# The Coronas-F Space Mission

For further volumes:
http://www.springer.com/series/5664

Vladimir Kuznetsov
Editor

# The Coronas-F Space Mission

## Key Results for Solar Terrestrial Physics

Springer

*Editor*
Vladimir Kuznetsov
IZMIRAN
Troitsk, Moscow
Russia

ISSN 0067-0057
ISBN 978-3-642-39267-2 ISBN 978-3-642-39268-9 (eBook)
DOI 10.1007/978-3-642-39268-9
Springer Heidelberg New York Dordrecht London

Library of Congress Control Number: 2014931378



Printed on acid-free paper

Springer is part of Springer Science+Business Media (www.springer.com)

# Preface

This book presents the results of the CORONAS-F experiments on observation of solar activity and its manifestations in near-Earth space carried out from July 2001 to December 2005.

The study of the Sun and solar–terrestrial coupling is of fundamental importance to the astrophysics and, in general, to the human life on the Earth. The Sun is a typical star like many others in the Universe. However, when studying the Sun, we do not observe it as a distant point object, but we can investigate the solar atmosphere and solar phenomena in detail. Quite a number of fundamental scientific problems of heliophysics and solar–terrestrial physics still remain unsolved. The disturbances in the near-Earth space associated with solar activity (so-called space weather) cause an ever growing impact on various aspects of the human life. The observation of solar phenomena and their terrestrial effects, as well as the long- and short-term forecast of these effects have become important elements in securing the human activity on the Earth and in space. In the context of the future programs of flight to the Moon and Mars, the significance of investigations in this field increases, in particular, as concerns the radiation safety of astronauts on interplanetary routes.

The chapter "CORONAS-F Project: The Study of Solar Activity and Its Effects on the Earth" contains a brief description of the CORONAS-F mission, general characteristics of the scientific payload, and the main observation results.

The solar activity originates in the solar interior, which is only accessible to the methods of neutrino astronomy and helioseismology. The chapter "Brightness Fluctuations and Global Oscillations of the Sun (DIFOS Experiment)" describes the DIFOS helioseismic measurements of the solar global oscillations. These results were obtained in spite of the difficulty of high-precision near-Earth observations associated with the reflected light from the Earth's atmosphere.

Active regions in the Sun and the related dynamics of the magnetic field in the outer solar atmosphere (from the photosphere to the corona) generate the variety of active events, such as solar flares, coronal mass ejections, and accompanying phenomena (dimmings, coronal holes, shock waves). These phenomena were detected and observed on the solar disk in the course of the SPIRIT experiment, whose results are presented in the chapter "Study of Active Phenomena in the

Solar Corona in the 8–350 Å Range by Imaging Spectroscopy Methods (SPIRIT Experiment)." The corona heating is one of the unsolved fundamental problems of the solar and stellar physics. The SPIRIT X-ray telescope was used to detect and investigate one of the corona heating mechanisms—extended hot plasma features with the temperature up to 20 million degrees.

The major manifestations of solar activity—solar flares—are of considerable interest as a mechanism of particle acceleration and transformation of magnetic energy to the energy of plasma. This is the most powerful explosive process in the Solar System. It is based on the reconnection of the magnetic field in the highly conductive solar plasma. High-resolution, broadband (from optical to gamma) instruments onboard the CORONAS-F mission recorded flare emissions in the continuous spectrum, in the ion lines of solar plasma and in the gamma-range nuclear lines. These observations enabled the study of atomic and nuclear processes in solar flares and the particle acceleration processes, as well as an extensive diagnostics of the flare-generated plasma and nonthermal processes in flares. The results of these investigations are described in the following chapters: "Study of Active Phenomena in the Solar Corona in the 8–350 Å Range by Imaging Spectroscopy Methods (SPIRIT Experiment)" (SPIRIT experiment comprising the SRT-K X-ray telescope and RES-K X-ray spectroheliograph); "Experiment with the SPR-N Instrument Onboard the CORONAS-F Satellite: Polarization, Temporal and Spectral Characteristics of the Hard X-Ray of the Solar Flares" (SPR-N solar spectropolarimeter); "Observations of Doppler Shifts of X-Ray Lines in Solar Flare Spectra Based on DIOGENESS Spectrometer Data" (DIOGENESS spectrophotometer); "Investigations of Physical Processes in Solar Flare Plasma on the Basis of RESIK Spectrometer Observations" (RESIK X-ray spectrometer); "The Study of the Cosmic Gamma-Emission Nonstationary Fluxes Characteristics by the AVS-F Apparatus Data" (experiment with the AVS-F time-amplitude spectrometer); "Variability of Extreme Ultraviolet Fluxes at Various Timescales as Measured Onboard the CORONAS-F Space Mission (SUFR-SP-K and WUSS-L Experiments)" (experiments with the SUFR-Sp-K solar UV radiometer and VUSS-L solar UV spectrophotometer); "Scientific Set of Instruments ≪Solar Cosmic Rays≫," "Protons Acceleration in Solar Flares: The Results of the Analysis of Gamma-Emission and Neutrons Recorded by the SONG Instrument Onboard the CORONAS-F Satellite," "Dynamics of the Relativistic Electrons Flux of the Earth Outer Radiation Belt Based on the MKL Instrument," "Dynamics of the Earth Radiation Belts During the Strong Magnetic Storms," and "Solar Protons in the Earth's Magnetosphere According to Riometric and Satellite Data During the Magnetic Storms of October 2003" (SCR experiments comprising the SONG spectrometer of solar neutrons and gamma rays, the cosmic ray monitor MKL, and the SKI-3 cosmic radiation spectrometer); "Spectrometer IRIS: Investigation of the Time Structure and Energy Spectra of X-Ray Emission from Solar Flares" (IRIS flare spectrometer); "Study of Solar Flares and Gamma-Ray Bursts in the Helicon Experiment" (HELICON gamma-ray spectrometer); and "RPS-1 Experiment" (RPS-1 X-ray spectrometer).

A series of outstanding events (powerful flares and ejections) occurred in the Sun in the epoch of maximum and in the declining phase of cycle 23. Simultaneous observations of these events on the solar disk and in near-Earth space with the CORONAS-F instruments contributed to better understanding of the solar–terrestrial coupling, e.g., the dynamics of the Earth radiation belts, deformation and restructuring of the magnetosphere, and penetration of high-energy particles. The results of these studies are presented in the chapter "Dynamics of the Earth Radiation Belts During the Strong Magnetic Storms."

The chapter "The Impact of Solar Activity on the Earth Upper Atmosphere as Inferred from the CORONAS-F Scientific Experiments" is devoted to the study of the Earth upper atmosphere using the SPIRIT X-ray measurements, which were taken as the satellite was entering and leaving the shadow. These observations provided us with new information on the altitude distribution of the atmospheric parameters and with the data necessary to construct an up-to-date model atmosphere of the Earth.

Regular precipitations of energetic particles from the Earth magnetosphere to the upper atmosphere determine the radiation background in the near-Earth space. The maps of such precipitations and their variations obtained with the aid of the AVS-F and RPS-1 instruments are provided, respectively, in the chapters "The Study of the Cosmic Gamma-Emission Nonstationary Fluxes Characteristics by the AVS-F Apparatus Data" and "RPS-1 Experiment."

The solar UV radiation exerts a direct influence on the Earth atmosphere and ionosphere determining their energy balance, ionization conditions, and disturbance. The results of measuring the solar UV flux with the SUFR-Sp-K and VUSS-L instruments and its variations depending on the level of solar activity are described in the chapter "Variability of Extreme Ultraviolet Fluxes at Various Timescales as Measured Onboard the CORONAS-F Space Mission (SUFR-SP-K and WUSS-L Experiments)." Along with the observations of solar flares, the chapter "Study of Solar Flares and Gamma-Ray Bursts in the Helicon Experiment" describes also the observations of gamma-bursts with the HELICON gamma spectrometer, including the unique event of 27 December 2004—an exceptionally powerful gamma-burst from a gamma repeater reflected from the Moon.

The topics concerning the control of the CORONAS-F scientific payload, collection, preliminary processing, and archiving of the scientific information are briefly dealt with in the chapters "On-Board and Ground-Based Complexes for Operating the Science Payload of the CORONAS-F Space Mission"; "CORONAS-F: Infrastructure and Organization of the Information Exchange"; and "Organization of a Unified Data Archive and Accessories for Processing Solar Images."

Throughout the book, the Universal Time (UT) is used when indicating the recorded events and analyzing the data obtained.

This book summarizes the results of a long work and broad cooperation of the scientific institutions involved in the CORONAS-F Project. It will make the data obtained accessible to the world scientific community and will demonstrate their significance in the general context of the present-day progress in space research and the future solar missions. This book is intended for the scientists working in the field of heliophysics and solar–terrestrial physics. It will be also useful to astrophysicists, students, and postgraduates of the corresponding specialties.

The edition was supported by the Russian Foundation for Basic Research, Project No. 08-02-07008. I am grateful to A.I. Rez, S.I. Boldyrev, L.V. Baibakova, and V.A. Koutvitsky for a large scope of preparatory work.

The English version of the book will acquaint the broad readership with the results of the CORONAS-F mission supplemented with data that became available after the publication of the original version. It will also help in relating the data obtained during the satellite operation period to the particular phases of the activity cycle, which may be important in the post-mission complex analysis of the solar and solar–terrestrial phenomena. The English edition, though a bit overdue, will meet these requirements to the best advantage.

Troitsk, Moscow, Russia V.D. Kuznetsov

# Contents

# CORONAS-F Project: The Study of Solar Activity and Its Effects on the Earth

**V.D. Kuznetsov**

## Introduction

The study of the Sun and solar–terrestrial coupling faces quite a number of unsolved scientific problems, such as the heating of the solar corona up to one million degrees, while the surface temperature in the Sun is only about 6,000°; the origin and acceleration of the solar wind, which exerts direct influence on the Earth and near-Earth space; the mechanisms by which various types of the solar energy are transformed and transported to the Earth's magnetosphere and ionosphere and influence the atmospheric processes, etc.

Unlike the remote stars, the solar atmosphere and phenomena therein can be observed in detail. This allows us to determine the structure of the Sun and solar atmosphere, as well as the origin and mechanisms of various active phenomena that are important to the astrophysics and to understanding the solar–terrestrial interactions.

The Sun and solar–terrestrial processes are studied by using both ground-based and space-borne means of observation. Many types of observations are only possible to conduct from space. Therefore, the most remarkable progress achieved in this field of research in the recent years is due to the space experiments.

The applications that ensue from the solar and solar–terrestrial studies are numerous.

The effects of solar activity on the Earth, which are studied within the concept of "space weather", are becoming increasingly important in day-to-day life. They cause regular disturbances in the near-Earth space and affect various aspects of the human activity. The Sun is never absolutely quiet. The active processes that occur in the Sun control variations in the geospace, i.e., in the Earth's magnetosphere,

V.D. Kuznetsov (✉)
Pushkov Institute of Terrestrial Magnetism, Ionosphere and Radio Wave Propagation RAS (IZMIRAN), Troitsk, Moscow, Russia
e-mail: kvd@izmiran.ru

V. Kuznetsov (ed.), *The Coronas-F Space Mission*, Astrophysics and Space Science Library 400, DOI 10.1007/978-3-642-39268-9_1,

ionosphere, and atmosphere and, eventually, in the natural environment of the man and biological objects. In the course of a 11-year cycle of solar activity, about 37,000 flares occur in the Sun. At the cycle maximum (in Cycle 23 it was 2001), one flare occurs every hour and a half, and from 5 to 10 mass ejections are recorded daily. The Earth suffers more than 600 magnetic storms during an activity cycle.

The effects of solar activity on the Earth are numerous, including the increased radiation hazard to astronauts during solar flares; abnormal deceleration or even premature fall of satellites due to the atmosphere bulging associated with solar activity; the failure of on-board equipment caused by the electric fields and high-energy solar and magnetospheric particles; disruptions of long ground-based electric power lines and pipe lines as a result of generation of geomagnetically induced currents during magnetic storms; radio blackouts; malfunction of the space navigation (precise positioning) and communication systems; and other effects.

The impact of solar activity on the Earth and various aspects of human life has become the object of detailed investigations at the junction of different sciences, such as geophysics and climatology, biology and medical sciences, and other disciplines, where a lot of new interesting and important results of social demand can be expected. This is due to the appearance of new experimental data on environmental variations caused by solar activity and to the technological progress—the development of space technologies such as GPS and GLONASS, long communication lines, etc., which make the technological medium increasingly sensitive to the solar activity.

Thus, the study of solar activity and its manifestations at the Earth and in interplanetary space, as well as its long- and short-term forecast are vital for securing the human activity on the ground and in space. This aim can only be attained as a result of complex ground-based and space observations of the Sun and in situ spacecraft measurements.

## The CORONAS Space Research Program

The program of Complex Orbital near-Earth Observations of Solar Activity (CORONAS) comprises a series of three space missions: CORONAS-I (active during 1994–2001), CORONAS-F (active during 2001–2005, see Figs. 1 and 2), and CORONAS-FOTON active during 2009. The CORONAS program forms part of a long-term research program of solar–terrestrial studies developed and implemented by the Division of Physical Sciences, Council on Space, and Scientific Council on Solar-Terrestrial Physics of the Russian Academy of Sciences within the frames of the Federal Space Program under the auspices of the Federal Space Agency.

The activity of the Sun has 11-year periodicity; i.e., the number of sunspots (or, more precisely, the Wolf number) changes from minimum to maximum for about 11 years. In the process, the general magnetic field of the Sun is reversed (changes its direction). It is the magnetic fields of sunspots and active regions (AR) that the observed solar activity is connected with. The magnetic fields in the solar

**Fig. 1** CORONAS-F satellite in orbit

atmosphere deformed continuously by the photospheric convective motions cause various active events in the upper atmosphere (flares, mass ejections, etc.).

Figure 3 shows how the number of sunspots was changing during the activity Cycle 23. The rise phase of the cycle was investigated by the CORONAS-I mission. CORONAS-F was launched at the cycle maximum and continued observations of solar activity and its effects in the Earth environment in the declining phase of the cycle.

## General Description of the Science Payload

The set of scientific instruments onboard the CORONAS-F mission was designed for a comprehensive study of solar activity and its terrestrial effects by observing active events in the solar atmosphere and recording their manifestations in

**Fig. 2** CORONAS-F satellite docked to the Tsiklon rocket at the Plesetsk Cosmodrome. From [12, p. 431]

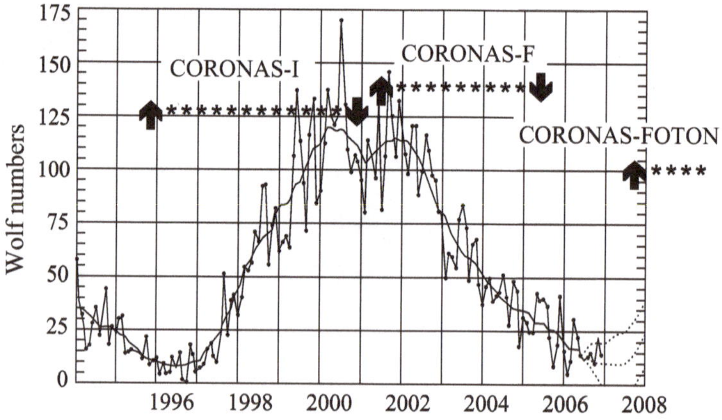


**Fig. 3** Solar studies onboard the CORONAS satellites. The figure shows the variation of sunspot numbers (Wolf numbers) over the 11-year activity cycle no. 23 (1996–2007). From [12, p. 431]

near-Earth space. The science payload comprised 15 instruments (Table 1), so that the satellite was operating as an autonomous solar space observatory controlled from the Earth. The on-board instruments observed the Sun in the entire electromagnetic spectrum from the optical to gamma-ray wavelengths, which enabled a complex analysis of the solar phenomena and solar–terrestrial relations. The measuring bands of different instruments are given in Table 2.

Figure 4 represents the exterior view of the CORONAS-F instruments.

The equipment layout on the satellite platform is shown in Fig. 5.

**Table 1** Science payload of the CORONAS-F mission

| Instrument | Destination |
|---|---|
| Helioseismology | |
| DIFOS multichannel photometer | Observation of global oscillations of the Sun |
| High angular resolution monochromatic imaging | |
| SRT-K solar X-ray telescope | The study of the spatial structure and dynamics of the solar upper atmosphere based on narrow-band images in the extreme ultraviolet (XUV) wavelength range |
| RES-K X-ray spectroheliograph | Diagnostics of hot plasma of the solar atmosphere based on X-ray and XUV images |
| DIOGENESS spectrophotometer | The study of X-ray emission from solar active regions and flares |
| Electromagnetic flux and polarization measurements (from ultraviolet to gamma wavelengths) | |
| RESIK X-ray spectrometer | Solar X-ray measurements with a high spectral resolution |
| SPR-N solar spectropolarimeter | Polarization studies of the flare-generated solar X-rays |
| IRIS flare spectrometer | Flare-activity observations in the X-ray wavelength range |
| HELICON gamma-ray spectrometer | Flare-activity observations in the gamma- and X-ray spectral ranges |
| RPS-1 X-ray spectrometer | The study of X-ray emission and precursors of solar flares |
| SONG—solar neutron and gamma-radiation spectrometer | The study of X-ray and gamma-radiation of solar flares |
| AVS-F amplitude–time spectrometer | The study of X-ray and gamma-radiation of solar flares |
| SUFR-Sp-K solar ultraviolet radiometer | The study of variations in the solar integral UV flux |
| VUSS-L solar ultraviolet spectrophotometer | Solar UV measurements in the vicinity of the hydrogen resonance line $HL_{\alpha}$ |
| The study of solar corpuscular fluxes | |
| The Solar Cosmic Ray complex (SCR) comprising the SONG, SKI-3, and MKL instruments) | The study of solar cosmic rays and manifestations of solar activity in the Earth space environment |
| Control of the science payload | |
| On-board control complex, data acquisition and processing system (SSNI), and ground-based control complex | Operation of the on-board scientific equipment and acquisition of data |

Besides the scientific equipment, the CORONAS-F mission comprised as its integral part the on-board control complex (BCC), data acquisition and processing system (SSNI), and ground-based control complex (GCC).

The scientific equipment of the mission was operating in off-line mode and was controlled by the GCC at IZMIRAN via the on-board control complex. The SSNI system ensured the acquisition, storage, and transmission of the scientific

**Table 2** Measuring ranges of the scientific instruments onboard the CORONAS-F satellite

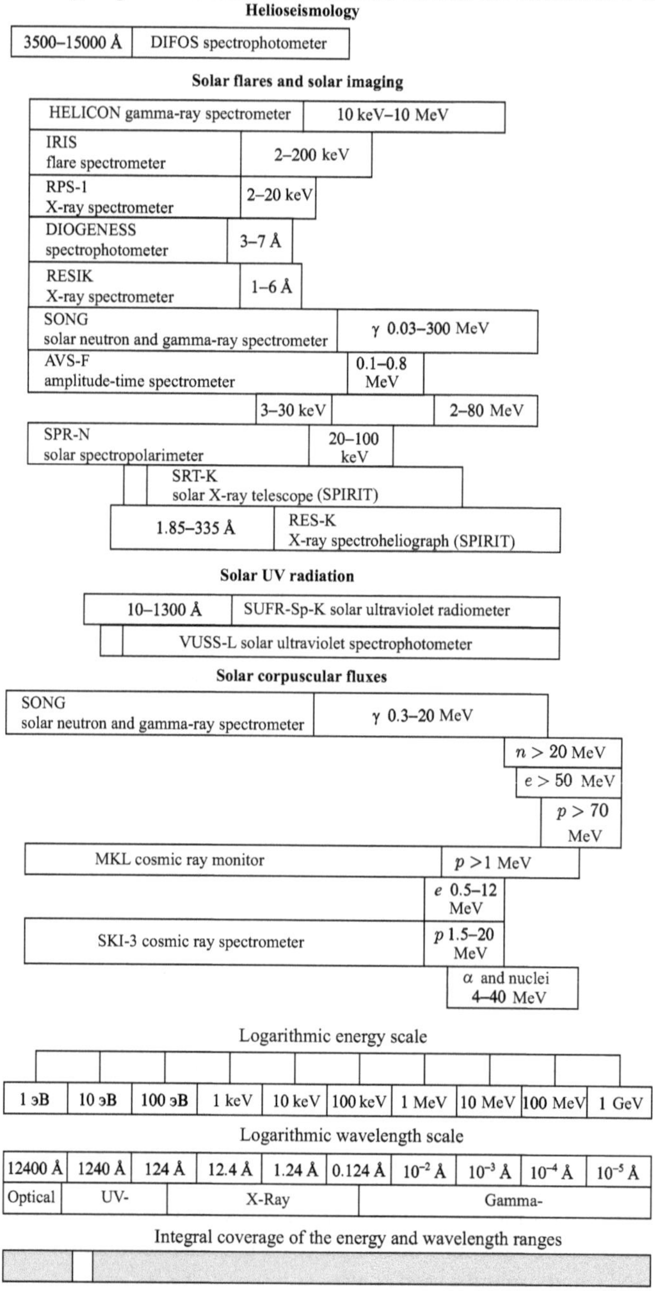

**Fig. 4** General view of the CORONAS-F scientific payload

information. The satellite itself and its service systems were controlled by a separate, special-purpose facility.

## *New Information Channels*

The measuring facilities of the CORONAS-F science payload were supplemented with new information channels for recording the solar corpuscular and electromagnetic emissions, which made it possible:

- to measure the emission in the magnesium line MgXII 8.42 Å (SRT-K solar X-ray telescope);
- to obtain solar images and spectra simultaneously in several (up to six) spectral channels with a high temporal (down to 1 s), spectral (down to 0.001 Å), and spatial (down to 3″) resolution (SRT-K solar X-ray telescope);
- to obtain simultaneous spectral and spatial records of the emission in two arbitrary lines in the spectral ranges of 280–335 and 180–210 Å (about 70 lines) (RES-K X-ray spectroheliograph);

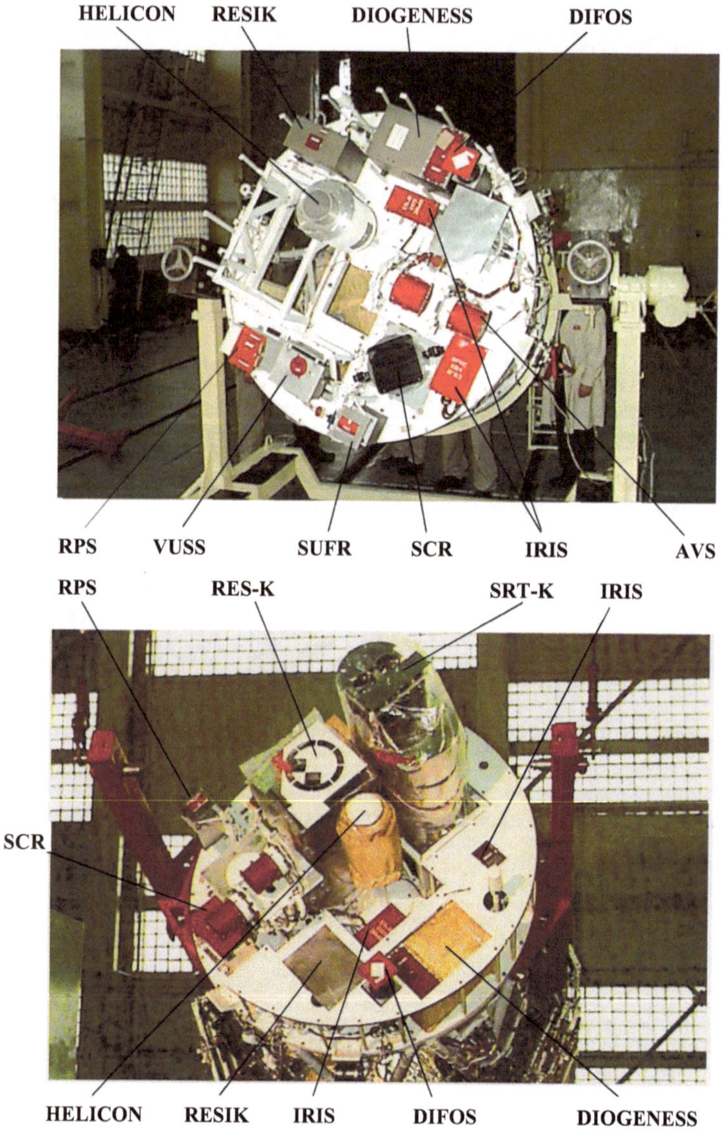


**Fig. 5** Arrangement of the scientific instruments on the CORONAS-F satellite platform

- to measure the linear polarization of hard X-rays (20–100 keV) (SPR-N solar spectropolarimeter);
- to measure emissions in the 3,500 and 10,000–15,000 Å wavelength ranges (DIFOS multichannel photometer);
- to measure gamma radiation with the energies of about 300 MeV (SONG solar neutron and gamma-ray spectrometer);

- to measure electrons and positrons in the energy range from 8 to 150 MeV (SONG spectrometer);
- to measure the 1–15 GeV protons (SONG spectrometer).

These channels provided quite a number of new results; some of them are described below and others in the following chapters.

## *CORONAS-F On-Board Observations*

The CORONAS-F satellite was launched on 31 July 2001 from the Plesetsk launching ground. On 6 December 2005, it ceased to exist as a result of the natural orbit evolution after having operated during 4 years and 4 months and having completed 24,623 revolutions around the Earth. The scientific program of the mission was fulfilled successfully. The initial orbital height was about 500 km; inclination was 83°. The orbit allowed continuous (without entering the shadow) observations of the Sun for about 20 days four times a year (the periods of solar-synchronous orbit).

The following sections give a general idea of the main results obtained in the course of the experiments onboard the CORONAS-F mission.

## Global Oscillations of the Sun

The clue to the 11-year periodicity of solar activity is to be sought for in the solar interior. At the center of the Sun, thermonuclear reactions are taking place, their energy moving up through the radiative zone. Then, as the solar surface approaches, the energy transfer by convection becomes more efficient. The convection zone, which is formed, determines the dynamics of the magnetic field over a solar cycle and the associated sporadic activity, such as flares and mass ejections.

Information on the solar interior can be obtained by observing global oscillations of the Sun as a gravitating plasma ball.

Continuous helioseismic observations onboard the CORONAS-F mission with the DIFOS multichannel spectrophotometer (differential photometer of solar oscillations) were started on 15 August 2001. The device detected low ($l = 0, 1, 2$) eigenmodes of the global oscillations of the Sun (acoustic oscillations with a period of about 5 min called $p$-modes) in a broad wavelength range (3,500–15,000 Å) by measuring minor variations in the solar radiation flux ($10^{-5}$–$10^{-6}$ of the total irradiance) associated with these oscillations [21]. Thus, it was shown that data from near-Earth satellites could be used for helioseismic studies. Special procedures were applied to fill in the gaps in observations due to the shadow entries of the satellite and to subtract the light reflected from the Earth's atmosphere to the detector. The results obtained were compared with other observations and theoretical calculations.

Figure 11 in the chapter "Brightness Fluctuations and Global Oscillations of the Sun (DIFOS Experiment)" represents the power spectrum of the $p$-modes of the solar global oscillations. Each peak in the spectrum corresponds to a harmonic global oscillation ($p$-mode) with a definite period and number of oscillation nodes in the azimuth $l$ and radius $n$ of the Sun. From 10 to 15 individual harmonics are recorded simultaneously at every instant of time. Such harmonics are excited at random at a certain time, live for a few days or weeks, and, then, disappear giving way to the new harmonics with different periods. Such a behavior is the result of influence of the outer convective shell with a broad noise spectrum on the solar global oscillations as a resonator in the solar interior.

Observations were carried out in a broad wavelength range, which allowed us to determine the dependence of the $p$-mode amplitude on the observation wavelength [21]. It was shown that the amplitudes increase significantly in the ultraviolet part of the spectrum (see Fig. 14 in the chapter "Brightness Fluctuations and Global Oscillations of the Sun (DIFOS Experiment)"). This makes the UV range very promising for the study of global oscillations and is to be taken into account when planning future helioseismic experiments. Note that the above-mentioned dependence agrees fairly well with the earlier ground-based observation and the Solar and Heliospheric Observatory (SOHO) data, but it was obtained for a much broader spectral range.

The analytical signal method applied to the DIFOS data [40] made it possible to carry out an amplitude–frequency–phase analysis of the global oscillations, in particular, to determine the instantaneous frequency, amplitude, and phase of individual oscillation $p$-modes at different levels in the photosphere and to show that the broadening of the $p$-mode spectral lines is due mainly to the amplitude fluctuations, while the oscillation frequency proves to be very stable ($\sim 10^{-4}$).

Many channels of the DIFOS spectrophotometer—the essential feature distinguishing it from the other space experiments on observation of the solar brightness fluctuations—allowed us to study the phase shift of the oscillations recorded in different optical channels of the device. A formerly unknown effect was revealed, which consists in the motion of oscillations from the upper and lower photospheric layers toward the middle layers. That is, apparently, where the oscillations damp most efficiently through nonadiabaticity, because the radiative heat exchange in these layers is most efficient and, hence, the relaxation time of temperature fluctuations is the smallest. These results provide experimental corroboration of the theory of nonadiabatic oscillations in the photosphere and cannot be explained in terms of the theory of adiabatic oscillations commonly used in helioseismology.

## X-Ray Imaging of the Sun

Unique studies were carried out with the aid of a multichannel X-ray telescope (SPIRIT experiment) [37]. Numerous high spatial resolution images of the Sun obtained in different spectral lines corresponding to different temperature layers

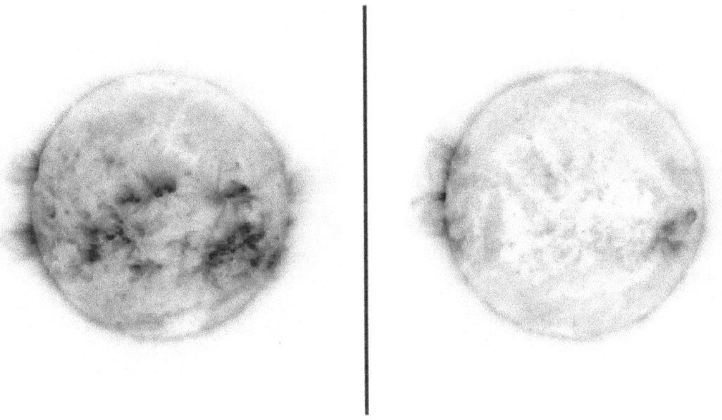

**Fig. 6** The state of two solar hemispheres in October–November 2003, when the most powerful in the past decades active events occurred in the Sun. One can see the longitudinal (hemispheric) asymmetry in the distribution of active regions over the solar surface. SPIRIT experiment

in the solar atmosphere made it possible to localize various active phenomena and to study their morphology. The telescope provided no less than 200 X-ray images of the Sun daily. Figure 6 illustrates the situation on the Sun in October–November 2003 when the activity was the highest recorded in the past few decades. The series of flares of unprecedented intensity was accompanied by powerful mass ejections at a speed up to hundreds of kilometers per second. The Sun was covered with active regions, numerous magnetic loops, and hot regions. The cause of the major flares and mass ejections observed in that period was, apparently, the global longitudinal (actually, hemispheric) asymmetry in the distribution of active regions over the Sun. The solar images in Fig. 6 obtained at half-rotation intervals show that one hemisphere is virtually free of active regions, while in the other, their number amounts to about 30. Large gradients of the magnetic field gave rise to the powerful events under discussion, which caused strong magnetic storms at the Earth and enhanced fluxes of cosmic rays in near-Earth space.

X-ray observations of these events were used to study the morphology and dynamics of the flares, mass ejections, and associated dimmings (darkenings) (Fig. 7) [4, 31].

It was found out that the dimmings observed were mainly formed as a result of full or partial opening of the magnetic fields in the process of coronal mass ejections and the associated matter outflow from large-scale magnetic features of the transition layer and corona accompanied by a decrease in the emission measure. The global nature of the dimmings in these and other events means that the process of eruption of CME involved a significant part of the solar atmosphere.

The homology of the eruptive events was observed. The pattern of the dimmings repeated in many details from event to event and, every time, the eruption affected approximately the same structures, which had managed to restore their magnetic field and luminosity in the lapse of time between the events.

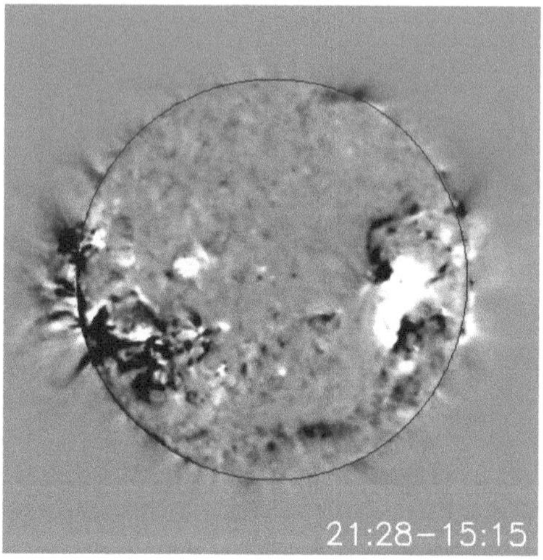


**Fig. 7** The major flares and mass ejections in the Sun recorded on 26 October 2003. SPIRIT experiment. Wavelength of 175 Å

Observations of the solar corona in the resonance line Mg XII (8.42 Å) with the RES-K spectroheliograph (SPIRIT experiment) revealed a new class of the phenomena—fast dynamic plasma features with temperatures up to 20 million degrees [3, 8, 36–38]. Their images were obtained and dynamics was studied for the first time. These features are likely to form when the hot plasma produced in solar flares fills the magnetic structures in the corona, and the plasma emission outlines their configuration. As a result, the features take various shapes (Fig. 14, chapter "Study of Active Phenomena in the Solar Corona in the 8–350 Å Range by Imaging Spectroscopy Methods (SPIRIT Experiment)"): ≪hot clouds≫, ≪spiders≫, loops, propagating wave fronts, and successively lighting magnetic arches. The hot plasma features observed in the corona are the manifestation of one of the mechanisms of the corona heating as a result of energy release in magnetic configurations and its transformation to the plasma energy. It was established that the hot plasma features in the corona are often accompanied by mass ejections from the solar atmosphere recorded with the coronagraph onboard the SOHO mission.

The X-ray coronagraph provided also the first maps of distribution of the hard ultraviolet radiation in the quiet and disturbed corona in the presence of bright active features on the limb, which visualize the magnetic heating of the corona over the active regions (see Fig. 8).

Observations with the X-ray telescope operated as a coronagraph made it possible to trace for the first time the dynamics of the solar corona up to the distances of three solar radii. This region, important to the understanding of many phenomena, is out of view of the other instruments, being intermediate between the limb regions observed with the telescopes and the remote corona observed in the white light with the coronagraphs. In this region, one can readily identify magnetic loops and arches, extended streamers and jets of the solar wind (see Fig. 9), as well as the mass ejections and eruptive prominences, which were observed for the first time at such altitudes.

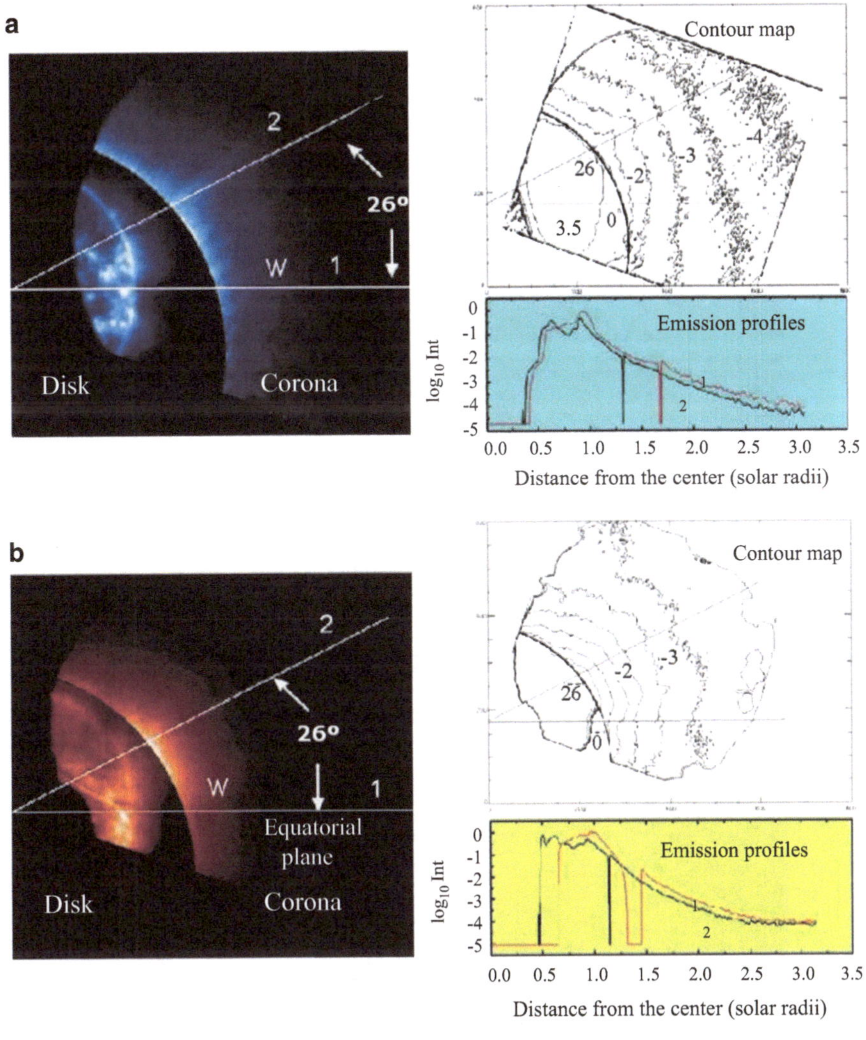


**Fig. 8** Distribution of the 175 Å (**a**) and 304 Å (**b**) hard UV radiation in the quiet and disturbed corona in the presence of bright active regions on the limb. SPIRIT experiment. *Left*—superposition of the disk and corona images. Adapted from [37]

A series of outstanding events occurred on the Sun in the period from 6 to 17 September 2005, i.e., 10 X-ray flares of importance X and 27 flares of importance M accompanied by powerful coronal mass ejections and strong geomagnetic storms. The CORONAS-F/SPIRIT experiment recorded 13 flares in the period under discussion. The analysis of the images obtained in the 175 Å coronal lines at the temperatures of about 1.2 MK showed that a dramatic reconstruction of the magnetic field occurred in the flare region during all flares (Fig. 10).

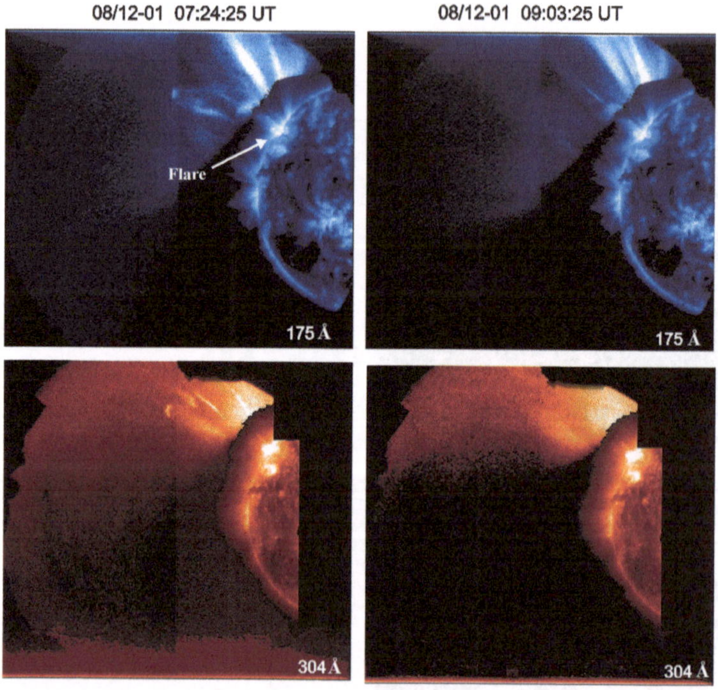


**Fig. 9** X-ray corona at a distance up to three solar radii. SPIRIT experiment. From [34]

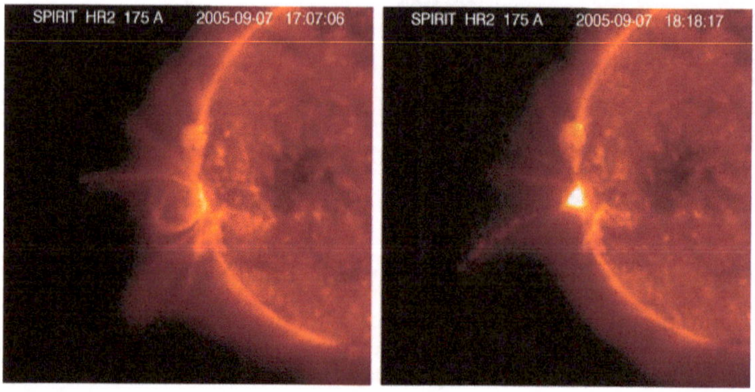


**Fig. 10** The coronal structure at the limb under pre-flare conditions (*left*) and at the X17 flare maximum (*right*) on 7 September 2005. SPIRIT experiment

A major limb flare on 7 September 2005 (the third intensive in the current cycle) was accompanied by a coronal mass ejection at a speed of about 550 km/s. As a result of evacuation of matter during the ejection, deep dimmings, areas of temporarily reduced brightness, appeared on the 175 Å solar images. After the extremely intense flare X 17 recorded at 17:40 on September 7 and the

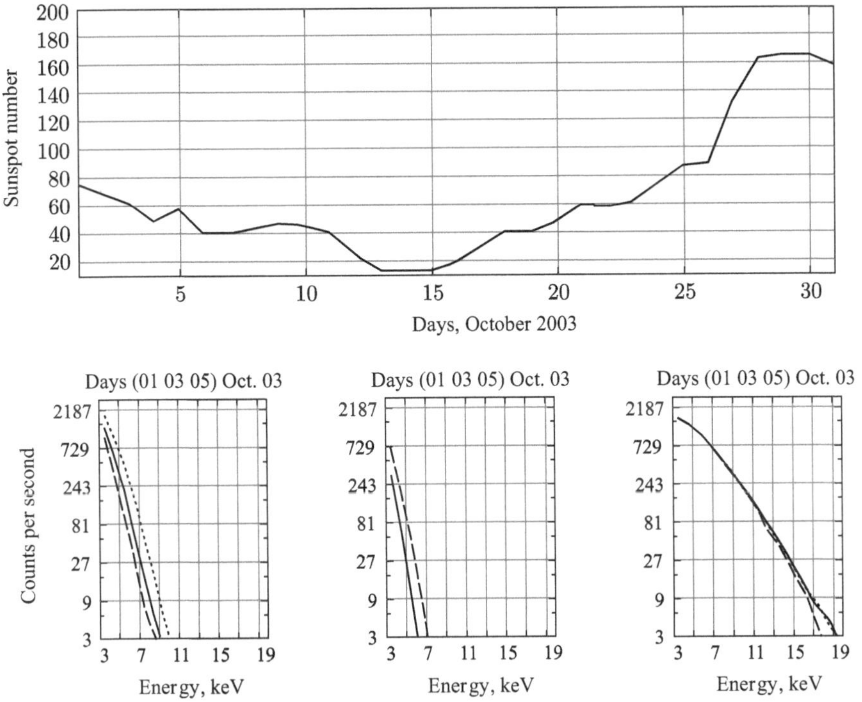


**Fig. 11** The background (without flares) X-ray spectra of active regions as a function of sunspot numbers illustrating the magnetic heating of the solar atmosphere. RPS-1 X-ray spectrometer data

accompanying CME, the decrease of the emission intensity in the 175 Å line in the dimming region amounted to ~7 % of the total solar brightness. On September 13, after the flares of importance X 1.5 (19:27) and X 1.7 (23:22), the brightness in the dimming region decreased by about 5 %. Simultaneous observations of the solar corona in two wavelength ranges 177–210 and 280–335 Å with the spectroheliometer of the CORONAS-F/SPIRIT experiment provided the line spectra corresponding to the pre-flare and maximum phases of the flare evolution, which revealed numerous ≪hot≫ lines and a general increase of line intensity at the flare maximum. The intensity ratio of the emission lines in the spectra obtained for the X 17 limb flare of 7 September 2005 was used to determine the differential emission measure. Its temperature dependence shows that the maximum luminosity of the flare-generated plasma falls on the temperature of about 10 MK.

The magnetic heating of the corona on account of active regions was also discovered from the solar X-ray spectra obtained with the RPS-1 X-ray spectrometer for different sunspot numbers. The more the sunspots are observed, the harder is the X-ray spectrum (Fig. 11), and, accordingly, the stronger is the effect on the Earth's atmosphere.

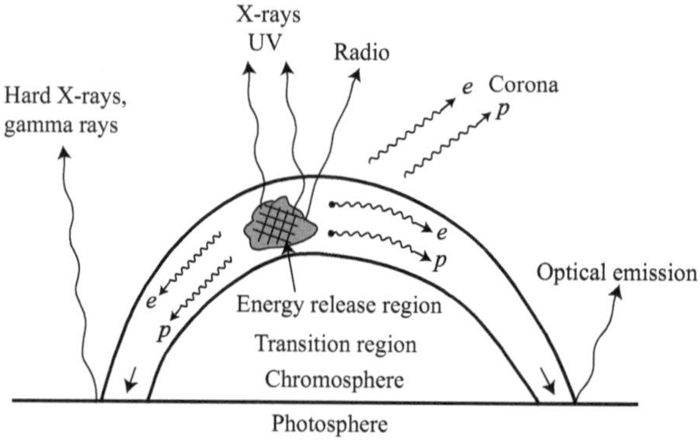


**Fig. 12** The types of emission generated by a solar flare

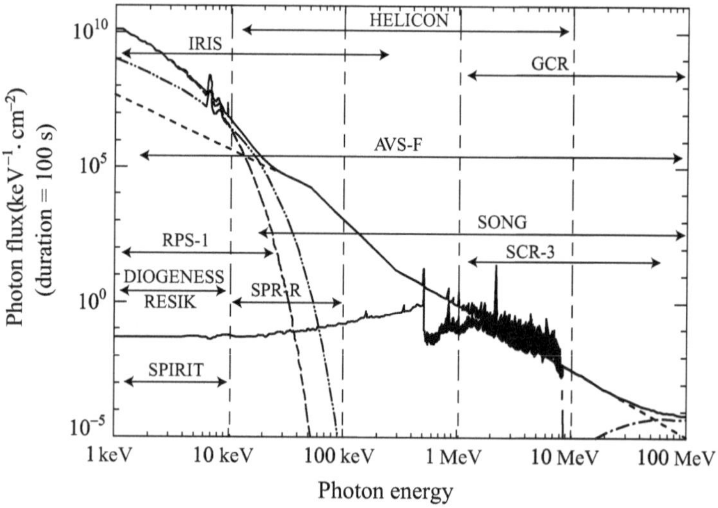


**Fig. 13** The integral spectrum of a flare and the energy channels of the CORONAS-F instruments

## Solar Flares

Virtually all types of electromagnetic and corpuscular emissions (except the radio emission) generated in solar flares (Figs. 12 and 13) were recorded onboard the CORONAS-F satellite (see Table 2). The data obtained were used for the diagnostics of the flare processes and analysis of the mechanisms of explosive release of the magnetic energy, the heating of solar plasma, and acceleration of particles up to high energies.

For the period of observations, the CORONAS-F instruments provided a wealth of new data on various physical processes in solar flares obtained with a high temporal and spectral resolution in a broad range of energies. These data involve the spectral, energy, polarization, and dynamic characteristics of the flare-generated emissions, the spectra of accelerated particles, gamma lines, etc.

The typical periods of plasma oscillations in active regions before, during, and after the flare were determined by the spectral analysis of high-resolution observations in the X-ray energy range (3–40 keV) with the IRIS flare spectrometer (e.g., see Fig. 6 in the chapter "Spectrometer IRIS: Investigation of the Time Structure and Energy Spectra of X-ray Emission from Solar Flares") [6, 29]. These periods vary from a few seconds to tens of seconds reflecting variations in the resonance properties of the AR magnetic configuration. Such signatures can be used for predicting solar flares from X-ray observations and for simulating the flare process.

## *The Flare Impulsive Phase*

The impulsive phase of a flare is characterized by the energy release, which, as seen from X-ray measurements, has an oscillation (saw-tooth) nature [5]. The flare-accelerated beams of charged particles penetrate the dense layers of the solar atmosphere and produce peaks of X-ray emission at characteristic intervals of about 20 s. A similar energy release pattern in the form of saw-tooth oscillations is observed in the tokomaks at the disruption instability caused by reconnection of the magnetic field lines [29].

The hard X-ray emission generated by interaction of the directed beams of charged particles with the dense solar atmosphere must be linearly polarized. A noticeable linear polarization at the flare maximum was measured for the first time with the SPR-N spectropolarimeter on October 29, 2003, in one of the most intensive flares (importance X10) [39] (see Fig. 4 in the chapter "Experiment with the SPR-N Instrument Onboard the CORONAS-F Satellite: Polarization, Temporal and Spectral Characteristics of the Hard X-Ray of the Solar Flares"). These measurements do not only provide a direct evidence for the existence of accelerated particle beams but also confirm that particles are accelerated by the pulsed electric field during the magnetic reconnection rather than by any stochastic mechanism.

The impulsive nature of energy release and acceleration of particles was also revealed in the time profiles and dynamic spectra of the flare-generated hard X-rays recorded in eight energy channels (26–380 keV) of the HELICON gamma spectrometer (see Figs. 1 and 2 in the chapter "Study of Solar Flares and Gamma-Ray Bursts in the Helicon Experiment"). These observations were used to determine the dynamics (i.e., the characteristic times and steepness) of the spectrum of hard X-rays in all phases of the flare. It was shown that the spectrum was the hardest at the flare maximum.

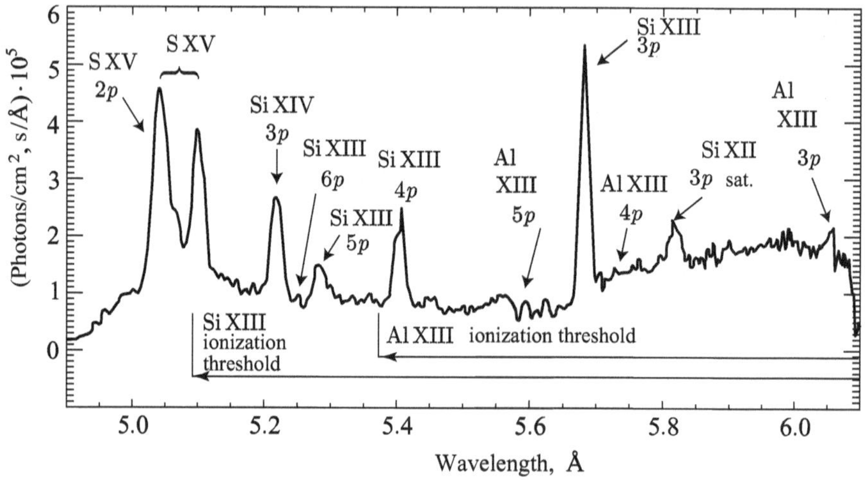


**Fig. 14** The spectral triplet of the S XV ion and the lines of the helium-like ions Si XIV and Al XIII corresponding to the atomic transitions at high quantum numbers: from $n = 3$ up to the ionization threshold. RESIK X-ray spectrometer data

## *Atomic Processes in Flares*

A detailed spectroscopic diagnostics of the flare-generated plasma and atomic processes in flares was carried out using the DIOGENESS spectrophotometer and RESIK X-ray spectrometer [32, 33]. This study made it possible for the first time to determine the absolute shifts of X-ray spectral lines in solar flares; to estimate the absolute content of potassium and chlorine in the solar corona; to reveal the spectral lines of ions of the astrophysical (solar) plasma at high quantum numbers $n$ (Fig. 14) offering new possibilities for temperature diagnostics of the coronal plasma; to detect and identify new spectral lines; and to measure hundreds of spectra of the helium-like ions Ca XIX, S XV, and Si XIII. About a million line spectra were obtained during the observation period, a catalog was compiled (http://www.cbk.pan.wroc.pl/resik_catalogue.htm), and an atlas of the solar spectral lines in the range of 3.4–6.1 Å was prepared.

## *Nuclear Processes in Flares*

The protons and nuclei accelerated in major flares to the energies above a few MeV collide with the nuclei of the surrounding medium giving rise to numerous nuclear reactions. The gamma emission, gamma lines of the excited nuclei, annihilation electron–positron lines, and neutrons produced in such reactions bear information of significant diagnostic value on the acceleration process and composition of the solar atmosphere.

Figure 40 in the chapter "The Study of the Cosmic Gamma-Emission Nonstationary Fluxes Characteristics by the AVS-F Apparatus Data" presents by way of example a record of gamma lines obtained with the amplitude–time spectrometer (AVS-F) during the flare of 29 October 2003, which shows what chemical elements and isotopes are present in the solar atmosphere [1]. In this particular case, they are iron, magnesium, silicon, neon, oxygen, and carbon. One can also see a feature in the spectrum corresponding to the line of the trapped flare neutrons. Since the gamma radiation of a flare is composed of the emission of individual gamma lines excited by the accelerated protons, the gamma spectrum represented in the figure gives an idea of the initial spectrum of the flare-accelerated protons, which is an essential characteristic of the acceleration process.

In the spectrum of the flare gamma radiation, the AVS-F recorded the annihilation line of electrons and positrons in the energy range of 0.5 MeV (see Figs. 6 and 37 in the chapter "The Study of the Cosmic Gamma-Emission Nonstationary Fluxes Characteristics by the AVS-F Apparatus Data"), which only appears during major flares, when sufficient amounts of positrons originate from the flare-generated radioactive nuclei.

The flare-generated neutrons can only be recorded at the Earth's orbit if their energy exceeds 30 MeV. At lower energies, the neutron decays into a proton, electron, and electron antineutrino (neutron beta decay, the decay time in the vacuum is 16 min) before reaching the Earth. Contrary to the protons and electrons, which are accelerated by pulsed electric fields of the flare, the flare-generated high-energy neutrons can only arise in nuclear reactions as a result of interaction of accelerated protons with nucleons (or nuclei) of the solar atmospheric plasma. Since the neutrons are uncharged particles and, thus, they are not affected by the magnetic field, they move from their origin to the detector along a straight line unlike the charged protons and electrons, which move along the magnetic field lines traveling a longer distance. Hence, at the same energies and velocities, it takes more time for the protons and electrons to reach the Earth; i.e., there must be a delay between the arrival of neutrons and protons generated simultaneously in one and the same flare. Figure 15 illustrates the records of gamma-ray and neutron fluxes made with the solar neutron and gamma radiation spectrometer (SONG) during the major flares of October 2003 [18]. Combined with other observations, these records make it possible to determine the moment at which the accelerated protons and high-energy neutrons leave the solar corona.

## *Ultraviolet Emission of Solar Flares*

The flare-generated UV fluxes were measured with the solar UV radiometer (SUFR-Sp-K) and solar ultraviolet spectrophotometer (VUSS-L) [9, 23–25]. Figure 1 in the chapter "Variability of Extreme Ultraviolet Fluxes at Various Timescales as Measured onboard the CORONAS-F Space Mission (SUFR-SP-K and WUSS-L Experiments)" shows an example of such measurements in the $L_\alpha$ line (at the

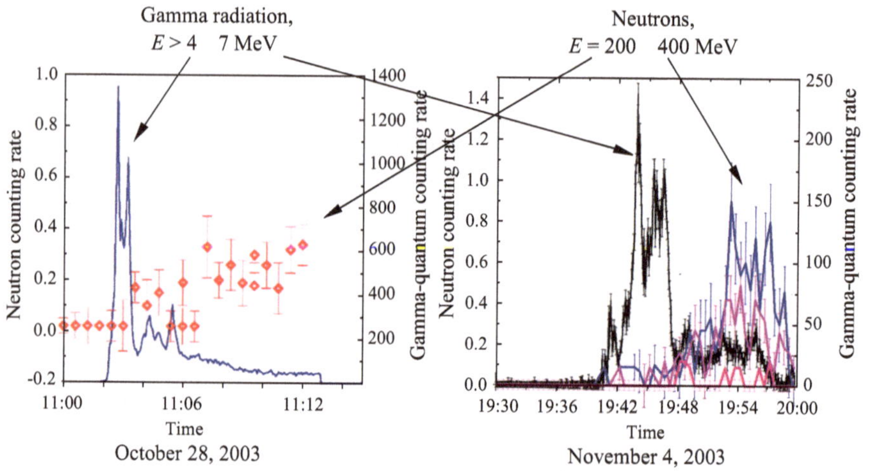


**Fig. 15** Gamma-ray and neutron fluxes from the solar flares of 28 October 2003. Data of the solar neutron and gamma-ray spectrometer (SONG)

wavelengths shorter than 130 nm) obtained with the VUSS instrument during the flare of 21 January 2003. For X-ray flares of importance C9-M1, the typical variation of ultraviolet emission in the $L_\alpha$ band is $\sim$0.5 %. The UV increase in the most intensive flares at the wavelength of 120 nm does not exceed a few percent.

## Solar–Terrestrial Coupling and Space Weather

### *Radiation Conditions in Near-Earth Space*

The measurement of solar cosmic rays along the mission trajectory with the CORONAS-F/SCR complex provided a continuous series of data on the fluxes of solar high-energy particles. The data on electrons with the energies more than 300 keV are unique, since no other measurements were taken in that period.

The SCR data obtained were used to study the dynamics of the Earth's magnetosphere and radiation belts and the change of boundaries of penetration of high-energy solar particles into the magnetosphere during strong geomagnetic disturbances [19, 20]. The periods of outstanding activity in the Sun, in particular, October and November 2003 when intensive flares caused two very strong geomagnetic storms, were most favorable for the study of solar–terrestrial interactions. Figure 16 illustrates the deformation of the magnetosphere and the dynamics of the Earth radiation belts in that period accompanied by penetration of high-energy particles to low $L$-shells.

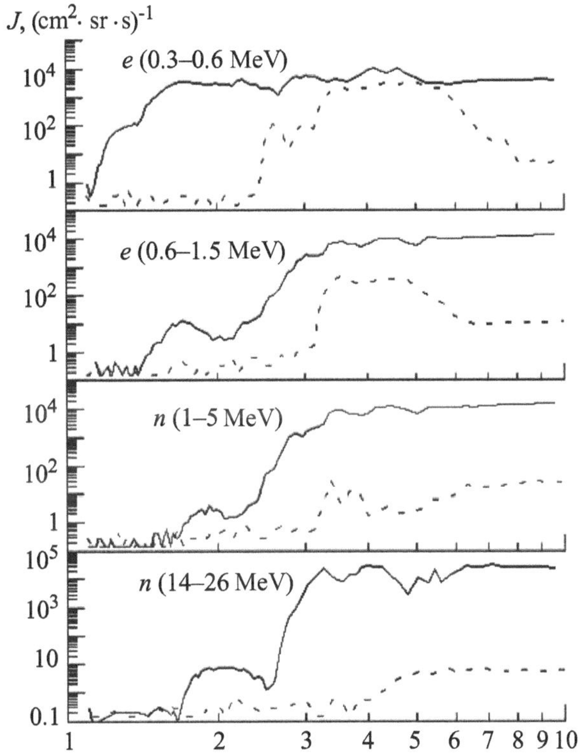


**Fig. 16** The distribution of fluxes $J$ of electrons $e$ and protons $p$ over the $L$-shells that illustrate the deformation of the magnetosphere and radiation belts and penetration of high-energy particles into the magnetosphere during strong magnetic storms of 28 October (*dashed line*) and 29 October (*solid line*) 2003. SCR experiment

Along with the well-known precipitations of high-energy particles from the magnetosphere to the ionosphere in the polar regions and in the Brazilian anomaly zone, the AVS-F device recorded localized (about 30° geographic latitude and about 10° geographic longitude) quasi-stationary low-latitude and equatorial precipitations (see Figs. 20 and 21 in the chapter "The Study of the Cosmic Gamma-Emission Nonstationary Fluxes Characteristics by the AVS-F Apparatus Data") [2]. These precipitations are characterized by a noticeable (20–30 %) increase in the background radiation. Their lifetime may reach 8 days.

## *The Earth Upper Atmosphere*

The maps of nightglow in the Earth upper atmosphere caused by solar radiation, galactic cosmic rays, and charged particles precipitating from the magnetosphere were plotted using RPS-1 X-ray data in the range of 3–31.5 keV (see Figs. 7–9 in the chapter "RPS-1 Experiment"). These data revealed important regularities in the behavior of the nightglow. A long-term variation of the nightglow depending on the season and the level of solar activity was found to exist in certain regions, such as the Brazilian magnetic anomaly, the outer radiation belt of electrons in the South and

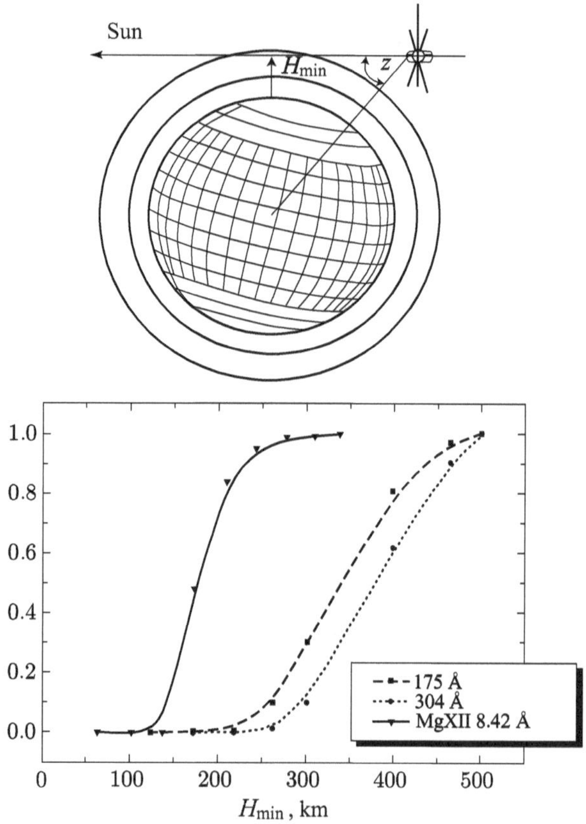


**Fig. 17** The X-ray absorption coefficient as a function of height $H_{min}$. SPIRIT experiment. Adapted from [35, Fig. 4]

North hemispheres, as well as in the regions of smaller world magnetic anomalies. The pioneering data were obtained on the low-energy electron fluxes of the Earth radiation belts, which precipitate into the atmosphere [10] and are responsible for its soft X-ray emission.

The Earth's upper atmosphere was studied using the absorption measurements of the solar hard X-ray radiation taken with the CORONAS-F solar X-ray telescope at the shadow entries and exits of the satellite [30] (Fig. 17). These observations were used to determine the height dependence of the X-ray absorption coefficients with a high resolution in altitude, to study the density and composition of the Earth's atmosphere up to 500 km as a function of solar activity, and to estimate the content of the molecular nitrogen and atomic oxygen. Similarly, the solar UV fluxes were measured with the SUFR radiometer and VUSS spectrophotometer.

Observations of the upper atmosphere carried out within the frames of the CORONAS-F project provided experimental data for constructing an up-to-date model of the Earth's atmosphere, which is important in light of the recent hypothesis of systematic trends of the upper atmosphere parameters and in connection with the problem of global warming.

## Observation of Cosmic Gamma Bursts with the HELICON Gamma Spectrometer

In addition to solar observations, the CORONAS-F gamma spectrometer (HELICON) recorded cosmic gamma bursts. On 27 December 2004, it recorded a unique phenomenon—the lunar reflection of a giant gamma burst from a neutron star residing deep in our Galaxy at a distance of tens of thousands of light years [22]. This is only the fourth case that such a giant gamma burst was recorded since the 1960s when the extra-atmospheric astronomy began to develop.

The direct signal of this gamma burst was so intensive that it could not be measured being off the scale of the detectors. This made it impossible to determine the time profile of the burst and to estimate the energy of explosion on the neutron star. The CORONAS-F satellite was shielded by the Earth from the direction to the burst source (see Fig. 7 in the chapter "Study of Solar Flares and Gamma-Ray Bursts in the Helicon Experiment") and recorded the reflected signal attenuated by a factor of 50000 (see Fig. 8 in the chapter "Study of Solar Flares and Gamma-Ray Bursts in the Helicon Experiment"), which could be used to estimate the explosion energy amounting to about $10^{44}$ erg [7]. This was the world's first observation of the cosmic gamma burst reflected from a celestial body or, in other words, the first location of a celestial body by means of a natural extra-powerful gamma-ray impulse.

## Concluding Remarks

CORONAS-F was an international Russian–Ukrainian mission implemented under the auspices of the Russian Academy of Sciences, National Academy of Sciences of Ukraine, Russian Federal Space Agency, and National Space Agency of Ukraine. The project was realized under the leadership of the Pushkov Institute of Terrestrial Magnetism, Ionosphere, and Radio Wave Propagation of the Russian Academy of Sciences (IZMIRAN) with the participation of Ukrainian, Polish, and German institutions, as well as scientists from France, England, Czech Republic, Slovakia, and the USA. In Russia, the cooperation involved the leading institutions in the field of solar and solar–terrestrial physics, such as the Lebedev Physical Institute of the Russian Academy of Sciences, Skobeltsyn Institute of Nuclear Physics of Moscow State University, Ioffe Physical Technical Institute of the Russian Academy of Sciences, Space Research Institute of the Russian Academy of Sciences, Moscow Physical Engineering Institute, Fedorov Institute of Applied Geophysics, and Space Research Center of the Polish Academy of Sciences.

During the operation period of the satellite:

- a unique archive of spectral images of the Sun and solar corona (more than one million images) was built up;
- unique data were obtained on X-ray, gamma-ray, UV, and corpuscular emissions of solar flares and their effect on the Earth's magnetosphere, ionosphere, and atmosphere;

- new spectral lines were revealed by a series of spectroscopic measurements, the absolute content of various elements in the solar corona was determined for the first time, and new methods were developed for the diagnostics of astrophysical plasma;
- the observation results were summarized in voluminous catalogs and data archives, which are used by scientists at many research institutions and will remain for years an important source of information for planning the future experiments in the field of solar–terrestrial physics.

The results of the CORONAS-F experiments are described in detail in the following chapters. See also the references therein and reviews [11–17, 26–28].

## References

1. Arkhangelskaya, I.V., Arkhangelsky, A.I., Kotov, Yu.D., Kuznetsov, S.N., Glyanenko, A.S.: The study of gamma-radiation from the solar flares of October–November 2003 recorded with the CORONAS-F/AVS-F device. Sol. Syst. Res. **40**(4), 302 (2006)
2. Arkhangelskaya, I.V., Amandzholova, D.B., Arkhangelsky, A.I., Kotov, Yu.D.: The properties of quasi-stationary equatorial precipitations as inferred from the CORONAS-F/AVS-F data. Sol. Syst. Res. **42**(6), 536–542 (2008)
3. Bogachev, S.A., Kuzin, S.V., Zhitnik, I.A., Urnov, A.M., Grechnev, V.V.: The dynamics of high-temperature plasma in the solar corona as observed with the SPIRIT telescope in the Mg XII 8,42 Å line. Sol. Syst. Res. **39**(6), 508 (2005)
4. Chertok, I.M., Grechnev, V.V., Slemzin, V.A., Kuzin, S.V., Bugaenko, O.I., Zhitnik, I.A., Ignat'yev, A.P., Pertsov, A.A., Delaboudiniere, J.-P.: Manifestations of coronal mass ejections in XUV-range as observed with the CORONAS-F SPIRIT telescope. Sol. Syst. Res. **39**(6), 462 (2005)
5. Dmitriev, P.B., Matveev, G.A., Charikov, Yu.E., Savchenko, M.I., Skorodumov, D.V., Kudryavtsev, I.V., Lazutkov, V.P., Nitsora, Yu.N.: Impulsive structure of energy release in the solar flare of 19 December 2001. In: Proceedings of the Scientific Conference of the CIS and Baltic countries ≪Active Processes in the Sun≫, St. Petersburg, 1–6 July 2002, pp. 195–198. NIIRF SPbGU, St. Petersburg (2002)
6. Dmitriev, P.B., Kudryavtsev, I.V., Lazutkov, V.P., Matveev, G.A., Savchenko, M.I., Charikov, Yu.E., Skorodumov, D.V.: The particularities of the solar flare X-ray radiation recorded with the IRIS X-ray spectrometer during the CORONAS-F mission. Sol. Syst. Res. **40**(2), 142 (2006)
7. Frederiks, D.D., Golenetskii, S.V., Pal'shin, V.D., Aptekar, R.L., Ilyinskii, V.N., Oleinik, F.P., Mazets, E.P., Cline, T.L.: Giant flare in SGR1806-20 and its Compton reflection from the moon. (2006). astro-ph/0612289
8. Grechnev, V.V., Kuzin, S.V., Urnov, A.M., Zhitnik, I.A., Uralov, A.M., Bogachev, S.A., Livshits, M.A., Bugaenko, O.I., Zandanov, V.G., Ignat'yev, A.P., Krutov, V.V., Oparin, S.N., Pertsov, A.A., Slemzin, V.A., Chertok, I.M., Stepanov, A.I.: On long-lived hot coronal structures observed in the MgXII line with the CORONAS-F/SPIRIT device. Sol. Syst. Res. **40**(4), 286 (2006)
9. Kazachevskaya, T.V., Nusinov, A.A., Katyushina, V.V., Gonyukh, D.A.: Variations of the solar HF-radiation flux over an activity cycle as inferred from the CORONAS satellite measurements. Sol. Syst. Res. **42**(4), 359–362 (2008)

10. Kolomiitsev, O.P., Zhitnik, I.A., Ivanov-Kholodny, G.S., Kuznetsov, V.D., Boldyrev, S.I., Kuzin, S.V., Surotkin, V.A.: Preliminary results of the comparison of satellite measurements and numerical simulations of some parameters of the Earth's upper atmosphere. Sol. Syst. Res. **40**(4), 326 (2006)
11. Kuznetsov, V.D.: The solar observatory CORONAS-F: three years of observation of solar activity. COSPAR Inf. Bull. **2004**(161), 90–93 (2004)
12. Kuznetsov, V.D.: Review of the results of solar and solar–terrestrial studies on board the CORONAS-F satellite. Sol. Syst. Res **39**(6), 433 (2005)
13. Kuznetsov, V.D.: Results of CORONAS-F observations of flares and active regions on the Sun. Izv. RAN, ser. fiz. **70**(1), 56–61 (2006) (in Russian)
14. Kuznetsov, V.D.: Results of CORONAS-F observations of the Sun. Izv. RAN, ser. fiz. **70**(10), 1449–1452 (2006) (in Russian)
15. Kuznetsov, V.D.: Observations of the Sun–Earth system within the CORONAS-F mission. J. Atmos. Sol.-Terr. Phys. **70**(2–4), 234–240 (2008)
16. Kuznetsov, V.D., Charikov, Yu.E., Kotov, Yu.D., Kuznetsov, S.N., Mazets, E.P., Nusinov, A.A., Pankov, V.M., Sobelman, I.I., Sylwester, J.: A review of the solar results from CORONAS-F satellite. In: Stepanov, A.V., Benevolenskaya, E.E., Kosovichev, A.G. (eds.) Multi-wavelength Investigations of Solar Activity. Proceedings of IAU Symposium 223, pp. 357–366. Cambridge University Press, Cambridge (2004)
17. Kuznetsov, V.D., Zhitnik, I.A., Sobelman, I.I.: CORONAS-F: contribution to solar–terrestrial physics. Vestnik RAN **75**(8), 704–714 (2005) (in Russian)
18. Kuznetsov, S.N., Kurt, V.G., Myagkova, I.N., Yushkov, B.Yu., Kudela, K.: Gamma-rays and neutrons from solar flares recorded with the SONG instrument in 2001–2004. Sol. Syst. Res. **40**(2), 104 (2006)
19. Kuznetsov, S.N., Yushkov, B.Yu., Denisov, Yu.I., Kudela, K., Myagkova, I.N.: The dynamics of the boundary of penetration of solar cosmic rays into the Earth magnitosphere as inferred from CORONAS-F data. Sol. Syst. Res. **41**(4), 348 (2007)
20. Kuznetsov, S.N., Myagkova, I.N., Yushkov, B.Yu., Denisov, Yu.I., Murav'yeva, E.A., Kudela, K.: The dynamics of the Earth radiation belts during strong magnetic storms as inferred from CORONAS-F data. Sol. Syst. Res. **41**(4), 338 (2007)
21. Lebedev, N.I., Kuznetsov, V.D., Oraevsky, V.N., Staude, J., Kostyk, R.I.: CORONAS-F/DIFOS helioseismic experiment. Astron. Rep. **48**(10), 871–875 (2004)
22. Mazets, E.P., Cline, T.L., Aptekar, R.L., Frederiks, D.D., Golenetskii, S.V., Il'inskii, V.N., Pal'shin, V.D.: The Konus-Wind and Helicon-Coronas-F detection of the giant gamma-ray flare from the soft gamma-ray repeater SGR1806-20 (2005). astro-ph/0502541
23. Nusinov, A.A., Kazachevskaya, T.V.: CORONAS-F observations of the high-frequency UV and X-ray emission of solar flares during 2001–2003. Sol. Syst. Res. **40**(2), 111 (2006)
24. Nusinov, A.A., Kazachevskaya, T.V., Katyushina, V.V., Svidsky, P.M., Tsigel'nitsky, Yu.N., Gonyukh, D.A., Afanas'yev, A.N., Boldyrev, S.I., Lisin, D.V., Stepanov, A.I.: The measurements of solar EUV radiation in different wavelength ranges on board the CORONAS satellites: instruments and principal results. Sol. Syst. Res. **39**(6), 470 (2005)
25. Nusinov, A.A., Kazachevskaya, T.V., Kuznetsov, S.N., Myagkova, I.N., Yushkov, B.Yu.: The ultraviolet, hard X-ray, and gamma-ray emission of solar flares recorded with the VUSS-L and SONG instruments in 2001–2003. Sol. Syst. Res. **40**(4), 282 (2006)
26. Oraevsky, V.N., Sobelman, I.I.: Comprehensive study of the solar activity on the KORONAS-F satellite. Astron. Lett. **28**, 568–577 (2002)
27. Oraevsky, V.N., Sobelman, I.I., Zhitnik, I.A., Kuznetsov, V.D.: Complex investigations of the Sun on board the CORONAS-F satellite: new results. Phys. Usp. **45**(8), 886–896 (2002)
28. Oraevsky, V.N., Sobelman, I.I., Zhitnik, I.A., Kuznetsov, V.D., Polishuk, G.M., Kovilin, P.N., Negoda, A.A., Dranovsky, V.I., Yatskiv, Ja.S.: CORONAS-F observations of active phenomena on the Sun. Adv. Space Res. **32**(12), 2567–2572 (2003)
29. Priest, E.R., Forbes, T.: Magnetic Reconnection, 600 pp. Cambridge University Press, Cambridge (2000)

30. Slemzin, V., Bugaenko, O., Ignatiev, A., Krutov, V., Kuzin, S., Mitrofanov, A., Oparin, S., Pertsov, A., Stepanov, A., Zhitnik, I.A.: Investigation of absorption of solar EUV-radiation in the Earth's atmosphere at altitudes of 100–500 km using solar images in the experiments TEREK-C (CORONAS-I) and SPIRIT (CORONAS-F). In: Proceedings of ISCS 2003 Symposium. ≪Solar Variability as an Input to the Earth's Environment≫, Tatranska Lomnica, Slovakia, 23–28 June 2003. ESA SP-535, pp. 389–392 (2003)
31. Slemzin, V.A., Chertok, I.M., Grechnev, V.V., Ignat'ev, A., Kuzin, S., Pertsov, A., Zhitnik, I.A., Delaboudiniere, J.-P.: Multi-wavelength observations of CME-associated structures on the Sun with the CORONAS-F/SPIRIT EUV telescope. In: Stepanov, A.A., Benevolenskaya, E.E., Kosovichev, A.G. (eds.) Multi-wavelength Investigations of Solar Activity. Proceedings of IAU Symposium 223, pp. 533–536. Cambridge University Press, Cambridge (2005)
32. Sylwester, J., Gaicki, I., Kordylewski, Z., Kowalinski, M., Novak, S., Iocieniak, P.S., Siarkowski, M., Sylwester, B., Trzebinski, W., Culhane, J.L., Whyndham, M., Bentley, R.D., Guttridge, P.R., Lang, J., Phillips, K.J.H., Brown, C.M., Doschek, G.A., Oraevsky, V.N., Kuznetsov, V.D., Stepanov, A.I., Lisin, D.V.: RESIK—bent crystal solar X-ray spectrometer for studies of Coronal Plasma Composition. Sol. Phys. **226**, 45–72 (2005)
33. Sylwester, J., Sylwester, B., Kordylewski, Z., Phillips, K.J.H., Kuznetsov, V.D., Boldyrev, S.I.: Observations of solar X-ray spectra with the RESIK and DIOGENESS spectrometers on board the CORONAS-F satellite. Sol. Syst. Res. **39**(6), 479 (2005)
34. Zhitnik, I.A., et al.: Observations of solar corona in the SPIRIT experiment on board the CORONAS-F satellite. Adv. Space Res. **32**(4), 2573–2577 (2003)
35. Zhitnik, I.A., et al.: Effects of the absorption of solar XUV radiation by the earth's upper atmosphere at altitudes of 100–500 km in the X-ray solar images obtained on board the Coronas-I (TEREK telescope) and Coronas-F (SPIRIT X-ray complex) satellites. Sol. Syst. Res. **37**(4), 296–301 (2003)
36. Zhitnik, I.A., Bugaenko, O.I., Ignat'ev, A.P., Krutov, V.V., Kuzin, S.V., Mitrofanov, A.V., Oparin, S.N., Pertsov, A.A., Slemzin, V.A., Stepanov, A.I., Urnov, A.M.: Dynamic 10 MK plasma structures observed in monochromatic full-Sun images by the SPIRIT spectroheliograph on the CORONAS-F mission. Mon. Not. R. Astron. Soc. **338**(1), 67–71 (2003)
37. Zhitnik, I.A., Kuzin, S.V., Sobelman, I.I., Bugaenko, O.I., Ignat'yev, A.P., Mitrofanov, A.V., Oparin, S.N., Pertsov, A.A., Slemzin, V.A., Sukhodrev, N.K., Urnov, A.M.: The main results of the CORONAS-F SPIRIT experiment. Sol. Syst. Res. **39**(6), 442 (2005)
38. Zhitnik, I.A., Kuzin, S.V., Urnov, A.M., Bogachev, S.A., Goryachev, F.F., Shestov, S.V.: The X-ray and XUV diagnostics of active plasma features with the RES spectroheliograph in the SPIRIT experiment on board the CORONAS-F orbital station. Sol. Syst. Res. **40**(4), 272 (2006)
39. Zhitnik, I.A., Logachev, Yu.I., Bogomolov, A.V., Denisov, Yu.I., Kavanosyan, S.S., Kuznetsov, S.N., Morozov, O.V., Myagkova, I.N., Svertilov, S.I., Ignat'yav, A.P., Oparin, S.N., Pertsov, A.A., Tindo, I.P.: The results of the CORONAS-F/SPR-N experiment on measuring the polarization, temporal, and spectral characteristics of the hard X-ray emission of solar flares. Sol. Syst. Res. **40**(2), 93 (2006)
40. Zhugzhda, Yu.D.: Analytical signal as a tool for the study of the properties of the solar oscillation $p$-modes. Astron. Lett. **32**(5), 329–343 (2006)

# Brightness Fluctuations and Global Oscillations of the Sun (DIFOS Experiment)

**Yu.D. Zhugzhda, V.D. Kuznetsov, and N.I. Lebedev**

Helioseismology is the basic tool for the study of the solar interior, which is impossible to explore using the traditional optical methods. The solar interior is accessible to the modern helioseismic methods almost entirely except for the innermost regions of the solar core, where the helioseismology is perfectly complemented by the neutrino astronomy. The main product of helioseismic observations of the Sun is the solar eigenmode spectrum. At present, we can observe reliably the spectrum of acoustic-type oscillations with a period of $\sim$5 min called $p$-modes. The accuracy of determining the solar eigenmode frequencies depends on the signal-to-noise ratio and the duration of continuous observations. To ensure round-the-clock observations, networks of observing stations have been set up all round the world. Besides, space-born helioseismic experiments have been carried out. One of them was the DIFOS experiment on board the CORONAS-F spacecraft aimed at the observation of global oscillations of the Sun in the solar irradiance in the wavelength range from 350 to 1,500 nm. At present, data for a period of more than 10 years have been accumulated. However, the accuracy of a few thousandth of a $\mu$Hz, which corresponds to the 10-year observation period, was not achieved in determining the $p$-mode frequencies. This is because the $p$-mode spectral line is split into many components, which makes it difficult to determine the mean frequency. The complex structure of the spectrum is the result of stochastic excitation of oscillations.

## Continuous and Discrete Oscillation Spectra

When solar eigenmodes are observed by measuring the brightness or radial velocity fluctuations, the signal is always sampled. On one hand, this improves the

Yu.D. Zhugzhda (✉) · V.D. Kuznetsov · N.I. Lebedev
Pushkov Institute of Terrestrial Magnetism, Ionosphere, and Radio Wave Propagation RAS (IZMIRAN), Troitsk, Moscow, Russia
e-mail: zhu@izmiran.ru

V. Kuznetsov (ed.), *The Coronas-F Space Mission*, Astrophysics and Space Science Library 400, DOI 10.1007/978-3-642-39268-9_2,

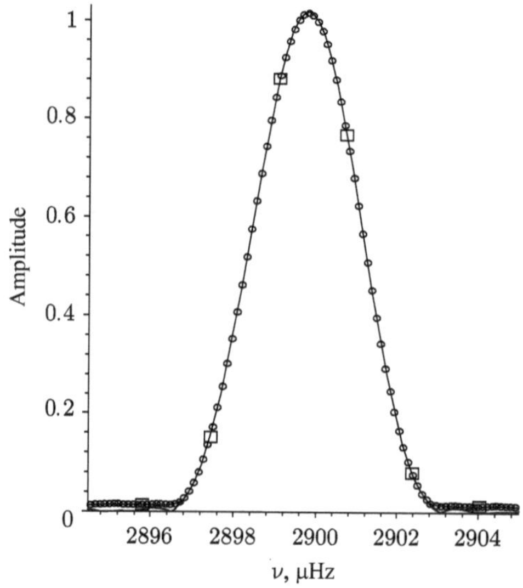


**Fig. 1** The spectral density of a model signal corrected for interruptions of observation due to the shadow entries of the satellite and telemetry problems is shown for three different processing methods. The *squares* indicate the discrete spectrum obtained using the fast Fourier transform; the *circles* show the discrete spectrum obtained likewise, but for a model signal, whose duration was increased by a factor of 14.5 by adding zero points; and the *solid line* indicates the continuous spectrum plotted with the use of the integral Fourier transform. From [14, p. 329]

signal-to-noise ratio through accumulation and, on the other hand, makes it possible to use the fast Fourier transform, thus reducing the data processing time. The discussion of this problem below is largely based on the material of paper [14].

In order to demonstrate the problems arising in the standard data processing method, let us first consider a model signal that reproduces in idealized form the actual observation data obtained during the CORONAS-F/DIFOS experiment, which are processed below. The model signal simulates observations for 7 days 50 min. The signal is interrupted every 94 min for about 40 min to imitate the shadow entries of the satellite. The interruptions due to the telemetry failures are also taken into account.

The sampling rate is 33.55 s. The squares in Fig. 1 indicate the discrete amplitude spectrum of the model signal obtained using the Hamming window, which is good at removing the effect of the realization ends. This is a typical situation where the true oscillation frequency falls somewhere in between the harmonics of the discrete spectrum, which makes it difficult to specify the true oscillation frequency. If instead of the model signal under consideration we take the corresponding analog continuous signal (even with the interruptions mentioned above), we can readily obtain the continuous spectrum using the integral Fourier transform. The continuous spectrum or, to be more precise, the spectral density is represented by the solid line in Fig. 1. For the continuous spectrum, the frequency of the spectral density maximum matches closely the frequency of the model signal.

In our case, it was easy to find the continuous spectrum, because we knew the continuous analog signal. The true spectrum of solar eigenmodes is difficult to determine, since the discrete spectrum recorded in the observations is used to obtain the discrete spectrum. This raises a natural question: How can the continuous spectrum of a signal be found from its values at discrete times? According to the Kotelnikov–Shannon theorem, a signal with the spectrum between the Nyquist frequency and the lowest frequency in the discrete spectrum can be fully restored from the values of the continuous signal at discrete times. The continuous signal $s(t)$ can be restored from its values at times $k\,\Delta t$ using the Kotelnikov–Shannon series:

$$s(t) = \sum_{k=-\infty}^{\infty} s(k\,\Delta t)\frac{\sin(t/\Delta t - k)}{t/\Delta t - k}. \tag{1}$$

As follows from this formula, an arbitrary continuous function $s(t)$ can be represented without loss of information by a sequence of discrete values $k\,\Delta t$, $k = 0, 1, 2, \ldots$ and can be unambiguously restored from this sequence if the highest frequency in spectrum of the function does not exceed its sampling frequency. This is the essence of the Kotelnikov–Shannon sampling theorem. Once the continuous signal corresponding to the discrete sequence is restored, its continuous spectrum can be found in principle by using the integral Fourier transform.

However, the integral Fourier transform for an actual signal is very difficult to perform. It takes much more computational time than obtaining the discrete spectrum by means of the fast Fourier transform. We are not aware of any attempts to restore the continuous analog signal and to obtain the continuous $p$-mode spectrum by using the scheme described above. Instead, the mean or maximum frequency of the $p$-mode lines in the solar eigenmode spectrum was found using various interpolations of the discrete spectrum based, for example, on the assumption of a Lorentz line profile.

It is quite clear that interpolation of the discrete spectrum cannot be used to find the spectral line profile without making additional assumptions, while passing to the continuous spectrum does not require additional hypothesis and is unambiguous. As a matter of fact, the continuous spectrum for a discrete signal can be readily found by a different method without passing to the analog signal via the Kotelnikov–Shannon series. Moreover, this can be done using the fast Fourier transform, which reduces the computational time essentially. The gradual passage from the discrete to continuous spectrum is performed through the increase of the realization length by adding points with zero signal values. When zeros are added, the number of harmonics in the spectrum increases. To eliminate the effect of decrease of the harmonic amplitude in the spectrum with increasing realization length $T$, the factor $1/T$ must be excluded from the discrete Fourier transform. Besides, factor 2 must be taken into account, because, unlike the discrete spectrum, the continuous one contains only positive frequencies. Given these corrections, the spectral density will be determined using the fast Fourier transform. Note that the increase of the realization length by adding zero points does not increase the frequency resolution; i.e., the spectrum width of the segment of a sine curve remains constant, while the number of harmonics in the spectrum increases.

The circles in Fig. 1 represent the discrete spectrum of our model signal extended in time by adding zero points up to the duration of 101 days and 19.38 h. One can see that the additional harmonics shown with the circles fall perfectly well on the profile of the continuous spectrum of the model signal, and the spectral line profile is well defined. Besides, it is evident from Fig. 1 that no increase of the realization length to infinity is required to find the continuous spectrum with a finite accuracy, but it will suffice to have a finite number of harmonics in the discrete spectrum. In principle, the continuous signal can also be restored if necessary using the inverse integral Fourier transform of the continuous signal found in this way. In the case of actual observations, this is absolutely unnecessary, since our goal is to study the individual $p$-modes that can be easily separated from the spectrum and only afterward can the inverse Fourier transform be applied to each of the modes separately. As a result, the analog signals will be isolated from the total signal for each of the oscillation modes of interest.

Below, the continuous spectrum will be used to construct the analytical signals for each of the oscillation modes. Thus, we pass to the continuous spectrum to be able to single out the continuous spectra for each of the oscillation modes present in the signal. By the way, the spectrum in Fig. 1, too, is only a part of the total spectrum of the model signal, which involves the side frequencies due to periodic interruptions of the signal when the satellite enters the Earth's shadow. Since the side harmonics are far enough from the central line, the spectrum of the "carrier frequency" can be isolated and studied separately. The inverse Fourier transform of the part of the total spectrum shown in Fig. 1 yields a continuous signal without interruptions due to the satellite shadow entries. This differs fundamentally from the restoration of the total signal including all $p$-modes and noise with the use of the Shannon–Kotelnikov series.

Of course, after the signal restoration according to the Shannon–Kotelnikov scheme and the subsequent determination of the continuous spectrum of the entire signal, one can, theoretically, also isolate individual spectral lines and find the continuous signal corresponding to each line by means of the inverse Fourier transform (Fourier inversion). However, this scheme for finding the continuous signal corresponding to an individual spectral line in the signal spectrum is very difficult to realize in practice. Our scheme, where the continuous spectrum is found by extending the realization through the addition of zeros, is very simple and does not take much computational time owing to the use of the fast Fourier transform.

Passing to the continuous spectrum makes the procedure of determining the maximum and mean spectrum frequency unambiguous and simple. As noted above, an assumption about the shape of the spectral line has to be made to find the spectral line maximum in the case of a discrete $p$-mode spectrum. For example, the hypothesis of Lorentzian shape is often used, though we are not quite sure that it is valid; moreover, the $p$-mode spectral lines were found to be asymmetric. Problems associated with the use of the discrete spectrum arise when the line in the spectrum is at the resolution limit.

In this case, only a few harmonics of the discrete spectrum fall within the line band as shown for our model signal in Fig. 1. One would think that it will suffice to use a longer realization, which will ensure higher spectral resolution; i.e., more harmonics will fall within the line band. However, in the case under consideration, the lines of the $p$-modes split as the resolution increases, and each of the resulting components again is at the resolution limit; i.e., the same problem arises.

The model signal was constructed taking into account the interruptions in observation in order to find out how much the latter affect the oscillation spectrum. For this purpose, we constructed the second model signal of the same duration, but without interruptions. Its spectrum turned out to differ insignificantly from that of the signal with interruptions. The continuous spectrum of the second signal is also shown in Fig. 1. However, it is difficult to distinguish in the plot because of insignificant difference between both spectra with the exception of the line wings, which lie below and virtually coincide with the abscissa axis for the signal without interruptions.

## Analytical Signal and the Instantaneous Frequency

At first sight, the oscillation frequency, amplitude, and phase are trivial notions that do not require special explanation. Actually, however, they were much discussed mainly among radio physicists, since the demands of radio physics prompted a deep insight into the subject.

The frequency of oscillations, in particular, the $p$-mode frequency is usually thought of as the spectral frequency determined by the spectral analysis. However, besides the spectral frequency, there also exists the so-called instantaneous frequency, which is a function of time. It may be suggested that this is one of those frequencies that are determined by the wavelet analysis. But this is not the case, because the wavelet analysis determines the frequency of the wave trains of finite duration. The notions of the instantaneous frequency, amplitude, and phase of oscillations cannot be defined without introducing the so-called analytical signal. It is well known that, in the case of monochromatic oscillations, all calculations are simplified significantly by passing to a complex signal $A\exp(i\omega t + \Phi)$. The real part of the complex signal is the physical signal, which is observed, measured, and subjected to other procedures. A natural desire arises to use the complex signal in the case of non-monochromatic signal, too, especially as all actual physical signals are not monochromatic. The monochromatic signal is a convenient idealization that is used to simplify the theoretical analysis.

It has been established [2, 8, 9] that the only consistent method for passing from the real signal $u(t)$ to the complex one $\omega(t)$ is to introduce the so-called analytical signal, which can be obtained through the inverse Fourier transform of the continuous signal spectrum by the method described in the previous section.

The analytical signal is

$$w(t) = u(t) + i\nu(t) = A(t)e^{i\Phi(t)}\frac{1}{\pi}\int_{0}^{\infty} W(\omega)e^{i\omega t}\,d\omega, \quad \omega(t) = \frac{d\Phi(t)}{dt}, \tag{2}$$

where the imaginary and real parts of the analytical signal are related by the Hilbert transform:

$$\nu(t) = H(\nu(t)) = \frac{1}{\pi}\int_{-\infty}^{\infty}\frac{u(s)\,ds}{t-s}. \tag{3}$$

Here, the integration is performed according to the Cauchy principal value; $A(t)$, $F(t)$, and $w(t)$ are the instantaneous amplitude, phase, and frequency of oscillations; and $W(\omega)$ is the spectral density. The analytical signal was first defined by Gabor [4]. The use of the analytical signal runs into certain difficulties [2, 8, 9] when broadband signals and fast amplitude variations are dealt with. But in the case of $p$-modes, no difficulties arise, since their bandwidth is small compared to their frequency, and the timescale of their amplitude variations is longer than the period by many orders of magnitude. To find the analytical signal of an individual oscillation mode, the integration in (3) is performed only within the frequency range that involves the selected spectral line. For our model signal, the integral was taken within the range indicated in Fig. 1.

Passing to the analytical signal, we shall be able to study the time evolution of the signal amplitude and frequency separately. Obviously, the instantaneous and spectral frequencies differ. Only their mean values $\overline{\omega(t)}$ and $\tilde{\omega}$ obtained by the time and spectral averaging are equal. The averaging is performed as follows:

$$\overline{\omega(t)} = \frac{\int_{-\infty}^{\infty}\omega(t)A(t)^2\,dt}{\int_{-\infty}^{\infty}A(t)^2\,dt} = \frac{\int_{-\infty}^{\infty}\omega|W(\omega)|^2\,d\omega}{\int_{-\infty}^{\infty}|W(\omega)|^2\,d\omega} = \tilde{\omega}. \tag{4}$$

Besides, there is a very important relationship between the second moments of the instantaneous and spectral frequency that is reduced to:

$$\overline{\omega(t)^2} + \overline{\left(\frac{dA(t)}{dt}\right)^2} = \frac{\int_{-\infty}^{\infty}\omega(t)^2A(t)^2\,dt}{\int_{-\infty}^{\infty}A(t)^2\,dt} + \frac{\int_{-\infty}^{\infty}\left(\frac{dA(t)}{dt}\right)^2 dt}{\int_{-\infty}^{\infty}A(t)^2\,dt}$$
$$= \frac{\int_{0}^{\infty}\omega^2|W(\omega)|^2\,d\omega}{\int_{0}^{\infty}|W(\omega)|^2\,d\omega} = \widetilde{\omega^2}. \tag{5}$$

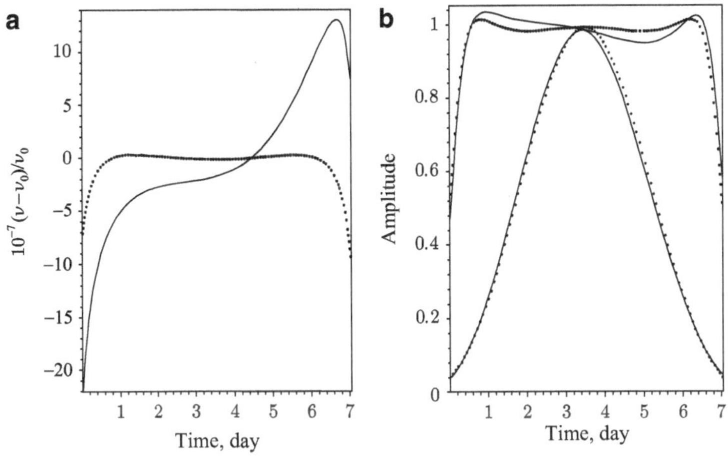


**Fig. 2** Relative variations of the instantaneous frequency (**a**) and amplitude (**b**) as a function of time for the continuous (*dotted curve*) and discontinuous (*solid curve*) model signals. The right-hand panel shows also the time dependence of the instantaneous amplitude of the two model signals corrected for the Hamming window function (7). In the latter case, the amplitude changes insignificantly, except for the ends of the time interval. From [14, p. 329]

As follows from this formula, the relationship between the deviations $(\delta\omega)^2$ and $(\Delta\omega)^2$ of the instantaneous and spectral frequency is:

$$(\delta\omega)^2 = \overline{\omega(t)^2} - (\overline{\omega(t)})^2, \quad (\Delta\omega)^2 = \widetilde{\omega^2} - (\tilde{\omega})^2,$$
$$(\Delta\omega)^2 = (\delta\omega)^2 + \overline{\left(\frac{dA(t)}{dt}\right)^2}. \qquad (6)$$

This formula reflects the important physical fact that the effective spectral width determined by the deviation of the spectral frequency is the sum of the broadenings caused by the frequency and amplitude modulations, which are equal, respectively, to the instantaneous frequency deviation and r.m.s. fluctuation of the time derivative of the instantaneous signal amplitude.

Thus, the introduction of the analytical signal allows us to separate two effects that influence the spectral width. The formulas given above were first derived by Fink [2, 3] for an analytical signal. It should be noted, however, that, long before Gabor [4] introduced the analytical signal, the same formulas had been derived by Rytov [6] for periodic signals.

Figure 2 shows the instantaneous frequency and amplitude of two model signals described in the previous section. The relative (with respect to the exact frequency $\nu = 2899.775626\,\mu\text{Hz}$) variations of the instantaneous frequency of the continuous model signal proved to be of the order of or smaller than $3\cdot 10^{-8}$ outside the zones of influence of the realization ends. The Hamming window used to analyze both model signals is rather efficient still it does not fully suppress the Gibbs effect, which gives rise to spurious fluctuations of the instantaneous frequency and amplitude.

The somewhat stronger fluctuations of the instantaneous frequency of the discontinuous model signal (solid curve in Fig. 2a) occur, probably, as a result of the Gibbs effect due to the presence of unsmoothed interruption ends in the signal that simulates the interruption in the DIFOS photometric observations because of the telemetry failure in the second half of the observation period. This interruption also manifests itself in the time dependence of the amplitude of the discontinuous model signal. This becomes obvious only after the amplitude dependence has been corrected by division by the Hamming window function, which equals to

$$H(t) = 0.54 - 0.46 \cos\left(\frac{2\pi t}{T}\right), \tag{7}$$

where $T$ is the realization length.

As seen in Fig. 2b, the amplitude of the interrupted signal corrected by the Hamming window function has a minimum in the second half of the interval. It should be noted that the amplitude by no means drops to zero as might be expected. This is because, actually, the instantaneous amplitude and frequency are not instantaneous, but are rather the means for a certain time interval. The Gibbs effect might, probably, be reduced by smoothing the ends of long interruptions in observations; however, we did not do it, since the accuracy achieved in reproducing the constant instantaneous frequency of the model signal proved to be very high. This manifests itself particularly clearly in the relative accuracy of the mean instantaneous frequencies $(\overline{\nu(t)} - \nu_0)/\nu_0$, which are, respectively, $-1,06 \cdot 10^{-7}$ and $-1,03 \cdot 10^{-9}$ for the continuous and interrupted signal; i.e., the mean instantaneous frequency is somewhat smaller than the true frequency of the periodic signal used to construct the model signals. Thus, the accuracy of the signal frequency was extremely high. This justifies the use of the method of analytical signal for the analysis of actual observations. The deviation of the instantaneous frequency $\delta\nu$ of both signals was zero. This suggests that using the analytical signal allowed us to establish that both model signals are exactly harmonic, which could not be established using the ordinary harmonic analysis of the signal.

Thus, introducing the Hamming window resulted in an insignificant shift of the mean frequency and did not lead to a broadening of the signal spectrum due to the spurious modulation caused by the Gibbs effect. Consequently, the analysis of the instantaneous frequency deviation shows that the spectral line width of the model signal is determined only by amplitude fluctuations. This agrees with the fact, since the model signals are segments of a sine wave. The use of the Hamming window introduces an additional amplitude modulation, which leads to a broadening of the spectrum: the widths of the spectra calculated with the analytical signal are 1.714 and 1.722 μHz for the continuous and interrupted model signals, respectively. The spectrum width of both model signals found by calculating the second moment of the continuous spectrum turned is 1.726 μHz; i.e., equality (5) holds to within the third decimal place. The correction of the analytical signal amplitude through division by the function $H(t)$ reduced the spectrum widths of the two model signals to 1.458 and 1.442 μHz, respectively, owing to the decrease of the signal amplitude fluctuations. The suggested amplitude correction does not cause undesirable effects, such as the Gibbs effect, which restricts significantly the capabilities of the spectral analysis.

In the next section, the analytical signal is used to analyze the observations of $p$-modes within the frames of the CORONAS-F/DIFOS experiment.

## Observation of the Solar Brightness Fluctuations

The DIFOS experiment on board the CORONAS-F satellite [5] was designed to observe the global oscillations of the Sun and to study its interior structure. The observations were made with a multi-channel photometer, DIFOS, which measured the intensity fluctuations of the solar radiation from the ultraviolet to near infrared range simultaneously in six optical bands. Two measuring channels (for 650 and 850 nm) were dubbed to estimate the proper noise contribution of the device. Figure 3 illustrates the functional diagram of the DIFOS photometer. Only three of the eight spectral measuring channels are represented for simplicity.

The CORONAS-F/DIFOS photometer was measuring the emission intensity simultaneously in six spectral channels: 350, 500, 650, 850, 1,100, and 1,500 nm with a bandwidth of 10 % of the central frequency covering a broad spectral range from the near ultraviolet to infrared radiation.

The relative intensity resolution of the photometer was $2 \cdot 10^{-6}$ of the total solar irradiance; the intensity sampling rate was 33.554 s; the spatial resolution was absent (accessible to observation were the lower-order oscillation modes with $l \leqslant 3$); the field of view of the photometer was 2°; the orientation of the photometer optical axis to the center of the Sun was no worse than 10′ (controlled by the satellite navigation system).

The CORONAS-F satellite was launched on 31 July 2001 to the high-latitude (82.5°) circular orbit at a height of about 500 km. The DIFOS photometer was brought into operation on 15 August 2001 and was functioning without interruption until the end of the CORONAS-F mission on 3 December 2005. During this period, the solar radiation intensity was measured in the vicinity of and after the maximum of Cycle 23. The total length of the records obtained amounts to 4 years, 3 months, and 11 days.

Figure 4 shows the relative position of the spectral channels of the CORONAS-F/DIFOS photometer and the SOHO instruments.

The DIFOS photometer started measurements on 22 August 2001. The data on radiation intensity were transmitted from the satellite via two independent telemetry systems. This allowed us to restore by comparison the data lost as a result of the failure in one of the systems. The analysis showed a good quality of the data obtained. The losses did not exceed 10 % of the photometer operation time and were associated mainly with the quality of reception of the telemetry at the receiving points.

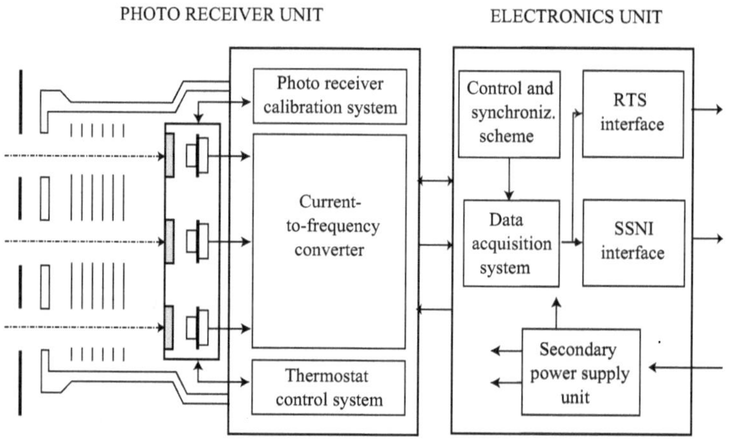


**Fig. 3** The functional diagram of the multi-channel photometer. From [5, pp. 871–875]

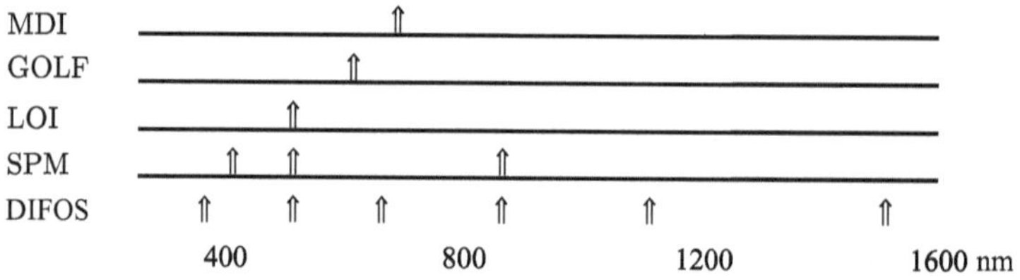


**Fig. 4** The relative position of the spectral channels of the DIFOS photometer and the SOHO helioseismic instruments (*MDI* Michelson Doppler Imaging of solar oscillations, *GOLF* Global low-degree velocity, *LOI* Luminosity Oscillation Imager, and *SPM* SunPhotoMeter). From [5, pp. 871–875]

Figure 5 represents a record obtained in one of the photometer measuring channels along the sunlit part of the satellite trajectory. The trend observed is attributed to the scattering of light reflected from the Earth surface that gets inside the instrument. The figure shows the analytical curve that approximates the trend.

After cleaning the record shown in Fig. 5 (subtracting the photo receiver dark currents and trends, normalizing, etc.), we obtained the variation of the solar emission intensity represented in Fig. 6. The intensity variation in this figure does not exceed 0.02 % of the total flux. The time series of the required length composed of such one-rotation records was used to obtain the amplitude spectra in the spectral range of 0–5 μHz.

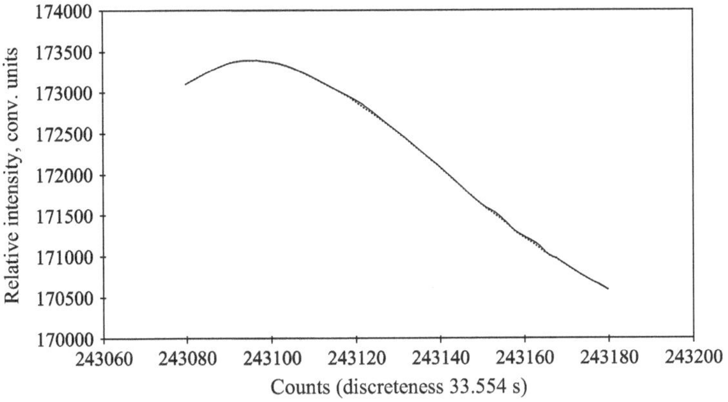


**Fig. 5** The intensity of solar radiation as measured in the 350 nm channel along the sunlit part of the satellite trajectory. The trend is due to the light scattered and reflected from the Earth's atmosphere. The analytical fitting curve is presented. From [5, pp. 871–875]

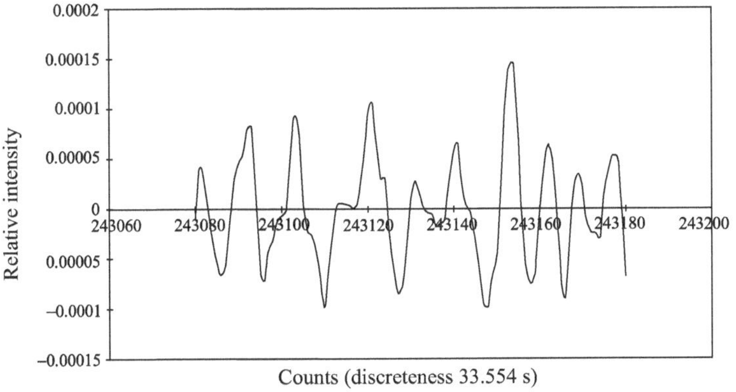


**Fig. 6** Variations of the solar emission intensity as measured in the 350 nm channel. From [5, pp. 871–875]

## The Instantaneous Amplitude, Frequency, and Phase of *p*-Modes at Various Levels in the Photosphere

The analytical signal method was used to analyze the solar brightness fluctuations observed within the frames of the CORONAS-F/DIFOS experiment.

The eight-channel photometer, DIFOS, was designed to observe the Sun as a star in six optical bands, namely, 350, 500, 650, 850, 1,100, and 1,500 nm. DIFOS had two independent photometers in each of the 650 and 850-nm bands, which enabled us to estimate the noise by comparing the signals in the parallel channels.

The data from the 1,100-nm channel will not be used because, unfortunately, they contain much noise, which is, apparently, due to the glare associated with reflections from the surrounding equipment.

The series of observations used for the analysis is a bit longer than a week, which is very short by today's standards. Nevertheless, it was not only sufficient to demonstrate the efficiency of the method based on the use of an analytical signal, but also provided some new results essential to helioseismology. Two modes with $l = 0$ and $n = 20, 21$ were chosen for the analysis. The modes with $l = 0$ were taken because they are not subject to splitting due to the solar rotation, which complicates the analysis significantly. The analytical signal for the chosen realization was constructed by exactly the same method as that used for the model signal with interruptions considered above.

Figure 7 illustrates the time dependence of the instantaneous amplitude. The instantaneous amplitude is corrected for the Hamming window. The amplitude variations differ in character for the two chosen modes. The amplitude for the $n = 20$ mode changes more smoothly than that for $n = 21$. The time variations of the instantaneous frequency of both modes also differ essentially. Compared to a smooth frequency variation of the $n = 20$ mode, the $n = 21$ mode displays narrow peaks of the instantaneous frequency. Particularly prominent are the instantaneous frequency maxima recorded in the 1,500-nm channel, which measures emission from the deepest photospheric layers. The maxima of the instantaneous frequency correspond to the minima of the instantaneous amplitude. This is particularly well pronounced in the 1,500-nm channel, whose amplitude is smaller than that of the other channels and is located in the lowest part of Fig. 7. The variations in the phase shift also correlate closely with the instantaneous frequency fluctuations: the phase variations are fastest at the instantaneous frequency maxima. It seems that the instantaneous frequency of the $n = 20$ mode also tends to reach the maximum immediately outside the time interval under examination; however, one cannot rule out that this is merely the effect of the realization ends. On the whole, the behavior of the analytical signal for the two chosen modes is different.

Figure 8 represents the results of calculation of the mean oscillation frequencies, dispersions, and phase shifts using both the oscillation spectra and the analytical signal. The left column illustrates the oscillation phase shifts with respect to the 1,500 nm channel. The phase shift between the channels was determined by two methods. The circles in Fig. 8 show the phase shifts determined from the spectra constructed for each channel. The phase shifts for the parallel channels do not always coincide perfectly. This can be attributed to the influence of noise or scattered light, which may differ in the parallel channels. The lines in Fig. 8 were plotted using the mean values in the parallel channels. The phases were also determined by time averaging of the phase difference between the analytical signals of the channels. In this case, the phase differences in the parallel channels did not always coincide either.

The phase differences determined by two methods proved to agree perfectly well. They clearly revealed the delay of oscillations in the 500-nm channel relative to

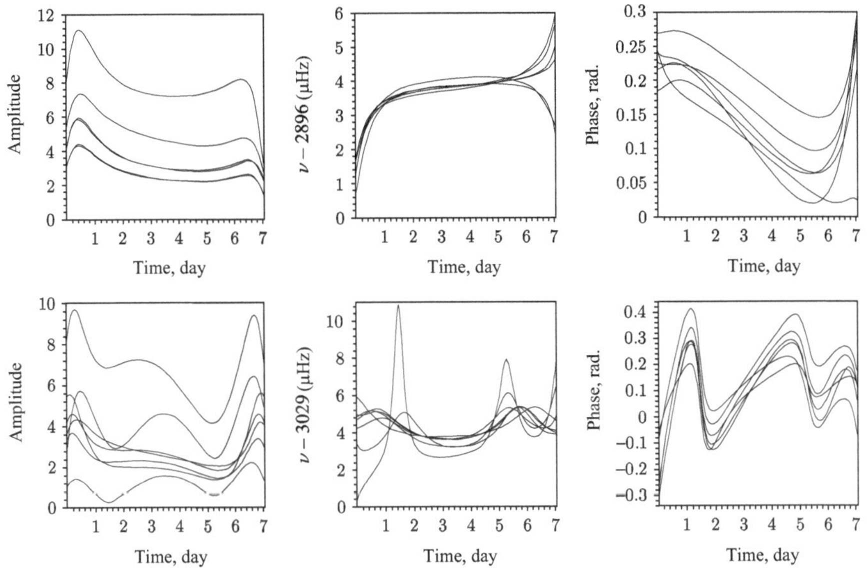


**Fig. 7** The instantaneous amplitude (in units $10^{-5}$ of the mean solar flux in the corresponding photometer channel), the excess of the instantaneous frequency (μHz) above 2,896 and 3,029, and the phase shifts (rad) in the optical channels of 350, 500, 650, and 850 nm relative to the 1,500-nm channel. The *upper row* of plots was constructed for the mode with $l = 0$, $n = 20$ and the *lower one*, for $l = 0$, $n = 21$. From [14, p. 329]

those in the 350- and 1,500-nm channels by ~30 s, which corresponds to the phase velocities of ~3 km/s.

The emissions at 350 and 1,500 nm come, respectively, from the uppermost and the deepest photospheric layers, while the emissions at 500, 650, and 850 nm originate in the middle photosphere. Hence, the waves from the upper and lower layers travel to the middle photosphere. A seemingly odd behavior of the waves can be naturally explained in terms of the nonadiabatic wave theory, which we shall consider in detail when discussing our results.

The middle column in Fig. 8 presents the frequencies of the two chosen modes determined both directly from the continuous spectrum and by averaging the instantaneous frequency. The frequencies obtained by two different methods differ. Being insignificant for the $n = 20$ mode, this difference can no longer be attributed to inaccurate calculation for the $n = 21$ mode. This is because the signal is not a single-component one as will be shown in the following section. Anticipating that discussion, we should emphasize that equality (4) is only valid for a single-component signal, i.e., for a quasi-monochromatic signal with smoothly varying frequency and amplitude.

It should also be noted that the *p*-mode frequencies change from channel to channel. The knowledge of the cause of this variation is essential to helioseismology, since the main objective of the latter is to determine the solar eigenmode spectrum

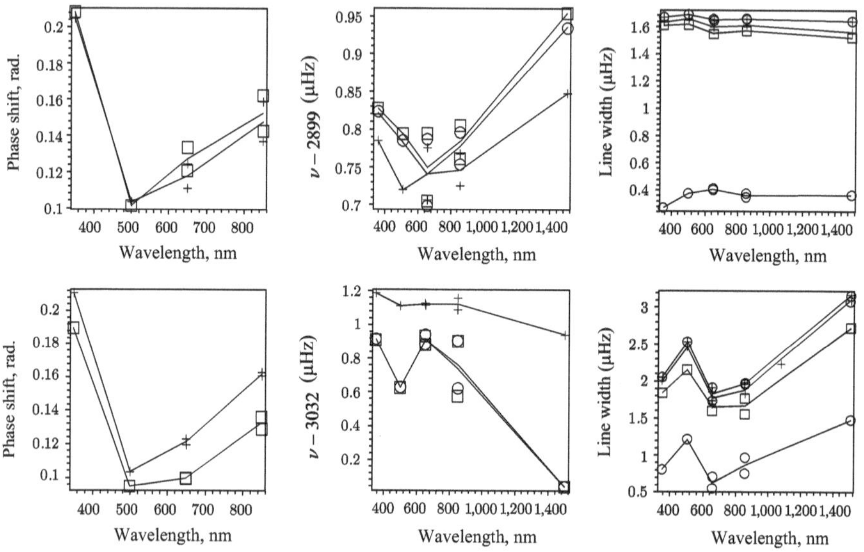


**Fig. 8** Mean parameters of the $l = 0, n = 20$ (*upper row*) and $l = 0, n = 21$ (*lower row*) modes. The *left column* represents the phase differences between the 350, 500, 650, and 850-nm channels and the 1,500-nm channel. The *squares* denote the differences of the time-averaged instantaneous phases $\overline{\Phi(t)}$. The *crosses* show the phase differences at the maxima of the continuous spectra in the corresponding channels. The *middle column* represents the mean instantaneous frequency $\overline{\omega(t)}$ (*squares*) and spectrum-averaged frequency $\tilde{\omega}$ (*crosses*). The *circles* denote the frequencies at the maxima of the continuous spectra in the corresponding channels. The *right column* represents the full line widths determined directly from the spectrum (*circles*) and by using the analytical signal (*crosses*), as well as the contribution to the full line width from the amplitude (*squares*) and frequency (*circled crosses*) modulation. From [14, p. 329]

as accurate as possible. Generally speaking, the instantaneous frequency of a signal can change when passing through linear systems, while the spectral frequency must remain constant.

The true spectral frequency can only be found by averaging over a sufficiently long time interval. In other words, averaging over a short time interval will yield but a smoothed instantaneous frequency. How long the averaging interval must be for obtaining the spectral frequency depends on the stability of the signal. We shall return to this question when discussing the observation results. Note, however, that the useful signal can be affected by various noise components, including those of solar origin. The first thing that calls our attention is a significant difference between the properties of the $n = 20$ and $n = 21$ modes as we noted above when analyzing the behavior of the instantaneous frequencies and amplitudes of oscillations. The range of channel-to-channel variations of the mean frequencies is much larger for the $n = 21$ mode. As shown in the next section, this is due to the presence of a signal close in frequency.

The right column in Fig. 8 shows the full line widths obtained both directly from the spectrum and with the use of the analytical signal, as well as the contribution to the spectrum width from the frequency and amplitude fluctuations. The full spectral widths determined by both methods virtually coincide as they should do according to the theory (5). The contributions to the line width from the amplitude and frequency fluctuations cannot be separated unless we use the analytical signal (5). Again, an essential difference can be noted between the parameters of the two modes.

The spectral width for the $n = 20$ mode is at the resolution limit depending on the realization length and does not virtually change throughout the photosphere, which is covered completely owing to the choice of frequencies of the optical channels. The line width is determined, primarily, by the amplitude modulation, while the contribution of the frequency modulation is insignificant. The spectral width for the $n = 21$ mode is much larger, although the contribution of the frequency modulation is also small. An important factor is that the spectral width is determined by the amplitude modulation rather than the realization length as is, apparently, the case for the $n = 20$ mode.

Indeed, in the latter case, the difference between the line widths determined by two different methods, as well as the width values themselves differ little from those found above for the model signals. This implies that the $n = 20$ line width may be even smaller, which can be revealed if longer observation series are processed. The analysis of the $n = 21$ mode will be continued in the next section where we shall propose another method for studying oscillations based on the analytical signal.

The relationship between the effects of amplitude and frequency fluctuations is of particular interest in the context of determining precise $p$-mode frequencies. At present, it is believed that the full line width is the measure of accuracy of the oscillation frequency. This follows from the assurance that the amplitude and frequency fluctuation effects responsible for the $p$-mode line broadening are impossible to separate.

True enough, these effects cannot be separated using the classical spectral methods. Actually, it is not the full line width $(\Delta\omega)^2$ but the instantaneous frequency deviation $(\delta\omega)^2$ resulting only from the frequency fluctuations that must be considered the true accuracy measure of the oscillation frequency. This obvious fact is not questioned, for example, when the methods for measuring the frequency of quantum frequency standards are developed. However, when applied to the results of our analysis, this approach runs into some fundamental difficulties associated with rather small frequency fluctuations of the $p$-modes under examination. For example, if the spectral line width for the $n = 20$ mode were determined by the frequency fluctuations alone, it would not exceed 0.4 μHz. This value is much smaller than the uncertainty in the frequency determination, 1.65 μHz, which follows from the uncertainty relation $\Delta\omega\,\Delta t \leqslant 1$.

On the other hand, with the model signal as an example, we could make sure that the accuracy of determining the frequency using the analytical signal is actually limited by the accuracy of our calculations. At the same time, the frequency of the maximum of the continuous spectrum of the model signal also coincides with the frequency of the model signal within the calculation accuracy. Likewise, the frequencies of the spectral maxima and the mean instantaneous frequencies of both modes under discussion are very close in all optical bands as is readily seen in Fig. 8 (middle column). However, it also follows from Fig. 8 that the difference between the spectrum-averaged frequencies and the time-averaged instantaneous frequencies is much larger because of the amplitude modulation.

We believe that the goal of the helioseismology must be determining the mean instantaneous frequency over long time intervals. These mean frequencies must, in principle, coincide with the frequencies of the maxima of spectral lines in the continuous spectrum. However, as noted above, the latter are difficult to determine because of the jaggedness of spectral lines in the spectra obtained by long-term observations. We hope that the use of the analytical signal will help us overcome this difficulty. An important advantage of using an analytical signal is the possibility to determine the instantaneous frequency deviation, which is the measure of stability and accuracy of the solar eigenmode frequency. It is to be emphasized that the reasoning above suggests implicitly that the signal under examination is monochromatic. Of course, this assumption is not always valid. In the next section, we show that using an analytical signal helps us overcome this difficulty in the data analysis.

## The Analytical Signal and Separation of a Two-Component Signal

In numerous applications of the method of analytical signal (the analysis of radio engineering circuits, nonlinear oscillations, random processes, parametric oscillations, etc.), it is always emphasized that the method is suitable, primarily, for the study of narrow-band monochromatic signals. The effects related to the influence of various types of noise were also considered, but it was specially noted that the presence of other narrow-band signals involves serious problems making the use of the analytical signal inefficient [1, 6].

Let us show that this opinion is erroneous, and the use of an analytical signal is quite efficient in the particular case of a two-component signal. This approach can also be applied to a three-component signal.

Let us consider the simplest two-component real signal

$$u(t) = A_1 \cos(\omega_1 t + \varphi_1) + A_2 \cos(\omega_2 t + \varphi_2), \quad \omega_{1,2} = 2\pi \nu_{1,2},$$
$$A_1, A_2 > 0, \quad \omega_1 > \omega_2. \tag{8}$$

The imaginary part of the analytical signal for this real signal is

$$v(t) = A_1 \sin(\omega_1 t + \varphi_1) + A_2 \sin(\omega_2 t + \varphi_2), \tag{9}$$

and its instantaneous frequency and squared instantaneous amplitude are, respectively,

$$\omega(t) = \frac{A_1^2\omega_1 + A_2^2\omega_2 + A_1A_2(\omega_1+\omega_2)\cos((\omega_1-\omega_2)t+\varphi_1-\varphi_2)}{A^2(t)}, \quad (10)$$

$$A^2(t) = A_1^2 + A_2^2 + 2A_1A_2\cos((\omega_1-\omega_2)t+\varphi_1-\varphi_2). \quad (11)$$

The amplitude and frequency of the analytical signal are periodic functions with a period $T = 2\pi/(\omega_1 - \omega_2) = 1/(\nu_1 - \nu_2)$.

The frequency and amplitude extrema of the analytical signal occur at the times

$$T_{\mathrm{ext}} = (n - \Delta\varphi)T, \quad n = 0, 1, 2, \ldots, \quad \Delta\varphi = \varphi_1 - \varphi_2 \text{ (rad)}, \quad (12)$$

and the frequency and amplitude values at the extremum are, respectively

$$\nu_{\max 1} = \frac{A_1\nu_1 + A_2\nu_2}{A_1 + A_2}, \quad \nu_{\max 2} = \left|\frac{A_1 - \nu_1 - A_2\nu_2}{A_1 - A_2}\right|, \quad (13)$$

$$A_{\max 1} = A_1 + A_2, \quad A_{\max 2} = |A_1 - A_2|. \quad (14)$$

If the instantaneous frequency at its extrema and the distance $T$ between its maxima or minima are known, then the harmonic frequencies of the two-component signal can be derived from the above formulas:

$$\nu_2^{(1,2)} = \frac{1}{2}\left(\nu_{\max 1} + \nu_{\max 2} - \frac{1}{T} + \sqrt{(\nu_{\max 1} - \nu_{\max 2})^2 + \frac{1}{T^2}}\right),$$

$$\nu_1^{(1,2)} = \nu_2^{(1,2)} + \frac{1}{T}. \quad (15)$$

Naturally, the harmonic frequencies of the signal do not depend on the amplitudes of the constituent harmonics. The first pair of frequencies $\nu_1^{(1)}$, $\nu_2^{(1)}$ corresponds to the case when the higher-frequency harmonic has smaller amplitude. In this case, the minima of the instantaneous frequency are sharper than the maxima. The other pair of frequencies corresponds to the case when the amplitude of the higher-frequency harmonic exceeds that of the lower-frequency one. In this case, the maxima of the instantaneous frequency are sharper than the minima as shown in Fig. 9 analyzed below.

Knowing only $\nu_{\max 1}$, $\nu_{\max 2}$, and $T$, we can obtain the harmonic amplitude ratio

$$\frac{A_1}{A_2} = \left|\frac{\nu_{\max 1} - \nu_2}{\nu_{\max 1} - \nu_1}\right| = \left|\frac{\nu_{\max 2} - \nu_2}{\nu_{\max 2} - \nu_1}\right|. \quad (16)$$

In order to find the absolute values of the constituent harmonic amplitudes of a two-component signal, it is necessary to determine additionally the variation range of the

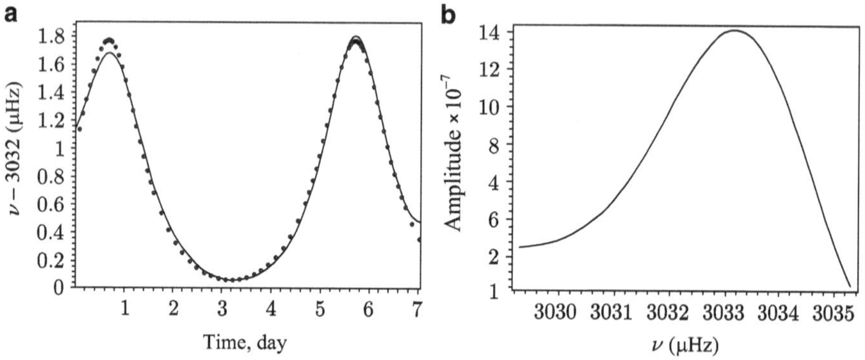


**Fig. 9** The instantaneous frequency vs. time (panel **a**, *solid line*) and the oscillation spectrum (panel **b**) in one of the 850 nm channels. The *dotted line* shows the calculated instantaneous frequency. From [14, p. 329]

instantaneous amplitude $A_{\max 1} - A_{\max 2}$, which is twice the amplitude of the smaller harmonics

$$|A_1 - A_2| = 2\min(A_1, A_2). \tag{17}$$

Figure 9a illustrates the time dependence of the instantaneous frequency for the $n = 21$ mode in one of the 850 nm channels of the DIFOS photometer. The same picture represents the instantaneous frequency calculated for a two-component signal with the frequencies $\nu_1 = 3033.23\,\mu$Hz and $\nu_2 = 3030.93\,\mu$Hz and the harmonic amplitude ratio $A_1/A_2 = 3.02$. The harmonic frequencies and their ratio were found from (15) and (16), and the maxima of the observed and calculated signals were matched by choosing the phase $\varphi_1 - \varphi_2$, whose variation makes the instantaneous frequency maxima shift along the time axis. The coincidence of the observed and calculated signals is nearly perfect. Thus, it can be established reliably that, apart from the principal harmonic, the signal contains a second harmonic with an amplitude three times smaller than that of the principal harmonic and a frequency differing by only 2.3 μHz.

In order to demonstrate an exceptional sensitivity of the instantaneous frequency to the presence of additional signal components, Fig. 9 (right-hand panel) shows the spectrum of the $n = 21$ mode for the 850 nm channel. It is clear that not only is such a weak additional component in the spectrum of the $n = 21$ mode impossible to isolate by any methods of the classical spectral analysis, but one cannot assert with confidence that the oscillation consists of two harmonics rather than being a single mode with an asymmetric spectrum. The signal separation into harmonics with the aid of an analytical signal is very efficient and allows us to establish reliably that the spectrum asymmetry of the $n = 21$ mode is due precisely to the presence of the second harmonic with a factor of three smaller amplitude.

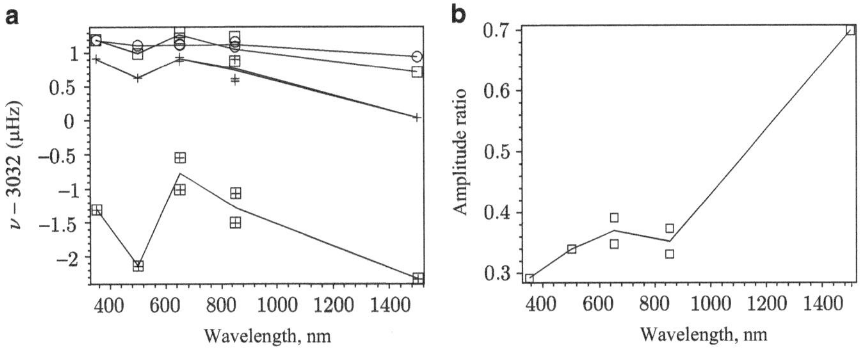


**Fig. 10** Panel **a**: *crosses* are the mean frequencies $\overline{\omega(t)}$ and $\tilde{\omega}$ of the $n = 21$ mode before the harmonic separation, *squares* and *crossed squares* stand for the high- and low-frequency harmonics of the two-component signal, and *circles* indicate the frequencies corresponding to the maxima of the continuous oscillation spectra in each channel. Panel **b**: the harmonic amplitude ratio as a function of wavelength. From [14, p. 329]

The separation into components for the $n = 21$ mode was performed for the brightness fluctuations in all optical channels of the DIFOS photometer. Not in all channels, the choice of the two-component signal was so ideal because of the slow, smooth frequency variations with time, which made the interaction of two modes differ from the ideal case of a two-component signal with constant harmonic amplitudes. As a result of slow variations of the instantaneous frequency, the maxima differ in height as seen, for example, in Fig. 7 (the lower middle panel). However, some uncertainty in choosing the height of the maxima turned out to have little effect on the frequencies and amplitudes of the components of the complex signal. Besides, a double maximum is sometimes observed suggesting the presence of a third component.

Figure 10 illustrates the results of harmonic separation for the $n = 21$ mode for all optical channels of the DIFOS photometer. The squares in Fig. 10a denote the frequencies of the $n = 21$ mode after the harmonic separation.

After the effect of the spurious harmonic has been removed, the $n = 21$ mode frequencies turned out to be similar to those determined from the position of the spectral line maximum and denoted by circles in the figure. For comparison, the $n = 21$ mode frequencies determined before the harmonic separation both by averaging the instantaneous frequency and by averaging over the spectrum are plotted in the same figure with crosses. The frequencies of the spurious harmonic indicated with crossed squares are seen to be, on average, lower than the $n = 21$ mode frequency by 3 μHz. Figure 10b shows the ratio of the spurious harmonic amplitude to the amplitude of the $n = 21$ mode. The relative amplitude of the spurious harmonic increases from 0.3 to 0.7 as we pass from the 350 nm to the 1,500 nm channel, which corresponds to the transition from the topmost to the deepest photospheric layers. Accordingly, the effect of the spurious harmonic on the frequency of the signal not separated into components increased with the increase of its amplitude.

The relationship between the mean frequency of the two-component signal $\overline{\omega(t)}$ and the frequencies of its components $\overline{\omega_1(t)}$, $\overline{\omega_2(t)}$ is expressed by a simple formula

$$\overline{\omega(t)} = \frac{\overline{(A_1(t))^2}}{\overline{(A(t))^2}}\,\overline{\omega_1(t)} + \frac{\overline{(A_2(t))^2}}{\overline{(A(t))^2}}\,\overline{\omega_2(t)}, \quad \overline{(A(t))^2} = \overline{(A_1(t))^2} + \overline{(A_2(t))^2}. \quad (18)$$

The mean frequency of the two-component signal can be easily calculated from this equation using the harmonic separation scheme described above.

Thus, it was established that the difference between the $n = 20$ and $n = 21$ modes is associated with the effect of the spurious signal. After the latter was eliminated, the channel-to-channel variation of the $n = 21$ mode frequency decreased significantly reaching approximately $\pm 0.25\,\mu$Hz. However, this amplitude of the channel-to-channel frequency variation is twice the variation amplitude of the $n = 20$ mode. This difference between the modes is due to the presence of another, higher-frequency spurious harmonic that distorts the $n = 21$ mode spectrum. The presence of such harmonic is suggested by the double maxima of the instantaneous frequency. The effect of this harmonic on the $n = 21$ mode frequency can also be eliminated, but we will not go into this problem here.

## Discussion of the *p*-Mode Properties

The channel-to-channel phase variations revealed for the $n = 20$ and $n = 21$ modes suggest that phase shifts exist between the oscillations at different depths in the photosphere. This oscillation property could only be established on the basis of data from the DIFOS experiment—the only helioseismic experiment that allows the solar eigenmodes to be observed simultaneously in various bands of the continuous spectrum formed at all levels in the photosphere from the temperature minimum to the deepest layers where the ~1,500 nm emission originates. The phase shift is the result of nonadiabatic nature of the oscillations. The 5-min oscillations in the photosphere are evanescent waves. The evanescent waves occur due to the total internal reflection in a medium in which the reflection takes place. In homogeneous media, they decrease exponentially with increasing distance from the boundary between the media where the reflection occurs. This is why they are called "evanescent." In contrast to the homogeneous media, in a stratified solar atmosphere, in which the density decreases with height, the amplitude of evanescent waves increases with height as we move away from the site where the total internal reflection of the waves coming from the resonance layers under the photosphere takes place. However, the energy density of evanescent waves in the photosphere decreases rapidly with height.

The phase of nonadiabatic evanescent waves must remain constant; i.e., the evanescent waves can only propagate along the surface from which the total internal reflection takes place.

The phase shifts between the evanescent oscillations at different distances arise only if the oscillations are nonadiabatic. The cause of the phase shifts between the oscillations at different distances from the site of the total internal reflection and, accordingly, the appearance of traveling evanescent waves is the wave absorption due to the nonadiabaticity of oscillations.

The traveling waves transport energy to the places where the absorption is the fastest. In the photosphere, the efficiency of the radiative heat transfer responsible for the nonadiabaticity changes with depth. It is for this reason that the waves at different photospheric depths do not coincide in phase as revealed from DIFOS data. There are no other reasons that could lead to such effect. It should be noted that the theory of nonadiabatic waves in the photosphere encounters serious difficulties associated with the dependence of the radiative heat transfer on depth and the effect of temperature waves in the photosphere that interact with the $p$-modes [11–13].

A more detailed comparison of theory and observations requires both further development of the theory and analysis of the effect under discussion using a larger sct of data from thc DIFOS cxperiment.

As noted in introduction, despite the availability of very long-term observations, the accuracy of determining the oscillation frequencies corresponding to so long data series could not be achieved. This is explained by the jaggedness of the $p$-mode spectra. The method of analytical signal must help us resolve this problem, since it allows the separation of fluctuations of the amplitude and instantaneous frequency. The following procedure can be suggested. Comparing the amplitude and instantaneous frequency fluctuations, we can determine which of the amplitude fluctuations are associated with the spurious signals or with the signal splitting into components. These amplitude fluctuations are accompanied by frequency fluctuations. After that, the spurious signals and the additional components of the main signal can be singled out and eliminated (the latter procedure is not described in this section). After such signal cleaning, the jaggedness of the spectral line must disappear, and its width will correspond to the true line width of the $p$-modes. The accuracy of the frequency determination must increase accordingly and, in our opinion, it will be determined by frequency fluctuations alone rather than by the full half-width of the spectrum, which involves amplitude fluctuations.

The study of the $p$-mode properties described above has revealed a high stability of the oscillation frequency. The line broadening on account of frequency fluctuations turned out to be no more than 0.4 μHz, which corresponds to the relative stability of the oscillation frequency of $10^{-4}$ or to a typical frequency variation timescale of the order of a month. This is in a certain conflict with the idea of stochastic excitation of oscillations, which is bound to lead to dephasing at every exciting "impact." This must be reflected in short-term variations of the instantaneous frequency, since the latter is a time derivative of the phase. Besides, there must be fluctuations of the instantaneous amplitude synchronized with the dramatic changes in the instantaneous frequency.

The method of analytical signal seems to be perfectly suitable for studying the stochastic excitation of oscillations. However, nothing of the kind is observed in the above results. This is, probably, due to the fact that the characteristic timescale of the stochastic fluctuations is of the order of 1 or 2 days. As a result, side frequencies must appear outside the main, central $p$-mode line. As emphasized repeatedly, the suggested method must be used to analyze long series of observation data, which will be done in the future.

The analytical signal concept described above differs from the method proposed by White and Cha [10] and applied later by Stebbins and Goode [7].

These authors found the analytical signal for the entire realization and not for a separate oscillation mode as we did. As a result, they determined the instantaneous frequency of the multi-component signal comprising all oscillation modes and the noise. Such a complex signal is, virtually, impossible to deal with. There are no methods for isolating the amplitudes and frequencies of individual modes from such a complex, multi-component analytical signal. Besides, because of an elementary mathematical error, the authors arrived at a conclusion that the purely amplitude modulation of the harmonic signal must be accompanied by the appearance of maxima of the instantaneous frequency. This is entirely in contradiction with the principle of separating the amplitude and frequency modulations.

## Dependence of the Oscillation Amplitude on the Observation Wavelength

One of the scientific problems that had to be approached in the course of the DIFOS experiment was the relationship between the $p$-mode amplitude and the observation wavelength. It is known that the solar radiation at different wavelengths originates at different subphotospheric levels, the 1,500 nm emission coming from the deepest layers.

The "amplitude–wavelength" relation was determined by two methods. The first one was as follows. We plotted the oscillation spectra for each of the spectral bands. The length of the data set was chosen such that it did not exceed the $p$-mode characteristic lifetime (from 3 to 10 days) but was sufficient for the individual $p$-modes to be reliably resolved and identified. The optimal length of the data set was, in our opinion, about 2 days, which corresponded to 30 orbital revolutions of the satellite. In order to increase the signal-to-noise ratio, we summed up these 2-day spectra over a sufficiently long time interval. An example of such summary spectrum for one of the observation channels is represented in Fig. 11.

The amplitudes of about ten reliably identified $p$-modes, the same for all observation channels, were derived from the spectra for each spectral band, and the results obtained were used to plot the "amplitude–wavelength" relation.

In the second method, the amplitude spectra were plotted using the time series corresponding to each particular orbital revolution. These spectra were summed up for the same time interval as in the first method. In this case, the individual oscillation modes could not be resolved, but the envelope of the 5-min oscillation spectrum was obtained for each observation band (Fig. 12).

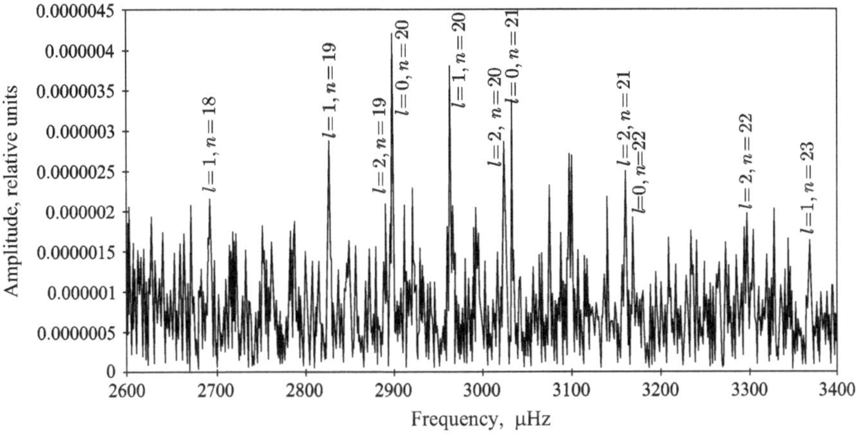


**Fig. 11** Amplitude spectrum of the 5-min solar oscillations in the 650 nm spectral band based on the 2-day series of CORONAS-F/DIFOS data (orbits 1831–1920, 144 h, 29 November to 5 December 2001). From [5, pp. 871–875]

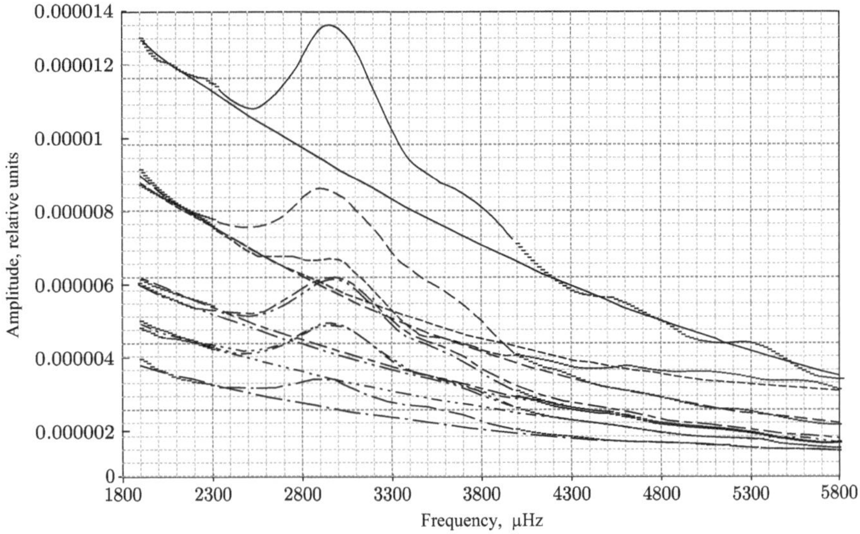


**Fig. 12** Amplitude spectrum of the 5-min solar oscillations based on the series of one-revolution CORONAS-F/DIFOS data (orbital revolutions 1831–1920, 29 November to 5 December 2001). The fitting curves for the noise component are presented

The "amplitude–wavelength" relation was plotted using the amplitude ratios that correspond to the maximum of the spectrum envelope (3,000 μHz).

The plots obtained by both methods are shown in Fig. 13.

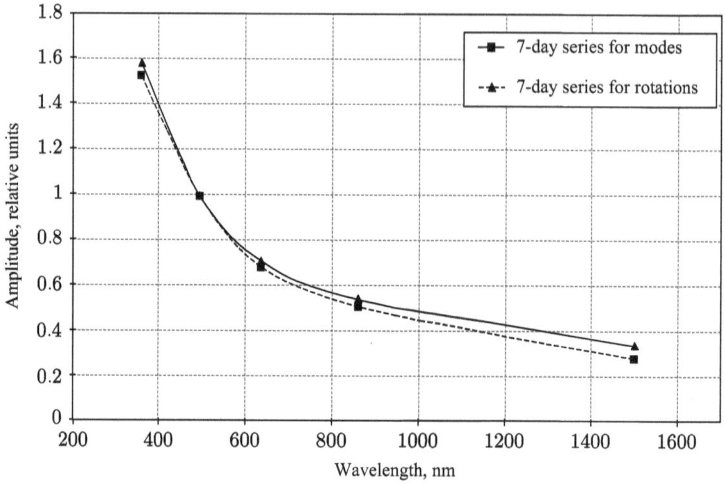


**Fig. 13** The $p$-mode amplitudes vs. the observation wavelength for individual oscillation modes and for the spectrum envelope

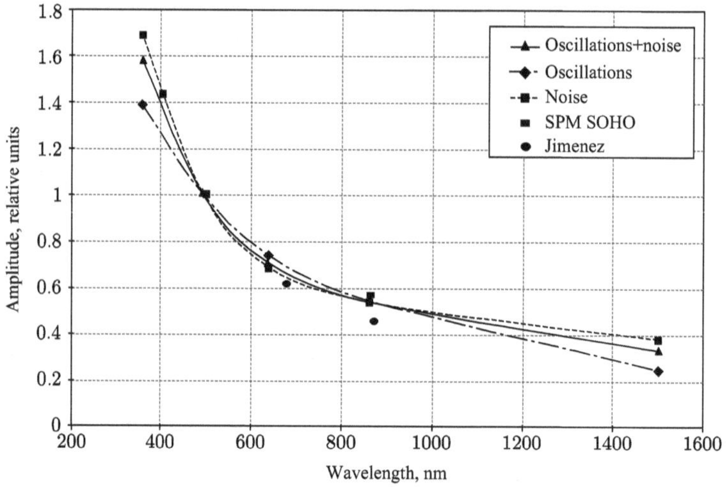


**Fig. 14** The amplitude of $p$-modes, solar noise, and total signal (oscillations + noise) as a function of the observation wavelength

One can see that the two curves agree perfectly well. A minor difference in the long-wave range is, apparently, due to the fact that the first method was applied to the range of 2,500–3,500 μHz and the second one, only to 3,000 μHz.

The spectral curves in Fig. 12 show that the $p$-mode amplitude can be separated from the solar noise by approximating the noise component to the 5-min oscillation range and, then, subtracting it from the total signal. The result of such calculation is represented in Fig. 14.

To interpret the results obtained, we need a theory of solar brightness fluctuations under the effect of the $p$-modes. It is necessary to find out how the oscillation amplitude and phase depend on the wavelength band the fluctuations are measured in.

Since the brightness fluctuations caused by the $p$-modes have small amplitudes, we can use the linear approximation and expand the expression for the solar radiation flux into minor disturbances of temperature and density. This yields the following expression for the brightness fluctuations at a frequency $\nu$:

$$\delta F_\nu = \int_0^\infty \frac{dB_\nu(\tau_\nu)}{d\ln T}\frac{\delta T(\tau_\nu)}{T_0(\tau_\nu)} E_2(\tau_\nu)\,d\tau_\nu$$
$$-\int_0^\infty \left(\frac{\partial \ln \chi_\nu(\tau_\nu)}{\partial \ln T}\right)_\rho \frac{\delta T(\tau_\nu)}{T_0(\tau_\nu)} \int_{\tau_\nu}^\infty \frac{dB_\nu(\tau'_\nu)}{d\tau'_\nu} E_2(\tau'_\nu)\,d\tau'_\nu\,d\tau_\nu$$
$$-\int_0^\infty \left(\frac{\partial \ln \chi_\nu(\tau_\nu)}{\partial \ln \rho}\right)_T \frac{\delta \rho(\tau_\nu)}{\rho_0(\tau_\nu)} \int_{\tau_\nu}^\infty \frac{dB_\nu(\tau'_\nu)}{d\tau'_\nu} E_2(\tau'_\nu)\,d\tau'_\nu\,d\tau_\nu, \qquad (19)$$

where $\tau_\nu$ is the optical depth at the frequency $\nu$, $B_\nu(\tau_\nu)$ is the Planck function as a function of optical depth, $E_2(\tau_\nu)$ is the integral exponential function, $\chi_\nu(\tau_\nu)$ is the absorption coefficient at a given frequency, $\delta T$ and $\delta\rho$ are the temperature and density fluctuations caused, in our case, by the $p$-modes, and $T_0$ and $\rho_0$ are the undisturbed temperature and density values at a given optical depth.

This formula takes into account two effects responsible for fluctuations of the solar radiation at a given frequency. The first integral in the formula accounts for the temperature dependence of the Planck function and appears as a result of expansion of the latter into minor temperature disturbances with allowance for the linear term alone. The other terms appear because the absorption coefficient depends on both temperature and density. In the linear approximation, this can be expressed as:

$$\delta\chi_\nu = \left(\frac{\partial\chi_\nu}{\partial T}\right)_\rho \delta T + \left(\frac{\partial\chi_\nu}{\partial\rho}\right)_T \delta\rho. \qquad (20)$$

If the absorption coefficient increases as a result of temperature and density fluctuations, the flux emerging from the photosphere decreases and vice versa. In order to simplify the general formula, we assumed that the $p$-modes do not differ significantly from adiabatic oscillations. In this case, the density fluctuations are expressed in terms of the temperature fluctuations as follows:

$$\frac{\delta\rho}{\rho_0} = \frac{\delta T}{(\gamma-1)T_0}. \qquad (21)$$

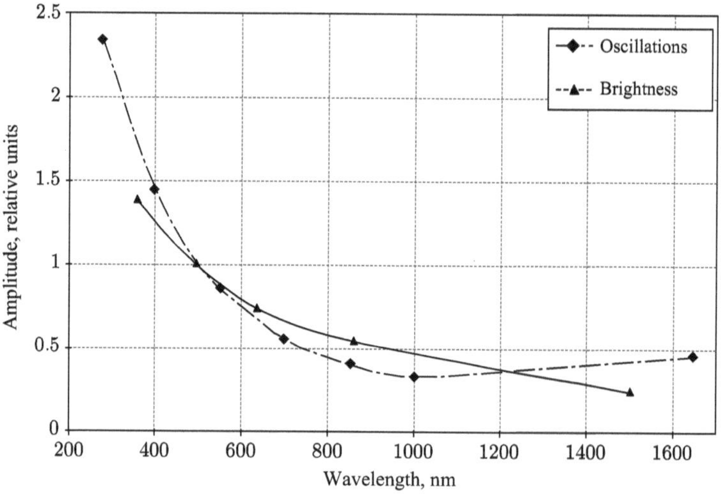


**Fig. 15** The dependence of the $p$-mode amplitude and the theoretical dependence of the solar brightness fluctuations on the observation wavelength

Using this formula, we can obtain an expression for the solar brightness, which depends only on brightness fluctuations. However, to calculate this integral expression, we need to know how the temperature fluctuations depend on depth. It may be assumed that $\delta T/T_0 = \text{const}$, i.e., is independent of depth. In this special case, we can remove $\delta T/T_0$ from under the integral sign and perform integration. Such calculations were made using the photosphere models [15]. The results are represented in Fig. 15.

A certain discrepancy between the theory and observations may be due to the assumption of nonadiabaticity of oscillations and $\delta T/T_0 = \text{const}$, as well as to the use of the mean inhomogeneous atmosphere model that takes no account of the granulation.

## References


1. Boashash, B.: Time–Frequency Signals Analysis: Methods and Applications. Wiley, New York (1992)
2. Fink, L.M.: Probl. Inf. Transf. **2**(4), 26 (1966) (in Russian)
3. Fink, L.M.: Signals, Noise, and Errors. Radio and Communications, Moscow (1984) (in Russian)
4. Gabor, D.J.: Inst. Electr. Eng. **46**(26), 429 (1946)
5. Lebedev, N.I., Kuznetsov, V.D., et al.: The helioseismological CORONAS-F DIFOS experiment. Astron. Rep. **48**(10), 871–875 (2004)
6. Rytov, S.M.: Trudy FIAN (Proc. Lebedev Phys. Inst.) **11**(1), 43 (1940) (in Russian)
7. Stebbins, R., Goode, P.R.: Waves in solar photosphere. Sol. Phys. **110**, 237 (1987)

8. Vainshtein, L.A., Vakman, D.E.: Frequency Separation in the Oscillation and Wave Theory. Nauka, Moscow (1983) (in Russian)
9. Vakman, D.E., Vainshtein, L.A.: Amplitude, phase, frequency-fundamental concepts of oscillation theory. Sov. Phys.—Usp. **20**(12), 1002–1016 (1977)
10. White, O.R., Cha, M.Y.: Analysis of the 5 min oscillatory photospheric motions. Sol. Phys. **31**, 23 (1973)
11. Zhugzhda, Yu.D.: Nonadiabatic oscillations in an isothermal atmosphere. Pisma v AZh. **9**, 631 (1983a) (in Russian)
12. Zhugzhda, Yu.D.: Non-adiabatic oscillations in an isothermal atmosphere. Astrophys. Space Sci. **95**(2), L255 (1983b)
13. Zhugzhda, Y.D.: Temperature waves in the solar photosphere. Sol. Phys. **124**, 205 (1989)
14. Zhugzhda, Yu.D.: Analytical signal as a tool for studying the solar $p$-mode properties. Astron. Lett. **32**(5), 329 (2006)
15. Zhugzhda, Y.D., Staude, J., Bartling, G.: Spectral darkening functions of solar $p$-modes—an effective tool for helioseismology. Astron. Astrophys. **305**, L33–L36 (1996)

# Study of Active Phenomena in the Solar Corona in the 8–350 Å Range by Imaging Spectroscopy Methods (SPIRIT Experiment)

**I.A. Zhitnik[1], S.V. Kuzin, S.A. Bogachev, O.I. Bugaenko, Yu.S. Ivanov, A.P. Ignatyev, V.V. Krutov, A.V. Mitrofanov, S.N. Oparin, A.A. Pertsov, V.A. Slemzin, N.K. Sukhodrev, I.I. Sobelman[2], A.M. Urnov, S.V. Shestov, and F.F. Goryaev**

## Introduction

Studies of the solar X-ray radiation with the use of spacecrafts started on the initiative of Professor S.L. Mandelstam in the Laboratory of spectroscopy at the Lebedev Physical Institute (LPI) of the USSR Academy of Sciences in the late 1950s of the twentieth century. The original purpose of these studies was to obtain general characteristics of solar short-wave radiation: flux intensity, spectral composition, temporal variations, and localization of radiation sources in the solar atmosphere. By mid-1970s more than 25 space-based experiments, including those on 15 high-altitude geophysical rockets and 7 specialized solar Cosmos and Intercosmos satellites, had been prepared and carried out. Large amounts of observational data and a number of priority findings for physics of the solar corona were collected during these studies.

The thermal nature of the X-ray emission of the "quiet" Sun was ascertained; flare dynamics and spatial structure were studied; X-ray line spectra of the "hottest" multi-charged ions of flare plasma were obtained at very high resolution. A great number of new lines, including shortest-wave resonance lines of H-like Mg XII (8.42 Å) and He-like Mg XI (9.17 Å) ions, were identified in the spectra. Polarization of X-ray flares was detected in both the continuous and line spectra, which is indicative of a dominant role of directed beams of accelerated electrons in the flare mechanism. The "satellite" spectral lines of multi-charged ions were recorded and

[1]Deceased

[2]Deceased

I.A. Zhitnik · S.V. Kuzin (✉) · S.A. Bogachev · O.I. Bugaenko · Yu.S. Ivanov · A.P. Ignatyev · V.V. Krutov · A.V. Mitrofanov · S.N. Oparin · A.A. Pertsov · V.A. Slemzin · N.K. Sukhodrev · I.I. Sobelman · A.M. Urnov · S.V. Shestov · F.F. Goryaev
LPI, Moscow, Russia

V. Kuznetsov (ed.), *The Coronas-F Space Mission*, Astrophysics and Space Science Library 400, DOI 10.1007/978-3-642-39268-9_3,

directly demonstrated for the first time that the dielectronic recombination process was essential in the ionization balance of the solar corona.

To a great extent, the study developed in the USSR was carried out in parallel with that in the USA and thereafter in England, complementing and improving each other. The success of solar X-ray emission study in the first 15 years of the Space Era was due to the high level of experimental and theoretical spectroscopy attained by the Lebedev Physical Institute of the Russian Academy of Sciences (LPI RAS). In this period the work schedule provided for annual launching one or two space vehicles with two new or improved X-ray instruments onboard. The use of innovations and non-standard engineering solutions in close coordination with spacecraft developers at all stages has become an important factor in the development of onboard scientific instrumentation.

The amount of data accumulated within the framework of rocket experiments depended on the time spent by the instrumentation outside the atmosphere (altitudes above 150 km). The total observation time for 15 launches amounted to 150 min. The net observation time for the satellite operating in the solar scan mode amounted to 4 days per 100 days on orbit.

The first stage of work has demonstrated that further progress in studying active phenomena of the solar corona necessitated primarily creation of a new type of spacecraft having long operation life (about 5 years). The spacecraft should be capable of carrying large-size telescopes (weighting up to 100 kg) providing for high-precision pointing of instruments toward the Sun and a telemetry system with transmission capacity of 150 MB/day and control capacity of about 10 kB/day. In 1975, on the suggestion of the Academy of Sciences of the USSR, the Government assigned the task of developing such a specialized satellite to the "Yuznoe" Special Design Bureau. The first new type satellite was planned to be launched during the period of the nearest solar maximum, in 1980–1981. However, the launch was rescheduled to be finally implemented in 1994.

At the same time LPI was assigned the task of designing and developing the onboard multi-wave X-ray telescope and spectrometer equipment. Based on the first stage outcome analysis, a scientific programme was adopted, and optical schemes of the future system of X-ray instruments were developed to comprise a six-channel telescope, four-channel spectroheliometer at 8.4 to 335 Å wavelengths and a polarimeter sensitive over an energy range of 20–100 keV. This system required the development of new X-ray optical components: the focusing optics based on multi-coated mirrors and concave crystals; diffraction gratings with $S = 100 \times 200\ \mathrm{mm}^2$ and $N = 3600$ lines/mm for 100–350 Å wavelengths; image detectors, i.e., large format 1000×1000 CCD arrays and electron-optical converters (EOCs) of high-resolution (over 40 lines/mm); free-standing ultrathin film X-ray filters ($\sim 200$ Å) of large diameter (up to 10 cm). All X-ray optical components have been designed in cooperation with institutions of the Russian Academy of Sciences, primarily with the Institute for Physics of Microstructures, and many other organizations in Russia. The components were tested for radiation resistance and temporal stability within the operating spectral range, subject to the period of solar maximum.

A new method which comprises abilities of solar experiments on space vehicles and achievements in design of short-wavelength imaging optical systems has been developed for studying the solar corona in the most informative 8–335 Å spectral range. The method known as "the full Sun imaging spectroscopy" consists in simultaneous obtaining images and spectra of the Sun with a high angular and spectral resolution in the XUV narrow spectral ranges and monochromatic lines being characteristic of different temperature layers of the solar plasma. This method was applied in the development of instrumentation for the subsequent experiments onboard the Phobos 1 interplanetary station (1988) and the CORONAS-I (1994) and CORONAS-F (2001–2005) solar space stations [1].

The SPIRIT experiment onboard the CORONAS-F satellite has made the most significant contribution to the investigation of the Sun's upper atmosphere. The experiment was aimed at studying the spatial-temporal structure of solar activity phenomena with the use of a set of X-ray instruments. For the SPIRIT experiment, LPI developed an instrumentation complex having no equivalent elsewhere in the world. This instrumentation allowed to simultaneously obtain spectral (ten channels) and monochromatic (in 160 spectral lines) X-ray full-Sun images with high spectral, spatial, and temporal resolution. More than 40,000 observation sessions were conducted and more than one million spectral images of the Sun and solar corona totalling over 250 GB were taken during the experiment.

This review presents the basic characteristics of measuring channels of the SPIRIT instrumental set, description of observation programmes, summary of observation sessions and scientific data obtained, as well as the main results in two areas: the research of hot coronal plasma and solar flares and the eruptive phenomena research. Other SPIRIT experiment results shall be further outlined in subsequent publications.

## SPIRIT Equipment Characteristics

The SPIRIT instrument comprised an assembly of solar SRT-C telescopes (six channels), a RES-C spectroheliograph assembly (four channels), and two optical sensors OD1 and OD2 to control the orientation of optical axes during the space flight [2]. The spectral regions recorded by individual channels, ion composition of the emission, and the relevant temperature range of the emitting plasma are shown in Table 1. The SRT-C and RES-C assemblies mounted on a platform of scientific instruments of the CORONAS-F satellite are presented in Fig. 1 (left panel).

### *Solar SRT-C Telescope Assembly*

Solar SRT-C telescope assembly (see right panel in Fig. 1) is designed to take EUV images of the Sun with high spatial resolution. The assembly consisted of a four-channel Ritchey–Chretien telescope (hereinafter referred to as T1), a two-channel Herschel telescope (T2), and two optical sensors to control the orientation of the

**Table 1** Spectral intervals of the SPIRIT instrumentation

| Instrument | Type of device | Channel | Spectral range, Å | Ions | $T$, $10^6$ K |
|---|---|---|---|---|---|
| SRT-C | Four-channel Ritchey-Chretien telescope T1 | 171 | $173 \pm 3$ | Fe IX–Fe X | 1.0 |
| | | 195 | $195 \pm 6$ | Fe XII, Fe XXIV | 1.6, 15 |
| | | 284 | $284 \pm 8$ | Fe XV | 2 |
| | | 304 | $304 \pm 8$ | He II, Si XI | 0.05, 1.6 |
| SRT-C | Two-channel Herschel telescope T2 | 175 | $175 \pm 5$ | Fe IX–Fe XI | 1.3 |
| | | 304 | $304 \pm 15$ | He II, Si XI | 0.05, 1.6 |
| RES-C | Two spectrohelio-graphs Mg XII | Mg XII | 8.418 – 8.423 | Mg XII | 5 – 20 |
| RES-C | Two two-channel spectroheliographs of the XUV range | XUV1 | 176 – 207 | O IV, Fe IX–Fe XXIV, Ca XIV–Ca XVII | 0.3 – 16 |
| | | XUV2 | 280 – 330 | He II, Si XI, Fe XV–Fe XVI, Mg VIII, Ni XVII, Ca XVII | 0.05 – 5 |

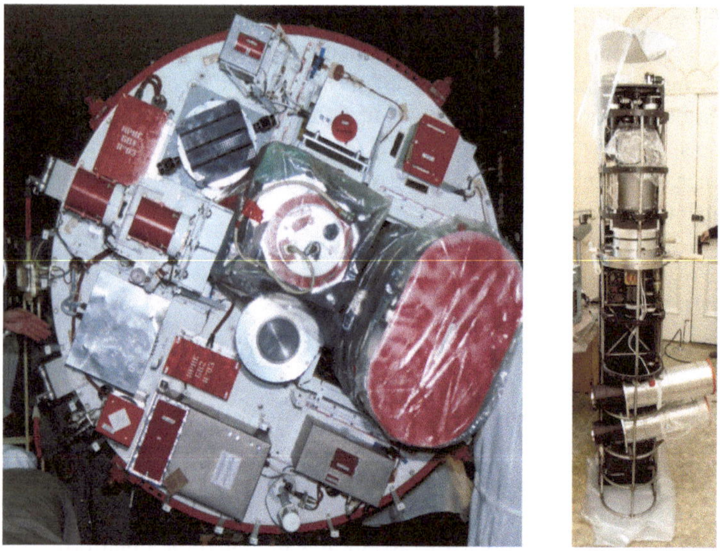

**Fig. 1** *Left panel*: view of the CORONAS-F scientific instrumentation platform with SPIRIT SRT-C and RES-C assemblies. *Right panel*: external view of the SPIRIT SRT-C telescope

telescope axes toward the stars (see Table 2). Optical schematic diagrams of SPIRIT telescopes are shown in Figs. 2 and 3.

The spectral ranges of SPIRIT telescopes have been selected to study the structure and dynamics of the Sun's transition region and lower corona within the temperature range of 0.05–2 MK. The optical scheme of the T1 telescope is identical with that of the telescope EIT (Extreme-Ultraviolet Imaging Telescope) on the SOHO (Solar and Heliospheric Observatory) spacecraft [3]. The main objective

**Table 2** Characteristics of the SRT-C telescopes of the SPIRIT instrumentation

| Type of telescope | Channel | Field of view, arc min | Angle scale, arc sec/pix | Imaging mode |
|---|---|---|---|---|
| Four-channel T1 | 171, 195, 284, 304 | 42 × 48 | 2.5″ | Full disk |
| Two-channel T2 | 175, 304 | 45 × 50 | 2.6″ | Full disk, corona up to 2 – 5 $R_{sun}$ |
| Optical sensors | 4000 – 8000 Å | Ø11° | 1′ × 2′ | Stars registration up to $12^m$ at the angle 90° to the direction of the Sun |

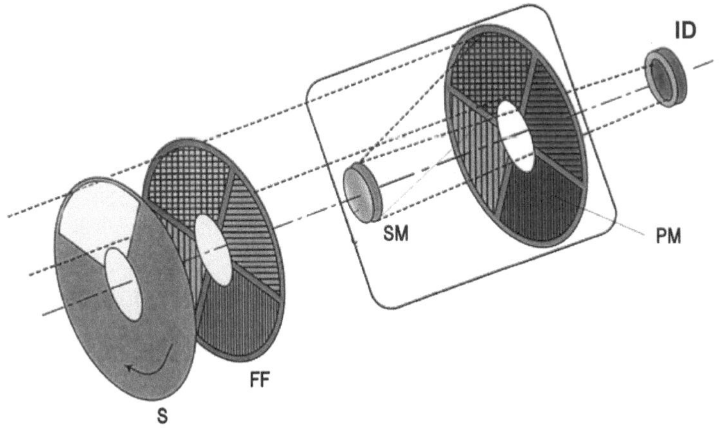


**Fig. 2** Schematic diagram of the SPIRIT Ritchey–Chretien telescope T1: *PM, SM* primary and secondary mirrors of the Ritchey–Chretien objective, *ID* image detector, *FF* front filters, *S* channel selection shutter. From [15, p. 489–500, Fig. 1]

of the T1 telescope was that of synoptic monitoring of solar plasma dynamics in the EUV range, joint observations with the RES-C spectroheliograph channels for the localization of solar structures and events, and coordinated observations with the EIT telescope onboard the SOHO spacecraft. A set of the T1 telescope mirrors, being identical to that of the SOHO/EIT telescope, was developed by the Institute of Space Astrophysics (Institut d'Astrophysique Spatiale, Orsay, France). The Ritchey–Chretien telescope with a diameter of 120 mm and a focal length of 1,660 mm consisted of two Mo-Si multi-coated mirrors having four sectors at wavelengths of 171, 195, 284, and 304 Å.

Telescope T2, with its normal incidence multilayer optical system and CCD-based detectors, is a third- generation solar telescope developed by LPI over the past 20 years. The telescopes TEREK onboard the Phobos 1 spacecraft [4] (1988) and TEREK-K onboard the CORONAS-I satellite [5, 6] (1994) became the prototypes of the above telescope. As compared to the TEREK telescope onboard the CORONAS-I satellite, the SPIRIT T2 telescope had a 2.6 times larger aperture as well as considerably higher sensitivity and angular and temporal resolution.

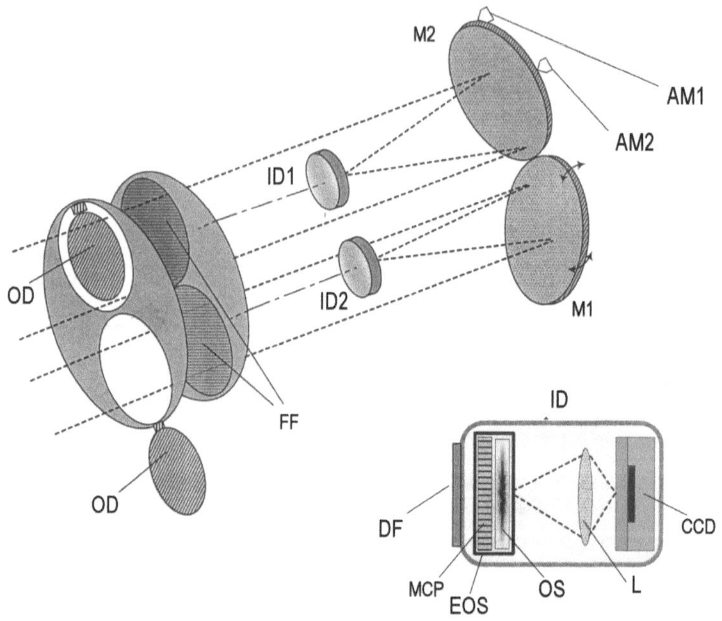


**Fig. 3** (**a**) Schematic diagram of the SPIRIT Herschel telescope T2: *M1, M2* off-axis paraboloid multi-coated mirrors, *ID* image detectors, *FF* front filters, *OD* occulting disks of the coronagraph, *AM1* and *AM2* angle and focus adjustment mechanisms; (**b**) schematic diagram of the image detector: *DF* detector filter, *MCP* microchannel plate of the electron-optical converter (EOC), *OS* output screen, *L* lens, *CCD* charge coupled device (array). From [15, p. 489–500, Fig. 1]

The Herschel telescope optical system consisted of off-axis paraboloids of 100 mm diameter and 1,500 mm focal length with a Mo-Si reflective coating. It has been jointly produced by LPI and the Institute for Physics of Microstructures of RAS by applying the original technology of aspherization by deposition of multilayer coating [7]. The Herschel telescope mirrors have angular and focal adjustment mechanisms to correct possible changes in adjustment during the satellite launch.

The main objective of the T2 telescope observations was to study solar structures in two spectral channels (175 and 304 Å) with the highest possible temporal resolution. Wide dynamic range of the instrument enabled observations during both high and low solar activity. It became the first instrument in the solar X-ray astronomy to implement the principle of taking simultaneous images with the coequal exposure in the spectral regions 175 Å and 304 Å, which correspond to the low corona and the transition region of the Sun. Besides, T2 telescope was used as a coronagraph with an external occulting disk to study EUV radiation of the middle corona at distances of up to $5R_{\rm sun}$ [8, 9].

Figure 4a illustrates the T2 telescope transition from the "disk" mode into the coronagraph mode. In the "disk" mode (position 1), the occulting disk in front of the telescope aperture is open. The solar disk image is focused by the telescope mirror onto the center of the detector. On the transition to the "corona" mode (position 2), the telescope mirror is tilted by radio command at an angle from 0° to

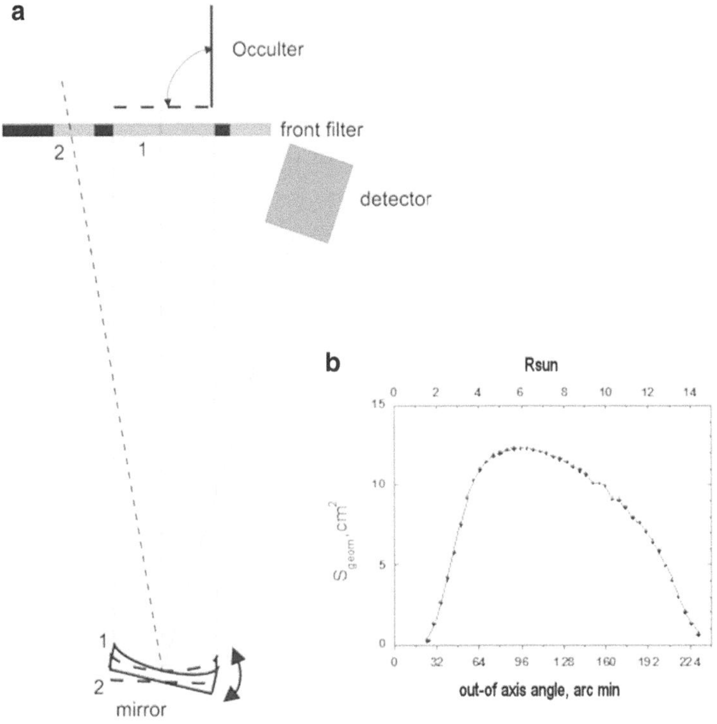


**Fig. 4** (**a**) T2 telescope operation scheme in the coronagraph mode; (**b**) vignetting function of the coronagraph. From [57, Fig. 1]

1.5°, so that the selected area of the solar corona is projected onto the center of the detector. In this position, an image of a displaced solar disk can be recorded; then the occulting disk covers the central part of the optical window upon command thus blocking out the disk emission, and solar corona images will be recorded through the lateral optical windows. Figure 4b presents the telescope geometric area in the coronagraph mode as a function of an off-axis angle (vignetting function). Geometrical area growth with the distance from the limb to 5 solar radii can to a great extent compensate for the fall of corona intensity, the effect being similar to that of a gradient filter in visible coronagraphs.

Image detectors of T1 and T2 telescopes are identical (Fig. 3b), consisting of electron-optical converters (EOCs) with an open microchannel plate (MCP), projecting lens, and CCD arrays. A modified EP-10 EOC converts EUV images into electronic images. The latter upon increasing their brightness are converted into visible images to be then focused by the lens onto the 1024×1152 CCD array. The EOC serves additionally as an electronic shutter to close the image when reading a frame from the CCD array. The EOC gain can be changed within about 200 times as much through MCP voltage commands (4 steps). Given the gain changes in the detectors, the dynamic range of brightness detected in EUV channels amounts to about $10^5$, and exposure times can vary between 0.01 s and 600 s.

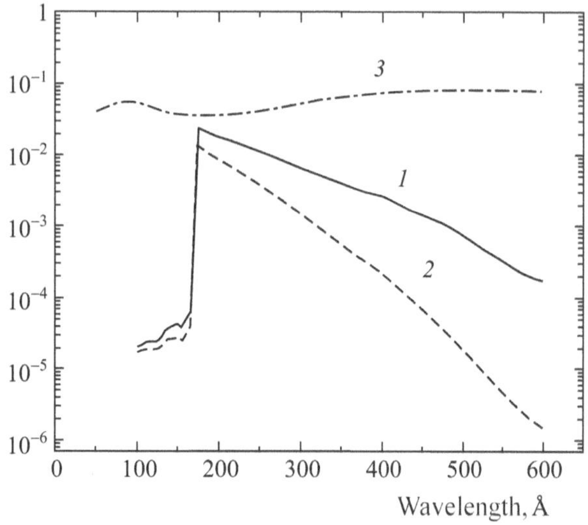


**Fig. 5** Spectral transmission functions of the filtering system (front filter + detector filter) used in all SPIRIT T1 and SPIRIT T2-304 telescope channels (1), the same of the SPIRIT T2-175 channel (2) and quantum efficiency of the MCP (3). From [15, p. 489–500, Fig. 1]

For small and medium light intensity values and moderate gain, the detector signal is a linear function of the incident flux. With high exposure levels and the maximum gain, the brightest elements in an image (active regions and flares) exhibit nonlinearity which on the verge of saturation equals up to 25–30 %. This property can be useful for the increase of the maximum registered brightness when bright flares are observed, thus making it possible to distinguish image details before the complete saturation.

Generally, the exposure times of 2 s or 9 s were used with an average EOC gain during the solar disk observations. In the coronagraph mode, the maximum gain was set, and exposure time amounted thus to 150 s or 300 s. The telescope sensitivity in this mode increased 140 times compared to that in the disk image recording mode in the 175 Å channel, and 320 times in the 304 Å channel.

The front filters, an aluminium foil about 0.25 $\mu$m thick on the supporting grid, were placed at the entry of channels of both telescopes to filter the X-ray flux and suppress the intense visible and UV solar radiation. In the channel T2-175, the foil is applied to a 600 Å base substrate of organic Formvar. There was no such base substrate in the other channels of T1 and T2 telescopes. Entry filters were similar by their design and properties to those previously used in the TEREK-K telescope onboard the CORONAS-I satellite [10]. Track-etched membrane filters were provided for additional protection from visible light at the entrance of detectors. The membranes consist of 0.15-$\mu$m-thick aluminium foil applied onto an 8.5 $\mu$m Mylar layer with a 1.5 $\mu$m pore diameter [11]. Figure 5 shows the total filter transmission and the MCP quantum efficiency [12] against the EUV radiation wavelength. Figure 6 presents spectral functions of the effective area of T1 and T2 telescope channels, where the net effective area is defined to be the product of the mirror geometric area, reflection coefficients, filter transmission, and detector quantum efficiency.

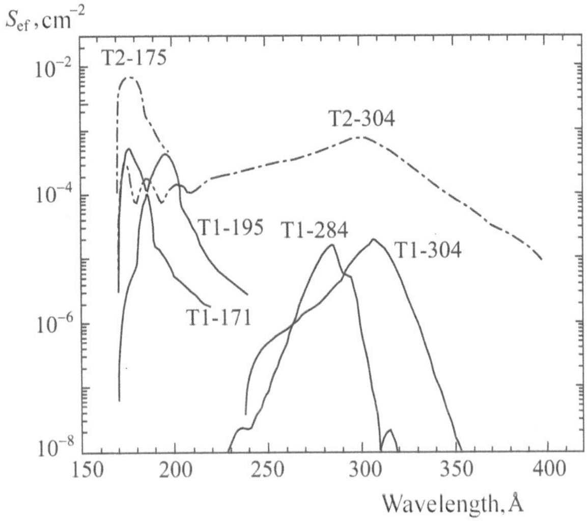


**Fig. 6** The net effective area of SPIRIT T1 (*solid lines*) and T2 (*dash-dotted lines*) telescope channels as a function of wavelength. From [57, Fig. 1]

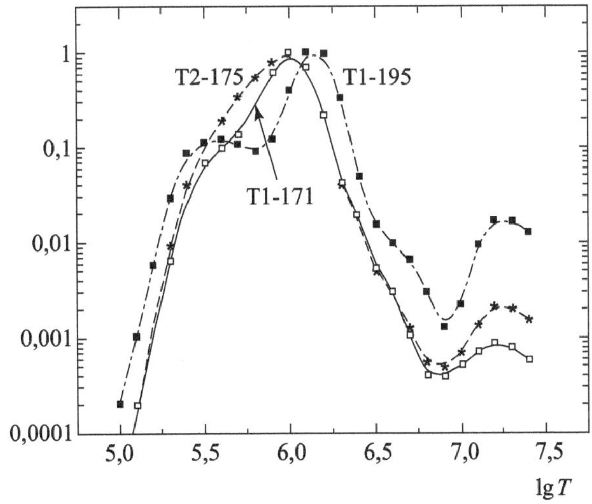


**Fig. 7** Normalized temperature response functions for SPIRIT T1-171, T1-195, and T2-175 channels. From [57, Fig. 1]

A solar plasma emission function $G(n_e, T, \lambda)$, calculated using the CHIANTI database [13] and the method described in [14], was applied to estimate the temperature response functions [15]. Normalized temperature response functions for different channels of the SPIRIT telescopes are presented in Figs. 7 and 8. For the T1 telescope channels, spectral composition of the detected radiation is almost the same as that in similar channels of the EIT telescope.

In the T2 175 Å channel, aside from the main lines Fe IX to Fe XI at wavelengths of 171, 174, 177, and 184 Å ($\lg T[\mathrm{K}] = 5.5 - 6.8$), there is a minor proportion (less than 1 %) of "cold", O IV–O V 182 and 172 Å lines at a temperature of $\lg T[\mathrm{K}] < 5.4$. In flares, "hot" Fe XX 173 Å and Fe XXIV 192 Å lines ($\lg T[\mathrm{K}] = 7.0 - 7.5$) make a considerable contribution. Comparison of the temperature response function

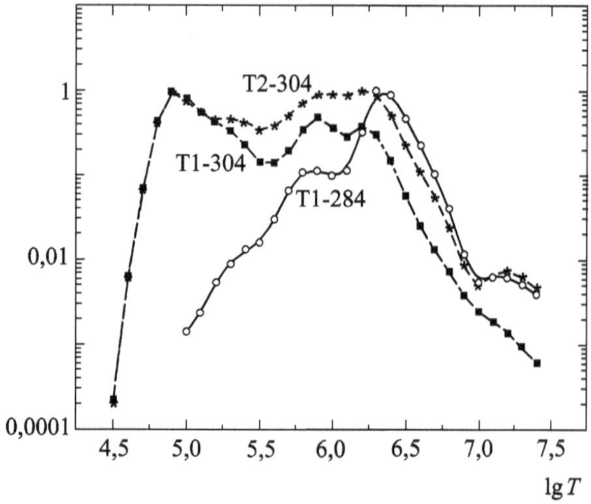


**Fig. 8** Normalized temperature response functions for SPIRIT T1-304, T1-284, and T2-304 channels. From [57, Fig. 1]

of the above channel versus that of T1 171 Å and T1 195 Å channels shows that the contribution of hot lines within the T2 175 Å channel is almost the same as that in the T1 171 Å channel and two times less than that in the T1 195 Å channel, with the maximum effective area of the T2 175 Å channel being, respectively, 12 and 15 times as much.

The largest contribution to the channel T2-304 is that of the He II 303.8 Å transition region line excited at $\lg T\,[\mathrm{K}] = 4.0 - 5.5$. At the same time a significant contribution (especially in active regions and flares) is made by a line of ions Mg VIII 315 Å, Si XI 303.3 Å, Fe XV 284 Å, Fe XVI 335 Å and Fe XXIV 255 Å emitted by the coronal plasma at a temperature of $\lg T\,[\mathrm{K}] = 5.8{-}7.5$. Due to the use of single reflection mirrors, the efficiency of the T2-304 channel is significantly (40 times) higher than that of the T1 telescope channel. However, for the same reason, spectral width of the T2-304 channel is much greater, and therefore the contribution of "hot" components is bigger.

## *RES-C X-Ray Spectroheliograph Assembly*

RES-C spectroheliograph assembly is designed to take Sun's images in spectral lines of highly ionized atoms of abundant elements (helium, magnesium, iron, and silicon) with high spatial and spectral resolution. RES-C spectroheliograph includes four channels: two X-ray MgXII channels and two channels of XUV vacuum ultraviolet range (see Table 3). Crystal optics and multilayer imaging optics as well as diffraction gratings were used in RES-C spectroheliograph channels to record images of the Sun. Image detectors in the XUV channels as well as in the SRT-C instrument were built on CCD arrays combined with EOCs. X-ray images in the MgXII channels were recorded directly with open CCDs.

**Table 3** Characteristics of RES-C spectroheliograph channels

| Channel | Spectral bandpass, Å | Spectral resolution, Å/cell | Angle scale over a cell | Working regime |
|---|---|---|---|---|
| Mg XII (two channels) | 8.41–8.43 | $10^{-3}$ | 4.1″ | Monochromatic images of the Sun in the Mg XII 8.42 Å line |
| XUV (2 channels) | 176–207, 280–330 | 0.03 | 6″ (⊥ dispersion) | Spectral images of the disk in all the lines (more than 150) |

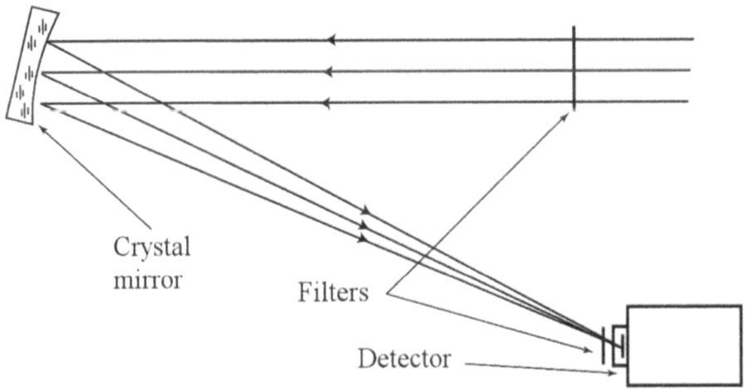


**Fig. 9** Optical scheme of Mg XII channels of the RES-C spectroheliograph

**Mg XII Channel Based on Crystal Mirrors.** Mg XII channel of the RES-C spectroheliograph is a unique scientific instrument in the world to obtain monochromatic images of the Sun in the Mg XII resonance line ($\lambda \approx 8.42$ Å). The spectral line emission occurs only in enough "hot" ($T > 5$ MK) coronal plasma.

It is virtually impossible to use normal incidence mirrors, usual for the visible range, to record solar images at wavelengths shorter than 10 Å due to a negligibly low reflection coefficient. Either pinhole cameras [16] or grazing incidence mirrors [17] have been used in instruments hitherto as focusing elements. The first type optics disadvantage is its low light gathering power. Grazing incidence mirrors are extremely difficult to manufacture as they require high optical quality of aspherical surfaces combined with ultra-smooth surface and a special control system. Besides, both types of optics are polychromatic, and spectral ranges are primarily determined by thin-foil filters.

Crystal focusing optics, with a high reflection coefficient and dispersion, is free from the above restrictions; it is comparatively simple to manufacture and control while enabling high light gathering power of the design. The modified Gamache scheme on a spherically bent crystal was used in Mg XII channels at 8.42 Å for infinitely remote sources [18, 19]. Optical scheme of Mg XII channels is shown in Fig. 9. Incident radiation reaches the pre-filter (aluminized mylar), then the quartz crystal mirror that focuses the image onto the detector. The detector is based on a front illuminated CCD array. There is an additional filter (aluminized mylar) in front of the array. The scheme parameters have been selected so that the image at a wavelength of $\sim 8.42$ Å is built on the detector. Since only Mg XII line falls within the narrow spectral range near 8.42 Å in the solar radiation spectrum, no special filtering of the adjacent spectral lines in this channel was required.

*Crystal mirror.* In the above scheme, a spherically curved crystal serves simultaneously as a focusing element and a dispersing element. Crystal parameters have been specially selected to ensure the diffraction angle as close to normal as possible in order to reduce the influence of field aberrations of the spherical mirror. At the same time, due to the use of spherical aberration inherent in spherical mirrors of an off-axis installation, the spatial and spectral image components of such a scheme could be separated.

The Bragg law lays the basis for the focusing crystal optics operation:

$$2d \cdot \sin\varphi = m \cdot \lambda\,, \tag{1}$$

where $d$ is a lattice constant, $\varphi$—an angle of radiation grazing incidence on the crystal, $m$—diffraction order, and $\lambda$—radiation wavelength.

Where a monochromatic beam is incident from a point source located at infinity on a spherically curved crystal (Fig. 10), the diffraction occurs in a narrow annular zone with a radius by

$$r = R\cos\varphi = R\sqrt{1-\left(\frac{m\lambda}{2d}\right)^2}\,, \tag{2}$$

where $R$ is a radius of curvature of the crystal. With the proper location of the recording plane, the diffracted radiation will be point-focused.

Varied wavelength radiation is diffracted in different annular zones of the mirror, with point focusing being possible only for one wavelength. As a result of spherical aberration inherent in this scheme, images at other wavelengths will take the form of concentric circles. The longitudinal spherical aberration for the image at the wavelength of $\lambda_0 + d\lambda$ is described by:

$$dl = -\frac{2d \cdot R}{2 \cdot m \cdot \lambda_0^2} \cdot d\lambda\,, \tag{3}$$

where $\lambda_0$ is the wavelength at which the optical system is focused.

The transverse spherical aberration is equal to the radius of the defocused image circle in the detector plane:

$$ds \approx dl \cdot \mathrm{ctg}\,\varphi\,. \tag{4}$$

Equation (4) determines the dispersion of the scheme in the detector plane.

**Fig. 10** Principal optical schematic diagram of the crystal mirror operation

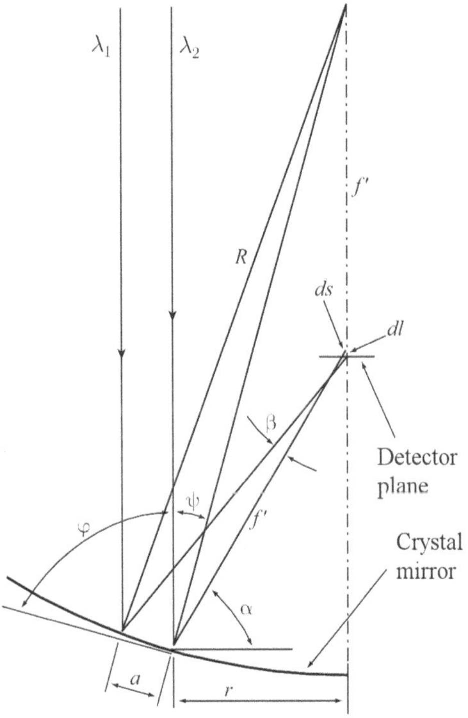


The spectral resolution for a point source will also depend on a spectral line width ($\sim$ 0.005 Å for the Mg XII doublet lines) and a crystal rocking curve width.

A comparatively small spherical crystal mirror, used in each Mg XII channel of the RES-C spectroheliograph, can be considered as a sector "cut out" of a full annular mirror. Therefore, when focusing a channel on one of doublet spectral lines, the point source in another line will be imaged in the form of arc segments. An uncertainty may arise in the separation of spatial and spectral image components for a spatially extended object, which is the case for the Sun. To address this phenomenon, two identical Mg XII spectroheliograph channels are placed with their dispersion being in mutually perpendicular directions.

Crystal mirrors of Mg XII channels are made of quartz crystal $2d_{\text{nom}} = 8.501$ Å (Y-cut, [10.0]). Such crystal orientation ensures Bragg reflection angle close to normal ($\psi \sim 8.0°$), which allows for a sufficiently compact scheme. Besides, the width of the quartz crystal rocking curve for the operating wavelength is less than $2'$, which can result in narrow diffracted regions in crystals and, as a consequence, in low aberration for narrow spectral lines. The mirror's bend radius $R = 1295$ mm, and the aperture of the crystal body is 60×40 mm. The main parameters of the optical system of Mg XII channels are listed in Table 4.

*Filters.* Dual radiation filtering system consisting of a pre-filter and detector filter was installed in the Mg XII channels of the RES-C spectroheliograph.

**Table 4** Principal optical characteristics of Mg XII channels

| Parameters | Wavelength<br>Mean value | $\lambda_1 = 8.418$ Å | $\lambda_1 = 8.423$ Å |
|---|---|---|---|
| $\varphi$ (°) | 81.92 | 81.80 | 82.04 |
| $r$ (mm) | 182 | 184.7 | 179.3 |
| $a$ (mm) | 5.4 | | |
| $f'$ (mm) | 654 | 654.192 | 653.804 |
| $\beta$ (°) | 0.24 | | |
| $dl$ (mm) | 0.381 | | |
| $ds$ (mm) | | 0.055 | 0.053 |
| $a_{30}$, working crystal zone for angular field 30′ (mm) | 11.3 | | |

Long-wavelength radiation was partially filtered by a crystal mirror with low reflection coefficient in this range (on the order of a few percent). Pre-filter consisted of two identical mylar film pieces 3.6 μm thick sputtered on both sides with aluminium layers of about 0.15 μm each. The detector filter was placed in front of the detector sensitive area. It consisted in a 2.5 μm thick uniform mylar film with five deposited layers of aluminium (with a total thickness of about 0.3 μm) deposited by magnetron sputtering (three layers to the outer side and two layers to the inner side). The filter system transmission amounted to $\sim 30\,\%$ for $\lambda = 8.4$ Å.

*Detector.* Images in the Mg XII channels of the RES-C spectroheliograph were recorded directly on the CCD array. A "Leader"-type CCD array (1024×1152 elements, 13 μm × 13 μm each) with frontal recording was implied. Conversion of X-ray emission into electrons occurred in the sub-electrode silicon layer. Conversion efficiency is equal to 3.2 eV/e, i.e., one photon transfers its energy of 1 KeV into about 300 photoelectrons, which is sufficient for reliable registration of single photons. In the instrument, CCDs without anti-blooming were used, which allowed using detectors as photometers with regard to strong signals, which are in order of magnitude greater than that of a signal required to fill the potential well of a single cell.

CCD arrays intended for direct detection of extreme EUV radiation in a long-term independent experiment must be highly resistant to radiation. During the preparation for the experiment, the applied arrays were tested for radiation resistance [20]. It was demonstrated that their resistance to space radiation was about $10^6$ rad, which appeared to be sufficient for more than 3 years of operation.

The basic parameters of Mg XII channels of the RES-C spectroheliograph are set forth in Table 5.

**XUV Channels of RES-C Spectroheliograph.** The primary objective of developing XUV channels in the range of 170–330 Å was to take full-disk monochromatic images of the Sun, with their width in terms of the wavelength scale not exceeding 1 Å [21–23]. It was necessary to reduce the superimposition of solar images in some spectral lines, which is determined by density of spectral lines in the relevant range and angular size of the Sun.

**Table 5** Mg XII channel parameters of the spectroheliograph RES-C

| Parameter | Parameter value |
|---|---|
| Entrance aperture for point (1′) monochromatic source | 1 $cm^2$ |
| Field of view | 70′ × 78′ |
| Resolution (over a cell) | 4.1″ |
| Spectral resolution (over a cell) | $1.2 \times 10^{-3}$ Å |
| Transmittance of the optical system for the working range of the spectrum | 10 % |
| Type of the detector | Open CCD |
| Format of the frame | 1024 × 1152 |
| ADC capacity | 14 |
| Time of exposure | 1–300 s |

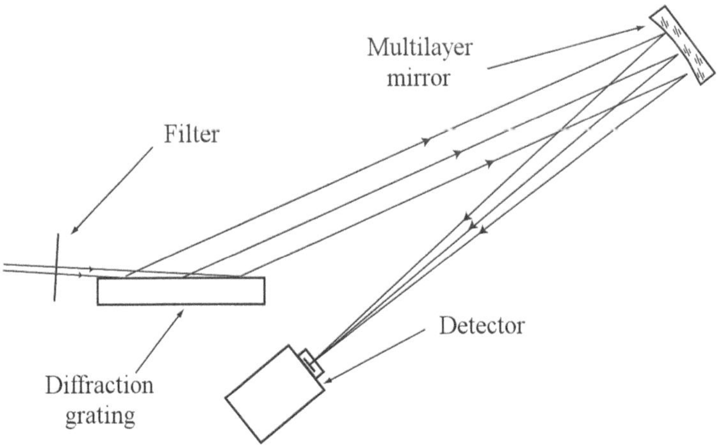


**Fig. 11** Optical scheme of XUV channels of the RES-C spectroheliograph. From [38, Figs.1–3, 4a, b, 7–10]

XUV channels of the RES-C spectroheliograph were built on a slitless optical scheme with a plane diffraction grating at the entrance [24] (Fig. 11). The solar radiation entered through the entrance filter and fell on a diffraction grating mounted at a grazing angle. Then, the diffracted radiation was focused by a normal-incidence multilayer mirror onto the sensitive area of the image detector.

*Optical scheme.* Diffraction of radiation on the reflection grating is described by the following relation:

$$m \cdot \lambda \cdot N = \cos\varphi - \cos\phi\,, \quad (5)$$

where $m$—the order of diffraction, $\lambda$—wavelength of the diffracted radiation, $N$—grating constant, $\varphi$ and $\phi$—grazing angles for the incident and diffracted light, respectively.

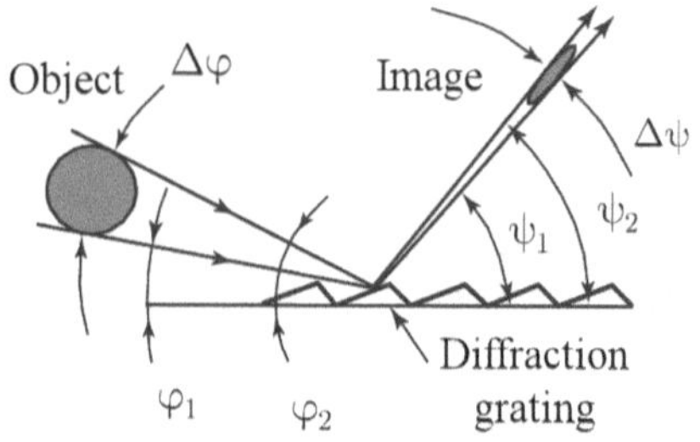


**Fig. 12** Effect of image "compression" on a reflecting grating in case of grazing incidence diffraction. From [38, Figs.1–3, 4a, b, 7–10]

From formula (5), we can obtain dependence of variation of the angular aperture of the beam due to diffraction:

$$\frac{d\phi}{d\varphi} = \frac{\sin\varphi}{\sin\phi}. \tag{6}$$

It follows from Eq. (6) that given a grazing incidence angle less than the radiation diffraction angle, the angular aperture of the diffracted beam in the dispersion plane is less than the original one (Fig. 12). In this case, there occurs an effect of spectral compression leading to the reduction in angular size of the image in the dispersion direction.

The angular size of an object that can be recorded without the superposition of wavelengths with a difference of $d\lambda$ is given by the expression

$$d\varphi = \frac{m \cdot N}{\sin\varphi} d\lambda. \tag{7}$$

Linear dispersion of the scheme in the detector plane is expressed by the relation

$$dl = \frac{f' \cdot m \cdot N}{\sin\varphi} d\lambda, \tag{8}$$

where $f'$ is a focal distance of the multilayer mirror.

The following optical elements were used in the XUV channels of the RES-C spectroheliograph:

- diffraction grating with a spatial frequency of 3,600 $mm^{-1}$,
- spherical multilayer normal incidence mirror with a focal length of 830 mm in the 176–207 Å channel,
- parabolic multilayer normal incidence mirror with a focal length of 570 mm in the 280–330 Å channel.

*Detector.* Two-axis sensors based on open microchannel plates (MCP) and CCDs with a scheme similar to those installed in T1 and T2 telescopes (see Fig. 3b) were used as image detectors in XUV channels [25]. As distinct from the telescope detectors, $MgF_2$ photocathodes were applied to the surface of the open MCP in XUV

**Table 6** XUV channel parameters of spectroheliograph RES-C

| Spectral band, Å | 176–207 | 280–330 |
|---|---|---|
| Grazing angle, ° | 1.16 | 1.40 |
| Diffraction angle for $\lambda_{mean}$, ° | 21.46 | 27.36 |
| Spectral resolution on a cell, Å | 0.024 | 0.043 |
| Spectral distance between re-superimposed images of the Sun, Å | 0.44 | 0.52 |
| Spatial resolution on a pixel in the direction perpendicular to the dispersion | 5.4" | 7.9" |
| Spatial resolution on a pixel in the direction of the dispersion | 97.5" | 148.6" |

channels, which increased the quantum efficiency for radiation with wavelengths of 200–300 Å up to ~ 30 %. Such a detector with low sensitivity to radiation with a wavelength exceeding ~ 1,500 Å provides additional long-wave radiation filtering.

*Filters.* Due to the use of a solar blind detector and to spectral filtering of entering radiation with the diffraction grating, only one filter installed at the entrance was used in the XUV channels of the RES-C spectroheliograph. It was made of 0.05±0.15 $\mu$m aluminium foil fixed on a rigid steel supporting grid. Transmission in the working spectral range amounted to ~ 20 % [26].

The basic parameters of XUV channels of the RES-C spectroheliograph are listed in Table 6.

## *SPIRIT Electronic System*

The SPIRIT microprocessor-based electronic system is designed to perform the following functions:

- reception of commands from the satellite's onboard systems to set an operating mode and to execute observational programmes;
- control over all electronic and mechanical units of the instruments;
- reception of data from detectors and operation information from the executing equipment;
- information coding and packaging it in the information digital array format to transmit through the *telemetry system*.

The electronic system of the SPIRIT assembly is based on a flexible design. It can operate both on the basis of pre-recorded cyclograms and those directly transmitted on board in the form of interactive files. The possibility of data compression without loss is implemented. The system includes an option of onboard reprogramming, which enables to enter, as may be necessary, onboard data processing elements and to use adaptive observation modes.

Structurally, the SPIRIT electronic system consists of two independent parts operating SRT-C (solar X-ray telescopes) and RES-C (X-ray spectroheliographs) instruments, respectively. Each subsystem had its own microprocessor unit, memory, power supply units, and interfaces with onboard systems and could be controlled independently. Quotas for the volume of information to be transmitted for each of the subsystems were set when planning the observations, and could be flexibly adjusted depending on the session type and solar activity status. In particular, several continuous Sun observation sessions have been carried out in the Mg XII line for 2 weeks, in the phases of fully illuminated orbit, involving virtually the entire telemetry volume.

## Solar Observations and Information Processing in the SPIRIT Experiment

### *Types of Observation Sessions and Observation Programmes*

The observation programme included the following types of sessions:

- synoptic observation sessions;
- series of observations in particular channels of SRT-C and RES-C instruments to study the dynamics of active regions, loops, arches, etc. in the telescope channels and channels of Mg XII and XUV spectroheliograph, in the full solar disk mode and within the selection window on the disk to increase the number of frames;
- corona observations with the use of a telescope/coronagraph at 175 and 304 Å;
- synchronous observations in several channels, e.g., telescope channels and channels of MgXII, MgXII and XUV spectroheliographs, within the periods of high solar activity in anticipation of flares;
- sessions of particular channel calibration and adjustment operations (T1 and T2 telescopes, Mg XII and XUV spectroheliographs);
- observation of the solar radiation absorption in the atmosphere at satellite rising and setting.

Stellar sky observations were carried out at each orbit with the use of optical sensors. Such observations are required to determine the orientation of optical axes of instruments and to analyze satellite dynamics.

Solar observations were carried out every day on the illuminated side of each of the 15 orbits (on average, within 10–48 min of a 94–92 min orbital period). Atmosphere observations were performed during 5–7 min in normal occulting orbits and during up to 45 min at non-occulting orbits, with the tangential immersing of the line-of-sight into the atmosphere. The number of observation sessions and amount of the information received was limited by restrictions on the quantity of radio commands transmitted on board and on the telemetry data volume.

The main solar observation modes with the use of SPIRIT equipment are summarized in Table 7.

**Table 7** Set of main regimes of observations of the Sun by means of the SPIRIT instrumentation

| Type of the seance | Duration of the seance $\Delta T$ and period of repetition $T$ | Channels | Quantity |
|---|---|---|---|
| Synoptic seances | $\Delta T = 30$ min<br>$T \sim 1.5$ h–1 day | All channels | 15 % of the total time |
| Observations with high temporal resolution (dynamics) | $\Delta T \sim 10$ s–30 min<br>$T = 1$–30 days | 175 Å, 304 Å, Mg XII | 40 % of the total time |
| XUV spectroscopy | $\Delta T \sim 10$ s–30 min<br>$T = 1$–30 days | 177–210 Å, 285–335 Å | 20 % of the total time |
| Observations of the corona under the coronograph regime | $\Delta T \sim 5$–30 min<br>$T = 1$–14 days | 175 Å, 304 Å | ~ 15 cycles |
| Monitoring of CMEs (SOHO keyhole) | $\Delta T = 15$ min<br>$T = 3$ weeks–2 months | 175 Å, 304 Å | 10 % of the total time |
| Patrol of flares in the Mg XII channel | $\Delta T = 5$–30 min<br>$T = 6$–24 h | Mg XII | ~ 200 cycles |
| Coordinated observations (MEDOC, EIT High Cadence, JOPs) | In accordance with the program | 175 Å, 304 Å, Mg XII | ~ 15 programs |
| Calibration seances | 1 time over 3–6 months | All channels | ~ 100 |
| Observations of the atmosphere | $\Delta T = 5$–90 min<br>$T = 12$ h–3 months | 175 Å, 304 Å | 5 % of the total time |

## *Coordinated Observations*

In 2002–2005, SPIRIT instrumentation was engaged in the coordinated observations with the use of SOHO observatory instruments and TRACE telescope under the "EIT Shutterless" programme. The programme coordinated by the Royal Observatory of Belgium (ROB) is aimed at studying temporal dynamics of solar structures with high angular resolution in several spectral regions simultaneously. The structures to be studied are active regions, coronal holes, filaments, quiet sun areas, etc. The EIT and TRACE telescope features enable observations of a part of the solar disk within 2 h with a frame repetition period of about 1 min. SPIRIT T1 and T2 telescopes provided technical facilities for observations of the full solar disk simultaneously at two wavelengths with a period of up to 7 s. Actually, due to telemetry data volume limitations, SPIRIT carried out observations of a part of the solar disk (up to $19' \times 19'$) with a cadence of about 30 s or the full solar disk with a cadence of 50 s during the illuminated part of the orbit (45 min).

Nine sessions of coordinated observations involving SPIRIT telescope were conducted with the use of SOHO and TRACE instruments: 13.10.2002 (17:48–19:44 UT), 08.01.2003 (17:37–19:44 UT), 27.03.2003 (14:07–16:25 UT), 20–24.06.2003 (17:00–19:11 UT), 9–17.09.2003, 14.12.2003, (12:00–14:42 UT),

02.06.2004 (21:00–21:40 UT), 30.11.2004 (17:36–19:39 UT), and 16.02.2005 (04:38–23:58 UT).

At the request of the Solar Data Centre MEDOC (Orsay University, France), the SPIRIT telescope was involved in 2-week observation campaigns in conjunction with SOHO instruments (CDS, EIT, LASCO, and UVCS) and the TRACE telescope: from 21.10.2002 to 01.11.2002 and from 19.05.2003 to 01.06.2003.

In June 2003, the positioning device of one of the SOHO transmitting antennas failed. As a result, the space station cannot transmit telemetry data quarterly for a period of 2 weeks. At the same time, these data are necessary for the continuous monitoring of the Sun by space weather services, in particular, SIDC (Solar Influence Data Centre) at the Royal Observatory of Belgium. Therefore, since 2003 and until the end of the CORONAS-F flight, during SOHO silence periods, the SPIRIT telescope carried out observations at the wavelength of 175 Å with a frequency of one frame per 15 min at the illuminated part of orbit. The unique solar activity data obtained from these observations, in particular, on powerful flares in September 2005, are absent in the SOHO database.

## *In-Flight Control over SPIRIT Instrumentation*

SRT-C and RES-C instrumentation design provided for two options of in-flight control over these instruments:

- using fixed and flexible (reprogrammable) cyclograms stored on board (16 fixed and 32 flexible cyclograms for each of RES and SRT instruments);
- using interactive command files (CF) downloaded into the SPIRIT electronic system immediately before an observation session.

The first control mode implied separate combinations of radio commands including a certain cyclogram to be transmitted through the satellite onboard control system (OCS). The second mode control implied digital command files to be transmitted from Earth via the interactive communication system. Command files are to be loaded into the SPIRIT instrumentation through the Scientific Data Collection System (SDCS, aka SSNI). It was an experimental mode, since the interactive communication system was installed within the CORONAS-F project for the first time ever in Russian extra-terrestrial solar observations, and there was no experience of its practical use. However, in October 2001, when due to an OCS technical malfunction it turned out to be impossible to transmit certain types of commands, solely interactive mode was used for controlling the SRT-C and RES-C instruments. Interactive communication system was also used for onboard processors reprogramming.

One week prior to sessions, operators compiled packages of one-time commands for switching the SPIRIT equipment on and off, including session times and switching commands for 3–4 days. Operating control (operating mode selection for each session) was performed through the CF prepared by the LPI Control Team and transferred to the Mission Control Centre (MCC) at the Institute of Terrestrial Magnetism, Ionosphere and Radio Wave Propagation (IZMIRAN) on the average thrice a week, and during heightened solar activity periods up to two times a day. The interval between the IF preparation and transmission to SPIRIT instruments amounted to several hours.

Within the framework of medium-term operational scheduling for SPIRIT instrumentation, LPI operators prepared a 2-week mission schedule. During the entire mission time, control over the SPIRIT instrumentation was very intensive: up to 2,000 commands per day was transmitted to the instruments, which in terms of data volume amounted to four Kbytes. In case of processor reprogramming, control data volume totalled up to 20 Kbytes per day. Commands were included to CFs: 30–40 CFs per day and up to 1,000 CFs per month.

In 2004, an integrated information system (IIS) was developed for more effective control over the SPIRIT experiment. IIS included two subsystems: control IIS (CIIS) and archiving and preprocessing IIS (AP IIS) [27].

Software package (SP), a component part of CIIS, provided for the semi-automatic generation of switching cyclograms, generation of control files (CFs) via the interactive communications link and scheduling CF delivery. SP was based on an Excel 2000 spreadsheet presenting the basic control procedures and events and an Access 2000 database automatically filled in with information about all newly generated CFs. Developers of special software for spreadsheet maintenance and database management used components of the instrument's software model and of the intelligent experimenter's assistant (expert system).

On 31 July 2001, the CORONAS-F satellite was launched from the Plesetsk launching site into a circular geocentric orbit at an average height of 530 km above the Earth's surface, 82.5° inclination, and initial orbital period of 94.7 min.

On 15 August 2001, the SPIRIT instrumentation started its operation. Within the period of August through October, the primary objective was to test various instrument channels in different modes, i.e., including switching on during shadow and light periods, with different sensitivity and exposure times, calibration of mechanical devices, etc. Regular observations and the scientific programme implementation started in November 2001 and continued until the exit of CORONAS-F from orbit in early December 2005. More than 23,000 observation sessions were carried out in different spectral channels during the mission. Over 2,000 equipment operating cyclograms were developed and over 45,000 interactive control files were transmitted on board to implement the scientific programme.

## *Experiment Data Transmission*

During the mission, the major portion of telemetry data was transmitted from the CORONAS-F satellite to the Neustrelitz ground station of the German Aerospace Centre (DLR). Technical capabilities of the receiving station enabled up to four downlink sessions per day with a total duration of 18–20 min, maximum daily

**Table 8** Number of frames in the FITS format over the channels of the SPIRIT assembly

| Channel | Number of frames | Channel | Number of frames | Channel | Number of frames | Channel | Number of frames |
|---|---|---|---|---|---|---|---|
| Mg 1 | 91595 | T 2-175 | 56598 | T 1-284 | 3522 | OD 2 | 6035 |
| Mg 2 | 98664 | T 1-171 | 2518 | T 1-304 | 3179 | XUV 1 | 24209 |
| T 2-304 | 52362 | T 1-195 | 4734 | OD 1 | 19556 | XUV 2 | 14616 |

telemetry data volume transmitted by SPIRIT instruments being about 150–160 MB. Within the first stage of flight, from August to December 2001, the daily information volume from the SPIRIT equipment was considerably lower, no more than 100 MB. Up to 50–55 % of the information received was corrupted, which required additional efforts to decipher it. The situation considerably improved when LPI specialists optimized monitoring of three-axis orientation of satellite on orbit based on data of optical sensors and preparation of forecasts of satellite's angular position for the next scheduling period, first in a trial mode in early 2002 and in a constant mode from April 2002. This permitted to preclude data transmission in situations of transmitting antenna signal shadowing caused by the satellite mass (reception error provoking situations) and to increase average daily volume of obtained information by 20–30 %, with a 100 % guarantee of data reception by earth facilities with an error level less than 1 %.

## SPIRIT Archiving and Database

85.5 GB of telemetry data comprising 377,260 frames of solar images and spectra, including more than one million of solar spectroheliograms in particular monochromatic lines, were obtained in terms of the SPIRIT experiment. With an average of about 400 frames per day, the number of frames per day amounted to 1,100 during the intensive observation periods. The volume of the electronic archive storing information obtained during 4.5 years of flight amounted to 250 GB. Alongside with backup copies, frames that were processed in different ways, project documentation, archived control files, and calibration and test data necessary for adequate interpretation of obtained images, the entire electronic archive volume of the experiment is about 1 TB. A three-server system was created at LPI to ensure safe and convenient archive storage, the most important archive portions being stored in RAID arrays.

Table 8 shows the distribution of data obtained during the SPIRIT experiment in various spectral channels.

Reviewing and primary processing of data received from SPIRIT was made by a single-module program KADIP.exe running under WINDOWS-2000 and XP. The KADIP program has been developed for a number of purposes:

- viewing the contents of primary telemetry files and recording the session protocol;

- subdividing telemetry data by instruments, channels and individual frames;
- recording separate frames as files under the international standard FITS (Flexible Image Transport System);
- viewing interactive control files and creation of new control files.

FITS-file headers were made in the format as similar as possible to that used in the SOHO/EIT experiment, which facilitates matching and processing with the use of identical programs. At present, the SPIRIT data on 175 and 304 Å channels are accessible from ftp server:

- http://spirit.lebedev.ru/database/SPIRIT/R175.php for 175 Å channel,
- http://spirit.lebedev.ru/database/SPIRIT/R304.php for 304 Å channel.

# Studies of Hot Coronal Plasma and Solar Flares

## *Studies of Solar Plasma in the Mg XII Line*

A class of new phenomena associated with emission of high-temperature ($T = 5$–20 MK) plasma structures in the solar corona has been discovered for the first time and since then frequently observed with the use of the unique (not to be found elsewhere in the world) SPIRIT spectroheliograph for recording monochromatic images in the Mg XII line at 8.42 Å [28]. This instrument formed an intermediate data channel between EUV telescopes imaging moderate temperature coronal plasma (SPIRIT 175, 195, 285, and 304 Å; SOHO/EIT and TRACE) and RHESSI providing data on super-hot plasma.

As a result of regular observations in the Mg XII line at 8.42 Å, we have revealed new types of hot coronal structures, which are characterized by lifetimes from minutes to days and scales from several arcseconds to several tenths of a solar radius [29, 30] (Fig. 13).

Besides non-stationary (transient) phenomena studied earlier by the Yohkoh satellite, a whole new class of dynamic structures of various shapes, scales, duration, and temporal features has been discovered, in particular long-lived (up to several days) spider-shaped plasma structures located high in the corona (up to 300,000 km) [29]. Within the framework of long-term (more than 20 days) series of continuous observations performed with high temporal resolution (from 0.6 to 1.7 min), we have also discovered active "life" of such "spiders" accompanied with hot "clouds," flare phenomena, coronal mass ejections, etc. Figures 13 and 14 exemplify images of "spiders" observed on 12/13 November and 20 December 2001 in the Mg XII channel.

The following regularly observed structures could be identified:

- "hot clouds": variable shape diffuse formations of a typical size and height above the limb of up to $0.4R_{\text{sun}}$ and lifetimes up to several hours;

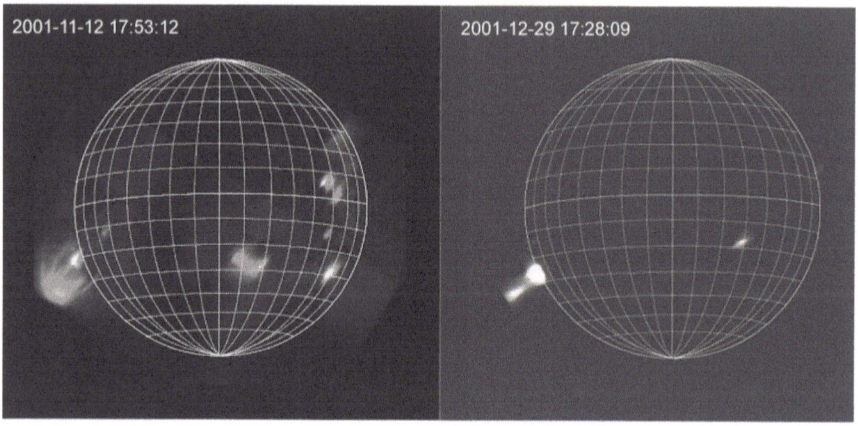


**Fig. 13** Examples of hot coronal structures observed in the Mg XII line. From [38, Figs.1–3, 4a, b, 7–10]

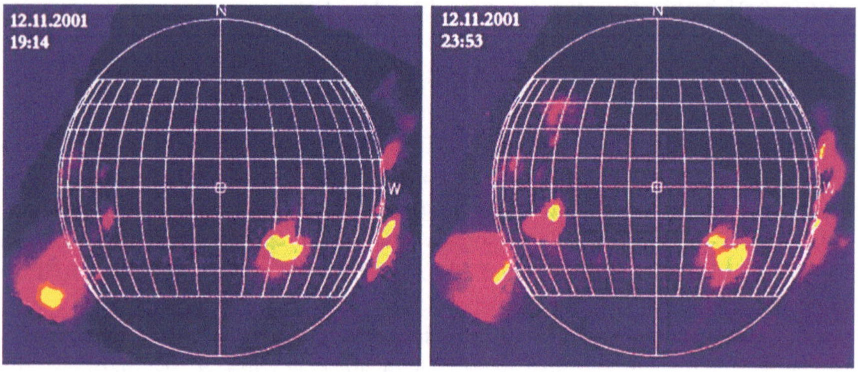


**Fig. 14** The evolution of hot plasma structures at 5–15 MK in the Mg XII line. From [58]

- "spiders": large-scale structures associated with active regions and characterized by having a spider shape for a long time (up to several days), a bright spherical body at 0.1–0.3 $R_{\rm sun}$ heights and less bright "legs," morphologically being a system of giant arches, which does not coincide with the cold magnetic loop system;
- complex events, in particular the appearance of clouds with the further formation of spiders and giant arches, accompanied by flares, coronal mass ejections, and other eruptive phenomena;
- "waves": phenomena in the form of divergent wavefronts or successive appearance of magnetic arches, which were observed during high solar activity.

Studies of the dynamics of active phenomena in the solar corona with a high temporal resolution of up to 7 s have also been performed in the Mg XII line. We have investigated temporal and spatial structure of flares in the Mg XII line

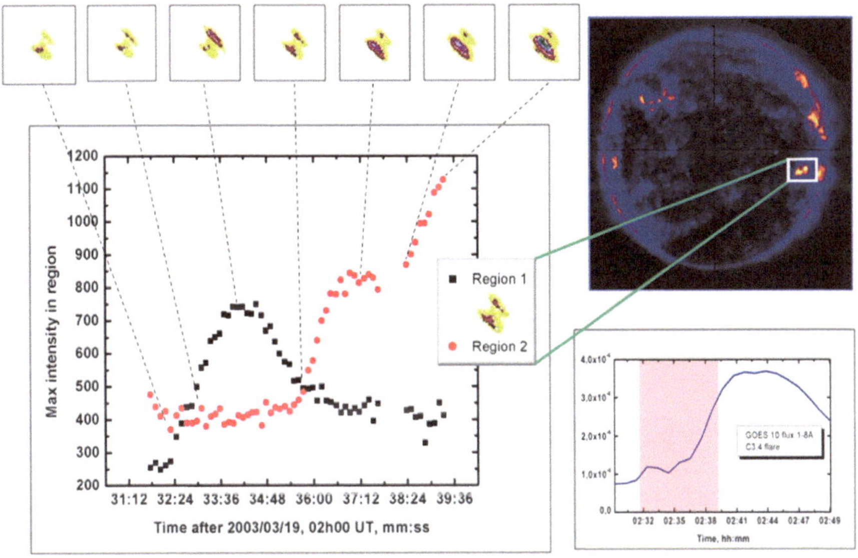


**Fig. 15** The dynamics of the active region during the impulsive phase of flare dated 19 March 2003 on the basis of measurements in the SPIRIT Mg XII 8.42 Å channel and in the GOES 1–8 Å channel. *Left* diagram shows integrated intensity in regions 1 and 2 with dynamics of images depicting flare regions in the Mg XII line at time points corresponding to positions on the plot; on the *right* of the figure the location of the flare regions on the disk image in the 175 Å (*top*) line and X-ray flux in the GOES 1–8 Å channel (*bottom*) are presented. From [31]

at 8.42 Å. Regular long-term observations have shown that the characteristic time of development of a great number of flares recorded in this range amount to a few minutes. Several class C and M flares may occur at the peak of activity during a day. To study such flares in detail, methods were developed and onboard software produced allowing capturing images every few seconds [31]. Within a few months of observations in the phase of maximum activity in 2002–2003, over 50 image sequences were recorded, covering preflare, initial, maximum phase, and flare decay phase. As compared to the GOES findings at 1–8 Å, SPIRIT data allow to observe and compare fine temporal structure of radiation in various areas of the flare (Fig. 15). We have revealed the energy transfer between the primary centers of activity for a part of flares, both in the transition layer and chromosphere, and in the higher layers of the corona.

In February and May 2002, programmes of continuous (within 10 days) solar imaging at 8.42 Å (Mg XII) with a temporal resolution of 100 s were arranged and implemented, with reported several C- and M-class flares. The correlation between the observed flares and data of the integrated X-ray flux observed by the GOES satellite has been analyzed.

Quasi-periodicity of X-ray bursts in the Mg XII line at 8.42 Å over particular active regions on the Sun from 19 February to 23 February 2002 was investigated [32]. Statistically significant quasi-periodic variations with periods in the intervals

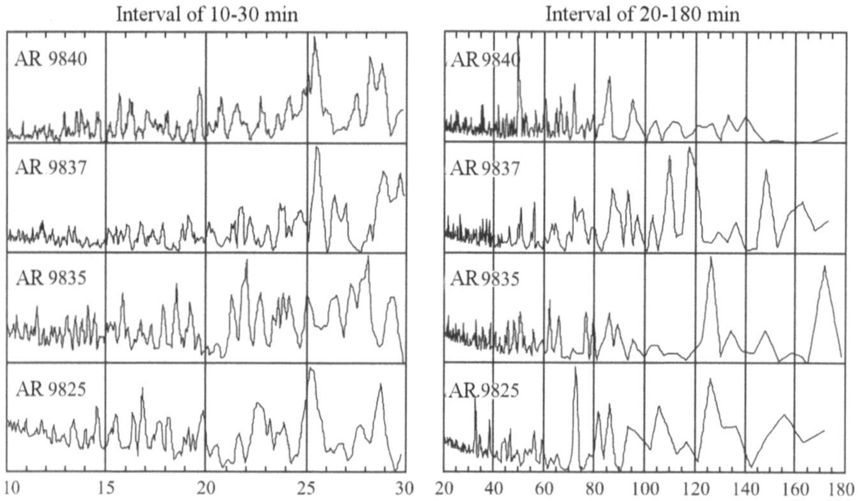


**Fig. 16** Periodograms of radiation intensity fluctuations in the Mg XII 8.42 Å line in active regions (according to the observations made on 19–23 February 2002)

of 12–30 min and 40–200 min were revealed (Fig. 16). Short-period oscillations seem to reflect the existence of periodic processes or waves in hot coronal loops. Our findings on changes in the power spectrum of any active region (AR) after the appearance of spots in this region suggest that the detected quasi-periods may depend on physical parameters of coronal loops. For example, after emergency of a new photospheric magnetic flux in the active region NOAA 9840 the power spectrum for these frequencies changed.

In the low-frequency region, power spectra peaks differ for different ARs. For example, quasi-periods characteristic of a compact growing region 9840 are 72 and 84 min, and those characteristic of the flare-active region 9825 are about 138 and 160 min. Thus, different positions of peaks in low-frequency regions associated with the incidence of large bursts in the different ARs appear to be determined by the structure and strength of magnetic fields in ARs.

For the first time the process of prolonged eruption of hot plasma (10 MK) into the distant corona was traced. In this phenomenon in the Mg XII line, hot structure altitude above the limb amounts to 0.7 solar radii, its existence being correlated with the phenomena recorded by the optical coronagraph onboard SOHO in this corona sector. A wide variation of intensity and nature of the dynamics is a distinctive feature related to the hot structures of eruptive phenomena (EP). Posteruptive arcades can be used to identify these structures. Such arcades usually form in a rapid manner with a subsequent transition to a more quiet stage. Long lifetimes of posteruptive arcades imply long-duration energy production in higher corona layers, which may be achieved through slow reconnection processes.

One such arcade was observed on 22 October 2001 by SPIRIT instrumentation, Yohkoh/SXT telescope and SSRT (Sibir Solar Radio Telescope) and RATAN-600

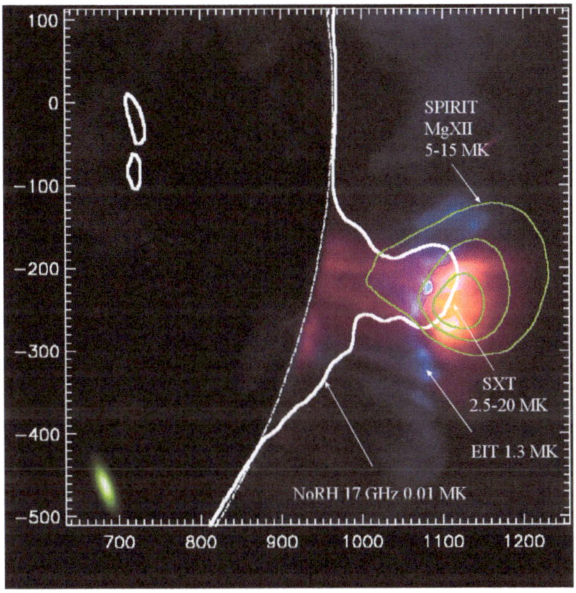


**Fig. 17** Posteruptive arcade during the event of 22 October 2001 as observed by SPIRIT (Mg XII 8.42 Å), Yohkoh/SXT, SOHO/EIT and NoRH radioheliograph, Nobeyama [33]. From [33, Fig. 1]

radio telescopes [33, 34] (Fig. 17). Estimates of the longitudinal component of a magnetic field led to $B_d > 75$ Gs. The temperature of a larger portion of the arcade, including its brightest region, amounted to 6 MK; the temperature of its upper edge reached 8 MK. Emission measure in the column $2.46'' \times 2.46''$ for the bright region was equal to $2.5{\cdot}10^{45}$ cm$^{-3}$, and for the resolved, less bright falling leg-loops $\sim 5{\cdot}10^{44}$ cm$^{-3}$, with the minimum value being $2.5{\cdot}10^{44}$ cm$^{-3}$. The plasma density in its bright portion, given the assumption that the region depth is equal to its lateral dimension, amounted to $5{\cdot}10^9$ cm$^{-3}$. High brightness of the central arcade may be accounted for an optical effect: it is possible that the brightness of the loops is the same throughout the height, but with their orientation along the line of sight differential emission measure for these peaks of loops accrues along their upper portions.

In the SPIRIT experiment, various programmes of observations using Mg XII channels were implemented, including simultaneous observations in different channels, long-term continuous observations in the heliosynchronous orbits, observation of high-speed processes, in particular flares. The accumulated experimental data allowed investigating the spatial structure, dynamics and physical mechanisms of emission of plasma structures observed in monochromatic X-ray images.

To simulate the temperature of the soft X-ray (SXR) plasma of the solar corona, we used data obtained simultaneously in the SPIRIT experiment with the use of RES-C instrumentation onboard the CORONAS-F, GOES (Geosynchronously Operational Environmental Satellite) X-ray monitor and softest X-ray emission on the RHESSI (Reuven Ramaty High Energy Solar Spectroscopic Imager) satellite. Absolute calibration was performed for fluxes in the magnesium channel based

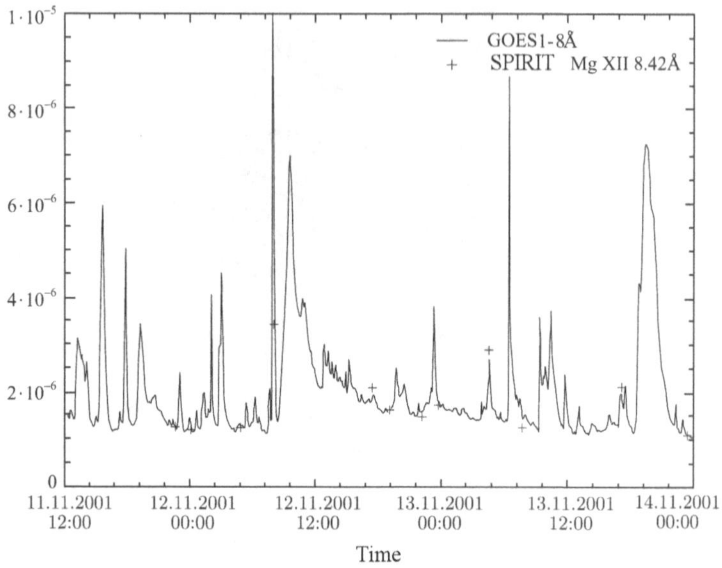


**Fig. 18** Time profiles of radiation fluxes (in units of $W/m^2$) in the Mg XII 8.42 Å channels and those according to the GOES (1–8 Å) data for the events dated 11–12 November 2001. From [38, Figs.1–3, 4a, b, 7–10]

on the data of emission time profiles in two channels (1–8 and 0.5–4 Å) of the GOES X-ray monitor as well as spatial correlation of flare sources according to the RHESSI satellite data.

The comparison of full emission fluxes (from the full Sun) in the 8.42 Å line (RES-C) and in the 1–8 Å range (GOES-8) performed during the analysis of monochromatic Mg XII images revealed good coincidence, within statistical error limits (on the order of 10 %), of their time profiles characterized by substantial changes in intensity within long time intervals, excluding relatively short periods during the maximum phase of the flare events (Figs. 18 and 19). The results of a series of continuous observations by the RES-C X-ray spectroheliometer for the period from 6 February to 28 February 2002 have been used for a more detailed study of correlation in the dynamics of X-ray fluxes. Within this period, flares and other transient events as well as the development of integrated large-scale and long-lived phenomenon, referred to above as "spider," have been observed in images in the Mg XII line.

The comparison of dynamic characteristics of flux time profiles in the 8.42 Å line and the 1–8 Å range allowed for conditional subdivision of all events into three types: short-term impulsive-type events (IE), long-term "gradient" events (LGE) related to the formation of spiders, and other complex events: e.g., pulse-type ones with a long decay (impulsive-type long-lived events, ILE) or LGEs accompanied with impulsive bursts [35]. Figure 20 demonstrates high correlation of fluxes for the three types of events in Mg XII and GOES (1–8 Å) channels. Deviation from the

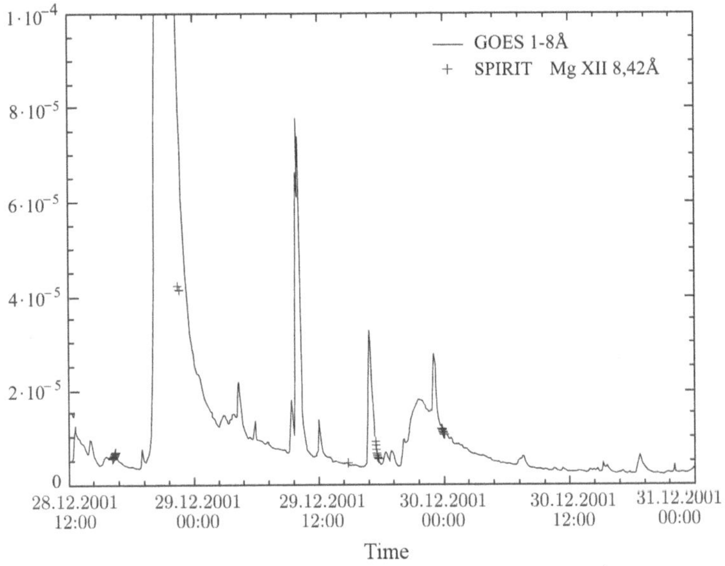


**Fig. 19** Time profiles of radiation fluxes (in units of W/m$^2$) in the Mg XII 8.42 Å channels and those according to the GOES (1–8 Å) data for the events dated 28–31 December 2001. From [38, Figs.1–3, 4a, b, 7–10]

linear relationship of these fluxes can only be observed during the peak phase of flare events, and it can vary depending on the event type within 5 % for the flare rise and decay phases, with its peak of 15–20 % for IEs.

A method based on the multi-temperature parametric (MTP) model was applied to restore the temperature distribution of the volume DEM of the hot flare plasma from spectral X-ray fluxes [35]. In addition, coronal abundances of elements were used.

Figure 21 presents flux time profiles for magnesium channels and GOES (1–8 Å) channels and their decomposition into components corresponding to three temperature ranges: relatively cold plasma (2–4 MK), an intermediary-temperature component (4–10 MK), and a hot component (10–20 MK). Figure 22 shows respective EM profiles and presents average-temperature and EM profiles calculated on the basis of this multi-temperature model as compared to the results provided by a single-temperature model (STM) on the basis of GOES data. A significant difference should be pointed out in the distribution of temperature components in the development of flare events of different types (IE and LGE). The contribution from the hot component is dominating for IEs, while the main contribution to LGEs is made by the transient intermediary-temperature plasma, with the intensity of the latter developing faster and lasting considerably longer than the hot component growth. A substantial difference between the temperatures and EMs calculated according to MTP model and STM must also be noted: EM buildup in the MTP model virtually coincides with the temperature growth as distinct from STM, which

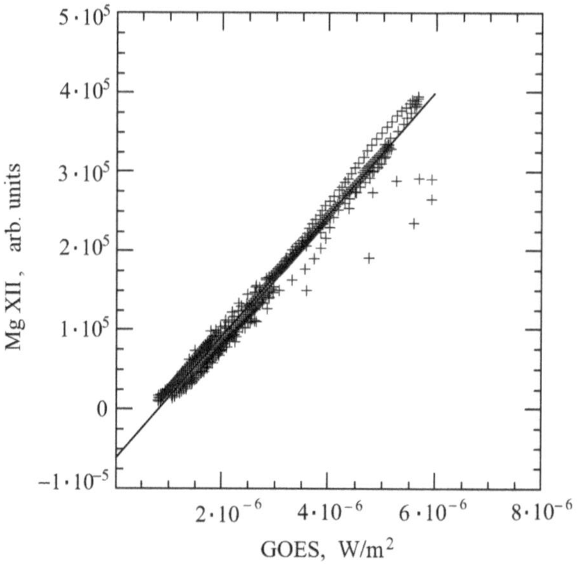


**Fig. 20** Correlation of X-ray fluxes in the Mg XII and GOES (1–8 Å) channels for a period of continuous observations of 26–28 February 2002. From [35, Figs. 3, 9]

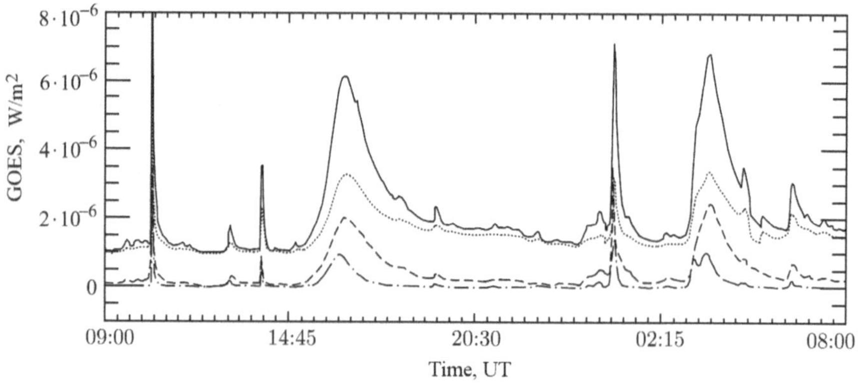


**Fig. 21** Temporal profiles of radiation fluxes in the GOES 1–8 Å channel (*solid line*) and their decomposition into components corresponding to temperature ranges of 2–4 MK (*dotted line*), 4–10 MK (*dash line*), and 10–20 MK (*dash-dotted line*)

is characterized by a considerable time delay in the EM growth as compared to the temperature. As Fig. 22 shows, the average temperature for many events is considerably lower, while EM exceeds the corresponding STM values.

Correlation between the results calculated by the MTP model and RHESSI data demonstrated the following: for impulsive-type events (IEs) at 10:26–14:00 UT, according to the MTP model and RHESSI data EMs amounted to $2.9 \cdot 10^{48}$ cm$^{-3}$

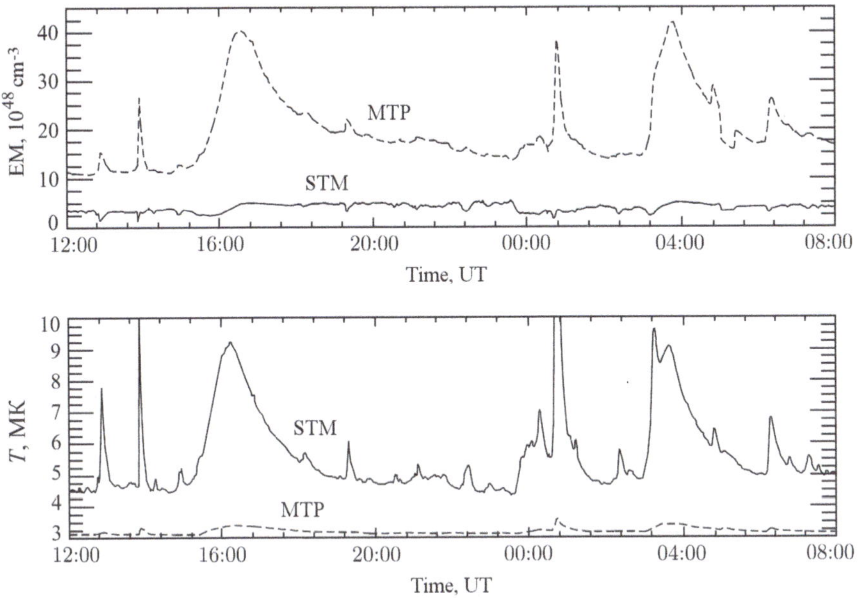


**Fig. 22** Comparison of temporal profiles of the emission measure EM (*at the top*) and temperature (*at the bottom*) for the MTP and STM models

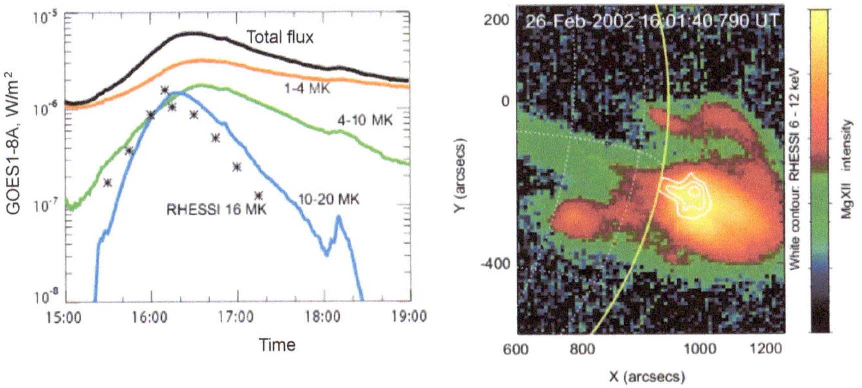


**Fig. 23** *Left*: temporal profile of the total flux from the "spider" according to the GOES 1–8 Å data (26 February 2002) and its decomposition into temperature components of 1.4 MK, 4–10 MK, and 10–20 MK as compared to the RHESSI data in the channel 6.12 keV (*asterisks*). *Right*: superimposed images in the Mg XII channel and RHESSI (*contours*). From [35, Figs. 3, 9]

and $0.5{\cdot}10^{48}$ cm$^{-3}$. Such agreement between MTP model results and RHESSI data is indicative of experimental data consistency, adequate modeling, and accurate determination of recalibration constants. Figure 23 compares temporal fluxes and images in the magnesium and RHESSI channels.

Another essential result is that the MTP model calculations using photospheric abundances cannot provide satisfactory agreement with RHESSI data.

The comparison of calculations of flux time dependencies with model-based calculations demonstrated that the deviation from a strictly linear dependence is based on the contribution from the luminosity of the cold component during the intensity rise and decay periods and the hot component at the maximum phase of flares, with these luminosities being associated with the proton bremsstrahlung. Thus, the observed interrelation of magnesium and GOES (1–8 Å) intensities was quantitatively interpreted based on the MTP model. Within the flare rise and decay phases the main portion of the source EM is contained in the transient plasma at 4–10 MK, which ensures the proportionality of the total GOES channel intensity to that of the Mg XII channel due to a weak dependence of the ratio of temperature coefficients in this range. The GOES channel intensity within the maximum phase markedly exceeds the magnesium channel intensity due to a significant contribution of the hot plasma component (10–20 MK) to the total intensity during this period, which is associated with the determining contribution of the proton continuous (bremsstrahlung) emission to the luminosity function in the GOES channel.

The aforementioned deviation from the linear law for fluxes in the line and spectral range may be entailed by the following physical causes: either a temperature change (e.g., when plasma heats up at the impulse stage of the flare or cools down at the decay stage, when the flux in the line is weaker than that in the 8–12 MK range at $T < 8$ MK or $T > 12$ MK) or a change in the luminosity function associated with the non-equilibrium mechanism of emission (e.g., in the presence of suprathermal electrons).

## *Analysis of the Solar Spectrum in the 176–207 Å and 280–330 Å Ranges*

During the SPIRIT experiment onboard the CORONAS-F satellite from August 2001 to December 2005, several thousand spectroheliograms were obtained in two spectral ranges, 176–207 and 280–330 Å, by XUV channels of the RES-C spectroheliograph (Fig. 24). Alongside with XUV-range spectroheliograms, monochromatic Mg XII 8.42 Å and XUV-telescope images were recorded for the investigation of structure and dynamics of the areas under study. Spectra of selected solar regions were obtained by scanning of spectroheliograms in the direction of dispersion with subsequent normalization for the spatial and spectral sensitivity of the instrument.

Typical intensity peaks interpreted as the spectral lines were detected based on the spectral analysis of 40 regions including various plasma structures (active regions, quiet areas on the solar disk, over limb structures). Lines were preliminary identified based on the CHIANTI database (version 4.2) through selection by intensity calculated in the range of densities $lg\, N_e = 8$–12, given the constant emission measure.

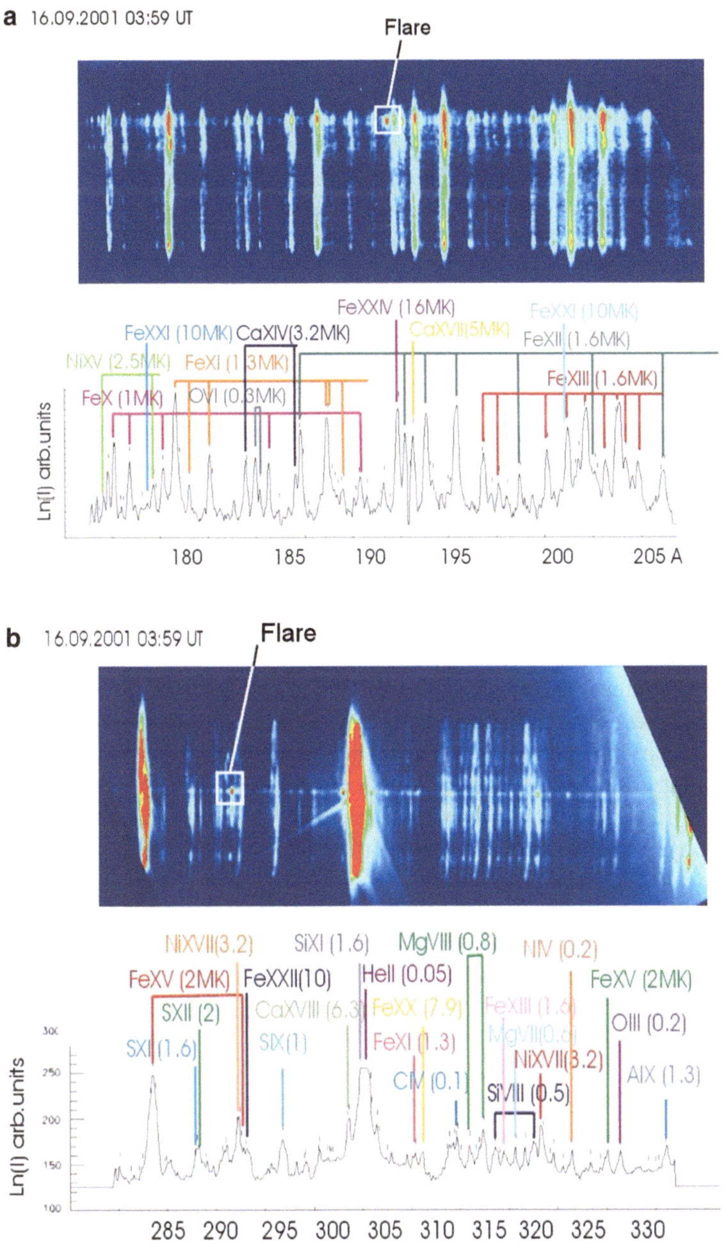


**Fig. 24** Spectral solar images in the 176–207 Å channel (**a**) and in the 280–330 Å channel (**b**) of the RES-C XUV spectroheliograph. From [38, Figs.1–3, 4a, b, 7–10]

A catalogue of spectral lines, which were observed by XUV channels of the RES-C spectroheliograph in the ranges 176–207 and 280–330 Å, was compiled [36, 37]. This catalogue includes 165 spectral lines. Identification of 105 lines is provided, with 15 lines being observed in flares only.

The diagnostics of electron density for various solar corona structures was carried out based on the data of XUV channels of the RES-C spectroheliograph and using the CHIANTI database. Calculated electron density values are as follows:

- for the quiet Sun (coordinates: N80-90 E016; 09.2001, 03.59.36UT)—$10^{8.7}$ cm$^{-3}$;
- for the active region NOAA AR 9765 (29.12.2001 14:48:53 UT)—$10^{9.3}$ cm$^{-3}$;
- for M5.6-class flares (AR 9608 16.09.2001 03:59:36 UT) > $10^{9.7}$ cm$^{-3}$;
- for M7.6-class flares (AR 9628, 25.09.2001 04:50:42 UT)—$10^{10}$ cm$^{-3}$.

Using lines with relative intensities, which are sensitive to temperatures in the range of 0.5–20 MK but do not depend on the electron density, temperature distributions of the differential emission measure (DEM) were calculated using inversion techniques in two ways: (1) by the method included in the CHIANTI package [13] and (2) by the Bayesian iterative method (BIM) (see [35, 38] for details). Figure 25 demonstrates the results of calculations made with the use of the most reliable line intensities in the spectra. DEM distributions calculated by the CHIANTI method yield greater dispersion in the ratio of observed intensities to calculated intensities, thus ensuring its average value, which leads to a smoother temperature dependence. Further studies concerning relative intensities of lines in spectra of various plasma structures and involving larger data sets are to show the reliability of the Bayesian method and the corresponding DEM distributions. Figure 26 compares such distributions for a number of active regions located in different parts of the solar disk during flare-free periods. We consider it of great importance that the dependence of DEM in these regions on the temperature shows similarity for a wide temperature range up to 8 MK; at a higher temperature, two of the six sources have got a pronounced maximum at a temperature on the order of 15 MK, which, to our knowledge, has not been previously observed.

Another important result of diagnostics of hot plasma in the active solar corona structures consists in the interpretation of mechanism of soft X-ray emission of the "spider" plasma based on the comparison of data obtained in the EUV and soft X-ray spectra. Figure 27 presents 28/29.12.2001 event images in the "hot" (8–12 MK) Mg XII channel of the RES-C spectroheliograph and "cold" (about 2 MK) channel near 195 Å captured with the use of the EIT telescope (Extreme Ultraviolet Imaging Telescope) onboard the SOHO satellite (Solar Heliospheric Observatory). Figure 27 displays a radially (along the solar radius) elongated "spider" structure in the X-ray (hot) Mg XII image and a posteruptive (cool) arcade of magnetic loops grouped along the perpendicular direction.

Proper orientation of the RES-C instrument (the "spider" was oriented along the axis perpendicular to the dispersion direction in the XUV 280–330 Å channel) made it possible to study the dependence of the DEM temperature distributions along the

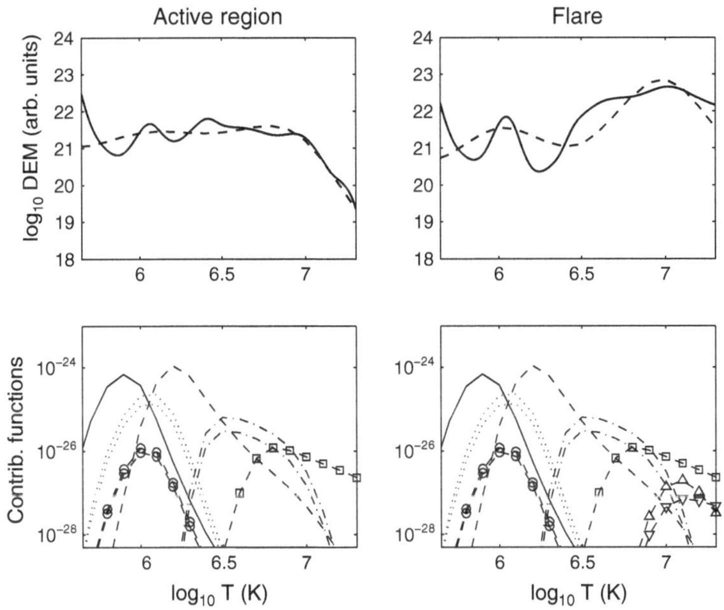


**Fig. 25** *Top* results of calculations of the DEM temperature distributions in the AR9765 active region and X34 flare region (29 December 2001) obtained by solving the spectral inverse problem: *solid line*—BIM, *dashed line*—CHIANTI procedure chianti_dem.pro. *Bottom*—spectral contribution functions from the CHIANTI database used for the DEM calculations: *solid line*—Mg VIII ($\lambda$ = 315.02Å), *dashed line*—Si XI ($\lambda$ = 303.33Å), *dash-dotted line*—Ni XVIII ($\lambda$ = 291.98 and 320.57Å), *dotted line*—Si IX ($\lambda$ = 292.82 and 296.14Å), *circles*—Al IX ($\lambda$ = 282.42 and 286.38Å), *squares*—Ca XVIII ($\lambda$ = 302.19Å), *top triangles*—Fe XXII ($\lambda$ = 292.49Å), *bottom triangles*—Cr XXII ($\lambda$ = 279.74Å)

solar radius. Figure 28 presents the DEM distribution for three distances from the solar limb and for the active region in the limb. The nonuniform distribution of hot material (DEM) depending on the solar radius in the temperature range of 8–12 MK can be easily seen in the figure.

Figure 29 shows the radial dependencies of intensities (normalized to the values at the edge of the solar disk) in particular lines of the EUV range as well as Mg XII X-ray line, which were recorded within the flare-free period about 14:40 UT. The principal finding here is that the observed intensities of the Ca XVIII EUV line and Mg XII X-ray line, while having the same temperature of formation, behave quite differently: the intensity of the first line falls, whereas that of the second rises. At the same time, behavior of the relative intensity in magnesium lines calculated based on Fig. 28 distributions (marked with crosses) is similar to that observed in the calcium line.

Owing to the relatively low threshold of excitation, the intensity of EUV lines is proportional to the emission measure formed by thermal (Maxwellian) electrons. Having a significantly higher excitation threshold, X-ray line of a multiply charged

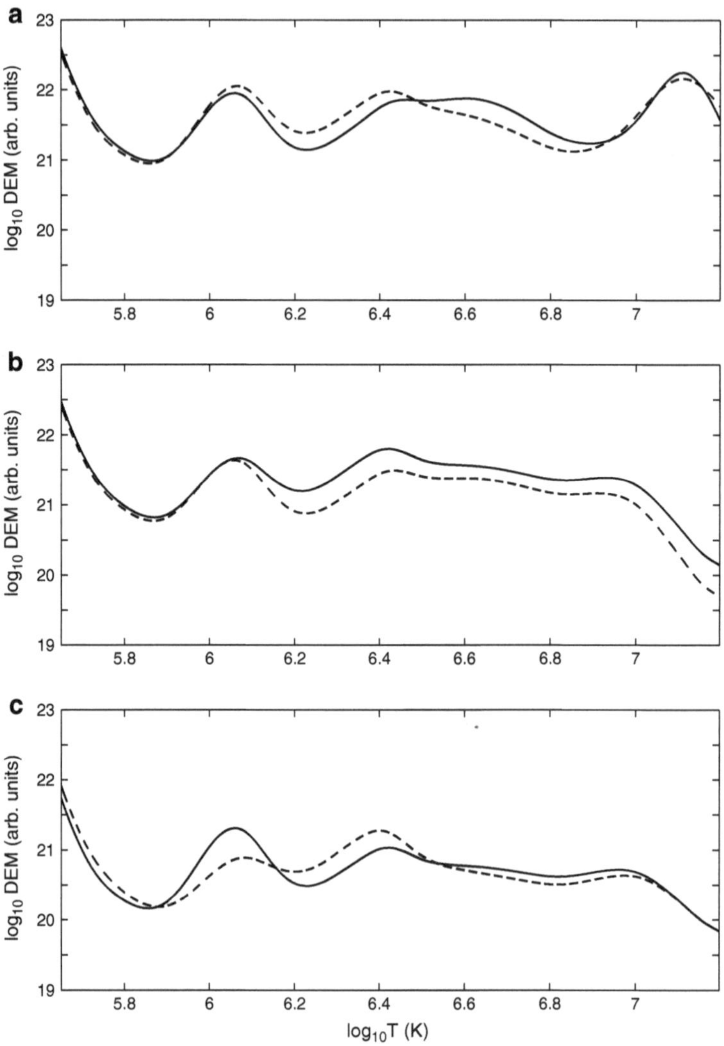


**Fig. 26** Temperature distributions (DEM) of plasma of different active regions. (**a**): *solid line*—01.10.2001, *dashed line*—03.10.2001; (**b**): *solid line*—29.11.2001, *dashed line*—03.11.2001; (**c**): *solid line*—21.10.2001, *dashed line*—03.11.2001. From [38, Figs.1–3, 4a, b, 7–10]

magnesium ion is produced by high-energy ($geg \geq 2$ keV) electrons, whose density in thermal plasma may reach a few percent of the total density of Maxwellian electrons. In this respect, the intensity of EUV lines does not depend on the slight infusion (about 5–10 %) of non-thermal (non-Maxwellian) electrons, and it is proportional to the thermal emission measure; whereas the intensity of X-ray lines is determined by the luminosity function, which may be in great excess over the thermal one because of the presence of non-thermal electrons. Thus, on considering

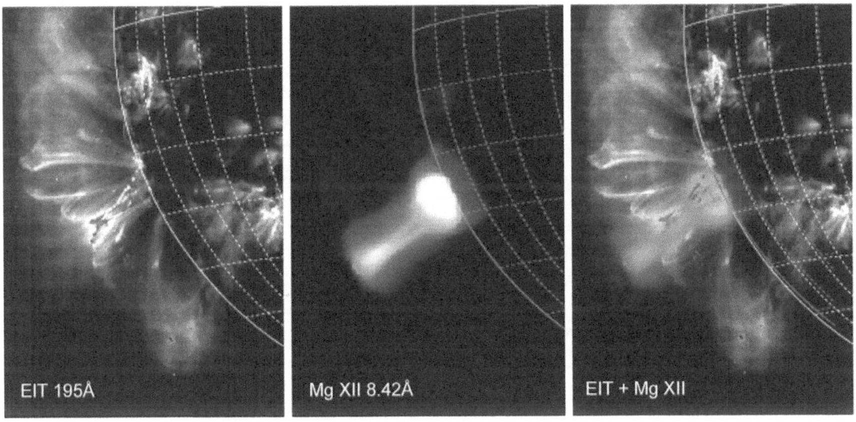


**Fig. 27** Images of the "spider" observed on 29.12.2001 in the Mg XII 8.42 Å line, in the spectral region near 195 Å (according to SOHO/EIT) and the combined image. From [38, Figs.1–3, 4a, b, 7–10]

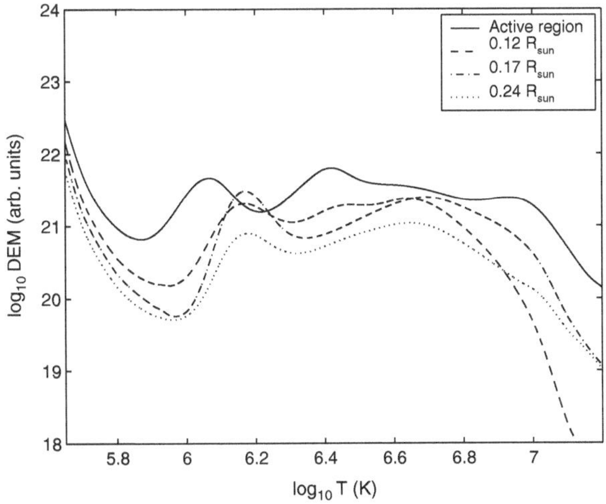


**Fig. 28** DEM distributions of the "spider" observed on 29.12.2001 and obtained at different altitudes over the solar limb. From [38, Figs.1–3, 4a, b, 7–10]

the RES-C channels' observation data, we can definitely make a conclusion on the non-thermal (non-stationary) nature of emission of hot plasma above the solar limb, to the intensity of which fluxes presented in Fig. 29 were normalized. Such a conclusion is in line with the conclusions based on the analysis of temporal profiles of the radiation in the Mg XII channel.

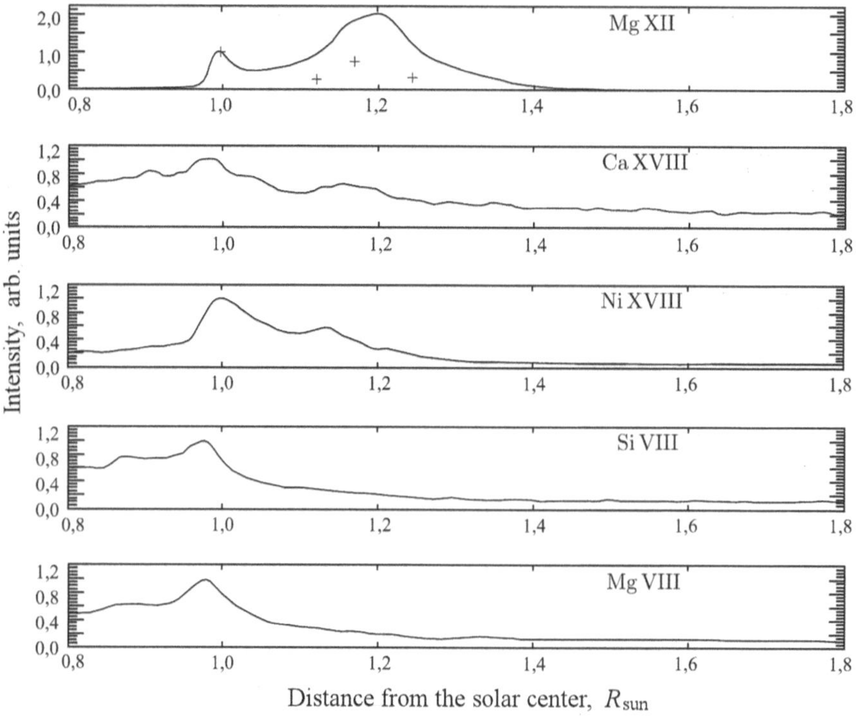


**Fig. 29** Radial dependencies of intensities of particular EUV and X-ray spectral lines (normalized to the solar limb values) in the "spider" region (observed on 29.12.2001). From [38, Figs.1–3, 4a, b, 7–10]

## *Analysis and Interpretation of the Results of Studying Hot Plasma of the Solar Corona According to the RES-C Data*

X-ray monochromatic images in the Mg XII resonance line obtained with high spatial resolution of about 8″ together with EUV spectra and data on temporal profiles and the soft X-ray (SXR) images from GOES, YOHKOH, and RHESSI satellites provided essentially new information on the structure, dynamics, and physical characteristics (electron density distribution, temperature, and pressure distribution) in the hot plasma structures at 4–20 MK or more. Multi-temperature analysis, performed on the basis of such integrated simultaneous observations in different spectral ranges, proved that temperature determined by the traditional filter method may be 2–3 times greater than, and the emission measure (EM) may be, by an order of magnitude or more lower than, respectively, the average temperature and total ME of the flare region. In particular, the maximum thermal matter distribution amounted to a value on the order of 10 MK in a uniquely powerful flare that took place on 7 September 2005, whereas by the single-temperature model temperatures about 40 MK were obtained in the flares of higher classes.

Important findings of studying temporal profiles of SX fluxes by the RES instrumentation and GOES monitoring equipment consist in the detection of transient plasma with intermediary temperature of 4–10 MK in all flare events, which contributes considerably, up to 70 % or higher, to the energy pool of active phenomena. Such plasma affects, in the most significant way, the dynamics of long-term gradient events (LGEs), e.g., "spiders," which were first discovered and studied due to a series of monochromatic images in the magnesium channel. The analysis of morphological peculiar features and diagnostics of spatial distributions of temperature and density proved the fundamental distinction of such flare phenomena from those previously detected in the Yohkoh experiment, which is indicative of different mechanisms of their development and evolution. Difference in time passage, spatial scale, density gradients and the maximum density (about two orders of magnitude) in impulsive-type events (IP) and the aforementioned LGEs can also testify to the specific nature of the mechanism of formation of the latter, as contrasted to the findings presented by the authors of Yohkoh data analysis publications concerning common origin of the two events. Therefore, obtaining new data from the RES-C instrumentation gives rise to the need to revise the interpretation of some results of quantitative modeling of flare phenomena formation and evolution. It is also worthwhile noting that a high correlation between LGEs (spiders) and coronal mass ejections (CMEs) is observed, which is of great practical importance for researches of solar-terrestrial relations and space weather.

## *Soft X-ray Sources in the Solar Corona*

As repeatedly noted above, regions of substantially hotter plasma with a temperature of 10 MK or higher, being sources of intensive X-ray radiation, can develop during flares in the solar corona, whose average temperature is only slightly higher than 1 MK. Numerous evidences of such heating were found as early as in the first decade of space researches of the Sun in the 1960s to early 1970s in experiments involving geodetic rockets as well as onboard the Soviet and American space vehicles. A detailed study of the structure and dynamics of high-temperature coronal regions, however, was not possible because of the low spatial resolution of observations (about one arcminute).

The possibilities for such study appeared in the 1990s with the development of X-ray optics as well as with the introduction of modern semiconductor radiation detectors permitting detection of short-wave solar radiation with a spatial resolution of up to $1''$. In 1992, the SXT telescope onboard Yohkoh, a Japanese satellite, provided first images of high-temperature X-ray sources in flares. The X-ray radiation sources were detected in the flare of 13 January 1992. They were located in the corona above the apices of flare loops and had a temperature of 15–20 MK, density of $\sim 10^{10}$ cm$^{-3}$, and the full emission measure of $\sim 10^{48}$ cm$^{-3}$. The most remarkable peculiarity of the radiation source was its upward movement during the

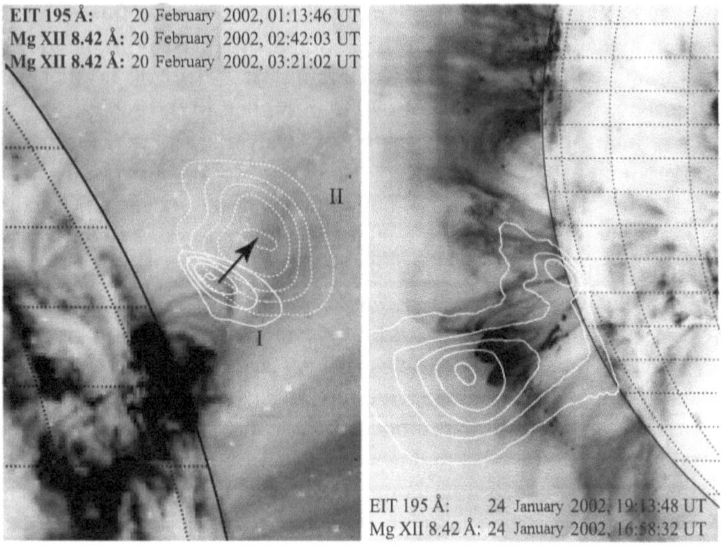


**Fig. 30** Impulsive and long-lived soft X-ray sources in the solar corona

flare at a speed of 3.5 km/s. Subsequently, similar results were obtained in the course of other Yohkoh observations, which have been continued until December 2001.

Since 2001, only the SPIRIT/MgXII imaging spectroheliometer onboard the Russian CORONAS-F satellite has been carrying out studies of coronal hot plasma. The most significant findings of these studies consisted in discovering a new class of high-temperature coronal objects: large regions of hot plasma with lifetimes far exceeding their typical times of conductive and radiative cooling. Observations revealed that the regions are located at a considerable altitude in the corona, up to 200,000 km, and have a temperature of 10–15 MK. Therefore, characteristics of radiation sources of this class differ essentially from those of impulsive-type sources detected by Yohkoh in flares.

Figure 30 demonstrates examples of impulsive and long-lived soft X-ray sources. It is based on the images captured by the SOHO/EIT telescope in the line of ionized iron Fe XII 195 Å ($\sim 1.5{\cdot}10^6$ K), with the position of high-temperature radiation sources, which were observed in each of the events by the SPIRIT/Mg XII spectroheliometer, contoured over it. For the impulsive-type source observed on 20 February 2002, two consecutive positions show its movement upwards. The first position corresponds to the time point 02:43:03 UT, when the center of the source was located at an altitude of $\sim$ 66,000 km, and the second one, to the time point 03:21:02 UT.

During this time interval, the source moved in the corona up to an altitude of $\sim$106,000 km. Therefore, an average upward velocity amounted to 17 km/s. The visible size of the high-temperature region within this period increased from $1.5'$ to $2.4'$.

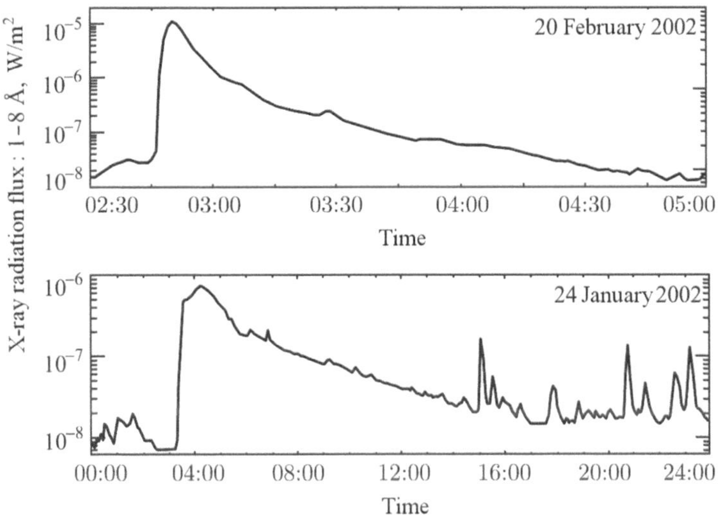


**Fig. 31** Temporal profiles of the impulsive and long-lived radiation sources based on the data of GOES-10 observations in the 1–8 Å wavelength range

Figure 31 compares temporal profiles of the impulsive and long-lived radiation source based on observation data of the GOES-10 satellite in the 1–8 Å wavelength range. Both temporal profiles have similarly shaped with a comparatively fast growth of the radiation flux at the beginning of the event and a prolonged radiation decline phase: 2 h for an impulsive source, and about 14 h—for long-lived one. In the first case, the radiation is decreased by three orders of magnitude from $10^{-5}$ W/m$^2$ to $10^{-8}$ W/m$^2$, while in the second case, only 50 times, from $10^{-6}$ W/m$^2$ to $2{\cdot}10^{-8}$ W/m$^2$. If we describe the decline phase using some characteristic time $\tau$, during which the radiation power decreases by a factor of $e$, we can conclude that for an impulsive source $\tau = 17.4$ min, and for a long-lived source $\tau = 214.7$ min, i.e., more than 3.5 h. Thus, plasma cooling patterns differ significantly in different types of high-temperature sources, at least by cooling times.

## Fast Shock Wave Heating of Impulsive Sources

Observations of high-temperature sources in the solar corona contribute to the improvement of the theory of solar atmosphere heating under the influence of flares and other phenomena.

Heating by thermal fluxes and energetic particles is often considered as the main mechanisms of solar atmosphere heating during flares. Particles and thermal fluxes can form in a high-temperature turbulent current sheet, where during flares, magnetic energy of current is converted into other forms of energy. Here, however, we discuss a different mechanism of gas-dynamic coronal plasma heating during

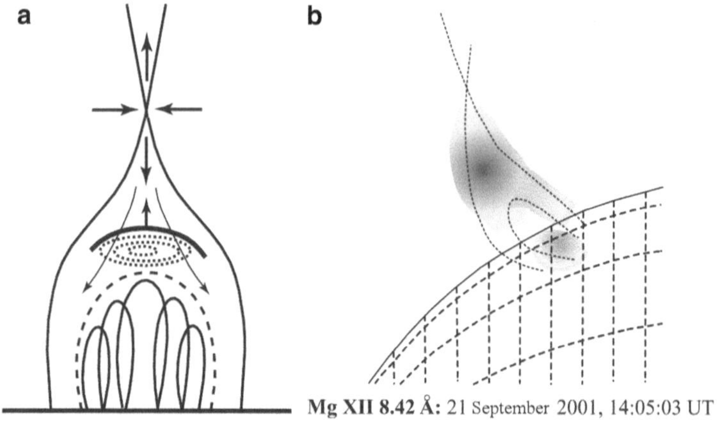


**Fig. 32** Formation of fast shock wave in the solar corona in the region of interaction between the plasma flux from the current sheet and a magnetic obstacle (a system of flare loops). *Left*: the scheme of the high-temperature radiation source formation. *Right*: the source image based on the SPIRIT/Mg XII data

a fast shock wave. Below, we provide arguments favoring the fact that exactly this process is the cause of impulsive flare occurrences that can be observed in soft X-ray range (see also [39, 40]).

## Adiabatic Shock Wave

Figure 32 shows a diagram of the evolution of a high-temperature radiation source in the corona. According to the model, a source forms above the flare loop system in the region of its interaction with the plasma flux from the reconnection region. Under certain conditions, the flux velocity exceeds the magnetosonic speed, and the forming of fast shock wave begins. In the reference system related to the loop apex, the wavefront propagates upward. This movement can be associated in the model with the upward movement of flare-related high-temperature radiation sources observed in the events of this type. To establish the relation between the plasma parameters before and after the shock wave, a system of continuity equations of mass, momentum, and energy fluxes is traditionally used. Implying this standard description method and having made several simplifying assumptions (a shock wave is considered perpendicular; the energy transfer equation is expressed with the use of adiabatic approximation; the solar corona plasma is considered ideal), we can theoretically calculate the main characteristics of a high-temperature source (temperature, density, and upward velocity) based on the specified parameters of undisturbed plasma (concentration, temperature, and flux velocity). Proceeding from the data of X-ray observations of undisturbed plasma, we can let $n_e = 2 \cdot 10^9\,\mathrm{cm}^{-3}$ and $T = 30\,\mathrm{MK}$. Flux velocity of plasma from the reconnection region depends primarily on the characteristics of reconnecting current layer, theoretical value of flux velocity being ~2,000 km/s. Note that plasma flux velocity is higher than the sound speed which amounts to ~900 km/s at 30 MK. It is sufficient to form a shock wave. Let us set magnetic field $B$ equal to null. In this case, the field behind the front will also be equal to null.

The calculation results show that, during the interaction between the loop system and plasma flux having the specified characteristics, a high-temperature radiation source must form with the following characteristics: temperature $T_2 = 130$ MK, density $n_2 = 4 \cdot 10^{10}$ cm$^{-3}$, and upward velocity $v_2 = 940$ km/s.

As noted above, the temperature of flare X-ray sources in the corona according to Yohkoh data is about 20 MK, electron density $n_e$ is $\sim 10^{10}$ cm$^{-3}$, and upward velocity of the source does not exceed 10 km/s. Save for the density, the above values essentially differ from those obtained from calculations. Thus, the actual conditions of high-temperature plasma heating in the corona are far from adiabatic ones.

The assumption of the magnetic field absence does not affect the general conclusion on thc inapplicability of the adiabatic approximation. It can be easily seen that the introduction of any nonzero values of B increases the front velocity, resulting in even greater divergence from observations.

As a result, there arises an issue of constructing a realistic model of coronal plasma heating due to fast shock waves which could explain the observed temperature and a low upward velocity of high-temperature radiation sources.

**Non-adiabatic Shock Wave Model with Cooling Behind the Wave Front.** Plasma heated behind the front will inevitably lose energy in course of conductive and radiative cooling. Expansion of the plasma along the field lines must also reduce its temperature. As it will be shown below, being essential in the energy balance, these losses lead to fast plasma cooling behind the shock wave front.

If we could apply classical thermal conduction, given the heated plasma parameters obtained, then the cooling time of the coronal source with a size of $10^9$ cm would amount to less than 0.1 s.

Coronal plasma, however, is collisionless, and within it heat is transferred more slowly. The turbulent heat front propagates with the velocity of waves interacting with thermal electrons of superhot ($T > 30$ MK) plasma. In case of ion-acoustic waves, this velocity is equal to $\sim 10^8$ cm/s. Therefore, the typical time of wave propagation is about 10 s, which is comparable with the time of the hydrodynamic expansion of plasma along magnetic field lines behind the shock wave front. For this reason, it must be considered as an upper bound estimate. In addition, the fastest thermal electrons can pass through the turbulent front, removing a significant part of thermal energy, thus reducing short characteristic cooling time still more.

Bremsstrahlung is known to provide determining contribution to plasma cooling by emission at temperatures above 10 million degrees. Its characteristic cooling time is about $10^6$ s, which is much greater than the conductive cooling time of a high-temperature source.

It should be noted that the role of radiative losses becomes more significant as the temperature decreases. First, radiative losses increase due to free-bound transitions,

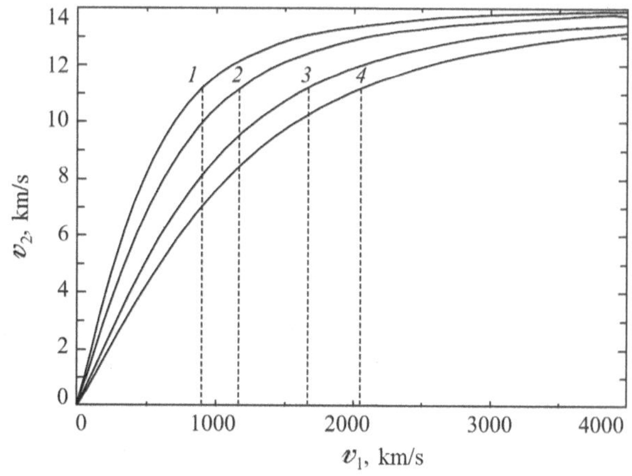


**Fig. 33** The ascending velocity of the high-temperature X-ray radiation source in the solar corona as a function of the reconnection region plasma velocity ($v_2 > v_1$), and its temperature T (30 MK (1), 50 MK (2), 100 MK (3), and 150 MK (4)). *Dashed lines* show velocity values, starting with which shock wave formation is possible. From [39, Fig. 3]

which leads to the formation of flare loops observed in the soft X-ray range. Then radiative energy losses reach their maximum values at temperatures $T_e \sim 10^5$, which causes rapid plasma cooling and compression, i.e., thermal instability. As a result, cold and dense plasma loops are formed; such loops can be observed in the $H_\alpha$ hydrogen line above the photospheric neutral line of magnetic field.

Generally, radiative losses are insignificant in the plasma heated behind the front. Cooling by heat conduction, on the contrary, is extremely effective. Its characteristic time is 10 s (upper limit), which is several times less than the lifetime of coronal hard X-ray sources.

We can suggest that the cooling processes in high-temperature coronal plasma are so effective that gas pressure can be completely neglected as compared to magnetic one. This allows us to consider a system of continuity equations in a non-adiabatic approach, i.e., differing from the one above.

The magnetic field can no more be assumed to be zero, since the problem formulation requires nonzero magnetic pressure. As an example, let us consider a flux with $B = 0.5$ Gs. It can be easily seen that in this case the velocity of the shock wave front, in fact, notably decreases, being equal to ~10 km/s (see Fig. 33); whereas in the adiabatic approximation, it amounted to about 1,000 km/s.

The obtained value is consistent with the upward velocity of a hard X-ray radiation source.

Thus, the formation of high-temperature X-ray sources in the corona may be described by a fast shock wave heating model. This heating model, which takes into account radiative and conductive cooling of hot plasma behind the front, is more consistent with actual solar corona conditions and observations than the traditional adiabatic approximation.

## Eruptive Phenomena Study

### *Study of Dimming Structures*

As is well known [41], coronal mass ejections (CMEs) and CME-associated magnetic field restructuring in the widespread solar atmosphere region are accompanied with a number of large-scale phenomena. In particular, the so-called dimmings and posteruptive arcades, with lifetimes ranging from hours to days, can be observed on the solar disk in the soft X-ray and extreme ultraviolet (EUV) ranges.

Dimmings, or transient coronal holes, are regions of reduced radiation intensity, which are formed due to the CME in the vicinity of the eruptive center and can cover a considerable part of the solar disk [42–44]. Dimmings that are adjacent to the eruptive center, as a rule, coincide by location and structure in multi-temperature lines, which gives us grounds for interpreting them as a result of full or partial opening of coronal magnetic fields resulting in the evacuation of matter and the corresponding reduction in radiation intensity. Direct evidence for the matter outflow from the dimmings in the vicinity of the eruptive center has been obtained in [45] paper through the Doppler shift of several EUV lines.

Regular solar observations by the SPIRIT telescope have provided important information on the powerful eruptive events of 2001–2005. An important advantage of the SPIRIT telescopes as compared to other instruments, in particular, to the SOHO/EIT telescope, consists in the possibility of simultaneous two-ranged imaging of the full solar disk: 175 Å range containing the Fe IX–Fe XI coronal lines ($T \sim 0.9$–$1.3$ MK), and 304 Å range with the dominating He II line of the transition layer ($T \sim 0.02$–$0.08$ MK). Correlation of temporal variations in the brightness of structures, which are visible in the different ranges, permitted us to study the dynamics of propagation of disturbances in active processes such as flares and CMEs. Other important tasks consisted in establishing global relations between the activity centers and searching for eruption precursors to forecast active phenomena and their geoeffectiveness.

The eruptive event, which occurred on 4 November 2001 after 16:00 UT, has been analyzed in detail. It was recorded in the channels of SPIRIT T1 (284 Å) and T2 (175 and 304 Å) telescopes [46]. There were two factors that preconditioned study of this particular event and the great interest that the event aroused. First, at this time no observations were carried out onboard SOHO/EIT because of preventive maintenance, and TRACE recorded active phenomena with a limited field of view aside from the eruption location. Secondly, this event is one of the largest and most geoeffective in the current 23-d activity cycle. It is comparable by its features to the widely known event of Bastille Day, on 14 July 2000.

A large active region AR 9684 located to the north-west of the center of the disk was an eruption source (Fig. 34). A flocculus visible in the $H_\alpha$ was crossed by an extended FF filament located above the polarities separating line of the photospheric magnetic field. As a Yohkoh/ XT soft X-ray image shows (Fig. 34a), a large high-temperature coronal arcade with a twisted (sigmoid) structure was observed

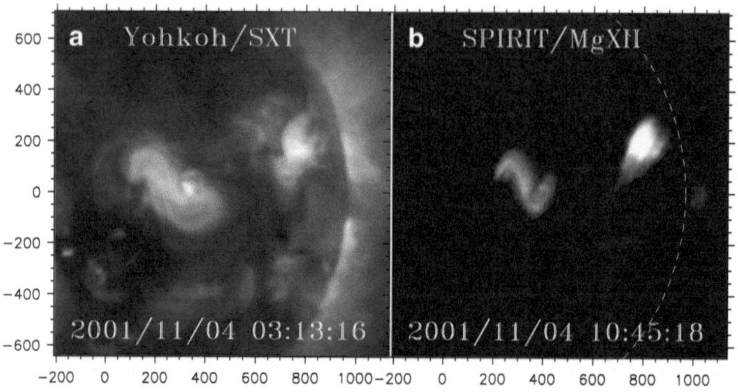


**Fig. 34** Sigmoidal structures prior to the event of November 4, 2001 recorded in (**a**) Yohkoh/SXT and (**b**) SPIRIT/MgXII images. From [47, Fig. 1]

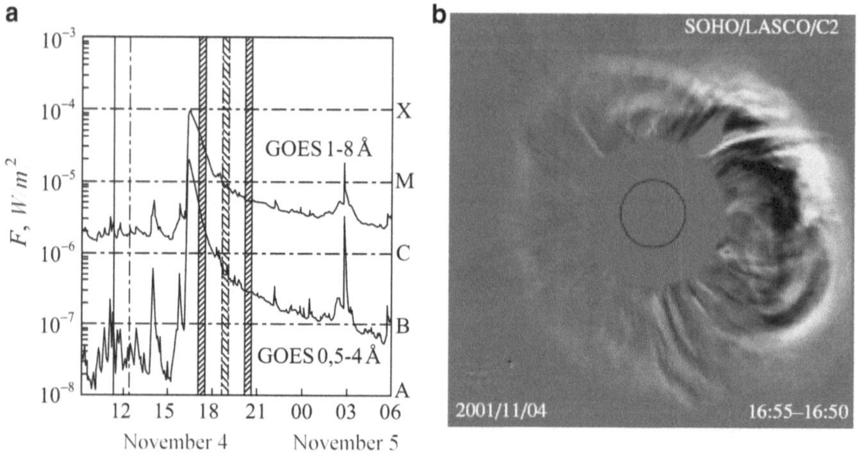


**Fig. 35** (**a**) Time curves of the signal in the GOES 1–8 and 0.5–4 Å channels and periods of observation by the SPIRIT telescope (*shaded areas*), (**b**) the CME image in the LASCO C2 coronagraph. From [46, Fig. 2]

above the filament. Such a structure is indicative of a sheared magnetic configuration, providing evidence for the excessive energy accumulated in the structure and a high probability of eruption. A similar structure was observed by SPIRIT in the hot Mg XII line before the event [47] (Fig. 34b): the sigmoid has been expressed more clearly as the Mg line emission corresponds to hotter plasma (5–15 MK).

The CORONAS-F/SPIRIT heliograms were recorded in 304 and 175 Å at 11:11 UT and in 284 Å at 12:19 UT, i.e., about 4–5 h prior to the event. The event began with the disappearance of the southern half of the aforementioned $H_\alpha$ filament and included a long-duration 3B/X1.0 flare (LDE) with its maximum at 16:20 UT (Fig. 35a). Halo-type CME observed with the LASCO/C2 coronagraph starting from

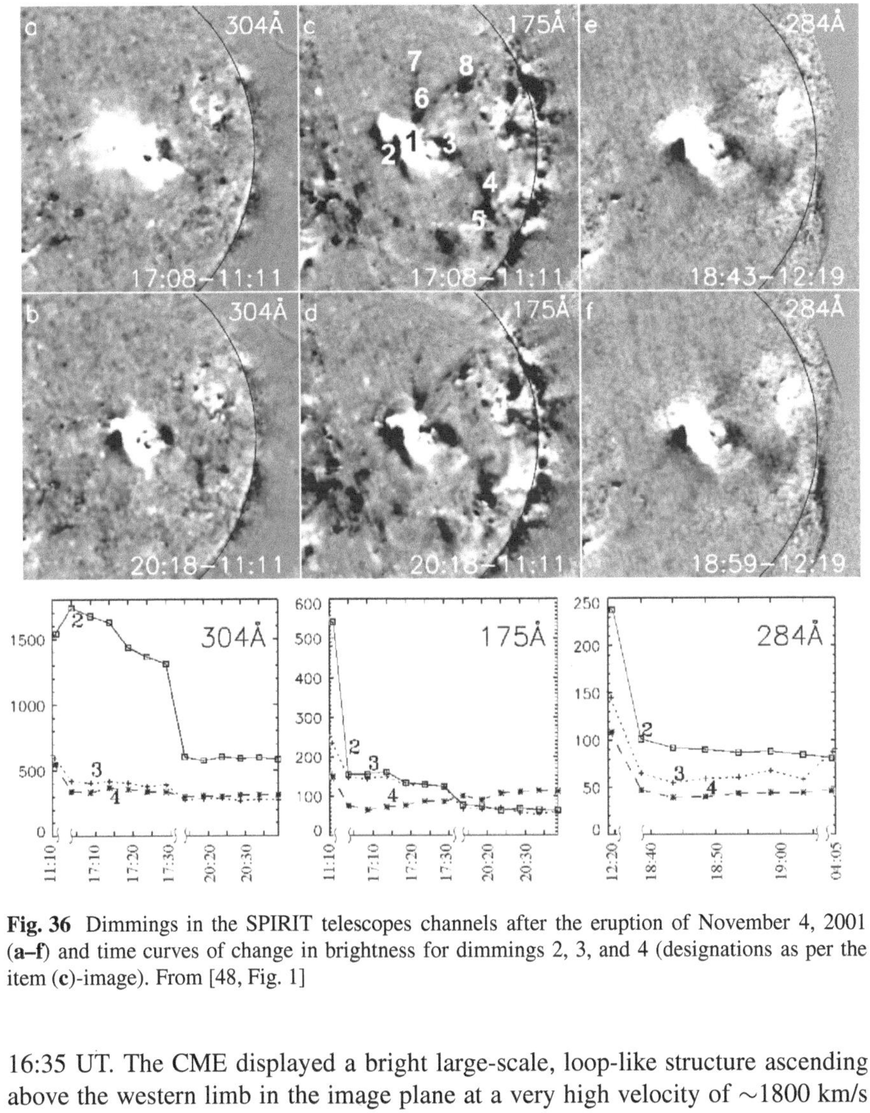


**Fig. 36** Dimmings in the SPIRIT telescopes channels after the eruption of November 4, 2001 (**a–f**) and time curves of change in brightness for dimmings 2, 3, and 4 (designations as per the item (**c**)-image). From [48, Fig. 1]

16:35 UT. The CME displayed a bright large-scale, loop-like structure ascending above the western limb in the image plane at a very high velocity of ~1800 km/s (Fig. 35b).

As a result of the eruption, long-term brightness decreases, i.e. dimmings, appeared in all the SPIRIT EUV channels (Fig. 36a–f). To identify these dimmings, the fixed-frame difference images were used, with subtraction being performed after compensation for solar rotation. The technique for obtaining such difference images is described in detail in [43,44]. The frames corresponding to 11:11 UT for 175 and 304 Å channels and to 12:19 UT for the 284 Å channel were used as the reference preeruptive images to be compared with all subsequent heliograms recorded by the SPIRIT telescope during this event. Solar derotation was made by rotating the heliograms to the times of the first frames obtained by the SPIRIT telescope after the event maximum phase, i.e., 17:03 UT for the 175 and 304 Å images and 18:38 UT for those at 284 Å.

Figure 36c, referring to the time moment of 17:03 UT in the 175 Å line, besides the brightly luminous arcade, displays relatively compact dimmings (2 and 3), which are immediately adjacent to the posteruptive arcade at its eastern and western sides. Alongside with the aforementioned dimmings, several narrow, extended dimmings were observed. These extended toward the south-western limb (3–4–5) and in the northern (6–7) and north-western (6–8) directions. Similar dimmings, though of different contrast, can be seen in a hotter 284 Å line and the 304 Å line of the transition region.

The lower part of Fig. 36 displays the extent of decrease in intensity in the central portion of the dimmings, which are identified by their corresponding numbers used in Fig. 36c, with the central dimmings being integrated over $22'' \times 22''$ areas [48]. Lower intensity in the dimmings, both in the coronal lines and in the transition-region line, persisted over several hours. The curves of Fig. 36 show that compact dimmings in the 175 and 284 Å coronal lines, near the flare localization, were the deepest (40–70 %). In the 304 Å transition-region line, the depth of these dimmings is also significant (30–60 %). We can see, however, that in the 304 Å band dimming 2 notably lags behind dimmings in coronal lines in both area (see above) and depth, which is indicative of the spread of disturbance from the corona to the solar transition region and the chromosphere.

Significant coronal mass ejections including halo-type CMEs recorded with the SOHO/LASCO white light coronagraphs occurred during the period of extreme solar activity during October–November 2003. Flares and CMEs accompanied with coronal waves and dimmings were observed by the SPIRIT instrumentation in the 175 and 304 Å channels [49]. Figure 37 depicts the structure of the dimmings, which were observed after the eruptions between 23 October and 2 November 2003 in the 175 Å channel. The large-scale nature of the dimmings in the aforementioned events demonstrated that a considerable portion of the solar atmosphere was involved in the CME process. It is worthwhile noting that a pattern of dimmings was to a great extent repeated in successive events. It is an evidence of eruptive events being homologous; it means that coronal structures involved in the formation of the CME were reproduced several times by the same long-lived configuration of the global magnetic field.

Observations of the powerful events of October–November 2003 have revealed two important advantages of SPIRIT telescopes over the SOHO/EIT telescope. Firstly, owing to differences in the spectral ranges 175 Å (SPIRIT) and 195 Å (SOHO/EIT) as well as the SPIRIT detector features, considerably more details can be seen in SPIRIT images during the brightest phase of the flare than in the corresponding EIT images. The latter are distorted by the detector saturation due to intense emission during plasma flares in a very hot Fe XXIV line at 192 Å as well as by a significant level of scattered light emitted by the flare regions. Such distortion effects were not observed in SPIRIT heliograms, as at very high signal levels microchannel plate detectors operate into a nonlinear mode, and thus are not saturated.

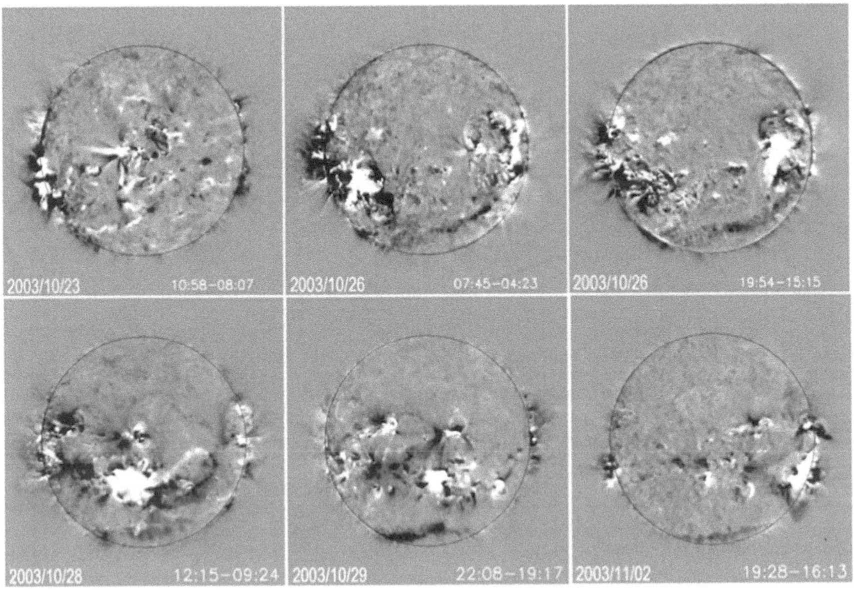


**Fig. 37** The dimmings' structure observed by the SPIRIT instrumentation during the events of October 2003 (channel 175 Å). Adapted from [49, Figs. 1, 2, 4]

Secondly, 20–30 min after the flare EIT images develop "snow," i.e., bright spots resulting from the detector exposure to the powerful flux of high-energy particles generated during the event (protons with the energy $E > 40$ MeV and secondary electrons [50]), which can often persists for several hours. SPIRIT heliograms remained unaffected throughout the event, since the CORONAS-F satellite in its comparatively low orbit was protected from the exposure to these particles by the Earth's magnetosphere. Figure 38 exemplifies difference images in the SPIRIT 175 Å channels versus EIT 195 Å channels [47] for the flare of October 28, 2003.

## *Study of the Temporal Characteristics of Dimmings and Their Relation to the Global Magnetic Field Structure*

To identify the relation of large-scale dimmings with the global structure of magnetic field, we have analyzed changes in the intensities of coronal structures within individual regions of dimmings prior to the eruption [51, 52]. For the events of October 28 and November 17, 2003, the maps of dimming regions and temporal profiles of integrated intensities within each region were built (see Fig. 39). Generally, dimming maps can be built by subtracting the base image recorded just before the eruption from the solar image taken from half an hour to 1 h after the eruption (estimated time of coronal mass ejection from the solar

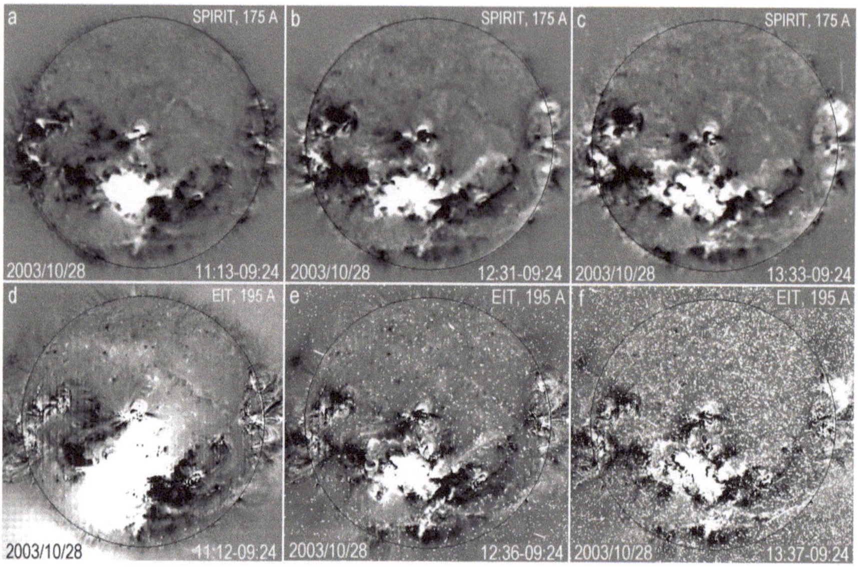


**Fig. 38** The effect produced by the charged particles background on SPIRIT 175 Å and EIT 195 Å difference images after the flare of 28 October 2003. Adapted from [49, Figs. 1, 2, 4]

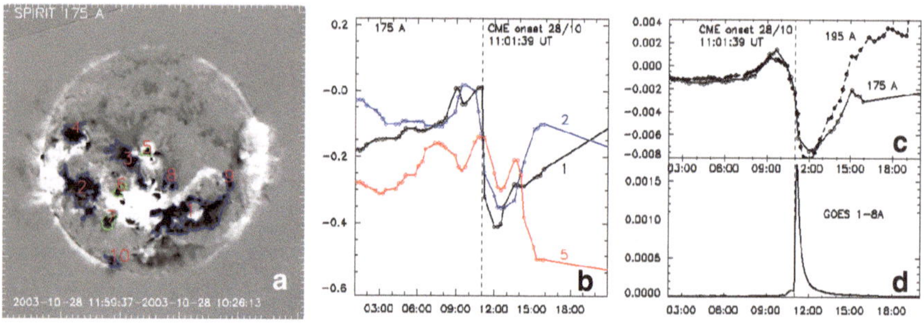


**Fig. 39** (**a**) Dimming maps for the event of 28 October 2003 built on the SPIRIT 175 Å data: 1–4, 8–10 are darkening with a high coefficient of intensity correlation with the main dimming area by time (1); 5–7 are darkening with a low correlation coefficient; (**b**) time profiles of the intensities for regions 1, 2 (high correlation) and 5 (low correlation); (**c**) profiles of the integrated intensity of the dimming in lines 175 Å and 195 Å vs. the profile of the GOES X-ray flux in the 1–8 Å channel (**d**). From [51, Figs. 1, 3]

surface, extrapolated for the CME propagation in a visible light coronagraph at distances longer than $2R_{\text{sun}}$, is usually assumed to be the beginning of an eruption). In this case, brightness variations of coronal structures within dimming regions were studied for the time point of several hours (up to 12 h) before the eruption. The analysis of temporal correlations of integrated intensity time profiles was performed for each of the dimming regions having the profile, whose contribution to the overall intensity of the dimming is the largest.

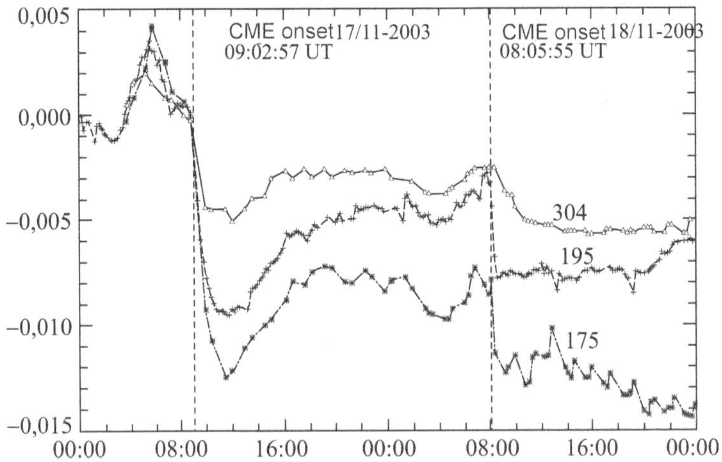


**Fig. 40** Time dependence of profiles for full dimmings in the 175 and 195 Å coronal lines and the 304 Å transition layer line. From [51, Figs. 1, 3]

The correlation coefficient was estimated based on the value of the linear Pearson's correlation calculated for the period of 4–6 h prior to the eruption. It was found that most of the regions show a high correlation, i.e., higher than the 50 % confidence level corresponding to the observation statistics; but this correlation is considerably lower for some regions. We can assume that in case of high correlation the intensity variations are caused by the same eruptive event, whereas in case of low correlation, such variations are due to the superimposition of other parallel processes. It is interesting that after the eruption starts, positive correlation in the regions of the first type drops sharply or even turns negative. The characteristic feature of the intensity of high-correlation regions associated with the main dimmings is a notable brightness increase (up to 10–20 %) 1–2 h prior to the eruption, which is followed by a decline, smooth first then turning into an abrupt one, reaching the minimum within 0.5–1 h after the eruption and slow recovery, sometimes to the initial brightness, within several hours (or even tens of hours). The integral intensities in the 175 Å and 195 Å coronal lines show virtually identical time dependencies. In the 304 Å transition region line, the minimum, which is notably less deep, is to be reached later, with the recovery being faster (Fig. 40).

The correlated variations of brightness in the remote regions of large-scale dimmings can be explained by the connections between spatially separated portions of the dimmings through the global magnetic structures, which are destroyed or completely disappear as a result of eruption. The so-called transequatorial magnetic loops, which can sometimes be observed in the X-ray range and detected by their manifestations in the form of type II bursts in the meter range, can exemplify such structures [53, 54]. Thus, the findings of the study of dimmings within the SPIRIT experiment testify to the global nature of CMEs during high solar activity.

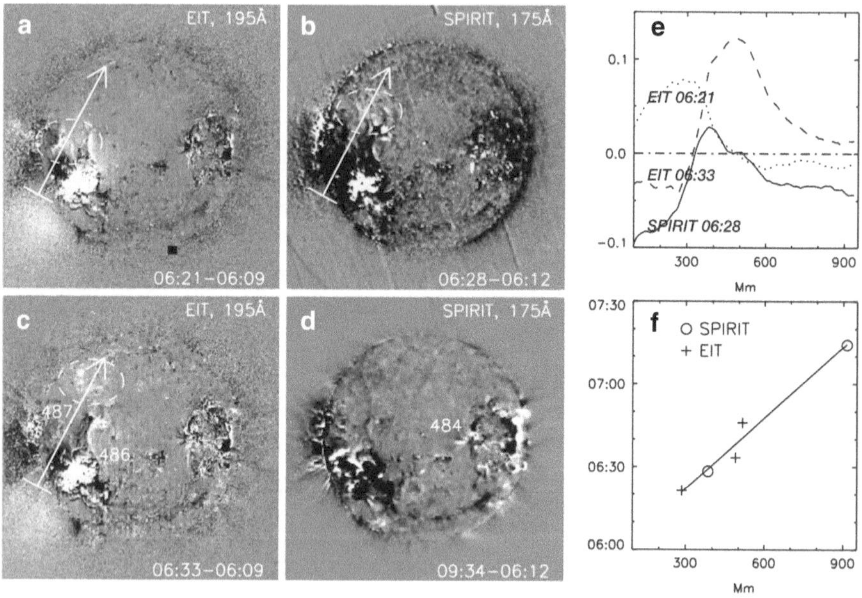


**Fig. 41** Coronal (EIT) wave observed by SPIRIT and EIT telescopes on 26 October 2003, (**e**) wave profiles at different time points, (**f**) distance-time dependence for the coronal wave distribution. Adapted from [49, Figs. 1, 2, 4]

## *Study of Coronal Waves and Propagation of Dimmings*

Around 6:00 UT on 26 October 2003, the SPIRIT telescope simultaneously with the SOHO/EIT telescope observed the 3B/X1.2-class flare that occurred in the active region 486 and was accompanied by a partial halo CME [49]. In the interval between 6:21 and 6:46 UT, both telescopes recorded a coronal wave (aka EIT wave [42]). Generally, coronal waves are recorded by the EIT telescope in the CME monitoring mode at a rate of 1 frame per every 12 min, and no more than 2–3 frames can be taken during the wave propagation (25–30 min), thus making it difficult to estimate the time of occurrence and propagation velocity of the wave. In that case, due to the difference in the time of recording, the wave could be observed with both telescopes in five frames, the data of both telescopes complementing each other well (Fig. 41). By combining these images, the average wave velocity was found to be about 190 km/s. The wave propagated from the active region 486 along the coronal structures toward the North Pole, without crossing other active regions. Displacement of the coronal wave front did not expanded the boundaries of dimmings. It means that the assertion made by Thompson et al. [42] that the coronal wave front is an outer boundary of the propagating dimming is not of a general nature.

Within the period of November 17–18, 2003, the SPIRIT telescope onboard the Coronas-F space station recorded solar images simultaneously in two channels:

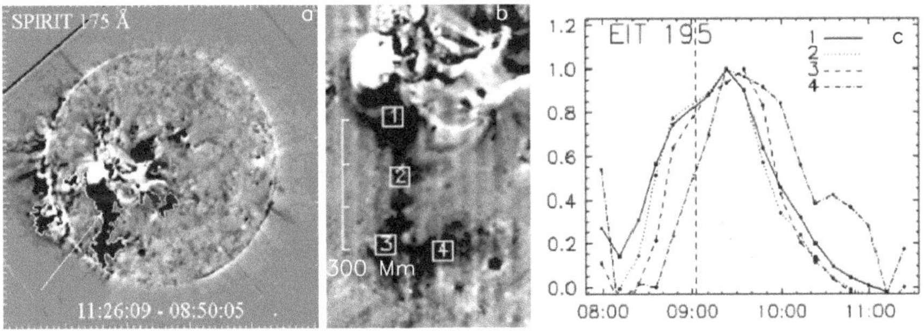


**Fig. 42** Propagation of a giant dimming during the eruptive event of November 17, 2003 in accordance with the SPIRIT and EIT data: (**a**) a difference image in the 195 Å EIT channel; (**b**) enlarged image of dimming No. 1 (sample areas of the dimming are marked with squares 1–4), (**c**) time derivative of brightness, normalized for the maximum, in squares 1–4. Adapted from [52, Figs. 3, 4]

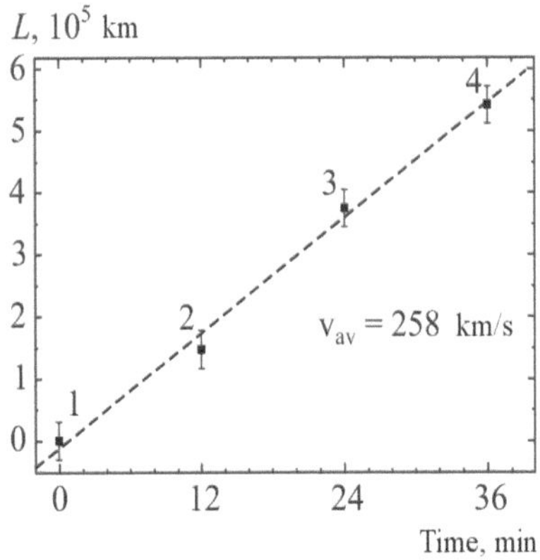


**Fig. 43** Estimation of the front propagation velocity for the giant dimming of 17 November 2003 over sample areas 1–4 according to the EIT (195 Å) data. Adapted from [52, Figs. 3, 4]

175 Å and 304 Å. The observations were made within 47 min at 15 min intervals in the unshadowed parts of orbits. CMEs were accompanied with dimmings that emerged in the vicinity of the eruption site on the disk. Based on the data obtained by the SPIRIT telescope, fixed-base derotated difference images have been built to study the structure of dimmings.

During the eruptive events of 17 November (the CME started at 08:50 UT), a giant dimming nearly 600,000 km long formed in the southern part of the solar disk (Fig. 42a, b). The boundaries of that dimming almost coincided in the difference images in the SPIRIT 175 Å channel and SOHO/EIT 195 Å channel. Figure 42c presents plots of the integrated intensity derivatives in the 195 Å line. By the time of disturbance passage through the areas marked with squares 1–4 at the 0.2 level, the average velocity of the intensity decrease front propagation (see Fig. 43) appeared to be 258 km/s, which is close to the average coronal wave velocity of 250–400 km/s.

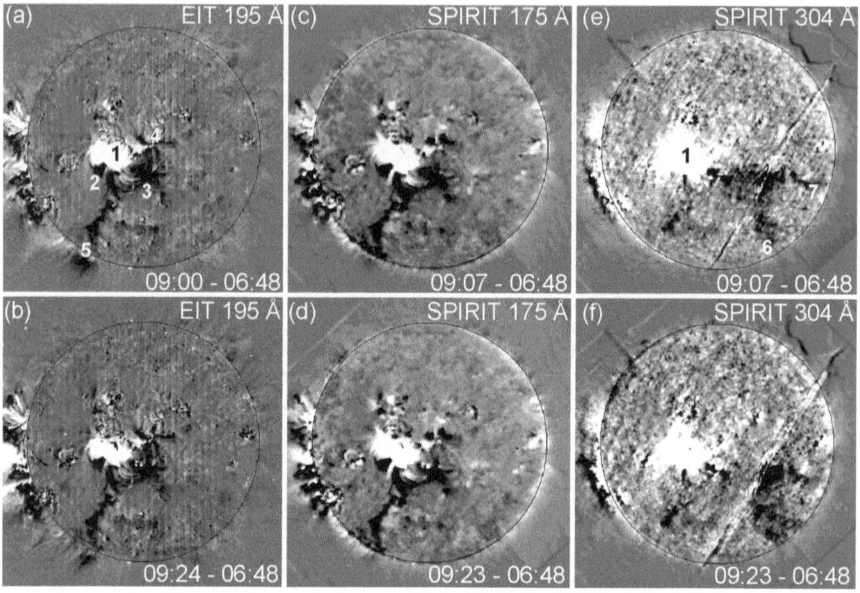


**Fig. 44** Drifting of the cold matter cloud during the eruption of November 18, 2003 observed in the difference images in the SPIRIT 304 Å channel and invisible in the SPIRIT 175 Å and EIT 195 Å coronal channels. From [48, Fig. 1]

Let us note here that no coronal wave propagation has been recorded during the event by EIT or SPIRIT telescopes, which substantiates the aforementioned conclusion that there is no unequivocal connection between dimmings and coronal waves.

## *Transient Phenomenon Observed During the Eruption of November 17, 2003*

An eruption-related transient phenomenon was observed during the events of 18 November 2003: an extending Y-shaped cloud passing through the solar disk and clearly visible as a dark shadow in the transition region lines (He II, $T = 0.08$ MK) but almost invisible in the 175 Å and 195 Å coronal lines [47,48] (Fig. 44). The large quiescent filament in active region AR 10501 was activated as a result of eruption at 07:41UT, destabilized and dispersed into a large-scale cloud of several fragments. The movement of different parts of the cloud was seen in images in the center, red and blue wings of $H_\alpha$ line registered by Kanzelhöhe Solar Observatory (KSO, Austria). The cloud has been observed within the disc between 08:23 and 08:54 UT as a "horseshoe" shadow with its top extended southwest from the active region. The horizontal velocity of cloud moving across the disk was estimated to be between 60 and 110 km/s. The main part of the cloud passed across the disk, but no CMEs were observed by the LASCO coronagraph afterwards. According to the analysis of possible absorption mechanisms in the He II 304 Å line (see [55]) and visibility of the cloud in different $H_\alpha$ channels, the most probable cause of absorption was non-resonant scattering. Estimations of the initial mass of the filament and the mass of the cloud from opacity in 304 Å as well as in 175 and 195 Å bands have shown that the cloud mass ($2–3 \cdot 10^{15}$ g) was of order the initial filament mass ($4–6 \cdot 10^{15}$ g). All observed phenomena suggest to be related to a return prominence, which was formed from erupted filament.

According to J.-P. Delaboudiniere's private communication, similar phenomena were observed earlier by the EIT telescope onboard SOHO, but only in the form of a single frame, since measurement interval in the 304 Å line is 6 h. In the SPIRIT experiment, the dynamics of the phenomenon in several wavelengths has been studied for the first time. In addition to the SPIRIT results, it should be noted that such phenomena were observed in several giant prominence eruptions by the TESIS telescope [56] during its operation from January to November 2009.

## Conclusion

Unique data on the solar atmosphere structure and dynamics in the temperature range of 0.05–20 MK were obtained with the use of the SPIRIT instrumentation set consisting of SRT-C solar telescopes and RES-C spectroheliographs, which operated onboard the CORONAS-F satellite from August 2001 to December 2005. These data represent a new stage in the study of fundamental processes of solar activity.

The method of *full-Sun multichannel monochromatic imaging spectroscopy* in the X-ray and extreme VUV wavelengths was implemented in the RES-C instrument for the first time. Unique database of spectroheliograms has formed the basis for studying the features of well-known solar atmosphere structures (active regions, coronal holes), and new ones, first observed in X-ray monochromatic images.

The following results obtained in the SPIRIT experiment can be referred to as the most important, top-priority ones:

A new class of phenomena, hot ($T = 5–20$ MK) plasma structures in the solar corona at heights of up to 0.6 solar radii, characterized by specific shape and dynamics, has been observed and studied in the Mg XII line (8.4 Å).

Spectra of more than 30 powerful flares have been obtained in the range of 177–335 Å; the spectral line catalogue database (from 100 lines up to 160 lines) has been significantly expanded.

Diagnostics of electron density, ion and chemical composition, and differential emission measure for various solar structures, including coronal holes and flares, has been performed.

For the first time, within the joint observations with the use of SPIRIT and SPR-N instruments, research teams of the LPI and D.V. Skobeltsyn Research Institute of Nuclear Physics (SINP of the Moscow State University) reliably recorded high (up to 80 %) degree of polarization of the continuous X-ray radiation (20 keV to 100 keV) during the powerful X10 flare on 29 October 2003; high degree of polarization is indicative of a non-thermal mechanism of energy release in the flare region.

A programme providing for the observations of a number of the most powerful solar flares in the recent 30 years was implemented: in the periods of 20.10.03 to 05.11.03, 01.01.05 to 05.01.05, and 04.09.05 to 19.09.05. Quasi-continuous series of full sun images were obtained in the 175 Å (Fe IX–Fe XI) and 8.42 Å (Mg XII) channels. Data analysis proved that the absolute flux for the flare of ~X30 importance, recorded at its maximum phase in the Mg XII line on 04.11.03, amounted to $5{\cdot}10^{-2}$ erg·cm$^{-2}$s$^{-1}$, and the emission measure was equal to more than $10^{51}$ cm$^{-3}$.

Based on simultaneous full solar disk imaging in the 175 Å and 304 Å channels, new data on the dynamics of eruptive processes in the transition layer and solar corona have been obtained, in particular, the data on the development of dimmings observed in the EUV spectral range and being indicative of plasma outflows during the evolution of coronal mass ejections (CMEs). It is shown that large-scale dimmings occurring during high solar activity periods indicate magnetic connections between remote centers of the activity, thus pointing to the global nature of CMEs. The multi-wavelength analysis of absorption phenomena during prominence eruption provided important information on kinematics and mass balance in this process.

It has been ascertained that the rapid solar plasma heating up to temperatures of about 10 million K, which can be observed in the solar corona during flares, cannot be explained in the context of the so-called thick-target model, which is traditionally applied to explain the energy release in flares. In accordance with this model, the effective plasma heating and intense X-ray radiation from flares can only develop in the dense layers of the solar atmosphere (photosphere and chromosphere) rather than in the rarified corona. SPIRIT observations have shown that such vision is inconsistent with the experiment. To resolve this inconsistency, modified models of energy release in flares, both effective in low-density coronal plasma and in good agreement with experimental results, have been suggested.

Thus, the data set obtained in the SPIRIT experiment allows us to analyze the structure and dynamics of the upper solar atmosphere plasma over a wide range of parameters for the first time: height (from the chromosphere up to 5 solar radii), temperature (from ten thousand to thirty million degrees), duration of phenomena observations (from seconds to several solar rotations), and spatial resolution (3–5 arcseconds).

The level and scale of the SPIRIT experiment findings are comparable with the achievements of milestone experiments, such as SOHO and Yohkoh experiments implemented due to the efforts undertaken by the USA, European, and Japanese international science consortia and space agencies. Before the next sunspot maximum, these records will provide a basic data source for national and international research and planning new experiments required to solve the fundamental problems of physics of solar atmosphere and solar-terrestrial relations.

**Acknowledgements** The authors of this paper are deeply grateful to a number of their colleagues in Russia and abroad for the participation in the equipment development, observations, data analysis, and interpretation, and especially to:

LPI Spectroscopy Department scientists V. K. Bardin, N. I. Potapov, E. A. Avetisyan, A. L. Goncharov, V. V. Korneev, V. V. Krutov, T. A. Shergina, V. M. Lomkova, and V. F. Sukhanov, who were actively involved in the equipment fabrication and experiment preparation;

J.-P. Delaboudinière (Institute of Space Astrophysics, Orsay, France) for supplying the Ritchey-Chretien lens for the T1 telescope and participating in discussions of the results;

N. N. Salashchenko and our colleagues from the Institute for Physics of Microstructures, RAS (city of Nizhny Novgorod) for the technology development, manufacture, and calibration of a multilayer X-ray optics for the SPIRIT telescopes and spectroheliographs;

I. L. Strulea and V. I. Maslov for the fabrication of high-precision optics; M. T. Pakhomov for the fabrication and testing of optoelectronic converters for SPIRIT light detectors;

V. O. Timofeev for the development and fabrication of CCD arrays; Yu. N. Tokaev for the fabrication of telescope T1 structural parts of carbon fibre plastic;

V. I. Dranovsky, S. A. Karapetyan, V. M. Brovko, A. M. Popel, and P. I. Lyagin for their assistance in the development of equipment, its placing on board the satellite, and carrying out ground-based tests;

A. I. Stepanov, D. V. Lisin, A. A. Afanasyev, and other IZMIRAN management group members for the organization of observational sessions, control over the scientific equipment and telemetry data delivery;

DLR Centre (Neustrelitz, Germany) specialists for their substantial assistance in telemetry data reception;

V. V. Grechnev and I. M. Chertok for their active participation in the observational data analysis and interpretation.

This paper was partially supported by the Russian Foundation for Fundamental Research (with regard to projects No. 99-02-17176, 02-02-17272, 05-02-17415, and 08-02-01301), Fundamental Research Programme of the Presidium of RAS No. 16 and Fundamental Research Programme of the Physical Sciences Division of RAS No. 16.

SOHO is an international cooperation project of ESA and NASA space agencies.

## References

1. Oraevsky, V.N., Sobelman, I.I.: Astron. Lett. **28**, 401 (2002)
2. Zhitnik, I.A., Bougaenko, O.I., Delaboudinière, J.-P., Ignatiev, A.P., Korneev, V.V., Krutov, V.V., Kuzin, S.V., Lisin, D.V., Mitrofanov, A.V., Oparin, S.N., Oraevsky, V.N., Pertsov, A.A., Slemzin, V.A., Sobelman, I.I., Stepanov, A.I., Schwarz, J.: SPIRIT X-ray telescope/spectroheliometer results. In: Solar variability: from core to outer frontiers. The 10th European Solar Physics Meeting, 9–14 September 2002, Prague, Czech Republic. Ed. A. Wilson. ESA SP-506, Vol. 2. Noordwijk: ESA Publications Division, ISBN 92-9092-816-6, 2002, p. 915–918
3. Moses, D., Clette, F., Delaboudinière, J.-P., Artzner, G.E., Bougnet, M., Brunaud, J., Carabetian, C., Gabriel, A.H., Hochedez, J.F., Millier, F., et al.: Sol. Phys. **175**, 571 (1997)
4. Sobelman, I.I., Zhitnik, I.A., Valnichek, B., Rybansky, M., Bernas, M., Gaponov, S.V., et al.: Pis'ma v Astronomicheskii Zhurnal **16**, 323 (1990)
5. Sobelman, I.I., Zhitnik, I.A., Ignat'ev, A.P., Korneev, V.V., Klepikov, V.Yu., Krutov, V.V., et al.: Astron. Lett. **22**, 539 (1996)

6. Zhitnik, I.A., Ignatiev, A.P., Korneev, V.V., Krutov, V.V., Kuzin, S.V., Mitrofanov, A.V., et al.: Current Russian research in optics and photonics: new methods and instruments for space- and earth-based spectroscopy in XUV, UV, IR, and millimeter waves. Proc. SPIE **3406**, 1–19 (1998)
7. Andreyev, S.S., Zuyev, S.Yu., Pozdnyakova, V.I., Salaschenko, N.N., Slemzin, V.A., Strulya, I.L., Shereshevskii, I.A., Zhitnik, I.A.: Poverkhnost'. Rentgenovskie, Sinkhrotronnye i Neitronnye Issledovaniya (Russian: Journal of surface investigation: X-ray, synchrotron and neutron techniques) **1**, 6 (2003)
8. Zhitnik, I., Kuzin, S., Afanas'ev, A., Bugaenko, O., Ignat'ev, A., Krutov, V., Mitrofanov, A., Oparin, S., Pertsov, A., Slemzin, V., Sukhodrev, N., Urnov, A.: Adv. Space Res. **32**, 473 (2003)
9. Slemzin, V.A., Bougaenko, O.I., Ignatiev, A.P., Kuzin, S.V., Mitrofanov, A.V., Pertsov, A.A., Zhitnik, I.A.: Ann. Geophys. **26**, 3007 (2008)
10. Mitrofanov, A.V., Pudonin, F.A., Starodubzev, N., Zhitnik, I.A.: Current Russian research in optics and photonics: new methods and instruments for Space- and Earth-based spectroscopy in XUV, UV, IR, and Millimeter Waves. Proc. SPIE **3406**, 35–44 (1998)
11. Mitrofanov, A.V., Zuyev, S.Yu.: Izvestiya of Russian Academy of Sciences. Physics **68**, 556 (2004)
12. Aynbund, M.R., Polenov, B.V.: Second-Order Electronic Multipliers of the Open Type and their Applications. Energoizdat, Moscow (1981)
13. Young, P.R., Del Zanna, G., Landi, E., Dere, K.P., Mason, H.E., Landini, M.: Astrophys. J. Suppl. Ser. **144**, 135 (2003)
14. Dere, K.P., Moses, J.D., Delaboudinière, J.-P., Brunaud, J., Carabetian, C., Hochedez, J.-F., et al.: Sol. Phys. **195**, 13 (2000)
15. Slemzin, V.A., Kuzin, S.V., Zhitnik, I.A., Delaboudinière, J.-P., Auchere, F., Zhukov, A.N., et al.: Sol. Syst. Res. **39**(6), 489–500 (2005)
16. Mandelshtam, S.L., Tindo, I.P., Voron'ko, Yu.K., Shurigin, A.I., Vasil'ev, B.N.: Studies of X-ray radiance of the Sun. I. Measurements by means of geophysical rockets. In: Artificial Earth Satellites, vol. 10, p. 12 (1961) (in Russian)
17. Tsuneta, S., Acton, L., Bruner, M., Lemen, J., Brown, W., Caravalho, R., et al.: Sol. Phys. **136,** 37 (1991)
18. Kuzin, S.V., Andreev, E.A., Korneev, V.V., Krutov, V.V., Mitropolosky, M.M., Pertzov, A.A., et al.: X-ray and ultraviolet spectroscopy and polarimetry, Silvano Fineschi. Proc. SPIE **2283,** 242–250 (1994)
19. Ignatiev, A.P., Kolachevsky, N.N., Korneev, V.V., Krutov, V.V., Kuzin, S.V., Mitrofanov, A.V., et al.: SPIE selected papers, New methods and instruments for space and earth-based spectroscopy in XUV, UV, IR and MM-waves. Proc. SPIE **3406**, 20–34 (1998)
20. Korneev, V.V., et al.: Proceedings of the workshop "X-rays optics 2002", Nizhnii Novgorod, Russia 2000, p. 175
21. Zhitnik, I.A., Urnov, A.M.: In: Uchida, Y., Canfield, R.C., Watanabe, T., Hiei E. (eds.) Proceedings of the international SOLAR-a science meeting 'flare physics in solar activity maximum 22' (Tokyo, Japan, 23–26 October 1990). Lecture Notes in Physics, vol. 387, p. 312. Springer, Berlin (1991)
22. Sobelman, I.I., Zhitnik, I.A.: X-ray telescopes aboard Fobos-I and Coronas. In: Hoover, R.B., Walker, A.B.C. (eds.) Multilayer and Grazing Incidence X-ray/EUV Optics for Astronomy and Projection Lithography. Proc. SPIE, vol. 1742, pp. 539–548 (1993)
23. Sobelman, I.I., Zhitnik, I.A., Kuzin, S.V., et al.: CORONAS-I information No 5, Preprint of Astrophysical Institute (Potsdam, 1993)
24. Kuzin, S.V., Zhitnik, I.A., Pertsov, A.A., Slemzin, V.A., Mitrofanov, A.V., Ignatiev, A.P., et al.: J. X-Ray Sci. Technol. **7**, 233 (1997)
25. Pertsov, A.A., et al.: Proceedings of the workshop "X-rays optics 2002", Nizhnii Novgorod, Russia 1999, p. 196

26. Mitrofanov, A.V., Zuyev, S.Yu.: Poverkhnost'. Rentgenovskie, Sinkhrotronnye i Neitronnye Issledovaniya (Russian: Journal of surface investigation: X-ray, synchrotron and neutron techniques) **1**, 81 (2002)
27. Ignat'ev, A.P., Slemzin, V.A., Kuzin, S.V., Bugaenko, O.I., Zhitnik, I.A., Pertsov, A.A., et al.: Instrum. Exp. Tech. **51**, 702 (2008)
28. Zhitnik, I.A., Kuzin, S.V., Bugaenko, O., Ignat'ev, A., Krutov, V., Lisin, D., Mitrofanov, A., Oparin, S., Pertsov, A., Slemzin, V.A., Urnov, A.M.: Adv. Space Res. **32**, 2573 (2003)
29. Zhitnik, I.A., Bugaenko, O.I., Ignat'ev, A.P., Krutov, V.V., Kuzin, S.V., Mitrofanov, A.V., Oparin, S.N., Pertsov, A.A., Slemzin, V.A., Stepanov, A.I., Urnov, A.M.: Mon. Not. R. Astron. Soc. **338**, 67 (2003)
30. Bogachev, S.A., Kuzin, S.V., Zhitnik, I.A., Urnov, A.M., Grechnev, V.V.: Sol. Sys. Res. **39**, 508 (2005)
31. Pertsov, A., Zhitnik, I., Bougaenko, O., Ignatiev, A., Krutov, V., Kuzin, S., Mitrofanov, A., Oparin, S., Slemzin, V.: Proc. ISCS Symposium (Tatranska Lomnica, Slovakia). In: Wilson, A. (ed.) Solar Variability as an Input to the Earth's Environment. ESA SP-535, pp. 239–241. ESA Publications Division, Noordwijk, ISBN 92-9092-845-X (2003)
32. Akimov, L.A., Beletskii, S.A., Belkina, I.L., Bugaenko, O.I., Velikodskii, Yu.I., Zhitnik, I.A., Ignat'ev, A.P., Korokhin, V.V., Kuzin, S.V., Marchenko, G.P., Pertsov, A.A.: Astron. Rep. **49**, 579 (2005)
33. Borovik, V.N., Rudenko, G.V., Slemzin, V.A., Stepanov, A.I., Shibasaki, K., Uralov, A.M., Zandanov, V.G., Zhitnik, I.A., Grechnev, V.V., Bugaenko, O.I., Bogachev, S.A., Grigorieva, I.Y., Kuzin, S.V., Lesovoi, S.V., Livshits, M.A., Pertsov, A.A.: Coronal and stellar mass ejections. In: Dere, K., Wang, J., Yan, Y. (eds.) IAU Symposium Proceedings of the International Astronomical Union 226 (13–17 September, Beijing), pp. 108–109, Fig. 1. Cambridge University Press, Cambridge (2005)
34. Grechnev, V.V., Uralov, A.M., Zandanov, V.G., Rudenko, G.V., Borovik, V.N., Grigorieva, I.Y., Slemzin, V.A., Bogachev, S.A., Kuzin, S.V., Zhitnik, I.A., Pertsov, A.A., Shibasaki, K., Livshits, M.A.: Publ. Astron. Soc. Jpn. **58**, 55 (2006)
35. Urnov, A.M., Shestov, S.V., Bogachev, S.A., Goryaev, F.F., Zhitnik, I.A., Kuzin, S.V.: Astron. Lett. **33**, 396 (2007); Fig.3 and Fig. 9 respectively
36. Zhitnik, I.A., Kuzin, S.V., Urnov, A.M., Beigman, I.L., Bozhenkov, S.A., Tolstihina, I.Yu.: Astron. Lett. **31**, 37 (2005)
37. Shestov, S.V., Bozhenkov, S.A., Zhitnik, I.A., Kuzin, S.V., Urnov, A.M., Beigman, I.L., Goryaev, F.F., Tolstihina, I.Yu.: Astron. Lett. **34**, 33 (2008)
38. Zhitnik, I.A., Kuzin, S.V., Urnov, A.M., Bogachev, S.A., Goryaev, F.F., Shestov, S.V.: Sol. Sys. Res. **40**, 272 (2006); Fig.1-left, Fig.1-right, Fig.3, Fig.4a, Fig.4b, Fig.2, Fig.7, Fig.8, Fig.10, Fig.9 respectively
39. Bogachev, S.A., Somov, B.V., Masuda, S.: Astron. Lett. **24**, 543 (1998); Fig. 3
40. Somov, B.V., Kosugi, T., Bogachev, S.A., Masuda, S., Sakao, T.: Adv. Space Res. **35**, 1690 (2005)
41. Hudson, H.S., Webb, D.F.: Soft x-ray signatures of coronal ejections. In: Crooker, N., Joselyn, J.A., Feynman, J. (eds.) Coronal Mass Ejections. American Geophysical Union: Geophysical Monograph series, vol. 99, pp. 27–38 (1997)
42. Thompson, B.J., Plunkett, S.P., Gurman, J.B., Newmark, J.S., Cyr, O.C. St., Michels, D.J.: Geophys. Res. Lett. **25**, 2465 (1998)
43. Chertok, I.M., Grechnev, V.V.: Astron. Rep. **47**, 139 (2003)
44. Chertok, I.M., Grechnev, V.V.: Astron. Rep. **47**, 934 (2003)
45. Harra, L.K., Sterling, A.C.: Astrophys. J. **561**, L215 (2001)
46. Chertok, I.M., Slemzin, V.A., Kuzin, S.V., Grechnev, V.V., Bugaenko, O.I., Zhitnik, I.A., Ignat'ev, A.P., Pertsov, A.A.: Astron. Rep. **48**(5), 407–417 (2004); Fig. 2
47. Slemzin, V.A., Grechnev, V.V., Zhitnik, I.A., Kuzin, S.V., Chertok, I.M., Bogachev, S.A., Ignatiev, A.P., Pertsov, A.A., Lisin, D.V.: Coronal and stellar mass ejections. In: Dere, K.,

Wang, J., Yan, Y. (eds.) IAU Symposium Proceedings of the International Astronomical Union 226 (13–17 September, Beijing), pp. 21–26. Cambridge University Press, Cambridge (2005); Fig. 1
48. Kuzin, S., Chertok, I., Grechnev, V., Slemzin, V., Bugaenko, O., Zhitnik, I., Ignat'ev, A., Pertsov, A.: Adv. Space Res. **38**(3), 451-455 (2006)
49. Grechnev, V.V., Chertok, I.M., Slemzin, V.A., Kuzin, S.V., Ignat'ev, A.P., Pertsov, A.A., Zhitnik, I.A., Delaboudinière, J.-P., Auchère, F.: J. Geophys. Res. **110**, A09S07 (2005); Fig. 4 – with different subscriptions, Fig. 2 – with different subscriptions, and Fig. 1 respectively
50. Grechnev, V.V.: In: Stepanov, A.V., Benevolenskaya, E.E., Kosovichev, A.G. (eds.) Proceedings of the IAU Symposium No 223 "Multi-Wavelength Investigations of Solar Activity", pp. 625–626. Cambridge University Press, Cambridge (2004)
51. Slemzin, V., Kuzin, S., Bogachev, S.: In: Danesy, D., Poedts, S., De Groof, A., Andries , J.(eds.) Proceedings of the 11-th European Solar Physics Meeting "The Dynamic Sun: Challenges for Theory and Observations", 11–16 September 2005, p. 166.1. Belgium (ESA SP-596) (published on CDROM); Leuven, Fig. 1 and Fig. 3 respectively
52. Slemzin, V.A., Grechnev, V.V., Kuzin, S.V.: In: Bothmer, V., Hady, A.A. (eds.) Proceedings of the IAU Symposium No 233 "Solar Activity and its Magnetic Origin", pp. 361–364. Cambridge University Press, Cambridge (2006); Cairo, Egypt, 31 March – 4 April. Combination of Figs. 3 and 4 and a part of Fig. 4 respectively
53. Khan, J.I., Hudson, H.S.: Geophys. Res. Lett. **27**, 1083 (2000)
54. Pohjolainen, S., Vilmer, N., Khan, J.I., Hillaris, A.E.: Astron. Astrophys. **434**, 329 (2005)
55. Grechnev, V.V., Uralov, A.M., Slemzin, V.A., Chertok, I.M., Kuzmenko, I.V., Shibasaki, K.: Sol. Phys. **253**, 263 (2008)
56. Kuzin, S.V., Bogachev, S.A., Zhitnik, I.A., Pertsov, A.A., Ignatiev, A.P., Mitrofanov, A.V., Slemzin, V.A., Shestov, S.V., Sukhodrev, N.K., Bugaenko, O.I.: Adv. Space Res. **43**, 1001 (2009)
57. Slemzin, V., et al.: Ann. Geophys. **26**, 1–10 (2008)
58. Kuznetsov V.D.: Sol. Syst. Res. **39**, 433 (2005)

# Supplement to the English Edition: New Results of the SPIRIT and TESIS Experiments Under the CORONAS Solar Program

**S.V. Kuzin, S.A. Bogachev, O.I. Bugaenko, A.P. Ignat'ev, A.S. Kirichenko, A.A. Pertsov, A.S. Ulyanov, A.A. Reva, S.V. Shestov, V.A. Slemzin, N.K. Sukhodrev, Y.S. Ivanov, L.A. Goncharov, A.V. Mitrofanov, S.G. Popova, V.A. Solov'ev, and I.A. Zhitnik**[1]

## Preface

This Supplement contains three parts. The first and the second parts describe results of two studies within the scope of the CORONAS-F/SPIRIT experiment completed after the issue of the Russian edition in 2009: an investigation of hot X-ray points (HXPs) from the data of the Mg XII Spectroheliograph and a study of the extended solar corona with the SPIRIT telescope in the 175 and 304 Å bands. The studies are based on the unique experimental data provided by SPIRIT and therefore the obtained results are important and greatly enhance our knowledge gained from other space-borne experiments.

The third part of the Supplement contains a brief description of the TESIS instrumental complex, operated during February–November 2009 onboard the last CORONAS-Photon mission of the CORONAS solar project and some principal results of the experiment. TESIS is a new generation instrument for investigations of the solar corona in the X-ray, XUV, and EUV spectral ranges by the method of imaging spectroscopy with considerably improved angular resolution, cadence, higher sensitivity, and extended spectral range. The TESIS instrumentation was designed using the experience of SPIRIT on the modern scientific and technological base which made possible to achieve parameters anticipating or even surpassing those of the SDO observatory. The scientific tasks of the TESIS experiment continue the tasks of the CORONAS-F for the period of extremely low solar activity in the minimum of 23–24 solar cycles.

---

[1]Deceased

S.V. Kuzin · S.A. Bogachev · O.I. Bugaenko · A.P. Ignat'ev · A.S. Kirichenko · A.A. Pertsov · A.S. Ulyanov · A.A. Reva · S.V. Shestov · V.A. Slemzin · N.K. Sukhodrev · Y.S. Ivanov · L.A. Goncharov · A.V. Mitrofanov · S.G. Popova · V.A. Solov'ev · I.A. Zhitnik
LPI, Moscow, Russia

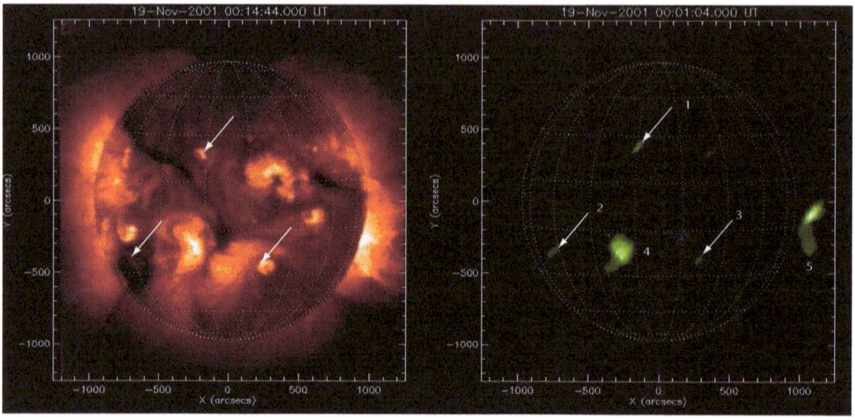


**Fig. 1** Images obtained by SXT/*Yohkoh* (*left*) and spectroheliograph Mg XII (*right*). 1, 2, 3: HXP; 4, 5: large hot structures (figure is taken from [1])

## Investigation of Hot X-ray Points (HXPs) with the SPIRIT/CORONAS-F Mg XII Spectroheliograph

The Mg XII spectroheliograph has revealed compact high temperature objects—hot X-ray points (HXP, Fig. 1) [1]. We have analyzed 169 HXP observed between 20 and 28 February 2002. During this period the satellite operated in completely illuminated orbits, and the spectroheliograph obtained images with cadence 40–120 s. HXPs have sizes less than 5 Mm, their typical lifetimes are 5–100 min. At maximum of solar activity HXPs occur with frequency of ten events per day. They are concentrated in active-region band.

Due to dispersion, the spectroheliograph produces images in two components of the Mg XII doublet shifted one from another. If we scan an HXP in the direction of dispersion, we will get the spectrum of Mg XII line 8.42 Å as it is shown in Fig. 2. In the coronal conditions the width of the component lines is determined by thermal Doppler broadening, and we can use it to measure HXP temperature.

Temperatures of HXPs lie in the interval 5–50 MK, emission measure is $10^{45}$–$10^{48}\,\mathrm{cm}^{-3}$, and electron density exceeds $10^{10}\,\mathrm{cm}^{-3}$. The HXP conduction cooling time is about 15 s which is significantly lower than its lifetime. This suggests that energy releases in HXP during all its lifetime. The HXP thermal energy is about $10^{28}$ erg, the power of energy release is about $10^{27}$ erg/s.

HXPs differ from X-ray bright points (XBPs [2]) and nanoflares [3] by lifetime, temperature, and size (see Table 1). HXPs have higher emission measure and electron density than nanoflares. HXPs differ from XBP by spatial distribution over solar surface: HXPs occur in active-region band, XBPs uniformly cover solar surface.

HXP and microflares [4, 5] have similar range of emission measure, electron density, and thermal energy (see Table 1). Also they have similar spatial distribution

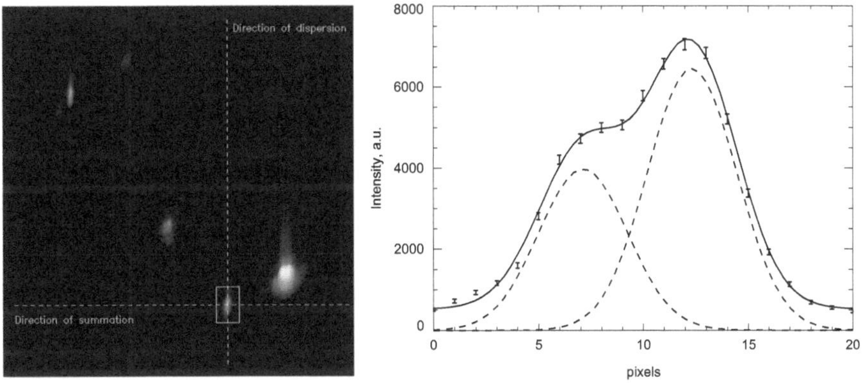


**Fig. 2** *Left*: Spectroheliograph image with HXP (direction of dispersion is *horizontal*, direction of summation is *vertical*. *Right*: Spectrum obtained after summation (figure is taken from [1])

**Table 1** Parameters of HXP and other microactivity phenomena

| | HXP | XBP | Microflares | Nanoflares |
|---|---|---|---|---|
| Lifetime | 5–100 min | 8–40 h | 1–10 min | 1–3 h |
| $T$ (MK) | 5–50 | 1–2 | 10–15 | 1–2 |
| $EM$ (cm$^{-3}$) | $10^{45}$–$10^{48}$ | $\approx 10^{47}$ | $10^{45}$–$10^{48}$ | $\approx 10^{44}$ |
| $n_e$ (cm$^{-3}$) | $\geq 10^{10}$ | $\approx 5 \times 10^9$ | $10^9$–$10^{11}$ | $10^8$–$10^9$ |
| Size (Mm) | $\leq 5$ | $\approx 25$ | 5–25 | 2–20 |

over the solar surface. But HXPs have longer lifetime, higher temperatures, and smaller sizes than microflares.

We conclude that HXP, XBP, and nanoflares are different phenomena. Differences between HXP and microflares are less clear, and we cannot rule out the possibility that they belong to the same class of phenomena.

## Observations of the Extended Solar Corona with the SPIRIT Telescope in the 175 and 304 Å Bands[2]

The solar corona is routinely observed with two types of instruments: the temperature sensitive space-based X-ray and EUV telescopes, e.g., SOHO/EIT [6], to study the inner corona from the limb to $R \sim 1.3$ (hereafter $R$ is the distance from the solar center in units of the solar radius) and the electron density sensitive space-based and ground-based WL coronagraphs to study the outer corona above $R = 2$. The inner corona being of special interest to study restructuring of global magnetic field from closed to open configuration, acceleration of matter in CMEs and forming the slow

[2]This chapter is an abridged version of the paper Slemzin et al. [7].

and fast solar wind, is not well investigated so far due to problems of separation its steeply falling down brightness from significant straylight.

The SPIRIT telescope-coronagraph T2 (see Chap. 2.1) was the first instrument for wide-field imaging of the corona to the distances of $R \sim 3$ in the 175 and 304 Å EUV spectral bands using the special coronagraphic mode [7]. It employed enhanced sensitivity and low straylight reduced by the use of an outer occulter. The instrument was able to observe wide segments (more than 90° in latitude) of the solar corona from the limb to $R = 5$ with high spatial (5″) and temporal (5 min) resolution, which has no analogues in the practice of the solar astronomy. In the period of high solar activity (2002–2003) the SPIRIT EUV telescope carried out 15 coronagraphic sessions of observations lasting from several hours to one week. As a rule, images of the disk were taken every 2–3 hours, between those from 4 to 12 coronal images were taken with a cadence of 15–30 min. The disk and coronal parts were combined at the distance of $R = 1.3$. To perform a photometric analysis, after subtraction of straylight, the coronal images were calibrated by dividing on the vignetting function.

## *Structure of the Corona in the 175 and 304 Å Bands*

Figure 3 shows the three selected frames from the movies demonstrating the common features observed by SPIRIT in the corona. In the 175 Å band (Fe IX–XI lines) the corona may be represented as a "diffuse" (unresolved) background with embedded bright structures like static/rising loops and radial/non-radial rays. The rays are evidently appearing above some active regions in the course of the solar rotation (e.g., a group of rays in the northwestern sector marked by the number 4 seen on 21–22 June 2002). Probably, these rays correspond to open magnetic structures associated with the solar wind streams [19]. Several large-scale rising loops (e.g., feature 3) have been identified with the CMEs in the LASCO catalogue http://cdaw.gsfc.nasa.gov (e.g., CMEs of 16 June 2002 15:50:07; 17 June 2002 21:20:06; 20 June 2002 22:30:32; 06 December 2002 13:31:47).

## *Comparison of the SPIRIT Coronal Images with LASCO and UVCS*

Figure 4 (a–d) shows the composite pictures of the SPIRIT and the Large Angle and Spectrometric COronograph (LASCO, [8]) images for 16 and 22 June 2002. The SPIRIT images of the corona were summed over 9 h intervals of observations. The rays seen in the SPIRIT 175 Å picture (Fig. 4a, c) between the limb and the inner boundary of the LASCO FOV ($R = 2.3$), marked by the numbers from 1 to 6, have counterparts in the streamers seen in the LASCO images. It is worth noting that the coronal brightness in the LASCO images is proportional to the total electron density $n_e$ of the plasma independently on its temperature, whereas in the 175 Å band it is proportional to $n_e^2$ of the plasma component with the temperature near

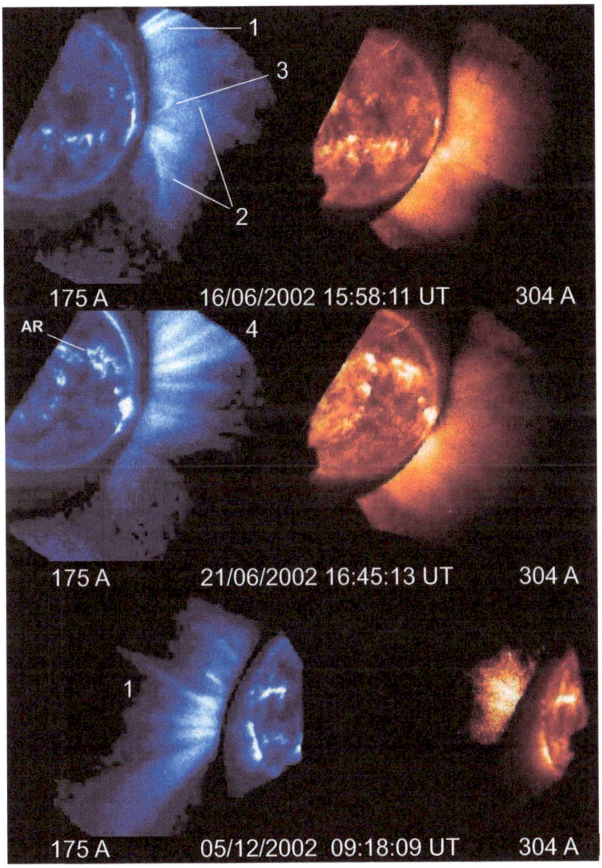


**Fig. 3** Specific coronal features observed by the SPIRIT EUV coronagraph in the 175 and 304 Å bands: 1 – radial rays, 2 – non-radial rays, 3 – a rising loop, 4 – a fan of rays linked with the active region at the disk

1 MK. The correlation between the latitudinal distributions of the coronal brightness in the SPIRIT 304 Å and LASCO images (Fig. 4b, d) is not high or even negative: some of the brightest features in 304 Å at $R = 1.1 - 1.5$ correspond to the regions of weaker brightness in the LASCO picture at $R > 2.3$ and vice versa.

Figure 5 demonstrates a comparison of the SPIRIT images in 175 and 304 Å bands and the images from the Ultraviolet Coronagraph Spectrometer (UVCS—[9, 10]) in the Ly $\alpha$ line (1216 Å) and in the O VI line (1032 Å) obtained on 16 June 2002. In the first case only some of the rays seen in 175 Å (e.g., features numbered as 1 and 2 in Fig. 5a) have counterparts in the UVCS images. The difference in contrast of the coronal structures seen in the SPIRIT and UVCS images can be explained by their specific mechanisms of excitation: whereas the Fe ion lines of the 175 Å band are excited by collisions, in the O VI line the collisional and radiative components have comparable rates [9]. Besides, emission of the Fe IX–XI lines

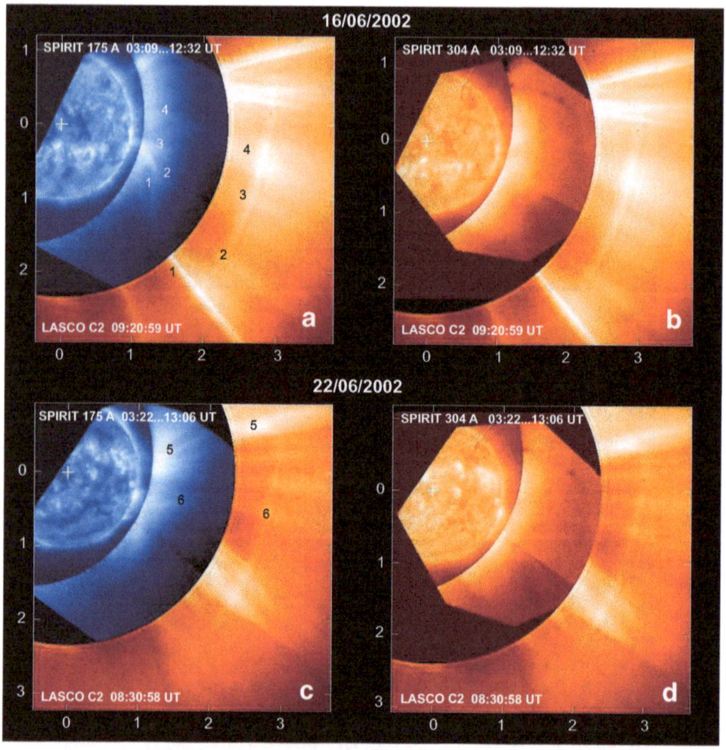


**Fig. 4** Composite SPIRIT–LASCO C2 images: (**a**) and (**c**) the SPIRIT 175 Å image at $R = 0{-}2.3$ with the LASCO C2 image (at $R > 2.3$), (**b**) and (**d**) the SPIRIT 304 Å image with the LASCO C2 image. The scales are in units of the solar radius (the origin marked by a *cross* corresponds to the solar center). The pair numbers 1–6 designate the features which probably belong to the same structural elements seen in EUV and WL

and the O VI line correspond to different excitation temperatures: $T_e \sim 1$ MK and 0.3 MK correspondingly.

The corona in the pair of the SPIRIT 304 Å and the UVCS Ly $\alpha$ 1216 Å images (Fig. 5b) do not show the structural elements seen in the SPIRIT 175 Å image. The angular distributions of brightness in both cases are roughly similar: the brightness in the sectors numbered by 1 is higher than those in the sector 2 in both bands. It corresponds to the suggestion that the coronal radiation in both cases is generated by the same mechanism of resonant scattering. The Ly $\alpha$ line is less sensitive to the line-of-sight velocity than the He II line [9].

## *Spatial Structure and Temporal Variability of EUV Brightness in the Corona at the Distances of $R = 1 - 2$*

Figure 6 shows the latitudinal distributions of the coronal brightness at different radial distances in the western hemisphere (a) and temporal variation of the coronal

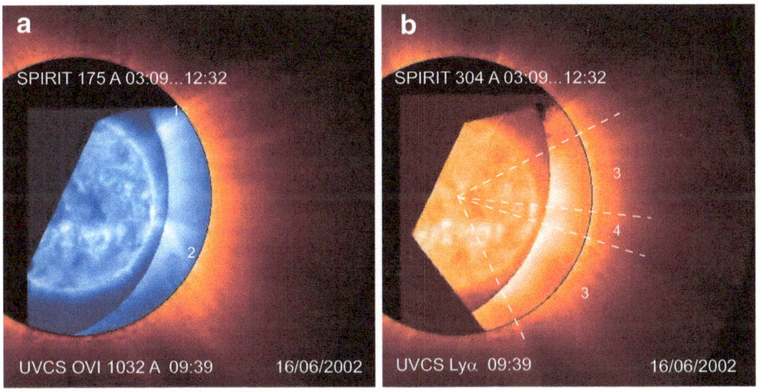


**Fig. 5** Composite SPIRIT (at $R = 1-1.6$) and the UVCS images (at $R = 1.6-3.5$): (**a**) SPIRIT 175 Å and UVCS O VI 1032 Å. 1 and 2: the rays in the SPIRIT having the counterparts in the UVCS image; (**b**) SPIRIT 304 Å and UVCS Ly $\alpha$. In both images the brightness in the sectors 3 is relatively higher than in the sector 4

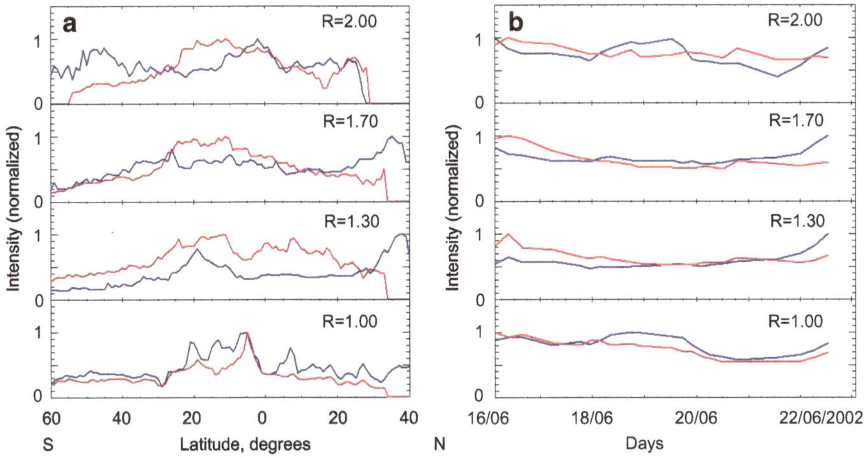


**Fig. 6** (**a**) Latitudinal distributions of the corona brightness above the West limb for 16/06/2002; (**b**) temporal variations of the corona brightness at different distances along the West equatorial line for the period 16–22 June 2002. The *blue lines* correspond to 175 Å, the *red lines* to 304 Å. The intensities are normalized to the maximum values for each distance

brightness in the western equatorial plane (b) in both SPIRIT bands for 16 June 2002. In order to enhance the signal-to-noise ratio, the initial SPIRIT images of the corona were summed over the period of 03:09–12:32 UT, and the resulting latitudinal distributions were averaged over 1° bins. The limb distributions were taken from the disk images at 09:19:51 UT.

The results may be interpreted in terms of the excitation mechanisms of the Fe and He ion lines. At distances of $R = 1-1.5$ due to dominated collisional excitation coronal brightness in the 175 Å band is the largest in relatively dense closed loops localized above the active regions. These structures cannot be resolved with our

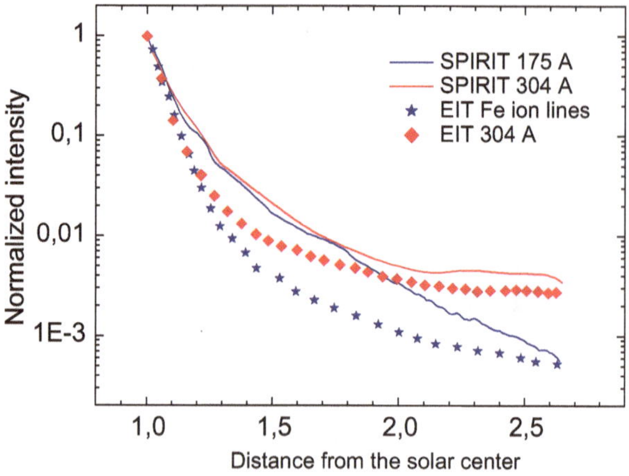


**Fig. 7** Normalized radial distributions of the EUV coronal brightness: *solid blue* and *red lines*—the SPIRIT data for the western equatorial plane taken on 16 June 2002 in the 175 and 304 Å bands correspondingly, *red diamonds* and *blue stars*—the EIT data for the eastern equatorial plane taken on 04 April 1996 in the 304 Å band and the averaged data obtained in the EIT 171, 195, and 284 Å bands

spatial resolution (1° bin corresponds to $1.2 \times 10^4$ km at the limb), but their averaged brightness is proportional to the brightness of active regions at the limb. At larger distances the major brightness is concentrated in non-radial rays, so the correlation of the coronal and limb brightness vanishes.

In the 304 Å band the dominating mechanisms of excitation of Helium are: collisional excitation near the limb with the minor contribution of the radiative excitation by the photoionization–recombination (P–R) process [11–14]. Above the limb, the emission in this band contains also a contribution of the collisionally excited Si XI and Fe XV lines. In total, they show a positive correlation between the coronal brightness in the 304 Å band with the brightness at the limb. At the distances $R > 1.5$ the electron density is low, and the radiative excitation, namely, the resonance scattering becomes dominant [15] which also provides a positive correlation with the brightness at the limb.

## *Radial Distributions of the EUV Brightness in the Quiet Corona*

We analyzed radial distributions of the EUV brightness in the quiet corona obtained by SPIRIT at the solar maximum and EIT at the solar minimum. Figure 7 shows the normalized (relative to the brightness at the limb) radial distributions in the SPIRIT 175 and 304 Å bands taken on 16 June 2002 in the western equatorial plane compared with the EIT data in the Fe ion lines and in the 304 Å band taken on 04 April 1996 in the eastern equatorial plane [15] when the SOHO spacecraft was in the off-set position.

The normalized radial distribution in 304 Å from the SPIRIT data coincides with the corresponding one in 175 Å in the region $R = 1 - 1.8$. The EIT distributions in 304 Å channel and in the coronal Fe lines also coincide in the region $R = 1 - 1.2$. It suggests that in these regions the emission in the 304 Å band (which consists of the major He II line and accompanying Si XI and Fe XV lines) has the same dependence on the coronal electron density as the corresponding emissions in the Fe bands which are excited by electron–ion collisions. The region of predominantly collisional excitation of the He II line is larger at the solar maximum (the SPIRIT case) in comparison with that at the solar minimum (the EIT case) because of higher electron density in the middle corona at the solar maximum.

## Experiment TESIS on the CORONAS-Photon Satellite

Based on the results of the SPIRIT experiment and actual tasks of solar physics, LPI has developed the next generation TESIS (TElescope-Spectrometer for Imaging Spectroscopy) instrumental complex [16] for the next CORONAS-Photon mission [17] of the CORONAS project. The TESIS experiment started in the deep minimum between the 23rd and the 24th cycles of solar activity: observations were performed from 30 January 2009 to 30 November 2009. The main goal of TESIS was observation of solar active phenomena from the transition region to the inner and outer solar corona with high spatial, spectral, and temporal resolution in the EUV and soft X-ray spectral bands.

The TESIS consisted of five instruments:

- Mg XII Spectroheliograph with spherical bent crystal mirror, for observations in the monochromatic Mg XII 8.42 Å line;
- EUV Spectoheliometer with grazing incidence diffraction grating, for the registration of full solar disk in monochromatic lines of the spectral band 280–330 Å;
- Two full-disk EUV Telescopes with multilayer mirrors (132/171 and 171/304 Å);
- Solar EUV Coronagraph, based on the Ritchey–Chretien scheme, to observe the inner and outer solar corona from 0.2 to 4 solar radii in the spectral band 290–320 Å;
- SphinX X-ray photometer-spectrometer designed by the Space Research Center of the Polish Academy of Science [18].

Main characteristics of the TESIS instruments are shown in Table 2.

TESIS investigated coronal hot plasma, phenomena in the extended corona, transition region dynamics, small-scale events, and oscillation processes. Below we list the major TESIS results.

### *Transient Events in the Solar Corona*

Cadence of the TESIS EUV telescopes was limited by signal readout time from the CCD-array (a readout time of the 2k*2k image is 16 s). In order to achieve higher cadence, a special observation program was carried out to take partial images each 3 s. The high cadence 171 Å images reveal numerous oscillating structures with sizes less than 10 Mm and periods of 50–350 s (see Fig. 8). These structures are believed to be origins of nanoflares–active events with energies $10^{24}$–$10^{26}$ erg.

**Table 2** TESIS instruments characteristics

| Channel | Pixel size | Field of view | $\lambda/\Delta\lambda$ | Major ions |
|---|---|---|---|---|
| 132/171 Å | 1.71″ | 1° | ≈26 (132 Å) ≈28 (171 Å) | Fe XIX, Fe XX, Fe XXIII (132 Å); Fe IX (171 Å) |
| 171, 304 Å | 1.71″ | 1° | ≈26 (171 Å) ≈30 (304 Å) | Fe IX (171 Å); He II, Si XI (304 Å) |
| 208–330 Å | 4.6″ | 2° | ≈7 | Fe XV, Ni XVIII, Si IX, Ca XVIII, Si XI, He II, Mg VIII etc. |
| 304 Å | 4.6″ | 2° | ≈7 | He II, Si XI |
| 8.42 Å | 2.3″ | 1.3° | ≈210 | Mg XII |
| Sphinx [18] | | Solid Si-spectrometer, 0.5–15 keV | | |

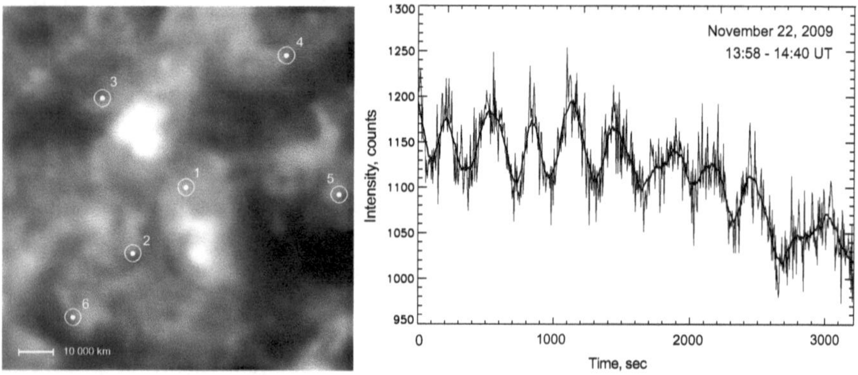


**Fig. 8** *Left*: TESIS 171 Å high cadence image. *Right*: Lightcurve of the region marked 1

Nanoflares are not associated with coronal bright points and uniformly cover visible solar surface.

## *Structure of Extended Corona*

The TESIS EUV telescopes have field of view of 1°. The images taken with 100 s exposure time display the corona and overexposed solar disk. The images with 1 s exposure display the normally exposed solar disk and faint corona. If we consequently take images with exposure times 1, 20, and 100 s and then merge them together, we will get an image where the corona and the solar disk are shown with equal contrast (see Fig. 9). Sometimes the TESIS images of the extended corona contained coronal rays or structures associated with CMEs as it was seen by SPIRIT at solar maximum. Using the TESIS data it was shown that large-scale coronal rays appeared above active regions in neighborhood of mid-latitude coronal holes at solar minimum indicate signatures of the slow solar wind [19].

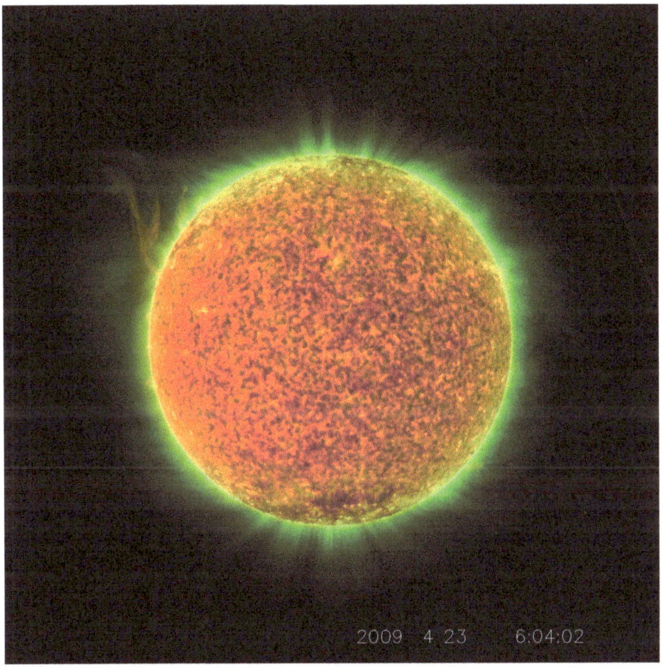


**Fig. 9** A combined TESIS image: *red*—304 Å, *green*—171 Å

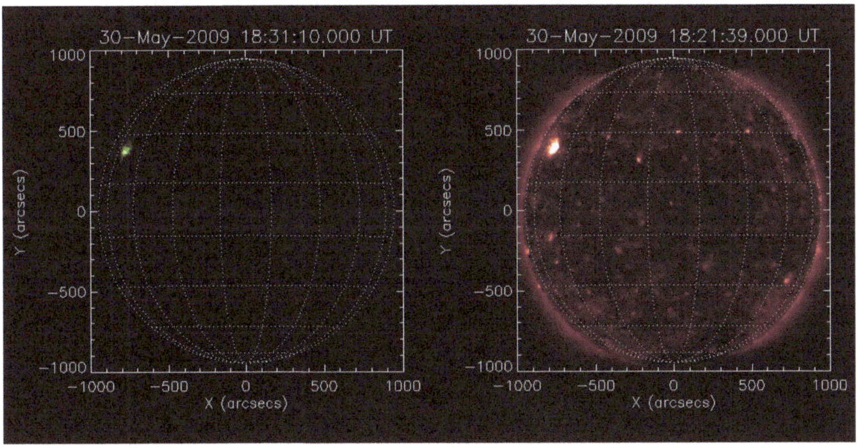


**Fig. 10** Image of active region NOAA 11019 obtained with TESIS Mg XII spectroheliograph (*left*) and XRT/Hinode telescope (*right*)

## *Studies of Hot Plasma with the Mg XII Spectroheliograph*

Observations of the Sun with the TESIS Mg XII Spectroheliograph have shown that even at solar minimum images the solar corona in Mg XII line 8.42 Å reveal

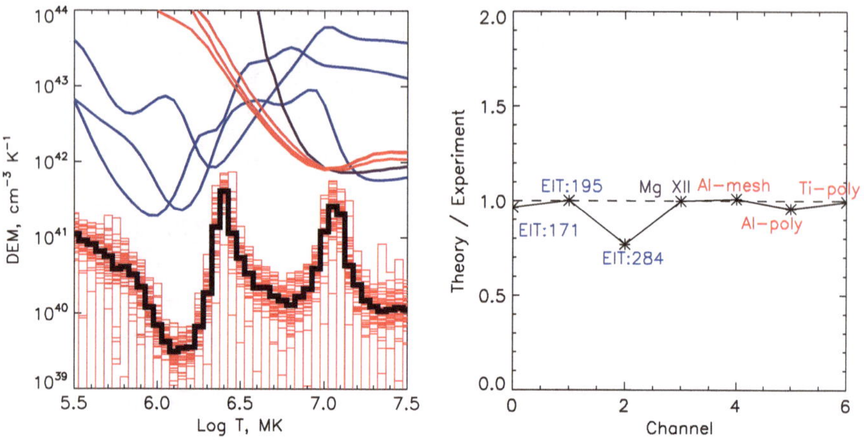


**Fig. 11** *Left*: Reconstructed DEM. *Right*: Ratio of theoretical to experimental fluxes (Al-mesh, Al-poly, Ti-poly—XRT channels, EIT:171, EIT:195, EIT:284—EIT channels, Mg XII—spectroheliograph Mg XII)

hot plasma with temperature higher than 5 MK. We carried out a multi-channel analysis of hot active region NOAA 11019 (28 May to 10 June 2012). We analyzed EIT/*SOHO* [6, 20], XRT/*Hinode* [21, 22], and the TESIS Mg XII images (Fig. 10). Using the genetic algorithm [23] we reconstructed differential emission measure (DEM) of this region. It was found that DME contained hot ($\approx$10 MK) and cool ($\approx$2 MK) components (Fig. 11). The hot plasma was observed during 9 days.

## References


1. Reva, A., Shestov, S., Bogachev, S., Kuzin, S.: Sol. Phys. **276**, 97 (2012)
2. Golub, L., Krieger, A.S., Vaiana, G.S.: Sol. Phys. **50**, 311 (1976)
3. Aschwanden, M.J., Tarbell, T.D., Nightingale, R.W., Schrijver, C.J., Title, A., Kankelborg, C.C., Martens, P., Warren, H.P.: Astrophys. J. **535**, 1047 (2000)
4. Christe, S., Hannah, I.G., Krucker, S., McTiernan, J., Lin, R.P.: Astrophys. J. **677**, 1385 (2008)
5. Hannah, I.G., Christe, S., Krucker, S., Hurford, G.J., Hudson, H.S., Lin, R.P.: Astrophys. J. **677**, 704 (2008)
6. Delaboudinière, J.-P., Artzner, G.E., Brunaud, J., Gabriel, A.H., Hochedez, J.F., Millier, F., Song, X.Y., Au, B., Dere, K.P., Howard, R.A., Kreplin, R., Michels, D.J., Moses, J.D., Defise, J.M., Jamar, C., Rochus, P., Chauvineau, J.P., Marioge, J.P., Catura, R.C., Lemen, J.R., Shing, L., Stern, R.A., Gurman, J.B., Neupert, W.M., Maucherat, A., Clette, F., Cugnon, P., van Dessel, E.L.: Sol. Phys. **162**, 291 (1995)
7. Slemzin, V., Bougaenko, O., Ignatiev, A., Kuzin, S., Mitrofanov, A., Pertsov, A., Zhitnik, I.: Ann. Geophys. **26**, 3007 (2008)
8. Brueckner, G.E., Howard, R.A., Koomen, M.J., Korendyke, C.M., Michels, D.J., Moses, J.D., Socker, D.G., Dere, K.P., Lamy, P.L., Llebaria, A., Bout, M.V., Schwenn, R., Simnett, G.M., Bedford, D.K., Eyles, C.J.: Sol. Phys. **162**, 357 (1995)

9. Kohl, J.L., Esser, R., Gardner, L.D., Habbal, S., Daigneau, P.S., Dennis, E.F., Nystrom, G.U., Panasyuk, A., Raymond, J.C., Smith, P.L., Strachan, L., van Ballegooijen, A.A., Noci, G., Fineschi, S., Romoli, M., Ciaravella, A., Modigliani, A., Huber, M.C.E., Antonucci, E., Benna, C., Giordano, S., Tondello, G., Nicolosi, P., Naletto, G., Pernechele, C., Spadaro, D., Poletto, G., Livi, S., von der Lühe, O., Geiss, J., Timothy, J.G., Gloeckler, G., Allegra, A., Basile, G., Brusa, R., Wood, B., Siegmund, O.H.W., Fowler, W., Fisher, R., Jhabvala, M.: Sol. Phys. **162**, 313 (1995)
10. Raymond, J.C., Kohl, J.L., Noci, G., Antonucci, E., Tondello, G., Huber, M.C.E., Gardner, L.D., Nicolosi, P., Fineschi, S., Romoli, M., Spadaro, D., Siegmund, O.H.W., Benna, C., Ciaravella, A., Cranmer, S., Giordano, S., Karovska, M., Martin, R., Michels, J., Modigliani, A., Naletto, G., Panasyuk, A., Pernechele, C., Poletto, G., Smith, P.L., Suleiman, R.M., Strachan, L.: Sol. Phys. **175**, 645 (1997)
11. Zirin, H.: Astrophys. J. **199**, L63 (1975)
12. Macpherson, K.P., Jordan, C.: Mon. Not. R. Astron. Soc. **308**, 510 (1999)
13. Andretta, V., Del Zanna, G., Jordan, S.D.: Astron. Astrophys. **400**, 737 (2003)
14. Judge, P.G., Pietarila, A.: Astrophys. J. **606**, 1258 (2004)
15. Delaboudinière, J.-P.: Sol. Phys. **188**, 259 (1999)
16. Kuzin, S.V., Zhitnik, I.A., Shestov, S.V., Bogachev, S.A., Bugaenko, O.I., Ignat'ev, A.P., Pertsov, A.A., Ulyanov, A.S., Reva, A.A., Slemzin, V.A., Sukhodrev, N.K., Ivanov, Y.S., Goncharov, L.A., Mitrofanov, A.V., Popov, S.G., Shergina, T.A., Solov'ev, V.A., Oparin, S.N., Zhukov, A.M.: Sol. Syst. Res. **45**, 162 (2011)
17. Kotov, Y.D.: Sol. Syst. Res. **45**, 93 (2011)
18. Gburek, S., Sylwester, J., Kowalinski, M., Bakala, J., Kordylewski, Z., Podgorski, P., Plocieniak, S., Siarkowski, M., Sylwester, B., Trzebinski, W., Kuzin, S.V., Pertsov, A.A., Kotov, Y.D., Farnik, F., Reale, F., Phillips, K.J.H.: Sol. Syst. Res. **45**, 189 (2011)
19. Slemzin, V., Harra, L., Urnov, A., Kuzin, S., Goryaev, F., Berghmans, D.: Sol. Phys. (2012). doi:10.1007/s11207-012-0004-y
20. Domingo, V., Fleck, B., Poland, A.I.: Sol. Phys. **162**, 1 (1995)
21. Golub, L., Deluca, E., Austin, G., Bookbinder, J., Caldwell, D., Cheimets, P., Cirtain, J., Cosmo, M., Reid, P., Sette, A., Weber, M., Sakao, T., Kano, R., Shibasaki, K., Hara, H., Tsuneta, S., Kumagai, K., Tamura, T., Shimojo, M., McCracken, J., Carpenter, J., Haight, H., Siler, R., Wright, E., Tucker, J., Rutledge, H., Barbera, M., Peres, G., Varisco, S.: Sol. Phys. **243**, 63 (2007)
22. Kosugi, T., Matsuzaki, K., Sakao, T., Shimizu, T., Sone, Y., Tachikawa, S., Hashimoto, T., Minesugi, K., Ohnishi, A., Yamada, T., Tsuneta, S., Hara, H., Ichimoto, K., Suematsu, Y., Shimojo, M., Watanabe, T., Shimada, S., Davis, J.M., Hill, L.D., Owens, J.K., Title, A.M., Culhane, J.L., Harra, L.K., Doschek, G.A., Golub, L.: Sol. Phys. **243**, 3 (2007)
23. Siarkowski, M., Falewicz, R., Kepa, A., Rudawy, P.: Ann. Geophys. **26**, 2999 (2008)

# Experiment with the SPR-N Instrument Onboard the CORONAS-F Satellite: Polarization, Temporal, and Spectral Characteristics of the Hard X-Ray of the Solar Flares

**I.A. Zhitnik[1], Yu.I. Logachev, A.V. Bogomolov, V.V. Bogomolov, Yu.I. Denisov[2], S.S. Kavanosyan, S.N. Kuznetsov[3], O.V. Morozov, I.N. Myagkova, S.I. Svertilov, A.P. Ignatiev, S.N. Oparin, and A.A. Pertsov**

During the experiment with a polarimeter SPR-N onboard the CORONAS-F satellite, polarization of the X ray of solar flares within the energy ranges of 20–40, 40–60, and 60–100 keV was measured according to Thomson scattering of X-ray photons in Beryllium plates with the following registration of the scattered photons by means of a system of six scintillation detectors based on Cs I(Na). As a result of observations for the period from August 2001 till December 2005, hard X-ray radiation was registered for 128 solar flares. During the event of 29 October 2003 degree of polarization of the radiation within the channels 40–60 and 60–100 keV exceeded 70 %, and within the channel 20–40 keV—50 %. Time profiles of the part of polarized radiation, orientation of the maximum polarization on the solar disc were obtained. The upper limit of the polarized radiation part for 25 events was estimated at the level of 8–40 %. For all registered flares time profiles (with resolution of up to 4 s) were registered, hard X-ray fluxes were determined, and spectrum factor was estimated. For the most powerful events which were observed during October–November 2003 and on January 20, 2005, the data on the dynamics of the characteristics of thermal and non-thermal components of X-radiation were obtained.

---

[1]Deceased

[2]Deceased

[3]Deceased

I.A. Zhitnik · A.P. Ignatiev · S.N. Oparin · A.A. Pertsov
P.N. Lebedev Physical Institute of RAS(PI), Moscow, Russia

Yu.I. Logachev · A.V. Bogomolov (✉) · V.V. Bogomolov · Yu.I. Denisov · S.S. Kavanosyan · S.N. Kuznetsov · O.V. Morozov · I.N. Myagkova · S.I. Svertilov
D.V. Skobeltsyn Institute of Nuclear Physics, M.V. Lomonosov Moscow State University (SINP MSU), Moscow, Russia
e-mail: aabboogg@rambler.ru

V. Kuznetsov (ed.), *The Coronas-F Space Mission*, Astrophysics and Space Science Library 400, DOI 10.1007/978-3-642-39268-9_4,

## Introduction

Degree of polarization of the hard X-radiation (~10–100 keV) of solar flares is one of the most important characteristics which allows to estimate the direction of the flux of accelerated electrons generating this radiation at their deceleration in the solar atmosphere [6, 12]. Very high degree of polarization can show high anisotropic electron beam which travels deep into the chromosphere along the power lines of the magnetic field which in turn requires presence of very big electric currents and powerful magnetic fields. Theoretic simulations predict the value of the degree of polarization of the hard X-radiation of solar flares within essentially wide ranges: from several percent by thermal model and non-thermal stationary model up to several dozen percent by non-thermal model with impulse injection of electron flux [3,6,12]. Concerning experimental data currently degree of polarization is measured only within the range of relatively soft X-radiation (wavelength of ~0. 8 Å) only for several flares [7,8,11]. Herewith variability of results is essential: from 2–3 % up to 20–40 %. Therefore the problem of the experiments on measuring of the polarization of hard X-radiation of the Sun remains currently central.

The SPR-N instrument onboard the CORONAS-F satellite was intended for measurements of polarization, spectral, and time characteristics of the hard X-radiation of solar flares [1]. The purpose of the SPR-N experiment is studying of the mechanism of the primary energy release in the flare, primarily, in order to estimate the relative role of non-thermal and thermal processes. It was supposed to determine degree of polarization of X-radiation with the energy within the ranges of 20–40, 40–60, 60–100 keV, location of polarization plane on the solar disc in relation with characteristic magnetic structures in the flare region, time dynamics of polarization by means of the SPR-N instrument.

Scientific program of the experiment also included monitoring of time profiles of X-radiation intensity within the energy range of 15–100 keV, determination of flux values, and estimation of spectrum rigidity within the mentioned energy range. In order to analyze spectra of X-radiation of solar flares data from SPR-N experiment were used along with the data from the SONG instrument which was also mounted onboard the CORONAS-F satellite. In order to determine the relative inputs of thermal and non-thermal radiation of the flares, the data from the SPR-N instrument were processed along with the data of the GOES satellites (http://swpc.noaa.gov) [4] on soft X-radiation.

## Description of the SPR-N Experiment

The method of measurement of polarization in the SPR-N instrument is based on Thomson scattering of solar photons in Beryllium plates. Five plates produced of metallic Beryllium are fixed inside a hollow hexagonal prism on the faces of which six scintillation detectors are located symmetrically. In order to exclude charged particles registration Phoswich was used: CsJ(Na)/plastic scintillator based on polystyrene. Geometric area of each detector is $\sim$ 8 cm$^2$, taking into account absorption in the protective covers, scattering efficiency, and registration in detector effective area varies $\sim$0. 3 cm$^2$ at $E_\gamma = 20$ keV up to $\sim$1.5 cm$^2$ at $E_\gamma = 100$ keV.

For non-polarized radiation probability of scattering at different angles relating to primary direction of the photon's direction is common; therefore it will scatter azimuth-symmetric and in this case detectors must detect the same intensity. In the case of plane-polarized radiation the main part of the photons scatters in perpendicular to the polarization plane direction. Considering that at orientation of the instrument to the Sun photons mainly fall perpendicularly to the scatterer's surface, obviously plane-polarized radiation will scatter in Beryllium plates essentially along the plates' plane. At this the pairs of opposite detectors conform the polarization plates turned for 120° relative to each other. As far as in the case of plane-polarized radiation photons scatter in azimuth-anisotropic way, counting rate of different pairs of detectors must differ from each other—counting rate of the detectors located near (within $\pm 60°$) to the plane perpendicular to the polarization plane must be the most high. In such a way difference of the detectors' counting rates allow to determine degree of polarization $P_{\text{meas}}$ of the detected radiation and in principle the angle $\varphi$ of the polarization plane's angular deflection relative to the pair of detectors.

In order to decrease the influence of systematic effects it was supposed to turn the Beryllium scatterer with detectors for $\pm 60°$ to the direction to the Sun during the measurements. At this if the detected radiation is polarized difference between the counting rates of the detectors apart from what pair of detectors is nearer to the polarization plane must be kept.

In order to determine the part of polarized radiation $P_{\text{meas}}$ and direction of polarization plane according to polarization detectors' counting rates $N_i$ preliminary input of each detector into the total counting rate (in percentage terms) was calculated for all parts of polarized radiation and all directions of polarization planes. In order to determine $P_{\text{meas}}$ according to the measured values $N_i$ Thomson scattering was simulated in Beryllium plates for the vertically falling linear-polarized radiation. For the given value of polarization of the falling radiation $P_0$ (which was chosen within the range of 0–100 %) the values of $N_i / \sum_{i=1}^{3} N_i$ which characterize the part of the scattered polarized radiation falling into the given pair of detectors relatively the total flux falling on the scatterer (Fig. 1a) were calculated depending on the angle $\varphi$. The found relations allow to solve the inverse problem—to determine degree of polarization $P_{\text{meas}}$ of the falling radiation according to the measured values taking into account the angle $\varphi$ (Fig. 1b). So the part of polarized radiation and direction of the polarization plane can be determined by means of solution of the inverse problem on the known counting rates of the polarization detectors.

A special "patrol" detector is also provided for the instrument. This detector allows to measure time profile of X-radiation within the energy range of 15–100 keV in detail. A spot-beam detector produced on the base of Phoswich CsJ(Na)/plastic

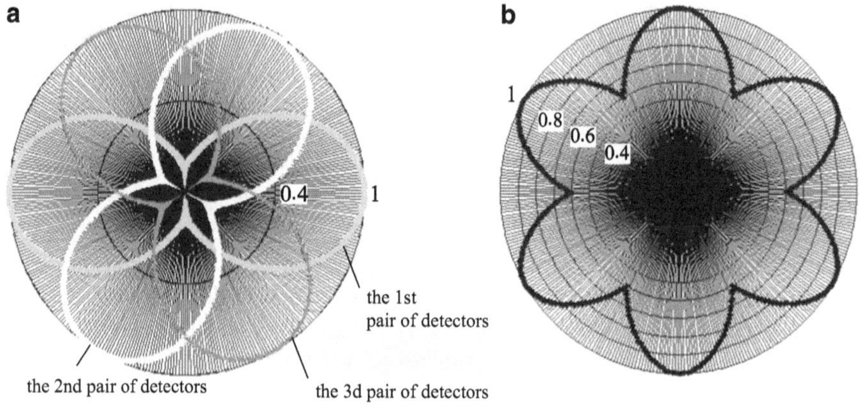


**Fig. 1** Diagrams characterizing the relative indications of polarization detectors (**a**) and degree of polarization determined according to these indications (**b**) depending on the location of the polarization plane on the assumption under 100 % polarization of radiation falling perpendicularly to the polarimeter's surface: (**a**) angular characteristic of detectors counting rate for 100 % polarization (maximum counting rate is taken as 1): the first pair of detectors, the second pair of detectors, the third pair of detectors; (**b**) dependence of degree of polarization $Q$ estimated by the relation $Q = (N_{\max} - N_{\min})/(N_{\max} + N_{\min})$, on the angle of the polarization plane for 100 % polarization. From [12]

scintillator (thickness of CsJ(Na) crystal is 3 cm, effective area is $\sim$1 cm$^2$ for the energy of the detected photons of $E_\gamma = 40$ keV) is used for it. All the range of the patrol detector is separated for subranges of 15–40 and 40–100 keV. The ratio of the indications in the given channels will allow to estimate parameters which define rigidity of energy spectra in the given spectral presentation: power-law for non-thermal component and "optically-thin plasma" for thermal component. Inputs of thermal and non-thermal components can be separated by means of attraction of the data of the GOES satellite's instruments which detect mainly the soft thermal X-radiation.

The SPR-N instrument and principles of its operation are described in [1] in detail.

Time resolution of the instrument (exposure duration of 4 or 8 s) was specified as duration of one measurement cycle (exposure at defined location of polarization detectors) and during the first months of instrument's operation was 8 s, during the rest time period—4 s.

Information of the SPR-N instrument included 28 output parameters, 15 of which are measurement results, the rest are service parameters. Information from the SPR-N instrument was transmitted to the Earth through the system of information acquisition (SIA) in the form of information-digital arrays of two types—polarization (data of polarization and patrol detectors) and patrol (more detailed data of patrol detector).

## Database of the SPR-N Experiment

During the observations from August 2001 till December 2005, the SPR-N instrument monitored solar activity actually continuously by means of patrol detector within energy ranges of 15–40 and 40–100 keV. During this period 3,750 observation sessions were obtained (672.16 MB of information in a form of archive files), including 163 sessions (42.11 MB) in 2001, 1,272 sessions (216.44 MB) in 2002, 840 sessions (165.86 MB) in 2003, 817 sessions (156 MB) in 2004, and 658 sessions (672.5 MB) in 2005.

In order of operative control of the incoming information special software was developed. It allowed:

- to sort arrays in polarization, patrol, and the rest;
- to convert information into counting rates and fluxes;
- to separate part of the session into separate file as giving time, as in graphic mode;
- to plot any combination of parameters versus time with any time resolution and within any intensity range;
- to determine if the Sun was in the field of vision of the instrument during the given time period including flare time according to light/shadow sensor indications;
- to plot counting rate of any parameter versus any other; such plot is suitable for correlation analysis of polarization parameters and for display of nonlinear effects at very big fluxes.

Answer to the question if the solar event was observed by the SPR-N instrument was obtained only several minutes after obtaining information on the event.

Information of the patrol detector of the SPR-N instrument is presented on a web site of Space Monitoring Data Center (Skobeltsyn Institute of Nuclear Physics, MSU, [13]) in the both forms of data and plots of X-ray fluxes ($cm^{-2}\,s^{-1}$) versus time for energy ranges of 15–40 and 40–100 keV. Appearance of one of the web pages is presented in Fig. 2.

User can select an interesting interval giving the primary time or selecting the flare from the catalogue of solar events. Counting rates of the protective cover of plastic scintillator (mainly counting rate of charged particles) is also presented in Fig. 2. Data of X-ray, gamma ray and charged particles fluxes obtained by SCR set of instruments (also onboard Coronas-F) can be plotted together with SPR-N data if necessary. The areas of shadow, i.e., time periods when the Sun was shaded with the Earth are marked with more dark colours. The period of absence of the shadow is presented in Fig. 2. Vertical lines show moments of the beginning, maximum and the end of each flare in soft X-rays according to the Solar events list taken from [4].

It is seen from Fig. 2 that the best background situation for flare observation was at the equator (threshold counting rate of $\sim 3\,s^{-1}$ within the range of 15–40 keV and $0.5\,s^{-1}$ within the range of 40–100 keV). In the area of polar caps (where counting rate is higher for an order of magnitude) it is also possible to observe relatively powerful flares (of M and X classes). Essential part of measurement time

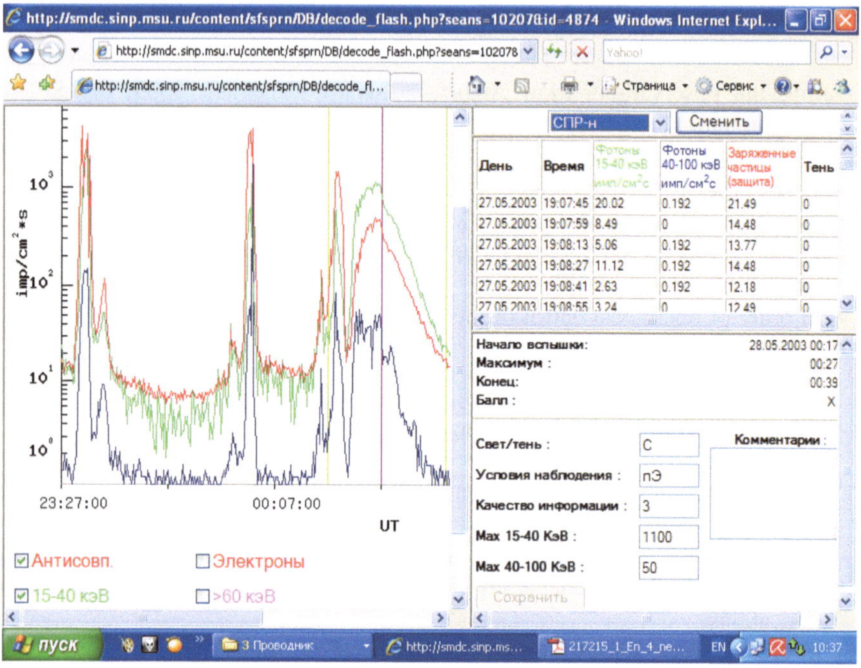


**Fig. 2** Web page http://smdc.sinp.msu.ru/content/sfsprn/DB/decode_flash.php?seans=10207&id=4874, with presented information of the patrol detector of the SPR-N instrument. From [13]

is set by the periods of satellite's passing of the radiation belt but for the most powerful flares of X-class counting rate of the flare radiation can be an order of magnitude higher than background in the outer radiation belt. As an example a period of the flare of X3.6 class (May 28, 2003, radiation maximum at 0:27) is presented in Fig. 2. The weakest flares for which the patrol detector of the SPR-N instrument detected hard X-radiation were of C2.5 class (for instance, the flare of 6 October 2004, 18:30).

Besides the data measured by the SPR-N instrument, short instrument description, information about other experiments onboard CORONAS-F, and other useful information are also presented on web site of CORONAS Project developing in IZMIRAN http://coronas.izmiran.ru/F/SPR.

Further processing of the data of the SPR-N instrument included first of all selection of the solar events and detailed analysis of their polarization, spectral, and time characteristics. During the whole observation period from August 2001 till December 2005, the hard X-radiation was registered from 128 solar flares. Time profiles with resolution of up to 4 s were plotted for all events. The results of the solar flares measurements by the SPR-N instrument in 2001–2002 were presented in [1, 2].

According to the data of the patrol detector of the SPR-N instrument a catalogue of solar flares recorded or not recorded by the SPR-N instrument is prepared.

**Table 1** Catalogue of solar events recorded by the SPR-N instrument during 2003

| Date | GOES | SPR-N | Session of SPR-N | Notes |
|---|---|---|---|---|
| 25.04.2003 | M1.2 | − | 09701 | Shadow |
| 26.04.2003 | M2.1 | + | 09716 | |
| 26.04.2003 | M7.0 | + | 09716 | |
| 26.04.2003 | M2.5 | − | 09725 | |
| 27.04.2003 | M1.7 | − | 09739 | Shadow |
| 29.04.2003 | M1.1 | − | 09762 | |
| 02.05.2003 | M1.0 | − | 09801 | Shadow |
| 26.05.2003 | M1.9 | + | 09724 | |
| 26.05.2003 | M1.0 | − | 10184 | Anomaly |
| 27.05.2003 | M1.4 | + | 10184 | |
| 27.05.2003 | X1.3 | + | 10184 | |
| 27.05.2003 | M1.6 | − | 10184 | |
| 28.05.2003 | ***X3.6*** | + | 10204 | |
| 29.05.2003 | X1.2 | + | 10222 | |
| 29.05.2003 | M1.5 | + | 10222 | |
| 31.05.2003 | M9.3 | − | 10245 | The Earth radiation belts |
| 02.06.2003 | M6.5 | + | 10275 | |
| 02.06.2003 | M1.0 | ? | 10282 | |
| 02.06.2003 | M3.9 | − | 10275 | Shadow/ERB |
| 08.06.2003 | M4.0 | + | 10376 | |
| 09.06.2003 | X1.7 | + | 10391 | |
| 09.06.2003 | M4.7 | − | 10391 | Shadow |
| 10.06.2003 | M5.6 | + | 10414 | |
| 10.06.2003 | M5.1 | − | 10406 | |

Information on all the events of M-class and over it happened during the operation time, and on the events of C-class which produced a response at least in the lowest channel of the instrument is included into the database. An approximate appearance of the catalogue is presented in Table 1.

In the "SPR-N" column signs ≪ + ≫ are hyperlinks to the plots of time code of X-radiation during the flare. As a rule all flares of the solar X-radiation registered by the SPR-N instrument within the range of 15–100 keV were accompanied by the increase of the soft thermal component measured onboard the GOES satellites [4] within the range of 1.55–24.8 keV. Due to the sensitivity of the patrol detector of the SPR-N instrument, the weakest flashes registered by the instrument correspond to the flares of M1-class according to the GOES classification. It means that X-radiation burst within the range of 1.55–24.8 keV of M-class and over it according to the GOES satellites can be a good indicator of X-radiation burst within the energy range of 15–100 keV.

During the experiment with the SPR-N instrument in 2001–2005 128 bursts of the solar X-radiation were recorded, including 20 bursts in 2001, 36 in 2002 and 2003, 14 in 2004, and 22 in 2005. The weakest flare with recorded hard X-radiation was of C2.5 class (6 October 2004). As seen from Fig. 3 the database contains

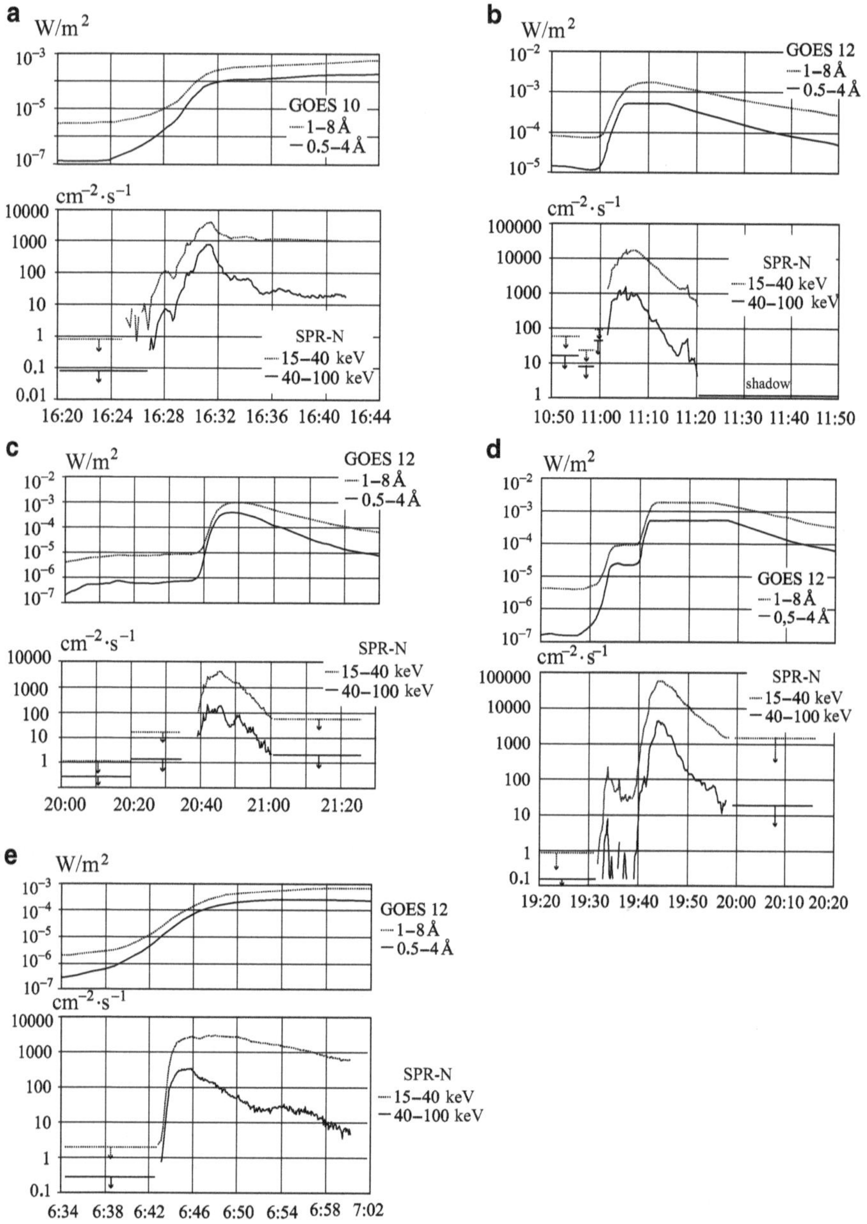


**Fig. 3** Time profiles of X-ray flux measured by the SPR-N instrument (channels of 15–40 and 40–100 keV) onboard the CORONAS-F satellite and onboard the GOES satellites (the data from [4]) (channels 1–8 and 0.5–4 Å) during five most powerful flares in 2001–2005: (**a**) August 25, 2001; (**b**) October 28, 2003; (**c**) October 29, 2003; (**d**) November 4, 2003; (**e**) January 20, 2005. From [12]

information on the power of the flare according to the GOES classification, the number of the corresponding session of the patrol detector. The database also includes the flares of > M1 class for which the SPR-N instrument did not record radiation as due to objective reasons (shadow, radiation belts, etc.), as under good observation conditions. If the patrol detector of the SPR-N instrument did not record the burst the reason for it is noted (South-Atlantic Anomaly, outer radiation belt, shadow or it is does not seen by the patrol detector of the SPR-N instrument). Database allows to visualize time dependence of the X-radiation burst intensity both according to the data of the SPR-N instrument and according to the GOES data. It is done by means of corresponding link.

The most interesting results during the time of the SPR-N instrument operation were obtained during five most powerful flares of August 25, 2001, October 28, October 29, and November 4, 2003, and January 20, 2005. Time profiles for these flares recorded by the patrol detector of the SPR-N instrument within the ranges of 15–40 and 40–100 kev and data on the flux of the soft X-radiation (wavelength of 1–8 and 0.5–4 Å), measured by the GOES satellites according to the site http://swpc.noaa.gov are presented in Fig. 3.

## Results of the Measurements of the Linear Polarization

Degree of polarization is estimated on the assumption that all hard radiation of the solar flare is a sum of some part of photons, polarized in one and the same plane and the rest part of photons polarized isotropically.

During the observation period from August 2001 about 25 solar events caused the response in the polarization detectors enough for the estimation of degree of polarization by means of the SPR-N instrument. For the most part of these flares it was possible to obtain only the upper limits of the part of polarized radiation at the level from ∼8 % up to ∼40 %.

The most interesting results of the measurements of the solar hard X-radiation polarization were obtained during the extreme events in October–November 2003 [10]. The last decade of October and beginning of November 2003 were marked with strong flare activity of the Sun. Nine flares of X-class happened on the Sun during this period. Six of them were detected by the SPR-N instrument, including three of four the most powerful. The results of the hard X-radiation (HXR) flares polarization are presented below for the following events: October 28, 2003 (X17/4B), October 29, 2003 (X10/2B), and November 4, 2003 (X28, 4B). One and the same active region 0486 was a source of all these flares; on October 28 and 29 it situated almost in the center of the solar disc (S15W02), on November 4—on the western edge of limb (S19W83). Data on gamma-radiation (up to the energies of 100 MeV) of these flares obtained onboard the CORONAS-F were presented in paper [10]. On October 29, 2003, the CORONAS-F satellite situated near geomagnetic equator where the background conditions were the best. The event of October 28, 2003, was observed during the passing of the polar cap. For the event of November 4, 2003, the results were obtained only for the primary flux increase because during the main increase the satellite was passing through the radiation belts.

During three flares polarization detectors have shown essential increase of counting rate over the background calculated according to the previous orbit passes in the points with geomagnetic coordinates closest to those conformed the moments of the flare registration. In order to diminish static error time profiles were smoothed by moving average method by five consequent exposures. For each consequent time interval 14 s long the data were accumulated during 4 s, i.e., exposure time was 4 s at pulse ratio of $4/14 \approx 0.28$. Polarization measurements data for the flare of October 29, 2003, processed in such a way are presented in Fig. 4 along with the data of patrol detector.

Three detectors at a distance of 120° were operated during the flares. For the flare of 29.10.2003, the ratios of the counting rates of polarization detectors of the SPR-N instrument varied essentially for all the channels comparing with the ratios of counting rates before and after the flare. Besides, these ratios differed essentially from the levels conforming non-polarized radiation. Counting rates of polarization detectors during the flare of 29.10.2003 are presented in Fig. 5. For the flare of 28.10.2003 and the first phase of the event of 04.11.2003 there were no noticeable variations of the counting rates' ratios. It allows to suppose that during the flare of October 29 hard X-radiation was polarized. In order to determine degree of polarization accurately it is important to take into account systematic dispersion of output indications of polarization detectors.

In-flight calibration of the detectors by light-emitting diodes and constant background counting rate near geomagnetic equator show stability of their threshold within the range of several percents. But counting rate of separate polarization detectors within the same energy ranges can vary (see, for instance, counting rates of polarization detectors during the flare of October 29, 2003 in Fig. 5). This difference can be caused by both different efficiency of polarization detectors and photomultipliers' noise level, and real flashing of the detectors with unequal X-radiation fluxes. As it was marked during registration of solar radiation this effect can be caused by its strong polarization, and in the case of background measurements in near geomagnetic equator and in the region of low and middle latitudes—by registration of the atmospheric X-radiation, because for different moments polarization detectors can be differently oriented comparing with it and, correspondingly, they detect atmospheric photons by different effective area.

For the measurements in the polar caps, precipitation zones, and radiation belts branches (as a rule in the radiation belts themselves the detectors were overloaded and their indications are unreliable), this effect can be caused by different angles between the detectors and the satellite's elements and construction details in which high-energy electrons produce secondary deceleration emission in the most effective way.

Therefore for the most accurate determination of degree of polarization of the solar flare radiation it is necessary to separate the dispersion of the polarization detectors' indications which can be caused namely by difference of their own effi-

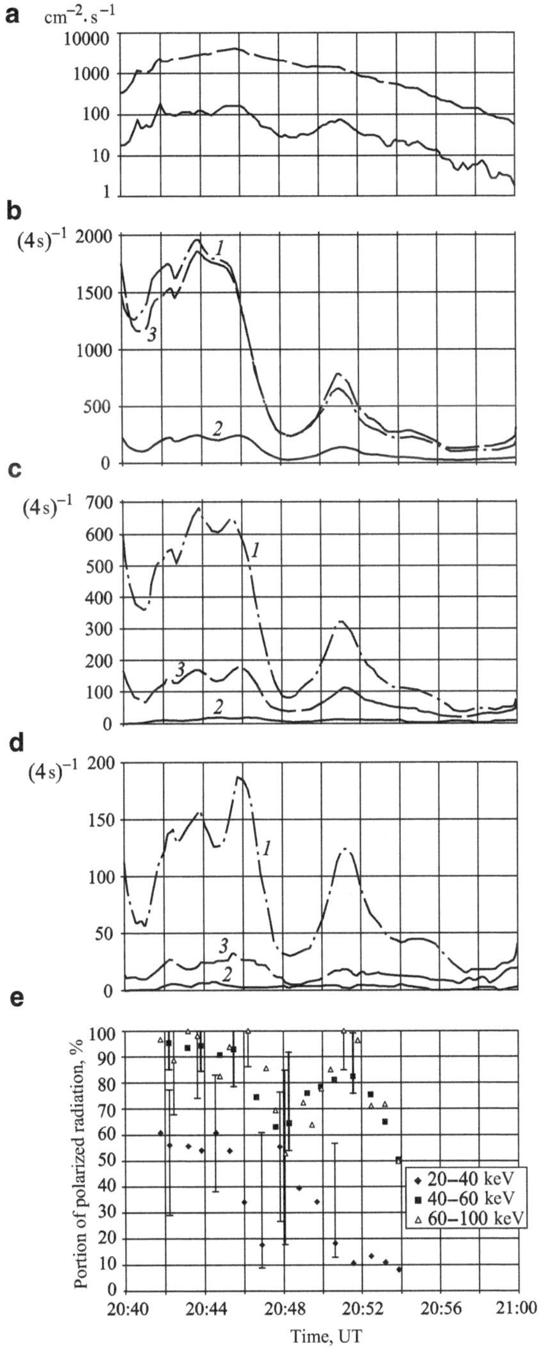


**Fig. 4** Data of the SPR-N instrument for the flare of October 29, 2003 (20:40–21:00): (**a**) patrol detector (*upper curve*—channel of 15–40 keV, *lower curve*—channel 40–100 keV); (**b–d**) polarization detectors within the following energy ranges: 20–40 keV (**b**), 40–60 keV (**c**), and 60–100 keV (**d**), 1, 2, 3: the numbers of polarization detectors groups; (**e**) dependence of the part of polarized radiation from time. From [12]

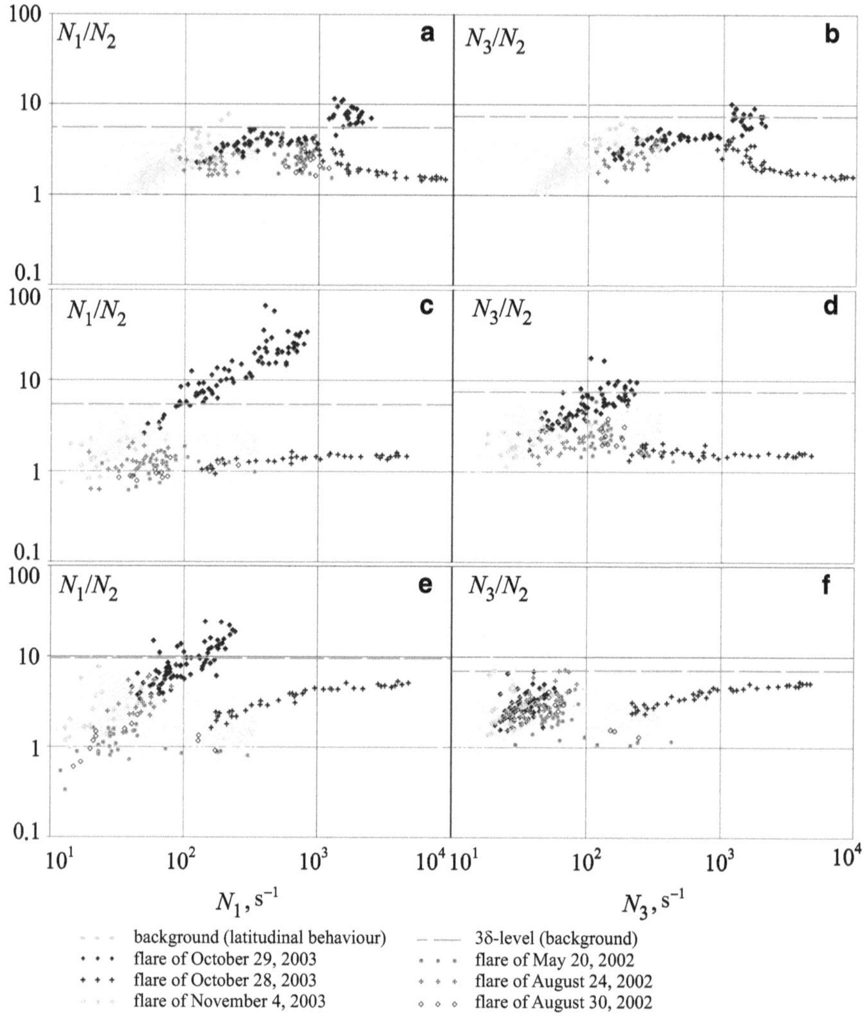


**Fig. 5** Diagrams representing dispersion of the ratios of the polarization detectors' indications ($N_1/N_2$, $N_3/N_2$) depending on full load within the channels of 20–40 keV (**a, b**); 40–60 keV (**c, d**); 60–100 keV (**e, f**). From [12]

ciency. Two groups of indications were chosen for this purpose. The first group consisted of the data of several most intensive solar events, when the increases of polarization detectors counting rates were statistically valid. The second group included the measurements near the geomagnetic equator, in the region of low and middle latitudes, i.e., for those parts of the orbit where latitudinal variation of the counting rate is clearly seen. It was supposed that during magnetically quiet periods input of atmospheric radiation into the latitudinal variation is negligible and the main input is made by X-rays and gamma-emission locally produced in the matter of the satellite as a result of interaction of galactic cosmic rays. Local gamma-emission is isotropic and therefore it is the best way to test detectors in order to compare their efficiency.

Diagrams in Fig. 5 present the ratio of the counting rates of different pairs of polarization detectors depending on their loading for two groups of measurements. It is seen that as a rule for all energy channels counting rates of polarization detectors 1 and 3 exceed counting rate of the detector 2. Apparently this difference of the detectors' counting rates has apparatus character. It can be caused by difference of dispersions and for the channel of 20–40 keV—additionally by the different level of noise of different photomultipliers. During preflight calibration tests and experiment itself photomultipliers' amplification factors were matched by tuning of the photomultiplier's supply voltage. So power voltages for each photomultiplier differed, indeed, just the value of the power voltage first of all determines both the level of photomultiplier's noise and its dispersion. And the latter in turn determines energy resolution of the detector, in consequence, its response function and efficiency.

From Fig. 5 it is seen that except for the flare of October 29, 2003, coefficients of spurious count for the detectors 1 and 3 comparing to the detector 2 do not depend on full load of the detectors. Dispersion of the ratios of corresponding counting rates $N_1/N_2$ and $N_3/N_2$ has both statistic and systematic (methodological) character. Statistic dispersion occurs essentially in the area of low loads: as a rule, background measurements, so-called latitudinal variation. Concerning dispersion at relatively high loads during solar flares, it can be caused by the dependence of detectors' efficiency from photons' energy which in principle can result in variation of coefficients of spurious count depending on radiation spectrum rigidity. But diagrams analogous to those presented in Fig. 5 plotted for spurious count coefficients versus spectral rigidity did not show up correlation between parameters $N_1/N_2$ and $N_3/N_2$ and spectral rigidity (the ratio of counting rate within the channels of patrol detector within the energy ranges of 15–40 and 40–100 keV was its measure). Besides as seen from Fig. 5 possible systematic dispersion of the values of $N_1/N_2$ and $N_3/N_2$ (except for the data of the flare of October 29, 2003) are within the ranges of statistic dispersion. So, statistic dispersion of the values $N_1/N_2$ and $N_3/N_2$ in the diagrams of Fig. 5 can be chosen as a measure of inaccuracy of normalization coefficients which is used for straightening of the separate polarization detectors. Exceeding by the ratios $N_1/N_2$ and $N_3/N_2$ of the level of dispersion $3\sigma$, marked in Fig. 5 for all energy ranges can be considered as a validity requirement of the presence of polarized component.

It is seen that the ratio $N_1/N_2$ for a part of measurements of the flare of October 29, 2003 exceed the level $3\sigma$ for all energy channels: $\sim 4\sigma$ (20–40 keV), $\sim 10\sigma$ (40–60 keV), $\sim 5\sigma$ (60–100 keV). Herewith the points for which the ratio $N_1/N_2$ lies within the level of $3\sigma$ or exceed it slightly (not more than $5\sigma$—for the channel of 20–40 keV) correspond to the final—thermal stage of the flare during which X-radiation hardly can be polarized. At the same time the points for which the ratio $N_1/N_2$ essentially exceeds level of $3\sigma$, corresponds to either the primary phase of the flare or separate maximum of intensity, i.e., the stages during which polarized radiation can be in principle observed [9]. Behavior of the ratio $N_3/N_2$ during the flare of October 29, 2003, is another argument in favor of the claim that significant falling of these measurements out of the ranges of statistic dispersion. As follows from Fig. 5 for the flare of October 29, 2003, although this ratio exceed the level of $3\sigma$ ($\sim 4\sigma$ for the channels of 20–40, 40–60 keV) it is essentially less than the ratio $N_1/N_2$, while for the rest measurements (as for the latitudinal variation, as during other flares and thermal phase of the flare of October 29, 2003) the range of these ratios is approximately the same. Presence of polarization naturally explains such behavior because polarization plane in general case must thread nonsymmetrically relatively the detectors 1 and 3.

But it is impossible to shut out the influence of systematic effects on the ≪anomalous≫ values of the ratio $N_1/N_2$ during the flare of October 29, 2003, completely. In particular, we can suppose that the increased count of the detector 1 can be caused by spectral specific characters of the detected radiation and possible dependence of the detectors' efficiency on photons' energy, which was noted above. But on their spectral characteristics within the range of 20–100 keV, in particular on the energy spectrum rigidity, the flare of October 29, 2003, does not differ significantly from other flares, including those related to the same active region and were registered by the SPR-N instrument on October 28 and November 4, 2003. Therefore it is appears to be unlikely that so essential deviation of the ratio $N_1/N_2$ can be explained by spectral dependence of the polarization detector efficiency. Apparently the behavior of polarization detectors' indications during this event can be considered as identification of the presence of polarized component of hard X-radiation.

In order to determine the part of polarized radiation $P$ and direction of polarization plane according to the detectors' counting rates for the vertically falling plane-polarized radiation, Thomson scattering in Beryllium was simulated. For given $P$ (taken from an interval of 0–100 %) and $\alpha$ parts of radiation coming to each detector in relation to the total flux were calculated. The obtained dependencies allow to solve the reverse problem—to find $P$ according to the measured values taking into consideration $\alpha$. In order to determine $P$ accurately it is important to take into account possibly unequal tuning of the detectors. In-flight calibration of the detectors by means of light-diodes and unchangeable background count in the region of geomagnetic equator show stability of detectors' thresholds within several percents. In order to exclude possible difference related to the spurious count of not scattered photons detectors' indications were normalized on their indications at the end of the flare, when polarization had to be low.

On behalf of quantitative assessment of polarization effect the indications of separate polarization detectors were normalized on the values of the ratios $N_1/N_2$ and $N_3/N_2$, related to the measurements for which beforehand non-polarized radiation was recorded, i.e., ratios $N_1/N_2$ and $N_3/N_2$ lie within the range of $3\sigma$ level. Herewith the value related just to the $3\sigma$ level was chosen as an upper limit of normalizing coefficient (which determines the lower limit of error for determination of polarization degree). The upper limit of error of polarization degree was determined by detectors' indications without normalizing because the lower limit of variance for $N_1/N_2$ and $N_3/N_2$ corresponds to $\sim 1$ (see Fig. 5).

The result of the measurement of polarization radiation part was presented in Fig. 4 (lower curve). It is seen that this part during the first maximum of hard

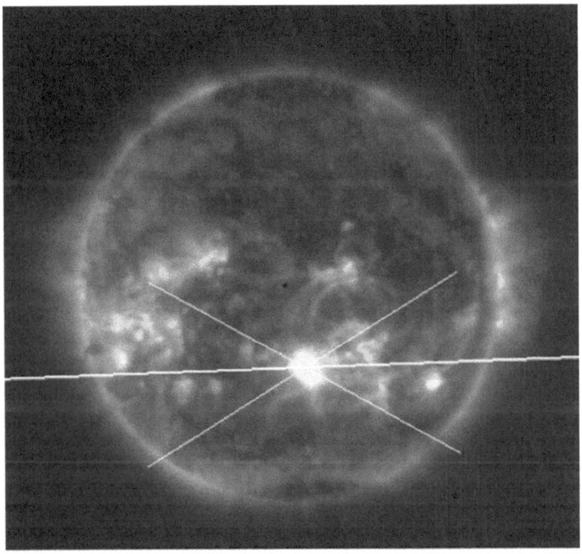

**Fig. 6** Location of the polarization plane of hard X-radiation on the solar disc according to evaluating of the SPR-N instrument. Image of the Sun is obtained during the flare of October 29, 2003, 20:50:21 in soft X-radiation (channel 175 Å) by means of the SPIRIT instrument onboard the CORONAS-F satellite. From [12]

radiation (~20:44) for the channel of 40–60 keV was ~85 %, and during the second maximum (~20:51)—~75 %. For the flare of October 28, 2003, and for the initial phase of the event of November 4, 2003, the ratios of the underlying indications of detectors did not differ significantly from those measured near the equator and the accuracy of determination of normalizing coefficients admits choosing them such as detectors' counting rates become equal. Therefore for these events it is possible to give only upper limits for polarization degree: $P < 25\,\%$ for the flare of October 28, 2003, and $P < 40\,\%$ for the flare of November 4, 2003. For the flares of October 29, 2003, and October 28, 2003, the accuracy of determination of $P$ is resulted from systematic error of determination of normalizing coefficients. For the flare of November 4, 2003, an essential input into the error is made by statistic variance of counting rate in the channels of polarization detectors.

In Fig. 6 location of the polarization plane of hard X-radiation on the solar disc during the flare October 29, 2003, is built according to the data of the SPR-N instrument. Image of the Sun in the moment of the flare was obtained by the X-ray telescope of the SPIRIT complex of the CORONAS-F satellite. It is seen that the polarization plane passes approximately in parallel to the solar equator. Measuring error is ±30°.

Unfortunately, the engine providing rotation of the polarization detector system on ±60° was used sufficiently fast, and taking into account of different tuning of the detectors using it was possible only for the flare of August 25, 2001. During this flare the engine was operated for two time intervals: ~16:27–16:30 and ~16:31–16:34 (see Fig. 3). Within the first interval (at the initial phase of the flare) an increase in the electron channels of the SONG instrument was recorded along with an increase of X-radiation detected by means of the SPR-N instrument. It indicates possible satellite passing the branches of the radiation belt and complicates taking background counting rate into account during this period of time. Therefore only the second interval which corresponds to the decrease of the flare radiation (beginning almost from the maximum of its intensity) was chosen in order to determine polarization degree.

If radiation is polarized during the operation of the rotating platform the counting rates observed for two consequent (even- and odd-numbered) expositions of one and the same pair of detectors must differ. Herewith for one pair indications must differ upwards and for other downwards. But although periodical change of location of detector system neutralizes variance of indications related to different detectors' efficiency for detection of radiation which was not scattered in Be-polarimeter, it does not eliminate influence of the effects related to the position of the detectors with respect to the satellite's mass (background flux of different detectors can differ due to this reason).

In order to find the part of polarized radiation for the flare of August 25, 2001, difference between detectors' indications for the even- and odd-numbered frames within the range of 10 % for the channels of 40–60 and 60–100 keV were used. Calculations have shown that the part of polarized radiation during the flare of August 25, 2001 (for the interval of ~16:31–16:34) was not over 8.5 % (at the level of $3\sigma$). The upper limit of the part of polarized radiation obtained for this flare without taking into account changes of detectors' position is 13 %.

## Dynamics of Thermal and Non-thermal X-Radiation of Solar Flare

In January 2005 a number of powerful flares accompanied with hard X-radiation and gamma-emission happened on the Sun. The strongest of them—a flare of January 20, 2005, approximately from 6:40 till 7:00— had X7.1 class. An active region 720 was a source of five X-class and several M-class flares, and on January 20, 2005, it situated in the western part of the Sun (N14W61). Analysis of the parameters of thermal and non-thermal components of hard X-radiation of the flare of January 20, 2005, was based on the data of the patrol detector of the SPR-N instrument. Total of nine flares was recorded by the SPR-N instrument from January 14 till January 20 (the list of flares is presented in the paper [12]). Time profiles of the solar X-radiation measured by the SPR-N instrument during the flare of January 20, 2005, are presented in Fig. 7. That time the satellite was situated at the equator. Background was taken into account according to the measurements of the previous orbit passes.

In order to determine X-radiation flux according to the counting rates of the SPR-N instrument, its response to the flux of gamma-radiation for power-mode and exponential spectra of falling photons with different parameters $kT$ and $\gamma$ was calculated. Calculations were made by means of Monte Carlo method and GEANT software.

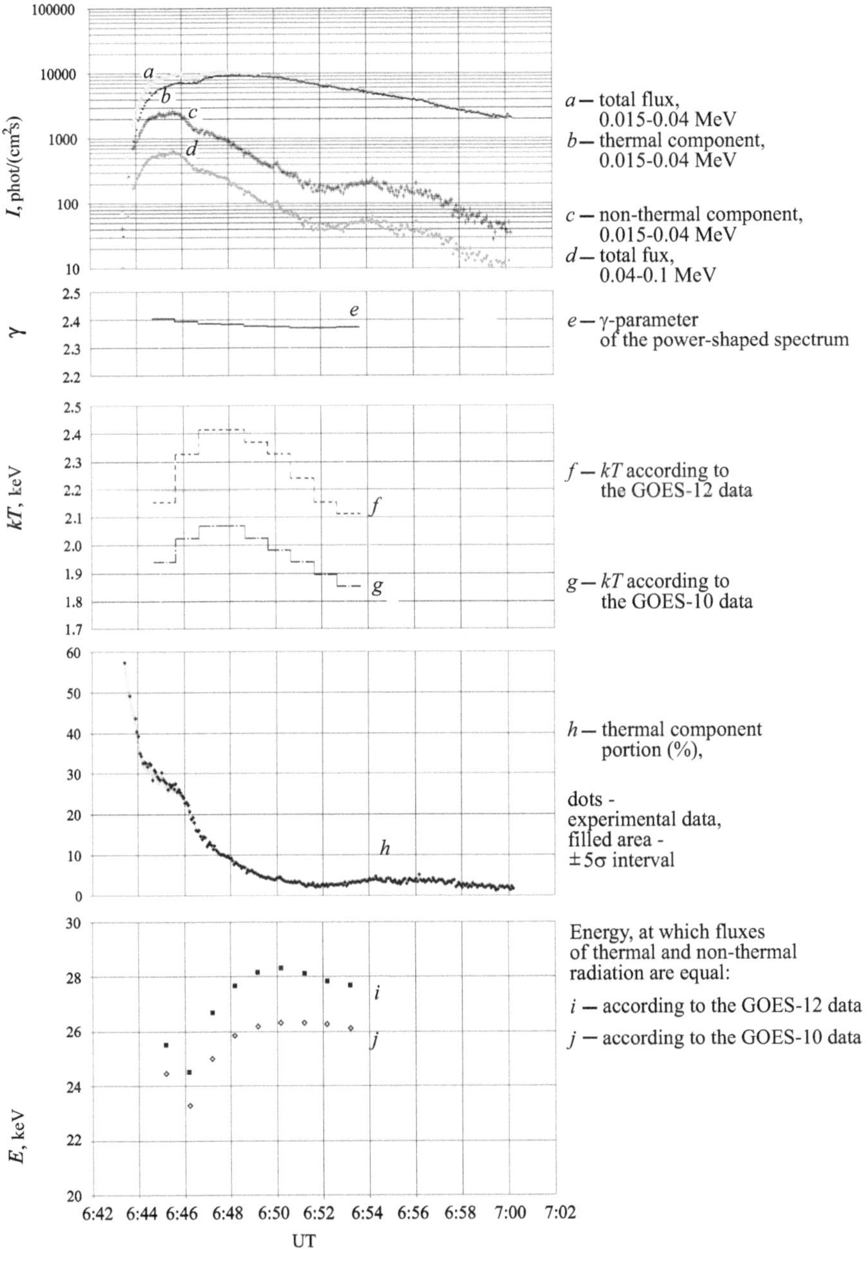


**Fig. 7** Time profiles of the solar X-radiation parameters measured by the SPR-N instrument during the flare of January 20, 2005 (Conference Proceedings, in Russian only)

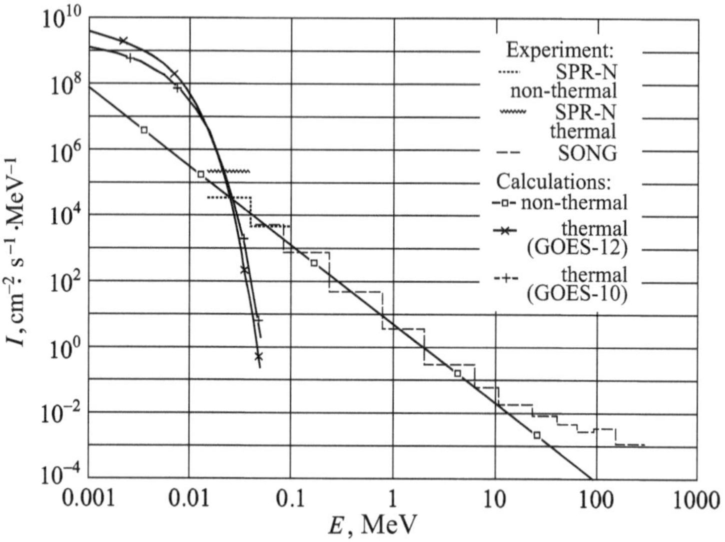


**Fig. 8** Energy spectra according to the data of the SONG and SPR-N instruments during the flare of January 20, 2005, along with calculated spectra of thermal and non-thermal component (Conference Proceedings, in Russian only)

In order to find an input of thermal and non-thermal processes into the hard X-radiation of the flare we supposed that there is only non-thermal component corresponding to the following power law: $I_{\rm n} = I_{\rm 0n} \cdot E^{-\gamma}$ for the channel of 0.04–0.1 MeV. In the channel of 0.015–0.04 MeV there are both non-thermal and thermal components corresponding to the exponential function: $I_{\rm t} = I_{\rm ot} \cdot e^{-E/kT}$. Other components (emission in lines, etc.) do not influence essentially on the form of the spectrum.

Factor $\gamma$ was determined on the basis of the data of scintillation $\gamma$-spectrometer SONG onboard the CORONAS-F satellite [6]. In order to improve statistical accuracy the indications of the SONG instrument were averaged out in nine 1-min intervals (beginning from 6:44:40). Spectra measured by the SPR-N and SONG instruments from 6:46:40 till 6:47:40 are presented in Fig. 8.

During the flare of January 20, 2005, high-energy component of $\gamma$-emission was observed. It is related to the decay of $\pi^0$, which was taken into consideration at reconstruction of the parameters of gamma-emission spectrum by counting rates of the SONG instrument. Time variation of $\gamma$ factor evaluated only by means of the channel 0.084–6.2 MeV of the SONG instrument is presented in Fig. 7. Evaluation has shown that during the flare $\gamma$ factor as a matter of fact stayed constant and stood at $\gamma = 2.38 \pm 0.012$. Constant $I_{\rm 0n}$ was determined by normalizing: the flux within the ranges of 0.04–0.1 MeV conformed the data of the SPR-N instrument. Then non-thermal spectrum within the ranges from 0.015 up to 0.04 MeV were integrated in order to calculate the flux of non-thermal component in the low channel of the SPR-N instrument and the part of non-thermal radiation in this channel. Time profiles of thermal and non-thermal components and the part (%) of non-thermal component in the total flux of hard X-radiation of the flare are presented in Fig. 7.

Parameters of thermal component were determined by means of data on the flux of soft X-radiation measured onboard the satellites of GOES series. The values of temperature during the flare of January 20, 2005, obtained by means of data of the satellites GOES-10 and GOES-12 are presented in paper (Garcia, private message, 2006). Time variation of $kT$ parameter is presented in Fig. 7. Along with it as far as the range of 0.015–0.04 MeV situated in the region of sharp decrease of thermal radiation determination of all parameters of thermal spectrum by means of only GOES data does not provide reasonable accuracy for the hard X-radiation. Therefore it was found to be more correct to calculate constant $I_{0t}$ by means of normalizing of exponential spectrum with known $kT$ at the experimental values of thermal component flux within the channel of 0.015–0.04 MeV.

In Fig. 8 along with experimental spectra measured by the SONG and SPR-N instruments calculated spectra of thermal and non-thermal components are also presented. Components' intensities are compared approximately at the level of $26 \pm 1$ keV. Time variation of location of the point of thermal and non-thermal radiation congruence is presented in Fig. 7.

Time profiles and spectra of thermal radiation and parameters related to it presented in Figs. 7 and 8 are shown separately for the calculations according to data of GOES-10 and GOES-12. The observed difference (tens of percent) denotes that the accuracy of the obtained results is determined better not by statistic dispersion, but by the accuracy of the instruments' tuning.

Detailed analysis of the dynamics of thermal and non-thermal X-radiation during solar flare 20 January, 2005, is presented in papers [5].

## Conclusions

In general scientific program of the SPR-N experiment was accomplished. During the experiment hard X-radiation was detected for 128 solar flares. For all events time profiles with time resolution of 4 s were plotted. For the period starting from August 2001 there were 25 solar events observed by means of the SPR-N instrument, and the response in polarization detectors was sufficient for evaluation of polarization degree. For the event of October 29, 2003, the measured significant level of polarization was over 70 % for the channels of $E$ = 40–60 keV and 60–100 keV, and about 50 % for the channel of 20–40 keV. Time profile of the part of polarized radiation and orientation of the plane of maximum polarization on the solar disc were obtained. For the rest flares the upper limits of the part of polarized radiation at the level from ~8 to ~40 % were found. For several most powerful flares relative input of thermal and non-thermal X-radiation was calculated.

The experiment has shown that it is possible to measure polarization by means of the instrument based on Thomson scattering in Beryllium, but it does not provide good accuracy without taking into account numerous factors which influence on the equal count of the detectors. Essential role can be played by the spurious count of the solar photons hit directly, i.e., without scattering. Therefore in the future experiments it is reasonable to use Compton scattering of photons by means of double coincidences method. It will lead not only to significant background decreasing, but also to assurance that only scattered solar photons are recorded.

## References


1. Bogomolov, A.V., Denisov, Yu.I., Kuznetsov, S.N., et al.: Hard X-ray radiation from solar flares in the second half of 2001: preliminary results of the SPR-N experiment onboard the Coronas-F. Sol. Syst. Res. **37**(2), 112 (2003)
2. Bogomolov, A.V., Denisov, Yu.I., Logachev, Yu.I., et al.: Time, spectral and polarization characteristics of hard X-radiation of solar flares observed during the SPR-N experiment onboard the orbital observatory CORONAS-F. Bull. Russ. Acad. Sci.: Phys. **67**(10), 1422 (2003)
3. Guzman, A.B., Kudryavtsev, I.V., Charikov, Yu.E.: The evolution of electrons accelerated in solar and stellar flares and the pulsed structure of hard X-ray radiation. Astron. Rep. **40**(2), 246 (1996)
4. http://swpc.noaa.gov. Web-site of the Space Weather Prediction Center
5. Kurt, V.G., Svertilov, S.I., Yushkov, B.Yu., Bogomolov, A.V., Grechnev, V.V., Galkin, V.I., Bogomolov, V.V., Kudela, K., Logachev, Yu.I., Morozov, O.V., Myagkova, I.N.: Dynamics and energetics of the thermal and nonthermal components in the solar flare of January 20, 2005, based on data from hard electromagnetic radiation detectors onboard the CORONAS-F satellite. Astron. Lett. **36**(4), 280–291 (2010)
6. Kuznetsov, S.N., Kurt, V.G., Myagkova, I.N., et al.: Gamma-ray emission and neutrons from solar flares recorded by the SONG instrument in 2001–2004. Sol. Syst. Res. **40**(2), 104 (2006)
7. Mewe, R., Gronenshield, E.H.B.M., van der Oord, G.H.J.: Calculated X-radiation from optically thin plasmas. Astron. Astrophys. Suppl. **62**, 197 (1985)
8. Skrynnikov, Yu.I.: About the polarization of hard X-radiation of the solar flare. In: Proceedings of the Second Annual Seminar ≪The problems of the solar flares physics≫ (Riga, December 1983), Moscow, 1985, p. 162 (in Russian)
9. Tindo, I.P., Ivanov, V.D., Mandelshtamm, S.L. et al.: Detection of polarization of X-radiation of solar flares. Cosmic Res. **9**(1), 116 (1971)
10. Veselovsky, I.S., Panasyuk, M.I., Avdyushin, S.I., et al.: Solar and heliospheric phenomena in October–November 2003: causes and effects. Cosmic Res. **42**(5), 435 (2004)
11. Zhitnik, I.A., Tindo, I.P., Urnov, A.M.: The studies of solar X-radiation in P.N.Lebedev Physical Institute. Trudy FIAN **195**, 3 (1989) (in Russian)
12. Zhitnik, I.A., Logachev, Yu.I., Bogomolov, A.V., et al.: Polarization, temporal, and spectral parameters of solar flare hard X-rays as measured by the SPR-N instrument onboard the CORONAS-F satellite. Sol. Syst. Res. **40**(2), 93 (2006)
13. Web site of Space Monitoring Data Center, developing in Skobeltsyn Institute of Nuclear Physics, MSU. http://http://smdc.sinp.msu.ru

# Observations of Doppler Shifts of X-Ray Lines in Solar Flare Spectra Based on DIOGENESS Spectrometer Data

**Z. Kordylewski, J. Sylwester, B. Sylwester, M. Siarkowski, S. Płocieniak, A. Kępa, M. Kowaliński, W. Trzebiński, and F. Farnik**

The idea of measurement of X-ray lines Doppler shifts in spectra of the Sun, applied in DIOGENESS spectrometer, was previously developed and verified in rocket experiment with RDR X-ray Dopplerometer (Vertical-11 Rocket, 1981) [3]. Upon the obtained results two X-ray DIOGENESS spectrometers have been manufactured; the first one was operated aboard the CORONAS-I satellite (launch in 1994) [6], while the second was operated aboard the CORONAS-F. The general view of the instrument is shown in Fig. 1.

The spectrometer with flat oscillating crystals is used for Doppler shifts measurement in DIOGENESS. According to Bragg law, the X-radiation, inciding crystal surface at $\theta$ angle, is reflected from the crystal at the same angle at a single wavelength. The crystal, due to the interference, quench all radiation except the radiation with $\lambda$ wavelength, defined by Bragg condition

$$k\lambda = 2d \sin\theta,$$

where $d$ is the crystal lattice spacing and $k$ is the order of reflection. By changing $\theta$ incidence angle, e.g., through crystal rocking back and forth, it is possible to receive X-ray source radiation spectrum within wavelength range, corresponding to scanning extreme angles.

The concept of spectrometer which can operate as the Dopplerometer system is based on mounting of two identical crystals fixed at the precise angular position against each other at the $\alpha$ angle and attached to rocking table. The crystals scan the spectrum in opposite directions, i.e., in the direction of increasing and

Z. Kordylewski · J. Sylwester (✉) · B. Sylwester · M. Siarkowski · S. Płocieniak · A. Kępa · M. Kowaliński · W. Trzebiński
Space Research Centre of Polish Academy of Sciences (SRC PAS), Wroclaw, Poland
e-mail: js@cbk.pan.wroc.pl

F. Farnik
Astronomical Institute of Czech Academy of Science, Ondrejov, Czech Republic

V. Kuznetsov (ed.), *The Coronas-F Space Mission*, Astrophysics and Space Science Library 400, DOI 10.1007/978-3-642-39268-9_5,

**Fig. 1** DIOGENESS X-ray spectrometer, designed for operation in the Copplerometer system. From [7]

decreasing wavelengths. When the incidence angle reaches the value $\theta = \alpha/2$, the same wavelength is measured in both spectra. Figure 2 shows a basic scheme of Copplerometer measurements.

The angle $\alpha$ between the crystals should be selected in a manner to ensure the simultaneous registration of selected intense spectral line from radiation source staying at rest relative to Dopplerometer. In case of the presence of a source moving along the line toward the instrument, the Doppler effect occurs, which causes the change in wavelengths. The spectral line in such a case will not be observed in both spectra simultaneously. The angle of table rotation between the detected lines in both spectra will represent a measure of Doppler shift and speed of the source along the line of sight.

There are four crystals in DIOGENESS spectrometer unit (two quartzes, beryl and ADP); their parameters have been adjusted with the purpose to receive the full spectrum in the vicinity of helium-like triplet lines Ca XIX (3.18 Å), S XV (5.04 Å) and Si XIII (6.65 Å) for table rocking by angle $140'$. Two quartz crystals, cut from the same single crystal block, were utilized in a system of the Dopplerometer adjusted accurately for resonance line of helium-like ion Ca XIX ($\lambda = 3.18$ Å). The spectrometer parameters are specified in Table 1.

The spectra from all crystals were registered by double proportional counters with beryllium entrance windows of the 145 μm thickness, filled with argon under the pressure of ~0.5 atm. Physical connection of two gas detector chambers (working and control ones) ensured the identity of gas parameters in the whole double counter. The window of detector control chamber was covered with $^{55}$Fe radioactive isotope emitting reference radiation at the energy of 5.9 keV. This ensured the constant energy gain of detector electronic system.

There was the multi-slit collimator attached to the rocking table. The narrow (FWHM $= 10''$) transmission window of collimator scanned the disc of the Sun

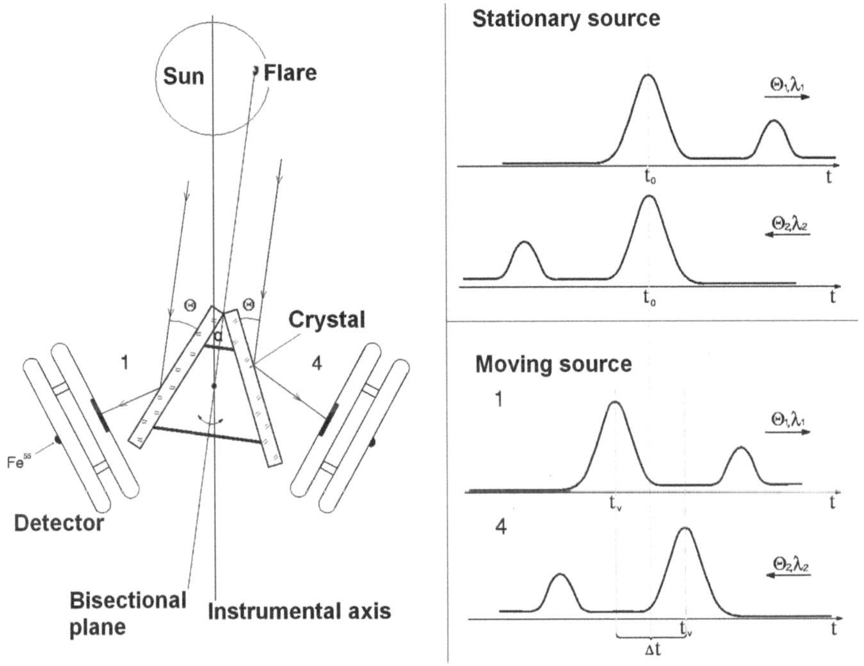


**Fig. 2** The scheme of X-ray Dopplerometer is shown in the *left part*. The simultaneous scanning of Sun spectra in opposite dispersion directions is obtained by rocking the table with two identical crystals. From [6]. The *right part* shows the schematic spectral records as obtained from the source at rest and moving (at the *bottom*). From [4]. For moving source the time interval between the registration of line centers in both spectra represents the measure of source speed along the line of sight

**Table 1** DIOGENESS spectrometer parameters

| Channel | 1 | 2 | 3 | 4 |
|---|---|---|---|---|
| Crystal | Quartz | ADP | Beryl | Quartz |
| Plane | 1011 | 101 | 1010 | 1011 |
| $2d$ spacing (Å) | 6.6855 | 10.5657 | 15.9585 | 6.6875 |
| $\lambda$ (Å) | 3.1781 | 5.0374 | 6.6488 | 3.1781 |
| Principal lines in range | Ca XIX | S XV | Si XIII | Ca XIX |
| $\lambda_{minimum}$ (Å) | 3.1436 | 4.9807 | 6.1126 | 2.9601 |
| $\lambda_{maximum}$ (Å) | 3.3915 | 5.3721 | 6.7335 | 3.2123 |
| Reflectivity (μrad) | 91 | 91 | 15 | 90 |
| Full Width at Half Maximum (FWHM) (arcsec) | 24.1 | 68.1 | 94.1 | 25.6 |

along the direction of spectrometer dispersion. Transmitted radiation was registered in two wavelengths ranges (2–4 keV and 4–8 keV), by the proportional counter. It was expected that the scans would give the localization of radiation source on the Sun, but, unfortunately, the operation of this detector terminated early during a flight.

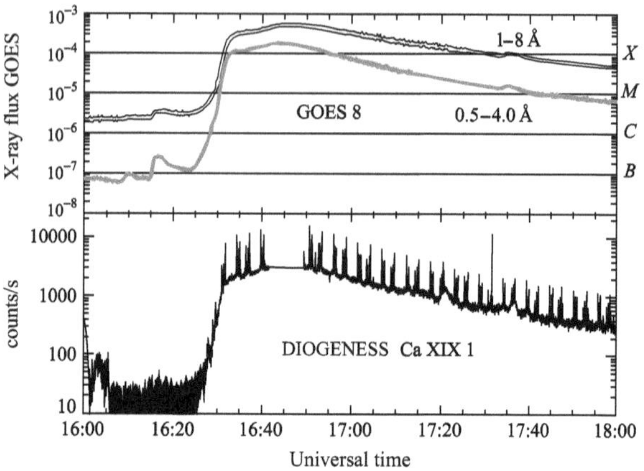


**Fig. 3** Observation of X-ray class X5.3 flare (importance 3B $H_\alpha$) of August 25, 2001. The X-ray radiation measured by GOES satellite detectors (two *upper curves*) and in DIOGENESS spectrometer first channel (*lower curve*) are presented. The peaks related to scanning of Ca XIX helium-like ion line spectra are seen atop the continuum level, corresponding to time changes of this level in the spectral band $\lambda = 3.1$–$3.2$ Å. From [2]

The rocking motion of table-collimator system with $1\,\mathrm{arcmin\,s^{-1}}$ angular velocity was actuated by the step motor which rotated back and forth the Archimedian spiral-shaped disc, against which the arm with collimator was pressed by a spring. The full scan in one direction required 40,000 motor steps.

Till mid September 2001, when the device stopped its operation, probably because of the mechanical damage of spectrometer drive, hundreds of spectra of solar flares above M1 class (according to GOES classification) have been obtained. The observations of the highest quality are that obtained on August 25, 2001 ($4 \times 30$ spectra), when a strong flare of 3B optical flare importance occurred on the Sun. According to X-ray radiation the intensity of this event has reached X5.3 class. Figure 3 (bottom) illustrates the time evolution of X-ray radiation of this flare. For comparison the X-ray measurements in two standard ranges measured by GOES satellite detectors are also provided. The continuum level evolution is similar as seen with DIOGENESS spectrometer: on the regular shape of the continuum the groups of strong X-ray emission lines, recorded in turn in two opposite directions along the dispersion plane, are notable in a form of narrow spikes.

As DIOGENESS device had very narrow instrumental width (better than in previous spectrometers), the observed spectral lines widths correspond to the "true" physical widths, caused by *thermal* and *nonthermal* spectral plasma widening. In the designed spectrometer the detailed laboratory calibration of mechanical drive has been performed, and the crystal rocking profiles have been measured with high accuracy. During the flight the control of the temperature of the crystal mount was constantly performed. Upon these data and with the consideration of effects caused by inaccuracy of pointing and stabilization of instrument axis relative to the Sun disc center, the obtained spectra could be accurately referred to wavelengths scale. This allowed to increase the reliability of identification of the observed spectral lines. Figures 4–6 show (separately for the "right" and "left" direction of rocking the crystals) the averaged spectra of solar flare of August 25, 2001. The detail coincidence of spectra received by scanning in opposite directions confirm the

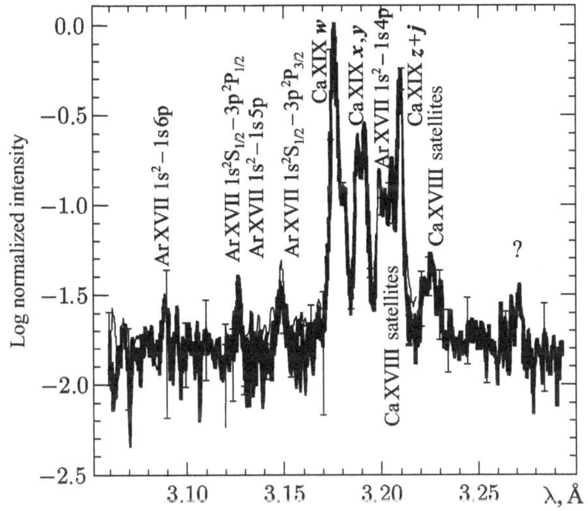


**Fig. 4** Averaged flare spectrum of August 25, 2001, obtained by DIOGENESS spectrometer in channels 1 and 4. There are two exactly coinciding spectra (*thick* and *thin curves*), averaged separately for the scans performed in opposite spectral directions. The *lines* identification are given. For the purpose of weak lines highlighting, the normalized spectra are shown in a logarithmic scale. From [5]

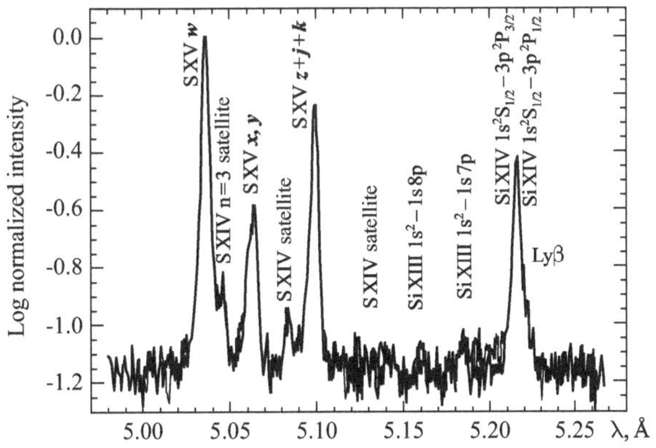


**Fig. 5** Averaged flare spectrum of August 25, 2001 flare, obtained by DIOGENESS spectrometer in the 4.98–5.26 Å range (shown, there are two well coinciding spectra averaged separately for the scans taken in two opposite directions). The spectrum covers the lines corresponding to indicated transitions. From [5]

reality of all spectral features. Many of them which have, according to atomic physics calculations, known intensities and wavelengths values have been identified for the first time. Identification of numerous spectral lines is in progress. The spectra, obtained by DIOGENESS spectrometer for the study of analyzed flare and other events, will be used for the analysis of temperature changes and emission measure analysis, with the purpose of formation of a complete image of time changes of principal thermodynamic characteristics of flaring plasma.

The good quality of spectra of August 25, 2001 flare also allowed to identify the Doppler shifts of spectral lines. Figure 7 shows two concurrent spectral scans embracing Ca XIX line triplet, which have been obtained by quartz crystals mounted in Dopplerometer configuration. It may be seen that resonance lines were not registered simultaneously. It was partially caused by small constant inaccuracy of adjustment of the angle $\alpha$ between atomic planes of Dopplerometer crystals pair, which has been noticed in course of device laboratory tests. Such inaccuracy

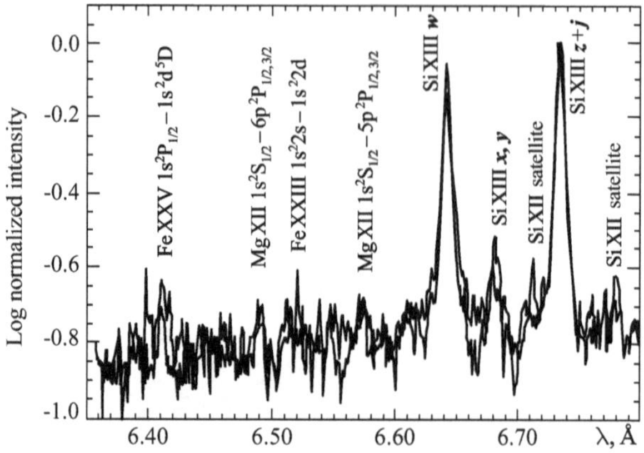


**Fig. 6** Two averaged X-ray flare spectra (two almost coinciding *curves*) of August 25, 2001 flare, obtained by DIOGENESS spectrometer in course of scanning in opposite spectral directions. The identification of principal lines in 6.35–6.80 Å range has been performed. From [5]

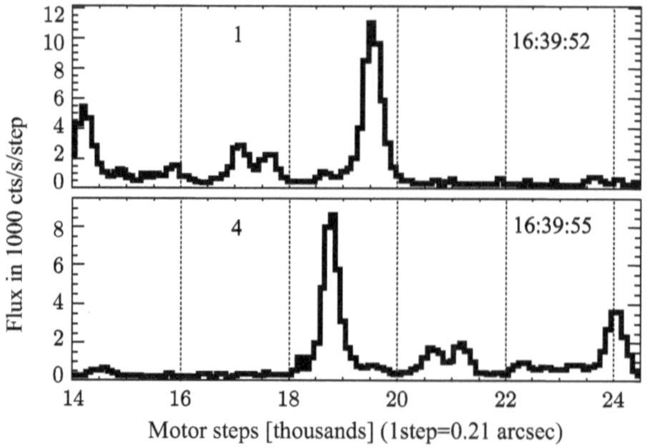


**Fig. 7** The example of two scans, covering spectral ranges of Ca XIX ion line triplets. The spectra have been obtained by the Dopplerometer system during the flare of August 25, 2001. The rotation angles are shown on the horizontal axis represented by the steps of the drive motor of the rotating table with crystals. It may be seen that the strongest resonance lines are not measured simultaneously. After elimination of the constant instrumental shift, the remaining difference of the positions of centers of the registered lines evidences the presence of a Doppler shift. From [1]

amounted to 20 arcsec, and it could be easily accommodated during the line shift determinations. After the correction for this instrumental shift the remaining displacement of lines may be interpreted only by Doppler effect, caused by the motion of flaring source along the line of sight. The results of such interpretation are shown in Fig. 8, which illustrates the change of radial component of flare hot plasma speed with time, i.e., in its development phase. The highest speed (100–150 km/s in the direction from the Sun) was observed in initial phase of the flare. The reliability of the obtained results is confirmed by the fact that similar velocity values have been

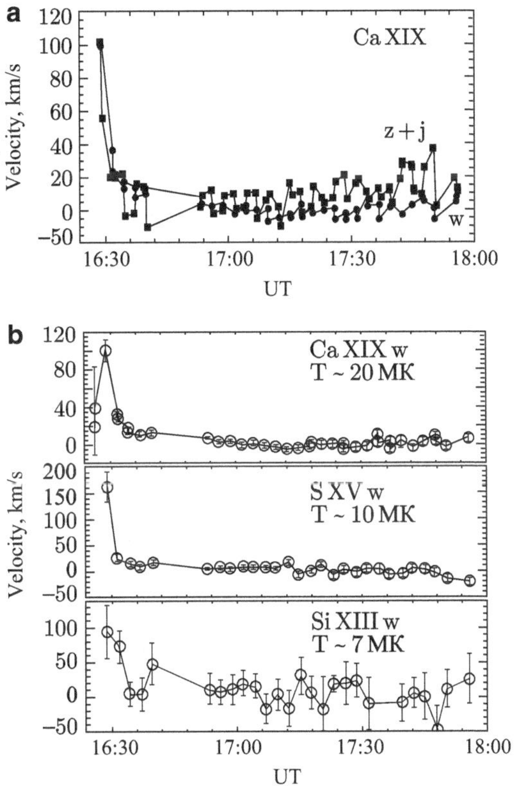


**Fig. 8** (**a**) The time dependence of plasma Doppler velocity for the flare of August 25, 2001. The "w" letter marks the values of velocity obtained from the measurement of relative shifts of Ca XIX resonance line ($\lambda = 3.178$ Å). The "z + j" letters mark the velocities for a line representing the blend of j satellite line and so-called forbidden line z of Ca XIX ion. From [5]. (**b**) Comparison of the time dependence of Doppler velocities obtained for resonance lines of Ca XIX, S XV, and Si XIII ions. The characteristic temperatures for the formation of these lines are specified. From [1]

independently obtained by measurements of both resonance (the strongest one) and forbidden lines of Ca XIX ion.

Ca XIX ion resonance line is effectively formed in a hot plasma with the temperature above 7 MK only. The maximum of efficiency occurs at ~20 MK. In the given flare the plasma had such a temperature in the very initial phase of the event, during the rise phase. The sulfur and silicon lines are formed in a lower temperature plasmas. DIOGENESS spectrometer was not designed for operation in Dopplerometer arrangement for a spectral range including the lines of these elements. However, the knowledge about device geometry has allowed to analyze, for S XV and Si XIII ion lines, their shift relative to the position of imaginary bisector of angle between the crystals. The results obtained in such a manner are also shown in Fig. 8, qualitatively they coincide with the data obtained for calcium ion lines.

**Acknowledgements** The works with DIOGENESS spectrometer have been carried out in the Solar Physics Laboratory of the Space Research Center of Polish Academy of Sciences in Wrocław (Poland), with the support of Polish Scientific Research Fund (project 1.PO3D.017.29 and 2011/01/B/ST9/05861).

## References


1. Plocieniak, S., Sylwester, J., Kordylewski, Z., Sylwester, B.: In: Wilson, A. (ed.) Determination of Wavelengths and Line Shifts Based on X-Ray Spectra from Diogeness Solar Variability: From Core to Outer Frontiers. The 10th European Solar Physics Meeting, 9–14 September 2002, Prague, Czech Republic. ESA SP-506, vol. 2, pp. 963–966. ESA Publications Division, Noordwijk (2002). ISBN 92-9092-816-6
2. Siarkowski, M., Sylwester, J., Plocieniak, S., et al.: In: Wilson, A. (ed.) Solar Variability: From Core to Outer Frontiers. The 10th European Solar Physics Meeting, 9–14 September 2002, Prague, Czech Republic. ESA SP-506, vol. 2, pp. 753–756. ESA Publications Division, Noordwijk (2002). ISBN 92-9092-816-6
3. Sylwester, J.: Experience of Wrocław team in construction of solar X-ray instruments. ESA SP **493**, 377–382 (2001)
4. Sylwester, J.: In: Battrick, B., Sawaya-Lacoste, H. (eds.) Solar Encounter. Proceedings of the First Solar Orbiter Workshop, 14–18 May 2001, Puerto de la Cruz, Tenerife, Spain. Scientific coordinators: Marsch, E., Marinez Pillet, V., Fleck, B., Marsden, R., ESA SP-493, pp. 377–382. ESA Publications Division, Noordwijk (2001). ISBN 92-9092-803-4
5. Sylwester, J.: In: Stepanov, A.V., Benevolenskaya, E.E., Kosovichev, A.G. (eds.) Multi-wavelength Investigations of Solar Activity, IAU Symposium, vol. 223, pp. 409–416. Cambridge University Press, Cambridge (2004). doi:10.1017/S1743921304006337
6. Sylwester, J., Farnik, F.: DIOGENESS soft X-ray spectrometer-photometer for studies of flare energy balance. Bull. Astron. Inst. Czech. **41**(3), 149–157 (1990). ISSN: 0004-6248
7. Sylwester, J., Kordylewski, Z.: Adv. Space Res. **30**(1), 67–72 (2002). doi:10.1016/S0273-1177(02)00173-4

# Investigations of Physical Processes in Solar Flare Plasma on the Basis of RESIK Spectrometer Observations

**Z. Kordylewski, J. Sylwester, B. Sylwester, A. Kępa, M. Kowaliński, and W. Trzebiński**

X-ray spectroscopy generally uses the data obtained with the help of spectrometers with crystals, acting as dispersive elements. According to Bragg's law, the wavelength $\lambda$ of X-radiation, reflected from the crystal, depends on $\theta$ incidence angle. This dependence is presented as

$$k\lambda = 2d \sin\theta,$$

where $k$ is the order of reflection and $d$ is the crystal lattice spacing. The change of angle of incidence of radiation at the crystal causes the X-ray spectrum is formed within the wavelengths range corresponding to the extreme values of crystal rotational change.

The X-ray spectrometers with the flat crystals have one deficiency, especially when they are used for observations of solar flare spectra. To obtain the spectrum through scanning with incidence angle change, some time is required—usually even few minutes. This complicates the interpretation of the obtained spectra for quickly varying radiation sources, such as solar flares. The obtained spectra are not homogeneous—individual parts are irradiated by plasma having various physical characteristics. This is caused by variation of physical conditions during the scanning.

Simultaneous registration of spectra of quickly varying sources may be obtained through the use of curved fixed crystals instead of the scanning flat crystals spectrometer. Illuminating such a curved crystal with parallel X-ray beam allows to obtain, after the reflection, the whole spectrum covering certain wavelengths range, as the incidence angle at curved crystal surface represents a monotonous function of incidence point position measured along the crystal. On the other hand, the spectrum registration becomes more complicated, as it requires the use of

Z. Kordylewski · J. Sylwester (✉) · B. Sylwester · A. Kępa · M. Kowaliński · W. Trzebiński
Space Research Centre of Polish Academy of Sciences (SRC PAS), Wrocław, Poland
e-mail: js@cbk.pan.wroc.pl

V. Kuznetsov (ed.), *The Coronas-F Space Mission*, Astrophysics and Space Science Library 400, DOI 10.1007/978-3-642-39268-9_6, 

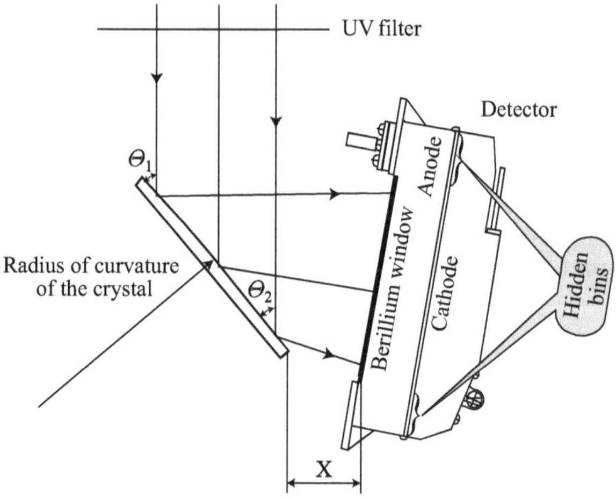


**Fig. 1** The scheme of spectrometer with bent crystals, used in RESIK experiment. The parallel rays bundle falls into the surface of cylindrically curved crystal (with indicated radius). The incidence angle smoothly changes along the crystal. As a result the single points on detector window are illuminated by radiation with different wavelengths, thus forming the X-ray spectrum. From [21]

so-called X-ray position sensitive detector. Such detector should be fitted with a known function of the detection of line coordinate of the points (along the dispersion direction), on which the radiation photons fall. In the same manner it is possible to measure simultaneously the radiation intensity in separate wavelengths. The scheme of the spectrometer with curved crystals is shown in Fig. 1.

Starting from the 1980s of the last century the Solar Physics Division of the Space Research Center of Polish Academy of Sciences in Wrocław conducted the works with X-ray spectrometers for the investigation of radiation of the Sun. The works with Polish spectrometers, which have been started on Vertical-8 and Vertical-9 geophysical rockets [12], were continued with so-called Dopplerometer spectrometers on Vertical-11 rocket [13] and CORONAS-I satellite [14], and have lead to the construction of RESIK spectrometer with bent crystals at CORONAS-F spacecraft [21]. During the design of RESIK instrument, the similar spectrometer BCS was used as a model. Such spectrometer operated on Yohkoh satellite [2]. The RESIK instrument design was performed jointly by the groups of experts of Naval Research Laboratory (NRL, USA), Rutherford Appleton Laboratory (RAL), and Mullard Space Science Laboratory (MSSL, UK). The works were also supported by scientists of Pushkov Institute of Terrestrial Magnetism, Ionosphere and Radio Waves Propagation of the Russian Academy of Sciences. General works supervision was executed by Solar Physics Division of the Space Research Center of Polish Academy of Sciences.

The RESIK spectrometer, shown in Fig. 2, consisted of two sections (A and B), closed in one unit, designed for operation on instrumental platform located outside the sealed satellite container. The electronic unit, which included the computer, was located inside the sealed CORONAS-F satellite container. In each of A and B sections of the spectrometer two crystals were used as dispersive elements—they were made of silicon (Si) and quartz (Qu) single crystals wafers. The application of materials with a low atomic number $Z$ helped to reduce the fluorescence effects, which, as is known, are proportional to $Z^4$ of the crystal material. Undesirable fluorescence radiation appears as a result of illumination of crystals with solar

**Fig. 2** The general view of RESIK spectrometer (without the cover) from the side of crystals support system. The figure contains the indications of sections A and B, crystal material, and spectrometer channels numbers. From [21]

radiation. The crystals, used in RESIK, have been manufactured, curved, and calibrated at the USA National Institute of Standards and Technology (NIST).

After the diffraction on crystals the X-radiation was registered by two linear double proportional detectors, manufactured according to those used in BCS device on Yohkoh satellite. In each of A and B sections of RESIK spectrometer there was one double detector with input window made of beryllium foil of 125 μm thickness. Each one of these detectors measured the radiation reflected separately by quartz and silicon crystals. The split of signals from two channels was performed in the counter with application of two anodes with the high voltage (power 1.5 kV), connected to preamplifier. The position of points of incidence of radiation at the window of the detector, defined in 256 bins, came out from the relationship of signal charges recorded from two parts of wedge and wedge cathode, located directly behind the anode. Due to small radius of curvature of applied crystals the spectral dispersion amounted to 0.02–0.04 Å/bin. These values are slightly higher than natural widths of spectral lines and represent instrumental width. The calibration of detectors with the use of $^{55}$Fe radioisotope source pulled out to the counters field of view was performed several times in course of the flight. The spare elements (detectors, high voltage sources, and electronic boards for detector signals processing) which remained after the BCS experiment at Yohkoh satellite were used in RESIK equipment.

The correct selection of crystal materials, their curvature radiuses, and optimal relative position of crystals and detectors inside the device has ensured the obtainment of nearly complete spectral coverage in a wide range, from 3.3 to 6.1 Å. Such range is especially valuable for analysis of elemental abundance of solar plasma. It includes the lines of elements with high and low first ionization potential (FIP). In particular, the spectra obtained by RESIK show the emission lines of the following elements: Ar (FIP = 15.8 eV), S (FIP = 10.4 eV), Si (FIP = 8.2 eV), and K (FIP = 4.6 eV). The detailed characteristics of four channels of RESIK spectrometer are presented in Table 1.

**Table 1** RESIK spectrometer parameters

| | | Channel 1 | Channel 2 | Channel 3 | Channel 4 |
|---|---|---|---|---|---|
| Detector | | A | B | A | B |
| Crystal | | Si | Si | Quartz | Quartz |
| Plane | | (111) | (111) | $(10\bar{1}0)$ | $(10\bar{1}0)$ |
| $2d$ spacing (Å) | | 6.27 | 6.27 | 8.51 | 8.51 |
| Bent radius (cm) | | 110.0 | 100.0 | 145.0 | 52.5 |
| Length (cm) | | 12.8 | 11.6 | 12.8 | 12.8 |
| Detector edge to crystal edge distance[a] (cm) | | 3.74 | 4.66 | 4.12 | 4.75 |
| Incidence angle range (°) | | 29.5–39.7 | 35.0–44.9 | 28.2–36.6 | 31.7–48.2 |
| Rc-integrated reflectivity[b] (μrad) | | 53 | 52 | 21 | 18 |
| Wavelength[c] (Å) | Nominal range | 3.40–3.80 | 3.83–4.27 | 4.35–4.86 | 5.00–6.05 |
| | Extreme range | 3.33–3.90 | 3.78–4.32 | 4.23–4.92 | 4.90–6.15 |
| Wavelength resolution (mÅ) | | 8 | 9 | 12 | 17 |
| Dispersion[b] (mÅ $\text{bin}^{-1}$) | | 2.49 | 2.28 | 2.85 | 4.99 |
| Principal lines in range | | Ar XVIII, K XVIII | Ar XVII, S XV | S XVI | S XV, Si XIV, Si XIII |

[a] Minimum distance between the crystal and detector measured along the line perpendicular to the direction of the source

[b] The values of reflection coefficient and the dispersion are given for the first-order reflection in the central part of the crystal

[c] Nominal range—for the radiation beam parallel to the optical axis of the instrument; maximum range—including the beams as inclined to the optical axis (slope angle is comparable with the angular radius of the Sun)

Owing to unprecedented sensitivity of RESIK spectrometer it became possible to make measurements not only of strong solar flares but also of the weak ones. The instrument used a dynamical method of on board adjustment of signal accumulation time (DGI). The RESIK on board computer analyzed the level of emission of observed X-radiation of the Sun and automatically matched the optimal DGI accumulation time. Thus, in the rise phase of a strong flare DGI time could amount to 5 min at the very beginning of the flare, decreasing to 1 s at maximum of the event.

The RESIK device operated successfully in a flight for almost 2 years. This period may be divided in several stages. After the initial control phase lasting for a short period in middle October 2001, the work activation was performed on August 24, 2001, and lasted till October 7, 2001. At this time period, on August 25,

2001, a very strong flare has occurred on the Sun. The analysis of first observations has shown the need for improvement of RESIK on board operation program. The power supply was henceforth disconnected from the device for the duration of improvement works, to protect the detector performance from deterioration. At the end of January and beginning of February 2002, the power supply was restored, and the device operated uninterruptedly till May 22, 2003. During the device active measurement phase the parameters of device operation have been changed several times upon the ground commands, in order to obtain the best possible results of observation. The optimal parameters have been matched at the end of December 2002. Starting from that date the most valuable material for many flares has been gathered, including the data for 14 phenomena (January–March 2003; middle level of solar activity), for which the average spectrum is shown in Fig. 3. The RESIK instrument has recorded nearly one million of X-ray spectra. Observation catalogue is accessible at the web site http://www.cbk.pan.wroc.pl/resik_catalogue.htm.

In course of the processing of observed spectra several factors were considered. One of them was the "notch" effect (FPS), which gives artificial peaks in spectra as a result of division of integer numbers which defined the positions of photons measured along the detector length. Another important fact in data processing was the correct accommodation of the results of ground calibration and in-flight calibrations with the use of $^{55}$Fe radioisotope. Many efforts were needed to eliminate the crystals fluorescence effect, caused by radiation of the Sun. This radiation contributed to the observed continuum level. This effect could be partially reduced by matching the relevant device operation parameters, while the remaining part of its impact could be eliminated by computational method only. It required the knowledge of detectors laboratory calibration parameters, as well as performance of special sequences of device operation taken in flight. Such actions required 34 h of measurements, which have been performed from August 30 to September 5, 2002.

The obtained spectra in 3.3–6.1 Å range are especially interesting, as in this wavelength range they have been observed for the first time with such high sensitivity and spectral resolution. Moreover, owing to the accurate calibration of the device in the laboratory (performed at RAL, MSSL) and during the flight, it was possible to reduce the observation and obtain the absolute error level not exceeding 20 %. Figure 3 shows the spectra from all four RESIK spectrometer channels, averaged for 14 flares of different classes. These averaged spectra have been obtained by summation of several tens of individual registrations covering the rise, maximum, and decay phase of the flares. The summation was performed with the purpose of increasing the statistics and revealing all spectral features characteristics for hot and cooler plasma.

The RESIK spectrometer measured the spectra in a wide intensity range, which covers four orders of magnitude. It allowed to obtain the spectra not only for maximum phases of strong flares, but also for their growth and decay phases, and even for quiet corona. The possible examples are the observations of a strong solar limb flare on January 21, 2003. This flare lasted almost 5 h and has reached M1.9 level according to GOES class. Figure 4 illustrates two solar spectra—one (on the left), for quiet solar conditions, has been obtained 5 h before the maximum of the

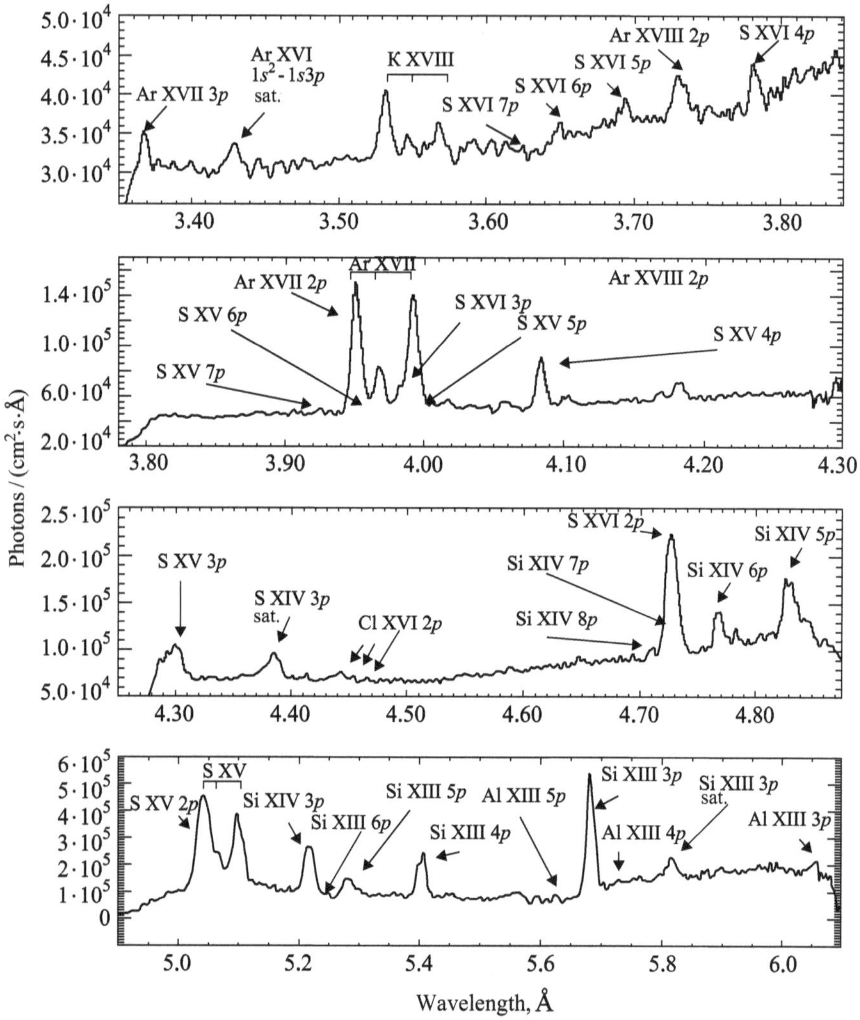


**Fig. 3** Averaged spectra of 14 flares observed in the period January–March 2003. The spectra have been obtained in four channels of RESIK spectrometer (the wavelength range 3.35–6.05 Å). The figure shows identification of lines—many of them have been observed in astrophysical plasma for the first time. From [6]

flare, the second one (on the right) corresponds to the flare rise phase. The difference between these spectra is clearly seen. It is caused not only by the growth of general intensity level, but also by the presence of different individual spectral lines and their varying intensities. For example, the helium-like Ar XVII ($\lambda \approx 3.95$ Å) triplet has a high intensity only during the flare development. The thin solid line below the spectra shows the continuum radiation level calculated using CHIANTI atomic code. The calculations have been performed with the use of characteristic

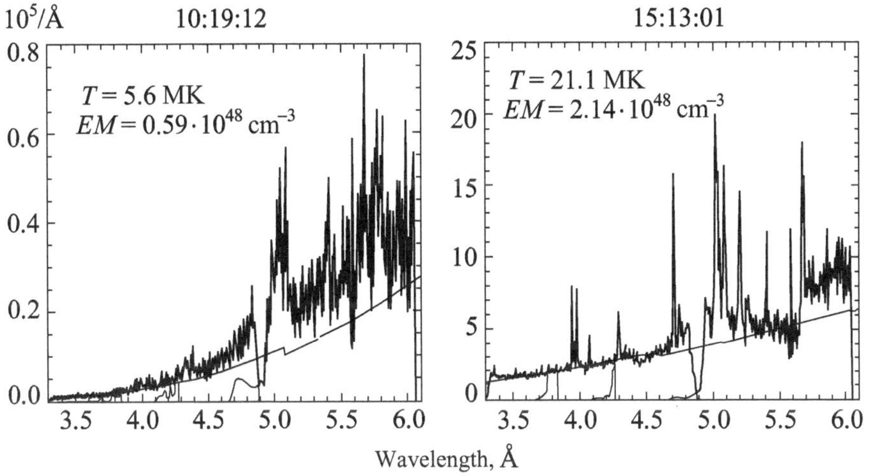


**Fig. 4** Comparison of spectra obtained by RESIK on January 21, 2003, in the conditions of quiet Sun (on the *left*) and during M1.9 class flare (on the *right*). The *thin line* below the spectra shows the level of continuous radiation, calculated with CHIANTI code for temperature and emission measure, indicated in the drawing. The spectra vary significantly in terms of intensity and ratios of the intensities in individual spectral lines. From [22]

parameters of plasma source (temperature $T$ and emission measure EM), indicated in the figure. These parameters have been determined based on ratio of fluxes in the first and fourth RESIK spectrometer channels, with the assumption that radiating plasma is isothermal. The time dependence of intensity changes recorded in a number of well-seen spectral lines is shown in Fig. 5. For clarity the zero levels of every curve have been vertically shifted in a manner to prevent their overlapping. It may be seen that while the radiation in "hot" lines decays faster (e.g., Ar XVIII, Ar XVII, K XVIII), it decays much slower in the lines formed in cooler plasma.

During ground calibration of the spectrometer, the accurate spectral reference wavelength scale has been established. The numerical model of every device sections has been applied in this respect. The accuracy of determination of incident positions along the detector and crystals in this model is better than 0.1 mm. The wavelengths scale accuracy is determined by such factors as crystal curvature radius, relative position of crystal and detector window, and linearity of electronic system against the photon incidence position. One of the most important physical factors influencing the wavelength scale accuracy is the inclination of emission source relative to diffraction plane. Depending on the position of active region or flare on the Sun, the observed positions of lines may move for several bins along the detector anode length.

The precise wavelength calibration as well knowledge of orientation of satellite axis relative to the center of the disc of the Sun has allowed the determinations of wavelengths of single spectral lines with accuracy of ~0.001 Å. It made possible to compare the values of wavelengths of observed lines with the data calculated

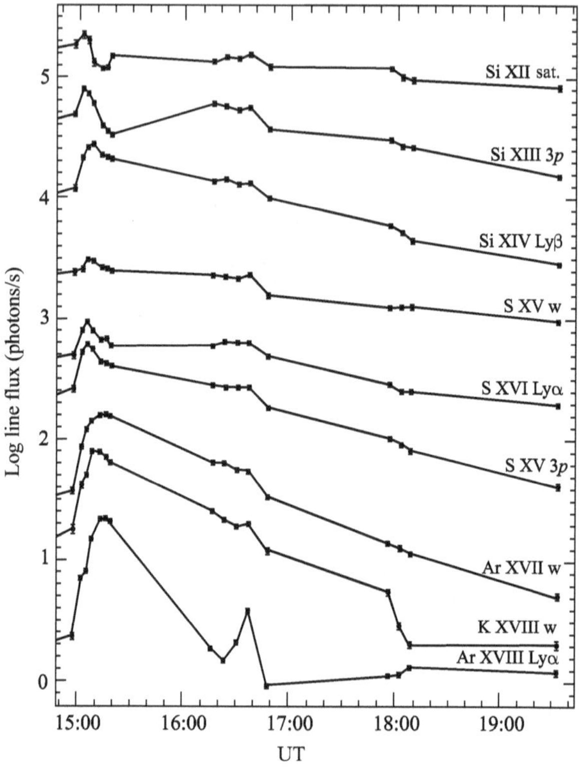


**Fig. 5** Time variations of fluxes in individual spectral lines, observed in RESIK spectra during the flare of January 21, 2003. This flare of M1.9 class occurred at the limb of the Sun. Individual curves are vertically shifted in order to increase visibility. From [22]

*theoretically* from atomic physics. The good coincidence of wavelengths between the two is the base for identification of new spectral features, presented in Fig. 3. The observed spectrum contains the lines corresponding to principal resonance transitions in hydrogen-like ions of Al, Si, S, and Ar and helium-like ions of Si, S, Cl, Ar, and K. The special achievement was identification for the first time of the new chlorine [20] and potassium [9] lines. The complete list of identified spectral lines is specified in Table 2 (for the first reflection order) and Table 3 (for the third reflection order). The table specifies the wavelength, temperature at maximum of emission function, the responsible ion, and relevant transition.

The obtained spectra show the lines corresponding to the transitions to low excited levels $n = 2$, which have been observed previously, and to high ones, $n = 10$, which have not been previously observed in the astrophysical plasmas. Moreover, the spectra show the forbidden and intercombination lines, corresponding to helium-like ions, as well as multiple satellites formed in the processes of dielectronic recombination and excitation of inner shell electrons. The values of relative intensities of lines of higher components of line series, for unknown reasons, are higher than theoretically predicted values for thermal plasma [6].

In the course of processing of spectra it appeared that the ratio of intensity of pair of Si lines (satellite line $\lambda = 5.82$ Å and resonance line $\lambda = 5.68$ Å), which are

**Table 2** The principal emission lines observed in the first reflection order spectra

| Number | $\lambda$ (Å) | $T$ (MK) | Ion | Transition |
|---|---|---|---|---|
| 1 | 3.36 | 22 | Ar XVII | $1s^2\ ^1S_0–1s3p\ ^{1,3}P_1$ |
| 2 | 3.43 | 19 | Ar XVI | Satellite to [1] |
| 3 | 3.53 | 26 | K XVIII | $1s^2\ ^1S_0–1s2p\ ^1P_1$ |
| 4 | 3.57 | 19 | K XVIII | $1s^2\ ^1S_0–1s2s\ ^3S_1$ |
| 5 | 3.59 | 17 | K XVII | Satellite to [3] |
| 6 | 3.70 | 26 | S XVI | $1s\ ^2S_{1/2}–5p\ ^2P_{3/2,1/2}$ |
| 7 | 3.74 | 36 | Ar XVIII | $1s\ ^2S_{1/2}–2p\ ^2P_{3/2,1/2}$ |
| 8 | 3.79 | 26 | S XVI | $1s\ ^2S_{1/2}–4p\ ^2P_{3/2,1/2}$ |
| 9 | 3.95 | 22 | Ar XVII | $1s^2\ ^1S_0–1s2p\ ^1P_1$ |
| 10 | 3.97 | 22 | Ar XVII | $1s^2\ ^1S_0–1s2p\ ^3P_{1,2}$ $(x+y)$ |
| 11 | 3.99 | 22 | Ar XVII | $1s^2\ ^1S_0–1s2s\ ^3S_1$ |
| 12 | 4.02 | | ? | |
| 13 | 4.09 | 16 | S XV | $1s^2\ ^1S_0–1s4p\ ^1P_1$ |
| 14 | 4.10 | 14 | S XIV | Satellite to $1s^2\ ^1S_0–1s5p\ ^1P_1$ |
| 15 | 4.18 | 35 | Cl XVII | $1s\ ^2S_{1/2}–2p\ ^2P_{3/2,1/2}$ |
| 16 | 4.30 | 15 | S XV | $1s^2\ ^1S_0–1s3p\ ^1P_1$ |
| 17 | 4.39 | 14 | S XIV | Satellite to [16] |
| 18 | 4.44 | 19 | Cl XVI | $1s^2\ ^1S_0–1s2p\ ^1P_1$ |
| 19 | 4.73 | 24 | S XVI | $1s\ ^2S_{1/2}–2p\ ^2P_{3/2,1/2}$ |
| 20 | 4.77 | | ? | |
| 21 | 4.83 | 14 | Si XIV | $1s\ ^2S_{1/2}–5p\ ^2P_{3/2,1/2}$ |
| 22 | 4.96 | 24 | Si XIV | $1s\ ^2S_{1/2}–4p\ ^2P_{3/2,1/2}$ |
| 23 | 5.04 | 13 | S XV | $1s^2\ ^1S_0–1s2p\ ^1P_1)$ |
| 24 | 5.10 | 14 | S XV | $1s^2\ ^1S_0–1s2s\ ^3S_1)$ |
| 25 | 5.22 | 14 | Si XIV | $1s\ ^2S_{1/2}–3p\ ^2P_{3/2,1/2}$ |
| 26 | 5.28 | 15 | Si XIII | $1s^2\ ^1S_0–1s5p\ ^1P_1$ |
| 27 | 5.40 | 11 | Si XIII | $1s^2\ ^1S_0–1s4p\ ^1P_1$ |
| 28 | 5.59 | | ? | |
| 29 | 5.68 | 10 | Si XIII | $1s^2\ ^1S_0–1s3p\ ^1P_1$ |
| 30 | 5.82 | 7 | Si XII | Satellite to [29] |
| 31 | 5.87 | | ? | |

clearly distinguished from the background, is a good indicator of the temperature of plasma, representing conditions in the quiet corona. The intensity ratio of these two lines decreases by an order of magnitude when plasma temperature increases from 3 to 10 MK [10]. It was predicted by the theory, but was for the first time observed on RESIK spectra. Sometimes the observed intensity of satellite line is higher than theoretical value for thermal plasma, which may evidence the existence of additional non-thermal processes contributing to satellite lines formation.

The solar plasma, besides the temperature $T$, is also characterized by so-called emission measure $\mathrm{EM} = N^2V$, which, with the assumption of constant volume $V$ of the flare during the evolutions, gives the density $N$ of radiating plasma. Another important characteristic parameter for radiating plasma is the so-called thermodynamic measure $\mathrm{ThM} = T \times (\mathrm{EM})^{1/2}$ [18], which is linearly related to the

**Table 3** The principal emission lines observed in the third reflection order spectra

| Number | $\lambda$ (Å) | $T$ (MK) | Ion | Transition |
|---|---|---|---|---|
| 1 | 1.457 | | ? | |
| 2 | 1.542 | 180 | Ni XXVIII | $1s\ ^2S_{1/2}$–$2p\ ^2P_{3/2,1/2}$ |
| 3 | 1.571 | 60 | Fe XXV | $1s^2\ ^1S_0$–$1s3p\ ^1P_1$ |
| 4 | 1.591 | 80 | Ni XXVII | $1s^2\ ^1S_0$–$1s2p\ ^1P_1$ |
| 5 | 1.750 | | Fe K $\beta$ | |
| 6 | 1.781 | 150 | Fe XXVI | $1s\ ^2S_{1/2}$–$2p\ ^2P_{3/2,1/2}$ |
| 7 | 1.850 | 54 | Fe XXV | $1s^2\ ^1S_0$–$1s2p\ ^1P_1$ |
| 8 | 1.910 | 30 | Fe | Multiple satellites |
| 9 | 1.980 | | ? | |
| 10 | 2.025 | | ? | |

amount of thermal energy of the flare plasma. The average values of $T$ and EM for the above-mentioned strong flare of January 21, 2003, were determined for 17 time intervals based on the ratios of total fluxes (including the continuous radiation and spectral lines) measured in two adjacent channels of RESIK spectrometer. The theoretical spectrum calculated with the application of CHIANTI code was used [3]. The comparison of time dependence of thermodynamic measure ThM and temperature $T$ shows that ThM maximum occurs 10 min later than observed maximum of plasma temperature $T$.

In the research works [5, 11, 17] was indicated that it is convenient to use the so-called diagnostic diagram (DD) for the investigation of energy release and cooling plasmas. In fact DD represents the temperature–density diagram. Figure 6 represents such a diagram for the analyzed flare; the diagram shows two theoretical straight lines: the first one (OFF), with inclination 2, corresponds to evolution in case of sudden switched-off of plasma heating in the flare, the second one (QSS), with inclination 1/2, corresponds to quasi-steady evolution with gradually decreasing heating. The zigzag line between OFF and QSS lines links the observation points. For flare decay phase it is located along the straight line with inclination $\sim$1.1. It indicates that during flare decay phase the significant heating of plasma was still operating.

Many interpretations in the area of solar plasma diagnostics use the so-called one-temperature model. The heavy line in Fig. 7 shows a small interval of spectrum in the range 5–6 Å, obtained by averaging of observations for nine selected flares, which occurred at the beginning of 2003 (from January to March). The thin line shows the theoretical spectrum calculated with CHIANTI code. The applied values of temperature $T$ and emission measure EM have been obtained in one-temperature approximation from the ratios of fluxes in the first (3.3–3.8 Å) and second (3.8–4.3 Å) spectral channels of RESIK instrument. As these two spectra do not coincide, the theoretical calculation has been repeated for $T$ and EM values, obtained from the ratios of fluxes of lines, sensitive to the temperature (shaded in black) [10]. In this case the calculated spectrum (light gray curve) and observed

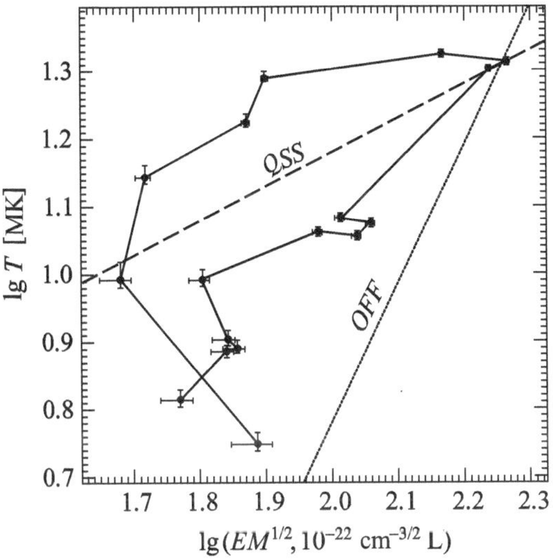


**Fig. 6** Diagnostic diagram (DD) for the flare on January 21, 2003. The location of points obtained from observations in relation to theoretical straight lines QSS (quasi-steady evolution) and OFF (sudden switched-off) defines the character of flare heating evolution. The *zigzag line* between QSS and OFF evidences the energy release in the flare, even during the phase of its late decay. From [22]

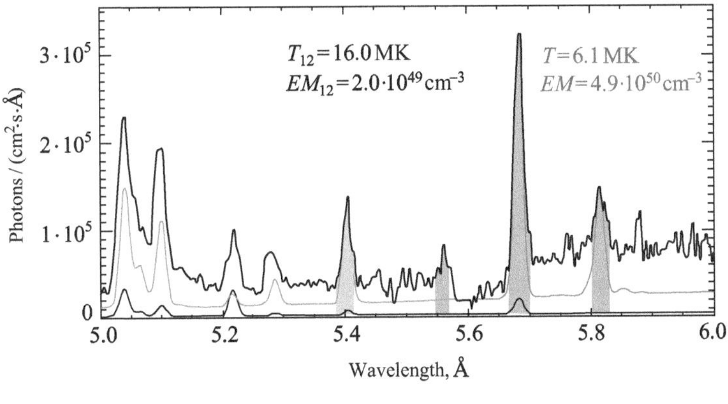


**Fig. 7** The fragment of spectrum (*heavy line*) averaged upon the observations of nine flares. The *thin lines* show the theoretical spectra, calculated with CHIANTI code for indicated parameters of emitting plasma. The disagreement between the theoretical curves and observations evidences the presence of multi-temperature plasma in the flares. From [22]

spectrum also did not coincide. These discrepancies indicate that radiating plasma is multi-temperature. It may be also seen that not all observed lines have their counterparts in the theory spectra calculated using CHIANTI code.

**Table 4** The spectral lines used for DEM($T$) calculations

| Number | Wavelength ranges (Å) | Main lines |
|---|---|---|
| 1 | 3.340–3.368 | Ar XVII $1s^2\ ^1S_0–1s3p\ ^1P_1$ |
| 2 | 3.688–3.699 | S XVI $1s\ ^2S_{1/2}–5p\ ^2P_1$ |
| 3 | 3.724–3.740 | Ar XVIII $1s\ ^2S_{1/2}–2p\ ^2P_{1/2,3/2}$ |
| 4 | 3.777–3.789 | S XVI $1s\ ^2S_{1/2}–4p\ ^2P_{1/2,3/2}$ |
| 5 | 3.944–3.962 | Ar XVII $1s^2\ ^1S_0–1s2p\ ^1P_1$ ($w$) |
| 6 | 3.963–3.980 | Ar XVII $1s^2\ ^1S_0–1s2p\ ^3P_{1,2}$ ($x+y$) |
| 7 | 3.981–4.004 | Ar XVII $1s^2\ ^1S_0–1s2s\ ^3S_1$ ($z$) |
| 8 | 4.076–4.091 | S XV $1s^2\ ^1S_0–1s4p\ ^1P_1$ |
| 9 | 4.288–4.315 | S XV $1s^2\ ^1S_0–1s3p\ ^1P_1$ |
| 10 | 4.720–4.743 | S XVI $1s\ ^2S_{1/2}–2p\ ^2P_{1/2,3/2}$ (Ly$\alpha$) |
| 11 | 5.030–5.061 | S XV $1s^2\ ^1S_0–1s2p\ ^1P_1$ ($w$) |
| 12 | 5.086–5.124 | S XV $1s^2\ ^1S_0–1s2s\ ^3S_1$ ($z$) |
| 13 | 5.204–5.232 | Si XIV $1s\ ^2S_{1/2}–3p\ ^2P_{1/2,3/2}$ (Ly$\beta$) |
| 14 | 5.267–5.290 | Si XIII $1s^2\ ^1S_0–1s5p\ ^1P_1$ |
| 15 | 5.396–5.417 | Si XIII $1s^2\ ^1S_0–1s4p\ ^1P_1$ |
| 16 | 5.669–5.697 | Si XIII $1s^2\ ^1S_0–1s3p\ ^1P_1$ |

For the purpose of detailed study of distribution of solar flare plasma with temperature, it is convenient to use the differential emission measure (DEM), which is described by the relation

$$\varphi(T) = \frac{N^2 dV}{dT},$$

where $T$, $N$, and $V$ are, respectively, the temperature, density, and volume of plasma in a flare. The $\varphi(T)$ distribution shows how much of the emitting plasma is characterized with given temperature $T$. DEM distribution characterizes actual physical conditions in plasma. The data collected by RESIK spectrometer are suitable for DEM distribution analysis, as they contain information about the fluxes of continuous radiation and fluxes of many spectral lines in range 3.2–6.1 Å, which are formed in different temperatures. Using the data for absolute calibration of RESIK device, the fluxes in 16 strong, well-visible lines have been determined. The list of these lines is shown in Table 4. On the basis of the observed fluxes in these lines, the distribution of DEM with the temperature for many flares in different evolutionary phases has been calculated. In fact, the fluxes in lines and the underlying continuum have been used. The iteration algorithm of Withbroe–Sylwester [15] was applied for calculations. The theoretical emission functions for the line and continuum have been calculated using CHIANTI code.

The example of DEM analysis is that performed for the strong flare of M1.9 class on January 21, 2003 at 15:26 UT. The DEM shapes for several phases of flare evolution are shown in Fig. 8. It may be seen that during the whole flare evolution the general shape of distribution may be considered as two-component. These components are clearly separated: the cold plasma with temperature of 5–8 MK and

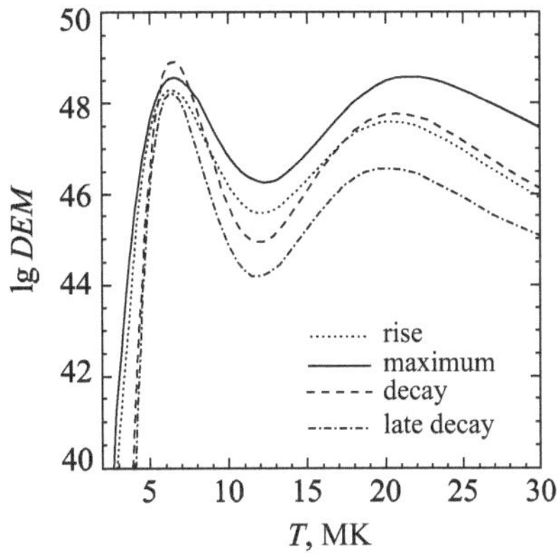


**Fig. 8** The distribution of differential emission measure (DEM) for the limb long-lasting flare on January 21, 2003 at 15:26 UT (M1.9 class). Each individual *curve* corresponds to various phases of flare development. It may be seen that at every evolutionary phase the flare contained both cold and hot plasmas. From [22]

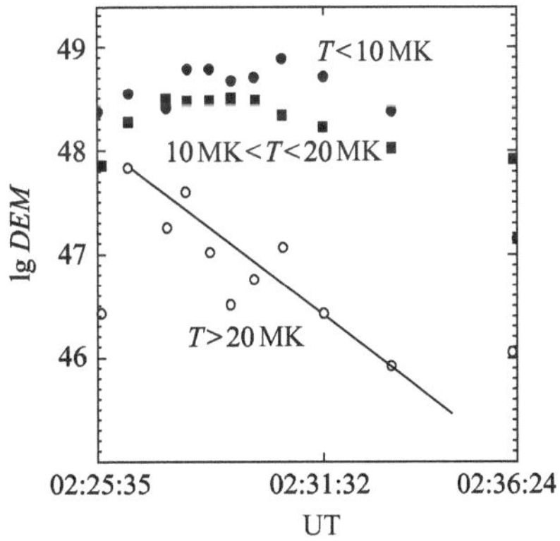


**Fig. 9** Time variations of the amount of plasma at various temperatures. The calculations were performed for the flare of January 21, 2003 at 02:28 UT (C8.1 class). The behavior of hottest component differs significantly from other components. From [7]

hot one with temperature of 18–25 MK. The cold component plasma temperature is almost constant in course of the flare; however, the amount of this plasma is subject to substantial changes. During the decay phase the amount of cold plasma is almost an order of magnitude larger than that of hot one. The amount of hot plasma in course of decay decreases much more (by two orders of magnitude), while the temperature maximum of this component drops from 22 MK at there maximum to 20 MK 3 h later.

The similar DEM distributions have been obtained also for other, weaker, and faster-evolving flares. In particular, there have been performed another calculations of DEM distribution for C8.1 class flare, which occurred on January 21, 2003 at 02:28 UT and lasted only 10 min. To obtain information about the time dependence of behavior of individual DEM components, three plasma component have been identified: the cold plasma with temperature $T < 10$ MK, the component with moderate temperature 10–20 MK, and hot plasma, with temperature above 20 MK. Figure 9 illustrates the time dependence of emission measure (amounts of plasma) in these separate temperature ranges. The hot component emission measure is monotonically decreasing almost from the very beginning of flare. The amount of cooler plasma is changing weakly with time.

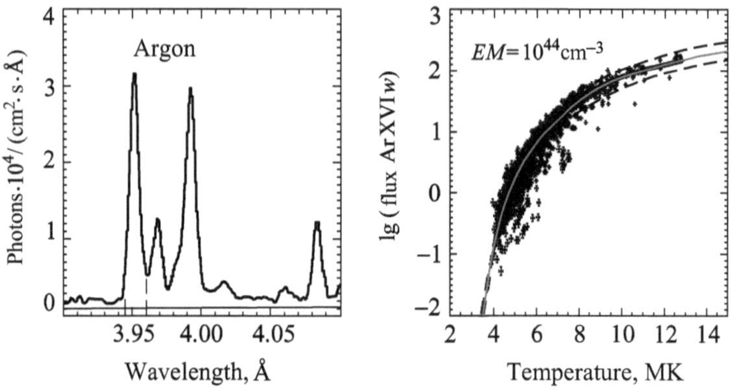


**Fig. 10** On the *left*: the fragment of spectrum obtained by summation of ~1,200 spectra of flares observed by RESIK instrument at the beginning of 2003. The spectrum involves the triplet of helium-like ion Ar XVII. On the base of observations of the fluxes of strong line $w$ of this triplet ($\lambda = 3.95$ Å) the dependence flux-temperature has been obtained, as seen on the *right*. The points are the values determined from observations, the *solid line*—theoretical dependence for ionization equilibrium and average coronal abundance of argon. The *upper* and *lower dashed lines* correspond to calculations for argon abundance, two times more and two times less than the average value respectively

The spectra obtained by RESIK instrument, besides spectral lines, also show clearly the continuous radiation, which is formed in free–free and recombination processes. The level of signal, measured by RESIK spectrometer in the continuum, exceeds significantly (up 100 times) the level of orbital noise. This allows to use the continuum in the analysis of physical parameters of radiating plasma. Thus, the measured relation of intensity of the lines to the level of adjacent continuum allows to determine directly the absolute (in relation to hydrogen) abundances of elements present in hot flaring plasma.

Determination of chemical composition based on the analysis of spectra, obtained by RESIK, leads to the conclusion that the abundance of elements with low first ionization potential (FIP $<$ 10 eV) in flare plasma is higher than abundances characteristic for the photospheric plasma. So far this fact was known from spectral observations only for potassium [16, 19]. The accuracy of obtained results is evidenced by the data presented in Fig. 10. At the left there is a section of spectrum obtained through summation of multitude (~1,200) of spectra, recorded during 400 h of observations, performed by the spectrometer from January to March 2003. This spectrum involves the triplet of argon Ar XVII helium-like ion, with resonance line $w$ corresponding to $1s^2$–$1s2p$ transition. Based on the ratios of total fluxes measured in spectrometer first and fourth channels for these 1,200 spectra there were determined temperature $T$ and total emission measure EM for every spectrum. These parameters have formed a base for the calculation of continuous radiation and normalization for unit emission measure of fluxes observed in argon triplet lines. By dividing the line emission by corresponding emission measure, this allowed to

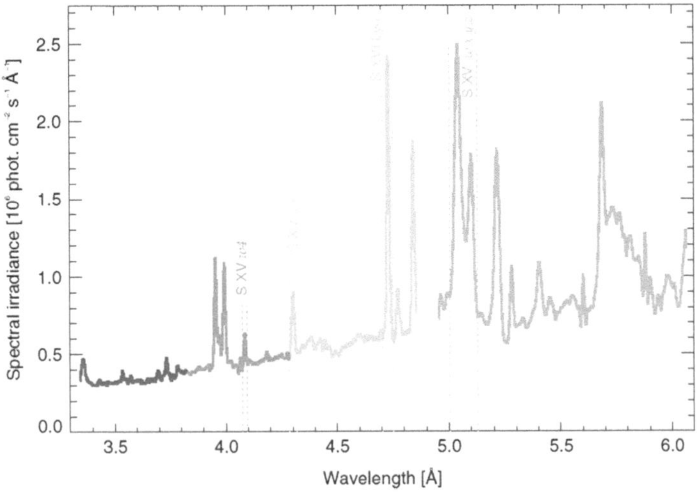


**Fig. 11** Complete RESIK spectrum (channels 1–4) taken during the M4.9 flare on 2003 January 7 over the period 23:29:00–23:32:35 UT. The principal S lines used in the analysis are indicated, with *vertical dashed lines* to show the interval over which the line fluxes were estimated. From [24]

accommodate the dependence of flux in a spectral line on the amount of plasma in the source. The dependence of thus normalized line flux on the temperature is shown at the right part of Fig. 10. The points correspond to individual spectra of flares and non-flaring active regions, measured at different physical conditions on the Sun. It may be seen that the points proceed along the characteristic curve. Three continuous lines presented in the figure have been obtained theoretically for ionization equilibrium [8] and assuming three various values of argon abundance: $5.64 \times 10^{-6}$ (upper), $2.82 \times 10^{-6}$ (medium-solid), and $1.41 \times 10^{-6}$ (lower-dashed line).

Using a similar spectroscopic method we determined the abundances of other elements for which the lines have been identified on RESIK spectra. In Fig. 11 the complete spectrum for the M4.9 flare on 2003 January 7 integrated over the 215 s is shown.

Different shades of gray indicate for the individual RESIK channels. The principal lines of the H- and He-like ions (S XVI and S XV) are noted and vertical dashed lines limit the wavelength intervals over which the line fluxes were estimated. For the sulfur abundance determinations we have used S XV *w4* line in the range 4.075–4.095 Å, S XV *w3* line in the range 4.285–4.320 Å, triplet ($w$, $x$, $y$, $z$) of S XV ion in the range 5.006–5.14 Å, and Ly-$\alpha$ of S XVI ion in the range 4.717–4.745 Å. The 1,448 spectra for 13 flares observed in 2003 have been used. As the most reliable we consider the value $A(\mathrm{S}) = 7.16 \pm 0.17$ (on a logarithmic scale with $H = 12$) as determined from *w4* line analysis ($1s^2$–$1s4p$ transition in S XV ion) seen in channel 2 of RESIK. In Fig. 12 the respective comparison of observations and theory is presented.

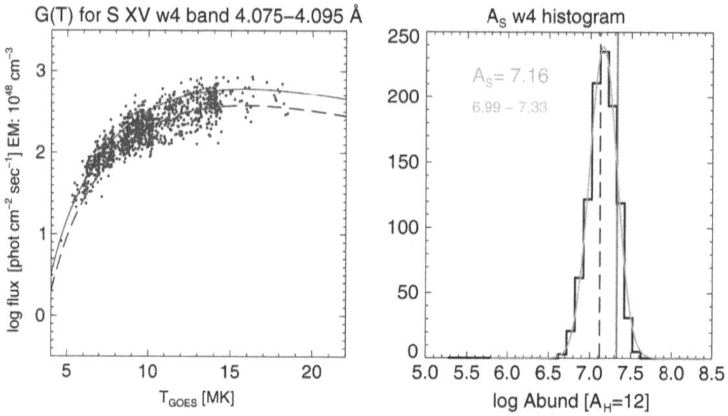


**Fig. 12** *Left panel*: measured S XV *w4* line emission (in RESIK channel 2 range 4.075–4.095 Å) divided by emission measure plotted against $T$. The *dashed curve* is the theoretical $G(T)$ function for the photospheric S abundance ($A(\mathrm{S}) = 7.12$ [1]), the *solid curve* is for a nominal coronal abundance ($A(\mathrm{S}) = 7.33$ [4]). *Right panel*: distribution of abundance determinations with best-fit Gaussian curve. From the peak and width of the Gaussian, the S abundance is determined to be $A(\mathrm{S}) = 7.16 \pm 0.14$. The *vertical dashed* and *solid lines* correspond to the photospheric and coronal abundance, respectively. From [24]

In the left panel the measured S XV *w4* line flux divided by the emission measure is displayed for each individual spectrum vs. the isothermal temperature $T$. The dashed curve is the theoretical $G(T)$ function for the photospheric S abundance $A(\mathrm{S}) = 7.12$ while the solid curve is for a nominal coronal abundance $A(\mathrm{S}) = 7.33$. In the right panel the distribution of S abundance determinations with the best fit Gaussian curve is shown. From the peak and width of the Gaussian, the value of S abundance and error limits are determined. The vertical dashed and solid red lines correspond to the photospheric and coronal abundances, respectively. Estimates from other sulfur lines seen by RESIK range from 7.13 to 7.24. The preferred S abundance estimate is close to photospheric and to quiet-Sun solar wind. Described abundance determination method has been also used for the analysis of potassium (K) abundance [23]. Potassium is of importance in the context of FIP effect as the value of FIP for K is very low (4.34 eV). The fluxes of the group of K XVIII lines at 3.53 and 3.57 Å ($w$, $x$, $y$, $z$ including unresolved dielectronic satellites) for 2,795 spectra have been analyzed in this respect. Based on the He-like K resonance line the value of K abundance $A(\mathrm{K}) = 5.86$ with the total range of variability by a factor of 2.9 has been obtained. This is much higher by a factor 5.5 than photospheric abundance estimates, a slightly greater enhancement than for other elements with FIP less than $\sim$10 eV.

The great number of spectra obtained by RESIK spectrometer will be analyzed in future with the purpose of identification of changes of plasma thermodynamic parameters and abundance of elements during individual flares.

**Acknowledgements** The works with RESIK spectrometer have been carried out with the support of Polish Scientific Research Fund (project 2.P03C.005.08 for technical design; project 1.P03D.017.29 and 2011/01/B/ST9/05861 for data processing).

# References


1. Asplund, M., Grevesse, N., Sauval, A.J., Scott, P.: The chemical composition of the Sun. Annu. Rev. Astron. Astrophys. **47**, 481–522 (2009)
2. Culhane, J.L., Hiei, E., Doschek, G.A., Cruise, A.M., Ogawara, Y., Uchida, Y., Bentley, R.D., Brown, C.M., Lang, J., Watanabe, T., Bowles, J.A., Deslattes, R.D., Feldman, U., Fludra, A., Guttridge, P., Henins, A., Lapington, J., Magraw, J., Mariska, J.T., Payne, J., Phillips, K.J.H., Sheather, P., Slater, K., Tanaka, K., Towndraw, E., Trow, M.W., Yamaguchi, A.: The Bragg crystal spectrometer for solar-A. Sol. Phys. **136**, 89 (1991)
3. Dere, K.P., Landi, E., Mason, H.E., Monsignori Fossi, B.C., Young, P.R.: CHIANTI—an atomic database for emission lines. Astron. Astrophys. Suppl. **125**, 149 173 (1997)
4. Feldman, U., Laming, J.M.: Element abundances in the upper atmospheres of the sun and stars: update of observational results. Phys. Scr. **61**, 222–252 (2000)
5. Jakimiec, J., Sylwester, B., Sylwester, J., Serio, S., Peres, G., Reale, F.: Dynamics of flaring loops II. Flare evolution in the density–temperature diagram. Astron. Astrophys. **253**, 269–276 (1992)
6. Kepa, A., Sylwester, J., Sylwester, B., Siarkowski, M., Pillips, K.J.H., Kuznetsov, V.D.: Observations of $1s^2$-1s(np) and 1s-np lines in RESIK soft X-ray spectra. Adv. Space Res. **38**(7), 1538–1542 (2006)
7. Kepa, A., Sylwester, J., Sylwester, B., et al.: Sol. Syst. Res. **40**(4), 294–301 (2006). doi:10.1134/S0038094606040058
8. Mazzotta, P., Mazzitelli, G., Colafrancesco, S., Vittorio, N.: Ionization balance for optically thin plasmas: rate coefficients for all atoms and ions of the elements H to NI. Astron. Astrophys. Suppl. **133**, 403–409 (1998)
9. Phillips, K.J.H., Sylwester, J., Sylwester, B., Landi, E.: Solar flare abundances of potassium, argon and sulphur. Astrophys. J. **589**, 113–116 (2003)
10. Phillips, K.J.H., Dubau, J., Sylwester, B., Sylwester, J., Culhane, J.L., Doschek, G.A., Lang, J.: Temperature-sensitive line ratios diagnostics of the corona based on satellite-to-resonance line ratios of $1s^2$-1s(np) transitions. Adv. Space Res. **38**(7), 1543–1546 (2006)
11. Serio, S., Reale, F., Jakimiec, J., Sylwester, B., Sylwester, J.: Dynamics of flaring loops I. Thermodynamic decay scaling laws. Astron. Astrophys. **241**, 197–202 (1991)
12. Siarkowski, M., Sylwester, J.: The analysis of MG XI ion X-ray spectra obtained from Vertical 9 rocket experiment. Artif. Satellites Space Phys. **20**, 63–70 (1985)
13. Sylwester, J.: Experience of Wrocław team in construction of solar X-ray instruments. ESA SP **493**, 377–382 (2001)
14. Sylwester, J., Farnik, F.: DIOGENESS Soft X-ray spectrometer–photometer for studies of flare energy balance. Astron. Inst. Czech. **41**(3), 149–157 (1990)
15. Sylwester, J., Schrijver, J., Mewe, R.: Multitemperature analysis of solar X-ray line emission. Sol. Phys. **67**, 285–309 (1980)
16. Sylwester, J., Lemen, J.R., Mewe, R.: Variation in observed coronal calcium abundance of X-ray flare plasmas. Nature **310**, 665–666 (1984)
17. Sylwester, B., Sylwester, J., Serio, S., Reale, F., Bentley, R.D., Fludra, A.: Dynamics of flaring loops III. Interpretation of flare evolution in the emission measure-temperature diagram. Astron. Astrophys. **267**, 586–594 (1993)
18. Sylwester, J., Garcia, H.A., Sylwester, B.: Quantitative interpretation of GOES soft X-ray measurements. Astron. Astrophys. **293**, 577–585 (1995)

19. Sylwester, J., Lemen, J.R., Bentley, R.D., Fludra, A., Zolcinski, M.C.: Detailed evidence for flare-to-flare variations of the coronal calcium abundance. Astrophys. J. **501**, 397–407 (1998)
20. Sylwester, B., Sylwester, J., Siarkowski, M., Phillips, K.J.H., Landi, E.: Detection of H- and He-like resonance lines of chlorine in solar flare spectra. Proc. Int. Astronomical Union **2004**(223), 671–674 (2004). Copyright © International Astronomical Union. doi:http://dx.doi.org/10.1017/S174392130400729X
21. Sylwester, J., Gaicki, I., Kordylewski, Z., Kowalinski, M., Nowak, S., Plocieniak, S., Siarkowski, M., Sylwester, B., Trzebinski, W., Bakala, J., Culhane, J.L., Whyndham, M., Bentley, R.D., Guttridge, P.R., Phillips, K.J.H., Lang, J., Brown, C.M., Doschek, G.A., Kuznetsov, V.D., Oraevsky, V.N., Stepanov, A.I., Lisin, D.V.: RESIK: a bent crystal X-ray spectrometer for studies of solar coronal plasma composition. Sol. Phys. **226**(1), 45–72 (2005). doi:10.1007/s11207-005-6392-5
22. Sylwester, B., Sylwester, J., Kepa, A., et al.: Sol. Syst. Res. **40**(2), 125-132 (2006). doi:10.1134/S0038094606020067
23. Sylwester, J., Sylwester, B., Phillips, K.J.H., Kuznetsov, V.D.: Highly ionized potassium lines in solar X-ray spectra and the abundance of potassium. Astrophys. J. **710**, 804–809 (2010)
24. Sylwester, J., Sylwester, B., Phillips, K.J.H., Kuznetsov, V.D.: The solar flare sulphur abundance from RESIK observations. Astrophys. J. **751**, 103–109 (2012)

# The Study of the Cosmic Gamma-Emission Nonstationary Fluxes Characteristics by the AVS-F Apparatus Data

**Yu.D. Kotov, I.V. Arkhangelskaja, A.I. Arkhangelsky, S.N. Kuznetsov[1], A.S. Glyanenko, P.A. Kalmykov, D.B. Amandzholova, V.T. Samoylenko, V.N. Yurov, A.V. Pavlov, O.I. Chervyakova, and I.V. Afonina**

The AVS-F apparatus (Russian abbreviation for Amplitude-Time Spectrometry of the Sun) [7, 8] is intended for the solar flares hard X-ray and gamma-ray emission characteristics studies and for the search and detection of the gamma-ray bursts (GRB). The experiment was carried out in the framework of the international program CORONAS(Complex ORbital Near-Earth Observations of the Active Sun) [56] on board the specialized automatic station CORONAS-F [74–76] successfully launched on July 31, 2001. The satellite operated until December 6, 2005.

The AVS-F apparatus [8, 44] provided information for the analysis of the following problems:

- definition of the solar flares gamma-emission variability minimum time;
- study of charged particles' acceleration dynamics during the flare by the energy spectra sequence analysis;
- studies of the particles acceleration and propagation mechanism during the flare basing on gamma lines from excited nuclei, neutron capture and positron annihilation analysis;

---

[1]Deceased

Yu.D. Kotov · I.V. Arkhangelskaja (✉) · A.I. Arkhangelsky · A.S. Glyanenko · P.A. Kalmykov · D.B. Amandzholova · V.T. Samoylenko · V.N. Yurov · A.V. Pavlov · O.I. Chervyakova · I.V. Afonina
National Research Nuclear University "MEPhI", Kashirskoe shosse, 31, Moscow 115409, Russia
e-mail: YuDKotov@mephi.ru; Irene.belousova@usa.net

S.N. Kuznetsov
D. V. Skobeltsyn Institute of Nuclear Physics M. V. Lomonosov Moscow State University (SINP MSU), Vorobyevy Gory, Moscow 119991, Russia

V. Kuznetsov (ed.), *The Coronas-F Space Mission*, Astrophysics and Space Science Library 400, DOI 10.1007/978-3-642-39268-9_7,

- search for temporal correlations with fine time resolution between solar flares emission in gamma-ray band and other energy ranges;
- analysis of the solar flares X-ray precursors behavior and their energy spectra;
- non-thermal emission lower boundary refinement;
- neutron fluxes measuring by analysis of a sequential sets of two-dimensional distributions of the relation between the slow light output component to total light output ratio and total energy deposition in the CsI(Tl) scintillation detector;
- background conditions analysis.

## AVS-F Apparatus Short Description

The AVS-F apparatus continued an experiment AVS conducted on board the satellite CORONAS-I [42, 71]. As well as for the AVS experiment of the CORONAS-I project, the apparatus AVS-F was intended for processing of the information from the SONG-D detector of the SCR complex, developed by SINP MSU (see description in the chapter "Scientific Set of Instruments "Solar Cosmic Rays"") and XSS-1 detector also named RPS-1 detector in Russian publications (see description in the chapters "RPS-1 Experiment").

The SONG-D detector (Solar Neutrons and Gamma-quanta) is a scintillation detector on the basis of CsI(Tl) crystal with a diameter of 200 mm and a height of 100 mm surrounded with plastic anticoincidence detector [83]. Using X-ray semiconductor spectrometer XSS-1 [90,91,105] on the basis of CdTe semiconductor with dimensions of $4.9 \times 4.9$ mm and thickness of 2.5 mm was conceptually new compared with the AVS experiment on board the CORONAS-I satellite. CdTe semiconductor detector does not need deep cooling (down to $\approx -30$ °C) in order to provide sufficiently high energy resolution. It provides high registration efficiency and radiation resistance, along with other advantages of the semiconductor detectors, and are operated at low power supply voltage. The XSS-1 was developed by MEPhI and IKI RAS in cooperation. The results of detailed data analysis for the XSS-1 detector are presented in the chapter "RPS–1 Experiment."

The AVS-F apparatus was the system of electronics for on board data acquisition from two detectors: SONG-D in the low-energy (0.1–10 MeV) and high-energy (2–80 MeV) bands of gamma-emission, and XSS-1 in the energy range of 3–30 keV of X-rays. The AVS-F apparatus operational ranges boundaries are presented according to the data of preflight calibration (about the changes of the operation ranges during the experiment see the section "In-Flight Calibration in $\gamma$-Bands").

The AVS-F apparatus had two operation modes: background mode for the analysis of radiation flux and the search for event of "burst" type, and burst recording mode switched in after such an event was identified (for the detailed description see p. 181).

The AVS apparatus was completely modified in order to provide complex processing of the data from two detectors simultaneously in three energy ranges [42, 71], and the part of the device providing acquisition of spectrometry information [8] was changed most strongly. The information from the low-energy channels (the XSS-1 and SONG-D detectors) was processed by analog-to-digital converters with 512 primary channels, and the data in the high-energy ranges and separation of the neutrons and gamma-quanta using the ratio of light output components in CsI(Tl) (from the SONG-D)—by specific ADC converters with 256 primary channels. In order to decrease the amount of the transmitted information ADC channels were compressed in a quasi-logarithmic algorithm. There were 114 channels in the low-energy range (32 for the XSS-1 and 82 for the SONG-D detectors) and 64 channels for the background mode and 114 channels for the burst mode for the SONG-D detector in the high-energy range. The more detailed description of the information obtained from the apparatus AVS-F is presented in page 178.

The discrimination between gamma-emission and neutron events was performed using the selection of events by the pulse shape analysis in the scintillation detector, based on the dependence of the ratio of intensities of light-output components in CsI(Tl) scintillator with different decay times to the average ionization density produced by charged particles in the detector material. Scintillation flare in CsI(Tl) consists of two basic components, and there is relation between the ratio of the slow component intensity $L_{sl}$ to the fast one $L_f$ and the mean specific ionization of the detected particle (measured in units MeV/g/cm$^2$) [41,46]:

$$L_{sl}/L_f = (0.96 \pm 0.03) - (0.11 \pm 0.01) \ln(dE/dx), \quad (1)$$

where $dE/dx$—the mean specific ionization.

The light flare energy is proportional to the total charge of the $RC$-chains capacitors, which compose the output cascade of the front-end electronics. The charge of the separate $RC$-chain characterizes the fast ($\tau_f \approx 0.7$ mks) or slow ($\tau_{sl} \approx 7$ mks) component of the CsI(Tl) scintillator light output. The method used in the AVS-F apparatus is based on the integration of the signal from the photomultiplier of the SONG-D in two different time intervals. The first interval corresponds to the period of the total charge $Q_{tot} = Q_f + Q_{sl}$ collection lasts from the pulse leading edge till 10 mks, and the second time interval was the time during which the slow fluorescence component was collected, and lasts from 1 mks after the leading edge of the pulse till 10 mks. For each detected event the values $Q_{tot}$ and $Q_{sl}/Q_{tot}$ were converted by 8-bit analog-to-digital converters and transmitted to the micro-controller KS-MP [30].

The data obtained by the AVS-F apparatus was transmitted by means of two telemetric systems: RadioTelemetry System RTS-9 and the system of information acquisition SSNI (see the chapter "On-Board and Ground-Based Complexes for Operating the Science Payload of the CORONAS-F Space Mission").

System control was accomplished by ten commands [7]. The telemetry system had following features:

- analogous channels—4;
- digital channels—9;
- quota SSNI—8 MB/day.

The AVS-F apparatus provided [8,44]:

1. acquisition and outputting of the following information to the radiotelemetry system (sample frame for RTS-9 is 2.5 s):
   - counts rate in four energy ranges of 3–5 keV, 5–10 keV, 10–20 keV, and 3–30 keV [105] from the XSS-1 detector;
   - total counts rate from the SONG-D detector in the low-energy and high-energy gamma-ranges;
   - total counts rate for the channel corresponding to the two-dimensional distributions of the relation between the ratio of the slow fluorescence component to the total light output and total energy release for the SONG-D detector;
   - total counts rate of the anticoincidence detectors of the SONG-D detector;
2. acquisition and outputting of the following information to the scientific telemetry:
   - in the background mode:
   - background energy spectra in the X-ray band (from the XSS-1 detector) and low-energy gamma-emission (from the SONG-D detector) (the integration time is 16 s, number of channels of amplitude analysis after compression—114, including 32 channels of spectra from the XSS-1 detector, 82 channels of spectra from the SONG-D detector);
   - a set of two-dimensional distributions of the relation between the ratio of the slow fluorescence component to the total light output and total energy release for the SONG-D detector, which are used for the investigation of the neutron flux (matrix size is $64 \times 64$, integration time is 128 s);
   - test arrays containing information about the available operation modes of the apparatus and the results of apparatus self-testing;
   - in the burst mode:
   - energy spectra in X-ray (from the XSS-1 detector) and low-energy gamma-ranges (from the SONG-D detector) (the integration time is 1 s, number of channels of the amplitude analysis after compression—114, including 32 channels of spectra from the XSS-1 detector, 82 channels of spectra from the SONG-D detector);
   - burst event temporal profile in the low-energy gamma-range with 1 ms time resolution (4,096 channels of temporal profile);
   - energy spectra in the high-energy gamma-range (from the SONG-D detector) (integration time is 4 s, number of channels of amplitude analysis after compression—114);
3. generation of a signal "burst" during the operation in the background mode in the case of the gamma-emission flux intensity sharp increase with specific time of 8–16 ms on the value more than 5 $\sigma$ and switching to the burst operational mode basing on the analysis of this information.

Users receive the description of the transmitted information formats along with the requested data. The information for the determination of the spacecraft position

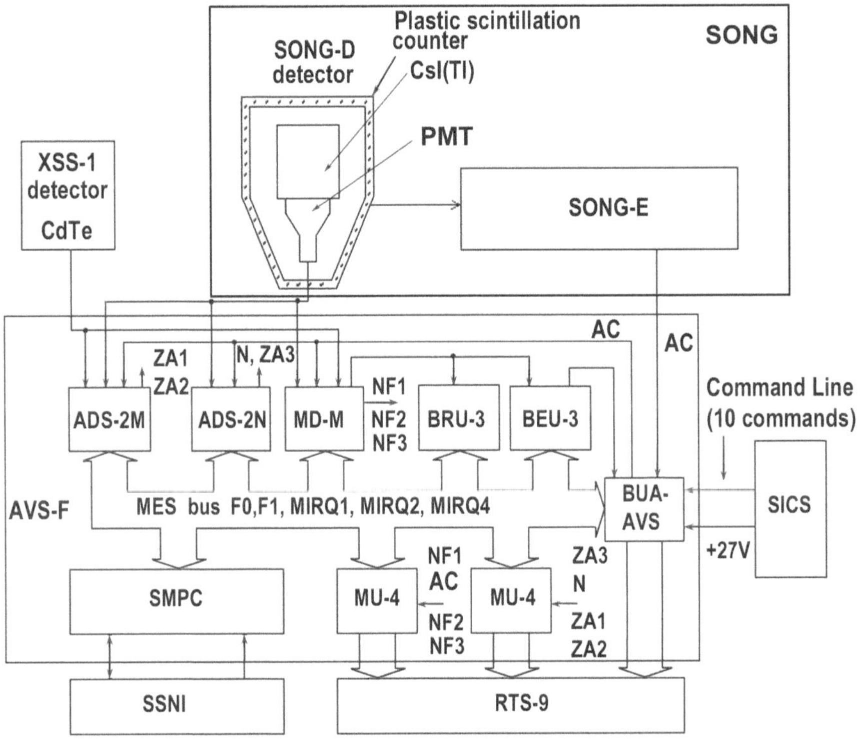


**Fig. 1** The AVS-F experiment functional diagram. From [30]

at the orbit according to the NORAD data (http://www.celestrak.com/NORAD/documentation/tle-fmt.asp) is also provided.

The AVS-F apparatus was developed basing on the block system of electronics [7, 8, 90] and included electronics crate with power supplies and a set of functional blocks [30]. Functional diagram of the experiment AVS-F is presented in Fig. 1.

Each block of electronics system is the plate with dimensions of $160 \times 180$ mm with the front panel 20 mm width (or multiple) intended for fitting of connectors and block's mounting in the crate.

Overall dimensions of the apparatus are $275 \times 400 \times 190$ mm, mass is $\sim 12$ kg. For the regular operation of the apparatus DC supply $27^{+7}_{-3}$ V was used. Power consumption did not exceed 22 W.

**KS-MP block** was a system's controller and was intended for the control, acquisition, and processing of the information, and for receiving of the digitized time label from the specific telemetry system SSNI and for data transfer to SSNI [30, 63]. Information was output to SSNI which serviced scientific apparatus according to the requests of electronic systems of the experiments. Data exchange was performed by the arrays of 960 bit frames through the synchronous sequence channel. Besides, the apparatus received 32-digit time label with the value of the lower order bit of 1/1024 s from SSNI. Microprocessor controller KS-MP provided acquisition, accumulation, primary processing of the information, generation of output data arrays, on board time referencing of the information, and data output to the SSNI system.

**ADC-2N block** was the part of modular electronic system intended for operation with CsI(Tl) detector. ADC-2N provided separation of the signals obtained from the detector in order to get information on the neutron flux ($n$) and gamma-emission flux ($\gamma$) by pulse shape analysis at the output of the SONG-D detector.

The objective of the ADC-2N included:

- outputting of counting impulses corresponding to the total number of the events ($n + \gamma$) and to the number of the events satisfying selection criteria according to the ratio $Q_{sl}/Q_{tot}$ (neutrons);
- outputting of the information (8 bit) on the status of full charge integrator and ratio of the charge corresponding to the slow light output component to the total charge to the modular electronic system data bus;
- identifying of signal overlapping and resetting in this case;
- exchange of service signals with modular electronic system in order to manage ADC-2N operation;
- resetting of process cycle according to the signal of the charged particle passing.

**ADC-2M block** was the part of modular electronic system intended for operation with SONG-D and XSS-1 detectors. Charge integration intervals for two different channels were selected individually and amounted to 7–10 mks to provide charge collecting from the detectors (for the SONG-D detector $\sim$ 10 mks; for the XSS-1 detector 7–8 mks). The objective of the ADC-2M included:

- outputting of the impulses corresponding to the total number of the events on each of two channels of ADC-2M in the case if there is no restricting signal from the anticoincidence shield of the SONG-D detector. Integration times for charge congregating were selected for each detectors and were 7–10 mks for optimal charge congregating from both detectors. (The main part both slow and total light output components in the CsI(Tl) crystal had de-excitation time $\sim$10 mks; the integration time for the XSS-1 was 7–8 mks);
- outputting the modular electronic system the information (8 bit) corresponding to energy deposition in the XSS-1 and the SONG-D detectors;
- identifying of signal overlapping and resetting in this case;
- exchange of service signals to manage the ADC-2M operation;
- resetting of process cycle according to the signal of the charged particle passing through the detector;

**The MD-M block** was intended for the operation with the XSS-1 detector. The objective of the MD-M was to provide a signal on entry of photons with energy in specified range (one of the following four ranges: 3–5 keV, 5–10 keV, 10–20 keV, and 3–30 keV) into the detector's volume. The data outputted to the block of intensimeters (BPN-4) for the following output to the RTS telemetry.

**The BUA-AVS block** included providing of different operation modes for the conducted devices according to the mode settings commands and using information from other experiments. According to the results of analysis of a set of starting and restricting commands coming to the BUA-AVS from other parts of the CORONAS-F complexes operated with the AVS-F apparatus (HELICON, IRIS, SONG, "Light/Shadow" sensors), individual signal "Burst" was produced in order to set the proper registration modes for the AVS-F apparatus. Also the signals managing the operation of the secondary power supplies of the AVS-F apparatus and control analogous signals from the secondary power supplies block to the service telemetry RTS were transferred through the BUA-AVS. Basing on the status of "switch on/ switch off" contacts of the secondary power supplies block, a set of control digital marks was formed in order to transfer them to the service telemetry system RTS.

**The BNP-4 block** was intended for the obtaining information on the flux of the input signals in the frequency bands up to 2 MHz, transformation of the measurements results into digital code and its outputting to the telemetry system through galvanic decoupler. The BNP-4 consisted of four identic measuring channels; each of them allowed to record the temporal profile of the input signals in the range of 0–15 V.

**The BRV-3 block** was intended for the writing of temporal profiles of gamma-emission with high time resolution. It was based on data storage of $4096 \times 8$ digits capacity.

Data acquisition interval was specified by the external fixed reference frequency.

**The BVV-3 block** was intended for identification of non-stationary events. Its operation principle was based on averaging of the background value for 128 collection intervals composing the cycle of the background collection, and using of the previous background average value as reference for the following collection cycle. Current value exceeding the reference value for a specified value (in this case—for 6 standard deviations) during one of the collection intervals was a mark of the event. In order to set an interval of collection the fixed reference frequency was used. Basic duration of one interval was 1 ms.

## AVS-F Apparatus Background Model in the $\gamma$-Bands

Detector background depends on various factors including satellite mass and configuration, orbit parameters, detector material, detector location on board satellite, and solar activity level. Effect of all these factors results to specific background value for each experiment. Background of gamma-detectors on board the spacecrafts at low near-Earth orbits are defined by the following components [45, 49, 54, 62]:

(a) cosmic diffuse gamma-rays;
(b) atmospheric gamma-rays;
(c) local gamma-rays and neutrons background produced via the direct interactions of cosmic rays and the materials of detector and spacecraft;
(d) $\gamma$-rays from decay of neutral pions produced via cosmic rays and atmosphere nuclei interactions;

(e) non-stationary events and electron precipitations;
(f) decay of nuclides produced by interactions of cosmic rays and magnetosphere trapped particles with the materials of the detector and satellite;
(g) discrete cosmic sources emission;
(h) bremsstrahlung of electrons and positrons accelerated in astrophysical objects or produced due decays of other leptons or charged mesons be generated via interactions of accelerated in astrophysical objects ions and nucleons.

Various background components relative contribution depends on detector composition, satellite mass, and construction and differs essentially for different energy bands. In general it is a function of time and spacecraft position on its orbit [5, 17, 49]. The (a) and (g) components do not depend on the position of the satellite in the Earth's magnetic field, while the other components contribution is specified by the value of cosmic rays flux, geomagnetic cutoff rigidity ((b)–(e) components), total dose accumulated during the satellite passing the regions of trapped radiation, and time interval duration from the moment of the trapped radiation regions passing till the moment of measurements ((f) component).

There are no requirements to construct detailed physical models of the abovementioned background components for events identifying and analyzing of their features. It is more convenient to consider background model as a sum of empirical relationships approximating background counts rate temporal variations in each channel. For the experiments on board satellites with orbit inclination up to 40° background counts rate temporal variations usually were approximated by periodic functions constructed during several sequential orbits for each spectral channel; but infrequent South-Atlantic Anomaly (SAA) region passages were taken into account by the activation analysis [62, 97].

For satellites with orbit inclination over 70° (inclination of the CORONAS-F orbit was 82.5°) the events in gamma-bands could be identified only in equatorial and polar-caps regions; therefore, background model rational should constructed only for these orbit parts. The detector counts rate background estimations were made due averaged counts rate values for several neighbour orbit parts with similar geomagnetic coordinates [17].

Typical dependance of the summarized AVS-F counts rate on geomagnetic latitude in the low- and high-energy gamma-bands is presented in Fig. 2. There are well-defined Earth radiation belts, polar caps, and SAA regions (1, 2, and 3 correspondingly) in the high-energy and low-energy bands. Also the areas of precipitated from the external Earth radiation belt electrons bremsstrahlung registration in the low-energy $\gamma$-band were identified (4). The equatorial regions (5) were more interest for study because of the most favorable conditions for the burst events (solar flares and gamma-ray bursts) registration and analysis.

In the low-energy gamma-band of the AVS-F apparatus only gamma-rays were registered including bremsstrahlung of charged particles in structural materials (charged particles could be detected in this band were screened by anticoincidence shield). AVS-F apparatus background for different spectral channels in the low-energy gamma-band was analyzed in details in [5, 17]. It has shown that background

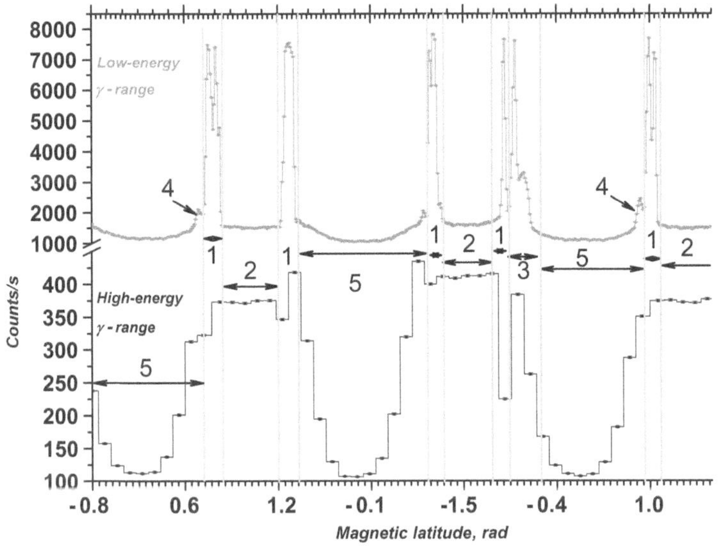


**Fig. 2** Typical example of latitudinal profile of the AVS-F counts rate in the low-energy and high-energy $\gamma$-bands

count rate temporal (latitudinal) profile for each of the 82 channels approximated by a fourth-or fifth-degree polynomials at the satellite orbit equatorial segments, and by parabolic curve, linear function or constant at the polar regions (see Fig. 3).

Such type background count rate profile approximation was used for solar flares spectra study and $\gamma$-lines identification. For instance, seven spectral lines complexes were separated by AVS-F data in the low-energy gamma-band at the October 29, 2003 solar flare spectrum using the above-mentioned method background subtraction [5, 17]. This event lasts during time interval 20:38–20:55 UT according to the data of the detectors on board the GOES-12 satellite. Its spectrum with and without background substraction and background spectrum is presented in Fig. 4. Four strong lines complexes were identified in the background spectrum and seven typical for solar flares $\gamma$-lines complexes were separated in this flare spectrum after background substraction.

Besides gamma-quanta, neutrons were detected via the secondary charged non-relativistic particles (isotopes of hydrogen and helium) in AVS-F high-energy band. Also relativistic protons were registered via ionization losses in this band due to limited efficiency of aniticoincidence shield.

As it was noted in the section "AVS-F Apparatus Short Description," ratios of the slow and the total light output components in the CsI(Tl) scintillator differ for the particles with different specific ionization, therefore their detection corresponds to different areas at two-dimensional distributions of the ratio of the slow light output component to the total one related to the total energy deposition in the SONG-D detector. Examples of the summarized background two-dimensional distributions

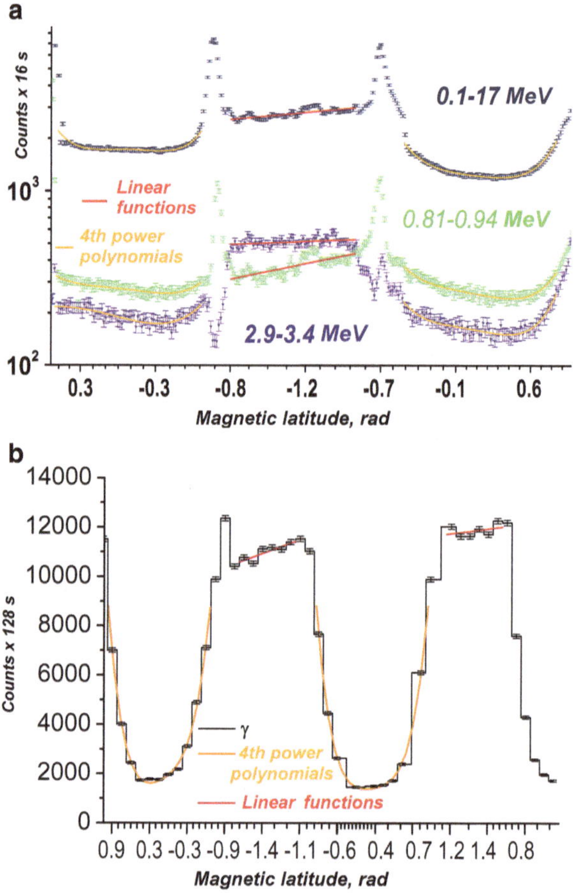


**Fig. 3** Typical latitudinal profiles of the AVS-F counts rate and their approximation in the low-energy (**a**) and high-energy (**b**) $\gamma$-bands. Profiles on the equatorial parts of the CORONAS-F orbit are approximated by fourth-degree polynomials (*orange curves*), on the polar parts by linear functions (*red lines*). From [5]

of the dependence of the ratio of the slow light output component to the total one from the total energy deposition in the SONG-D detector according to the data of the AVS-F apparatus in the equatorial and northern polar CORONAS-F orbit regions for undisturbed magnetosphere ($Kp < 4$) without burst events are presented in Fig. 5.

The data correspond to 296 orbits (N 19263–19559) in January 2005, total accumulation time was 69056.81 s (over 19 h) for the equatorial regions ($-30° < \phi < +30°$, except the SAA region), and 24750.30 s (over 6 h) for the North polar regions ($+83° > \phi > +65°$). The areas corresponding to registration of the secondary particles from neutron interactions (including non-relativistic protons and $\alpha$-particles), gamma-quanta, and relativistic protons are shown.

Detailed analysis of two-dimensional distributions of the ratio of the slow light output component to the total light output dependence on the total energy deposition in the SONG-D detector allowed to study in the high-energy range the behavior of

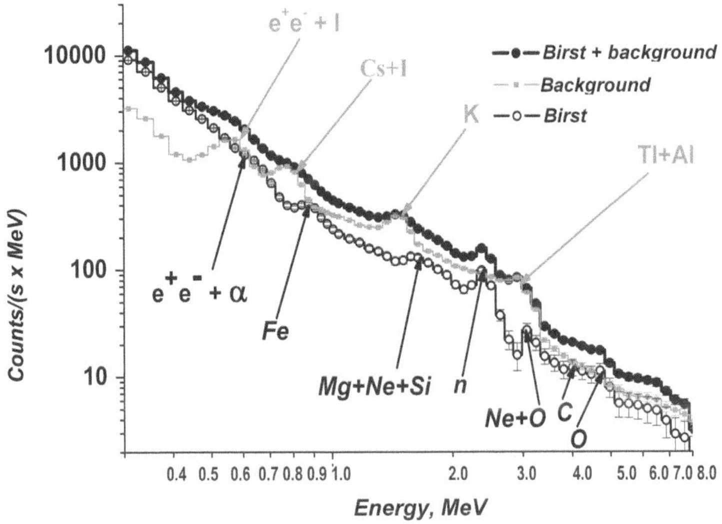


**Fig. 4** Energy spectrum with and without background subtraction and background spectrum for the solar flare of January 29, 2003. Specific spectral features are given by *gray arrows* at the background spectrum and by *black* ones at the flare spectrum with background subtraction. From [9]

the counts rate dependence on geomagnetic latitude separately for non-relativistic protons, $\alpha$-particles, relativistic protons, and gamma-quanta [69].

In the high-energy range of the AVS-F apparatus the latitudinal profile of the background counts rate in the areas with specific ionization corresponding to gamma-quanta on two-dimensional distributions of the ratio of the slow light output component to the total light output dependence on the total energy deposition in the CsI(Tl) detector can be described by the forth- or fifth-degree polynomials at the equatorial orbit segments and by linear functions or constants at the polar ones analogous to the low-energy band (see Fig. 3b).

Latitudinal profiles for the areas corresponding to non-relativistic protons and $\alpha$-particles can be approximated with the forth-degree polynomials in the equatorial region of the CORONAS-F satellite. One and the same polynomial (with normalization coefficient which takes into account difference of the counts rate in the analyzed regions) can be used for approximation of both profiles for non-relativistic protons and $\alpha$-particles at the significance levels $R_{p_{\text{nonrel}}\text{AVS-F}} = 0.99$ and $R_{\alpha\text{AVS-F}} = 0.85$, correspondingly. An example of such approximation for summarized latitudinal profiles normalized to the number of channels for Kp $= 2$ is presented in Fig. 6. Therefore the latitudinal profiles in the areas corresponding to non-relativistic protons and $\alpha$-particles were in a good agreement, as it must be for the secondary particles registration.

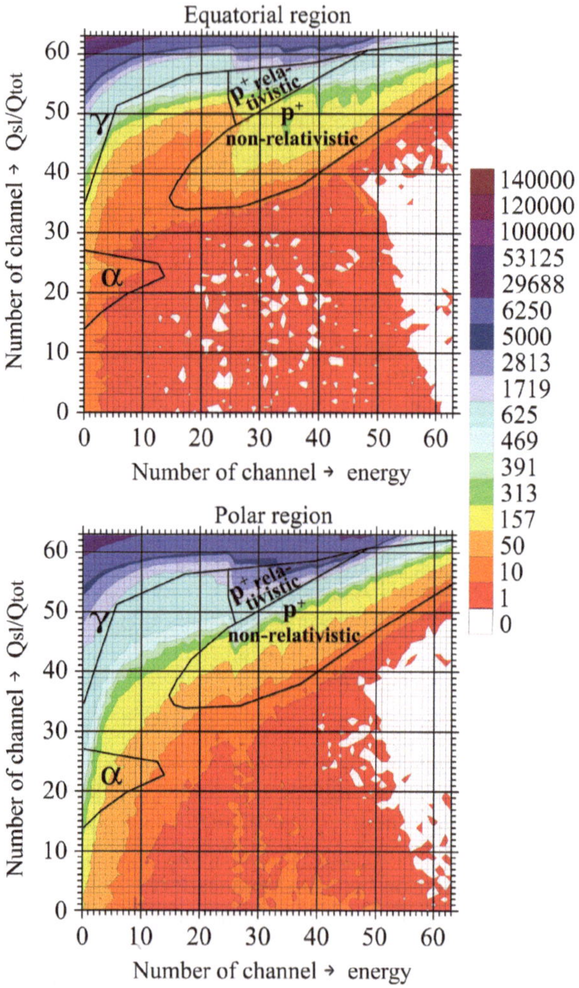


**Fig. 5** Summarized background two-dimensional distributions of the ratio of the slow light output component to the total light output dependence on the total energy deposition in the SONG-D detector according to the AVS-F apparatus data in the equatorial and polar regions of the CORONAS-F orbit at quiescent magnetosphere ($Kp < 4$)

For relativistic protons in the equatorial region of the CORONAS-F satellite orbit the dependance of the summarized counts rate from the latitude according to the data of January 2005 normalized to the number of channels is presented in Fig.7a. It can be approximated with the forth-degree polynomial at the level of significance of $R = 0.98$—thick grey curve.

For relativistic protons area identification confirmation at two-dimensional distributions of the ratio of the slow light output component to the total light output dependence on the total energy deposition in the CsI(Tl) detector the AVS-F data were compared with the cosmic protons flux latitudinal profiles obtained by the AMS experiment [1, 2] on board the "Discovery" spacecraft in 1998. The profiles were observed at the same altitudes (360–380 km) as for the CORONAS-F satellite

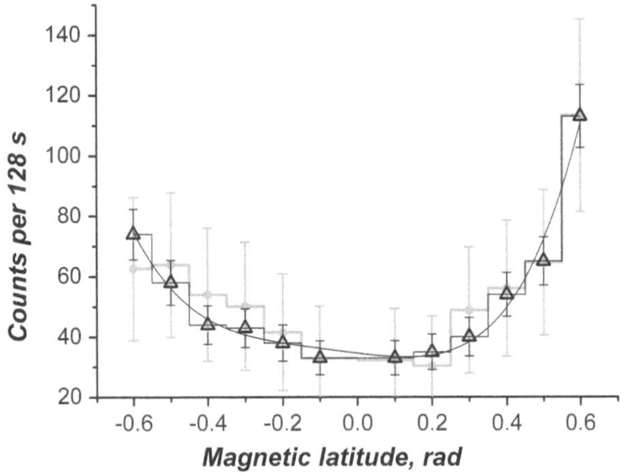


**Fig. 6** Summarized latitudinal profile of non-relativistic protons background counts rate normalized to the number of channels (*black* histogram) and summarized latitudinal profile of $\alpha$-particles background counts rate normalized to the average counts rate of non-relativistic protons (*gray* histogram) according to the data of AVS-F apparatus at the equatorial parts of the CORONAS-F satellite's trajectory (orbits 19263–19559 for January 2005) at $Kp = 2$. *Black curve* shows fourth-degree polynomial approximation of these profiles

in January 2005. Latitudinal distribution of protons according to the data of the AMS experiment is presented in Fig. 7b. It also can be approximated with the forth-degree polynomial [69] at $R = 0.99$. The data of the AVS-F apparatus corresponding to registration of relativistic protons (at $Kp = 2$) after renormalization taking into account difference of the response functions of the AVS-F and AMS detectors were approximated with the polynomial of the same degree with the same (in statistical errors) coefficients at the level of significance of $R_{p_{\text{rel}}\text{AVS–F}/p_{\text{rel}}\text{AMS}} = 0.89$ [69] (thin black curve in Fig. 7a). Moreover latitudinal profile corresponding to the relativistic protons area doesn't agree with the profiles corresponding to the areas of non-relativistic protons and gamma-quanta (black dotted curve and thin gray curve correspondingly in Fig. 7a): the level of significance of the approximation with conforming polynomials are $R_{p_{\text{rel}}\text{AVS–F}/p_{\text{nonrel}}\text{AVS–F}} = 0.64$ and $R_{p_{\text{rel}}\text{AVS–F}/\gamma\text{AVS–F}} = 0.56$ only.

So, it is possible to identify areas corresponding to registration of secondary particles from interactions of neutrons (including non-relativistic protons and $\alpha$-particles), gamma-quanta, and relativistic protons at two-dimensional distributions of the ratio of the slow light output component to the total light output dependence on the total energy deposition in the SONG-D detector; latitudinal profiles for the different types particles also differ due to the features of their origin and movement in the Earth's magnetosphere.

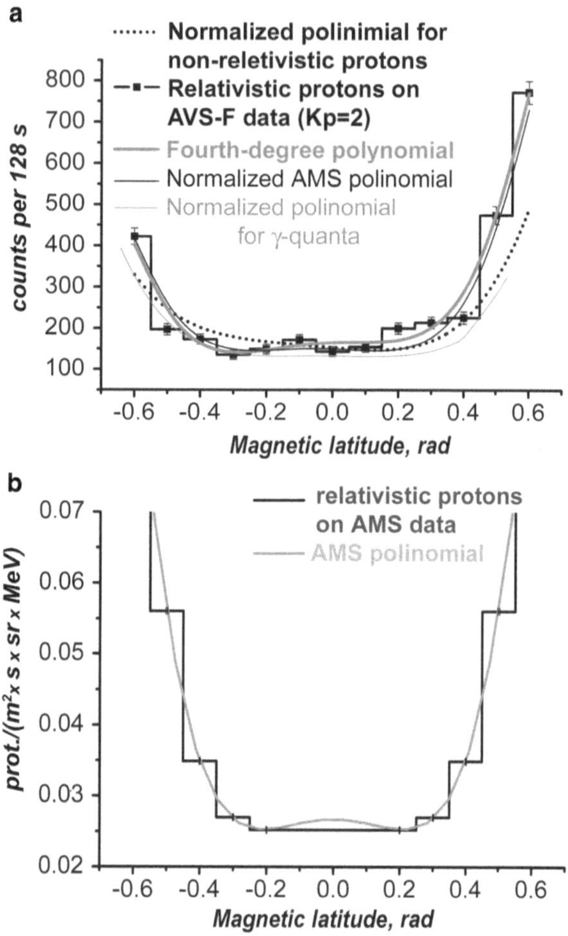


**Fig. 7** The relativistic protons counts rate dependence on geomagnetic latitude in the equatorial region by data of AVS-F (**a**) and AMS (**b**). AVS-F data were normalized to the number of channels. From [69]

# In-Flight Calibration in $\gamma$-Bands

As it was noted in the section "AVS-F Apparatus Background Model in the $\gamma$-Bands," background counts rate of the AVS-F apparatus in all energy ranges change essentially with time (depending on the spacecraft location in the orbit). During the movement of the satellite along the orbit in the Earth's magnetic field charged particles fluxes significantly varied. Satellite was periodically exposed to intensive fluxes of charged particles activated both detectors and structural materials of the spacecraft, and these effects led to changes of the AVS-F apparatus counts rate too.

Changes of the dependance of the observed energy from the number of the channel of the registering instrument during its total operation period resulted due to ageing of the detecting elements and electronics. It should be taken into account during database treatment both for the low- and high-energy gamma-bands of the AVS-F apparatus due regular calibrations.

For apparatus calibration (in this case for determination registered particle energy dependance from the number of the analog–digital converter channel), it was necessary to separate characteristic lines with specified energy from the obtained background energy spectra. Average AVS-F apparatus counts rates at the parts of the orbit outside the radiation belts and SAA were $\sim$ 1,000 counts/s in the low-energy gamma-band [72] and approximately ten times lower in the high-energy band [73]. Therefore it was necessary to accumulate spectrum during several hours at least to provide the reliable spectral lines identification. Orbit period of the CORONAS-F satellite was 90 min, so the accumulation time was not more than 20 min for one equatorial orbit part and not more than 15 min for one polar cap orbit region. The polar cap parts were selected from the shadowless orbit segments in the north polar cap in the latitudes range from $+65°$ up to $+82.5°$, and equatorial ones from the shadowless orbit segments in the latitudes range from $-30°$ to $+30°$; the SAA region and the parts of the orbit in 3 h before and after its passing were excluded from the analysis. Therefore, it was necessary to summarize sufficiently great amount of equatorial or polar spectra to obtain calibration spectrum with accumulation time of several hours.

The primary calibration spectrum was resulted from summarizing 765 parts of the orbit near the equator during the initial period of the apparatus operation and its accumulation time was approximately 80 h. Typical example of the calibration spectrum is presented in Fig. 8a. Spectrum shape was specified by the bremsstrahlung of the electrons (with power spectrum) passed through the detector and the spacecraft elements structural materials, Compton scattering, and (or) pair production during registration of the background gamma-emission. It also included positron line of 0.511 MeB and background lines characteristic for this apparatus formed via the protons and neutrons interactions with the material of the detector and satellite construction. Just these lines were used for calibration. For the real channel number definition it was necessary to subtract the base (power-law at a first approximation) from the spectrum corresponding to background line with specified energy, or to multiply the spectrum on the power-law function, and in this case the lines basic for the calibration will be distinctly identified.[2] The spectrum tends to the shape presented in Fig. 8b.

There were 11 background lines separated in the calibration spectrum [9,27,72]. Their energies are presented in Table 1.

Cesium, iodine, and thallium were composed in the scintillation detector crystal, aluminium was consisted in the structural materials of the apparatus and the

[2]At spectrum multiplying on a power-law function of the $r$-th order at $r > 0$ shift of the line center was taken into account. This shift was calculated by the following equation:

$$x = x_0/2 + \sqrt{(x_0/2)^2 + r\sigma^2} \tag{2}$$

for the features which can be approximated with a gaussian with the parameters $x_0$ and $\sigma$.

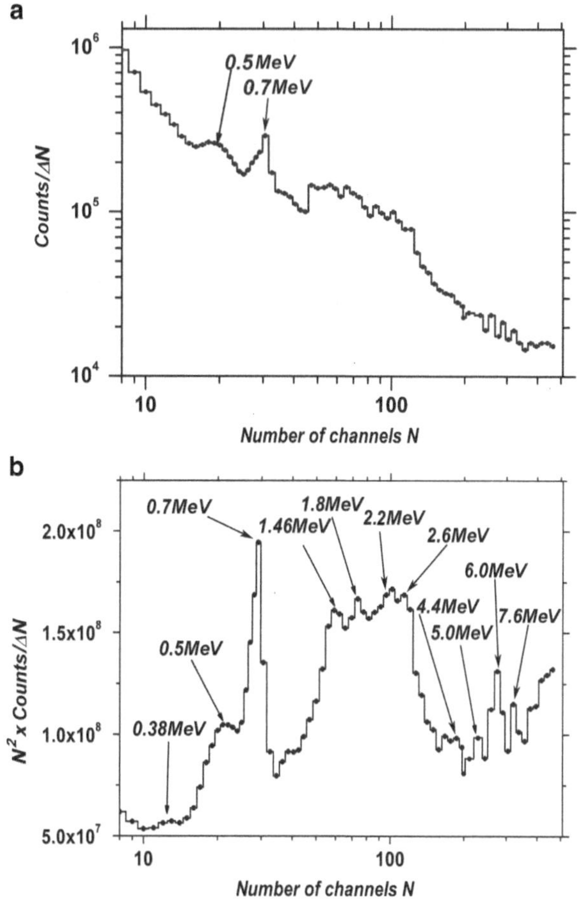


**Fig. 8** Typical example of the summarized calibration spectra: (**a**) spectrum in the equatorial region with total integration time over 80 h, (**b**) the same spectrum, as in (**a**), multiplied by the squared number of the channel for more demonstrable lines identification. From [28]

**Table 1** Parameters of the background lines [55, 89, 92, 93]

| $E$, MeV | Basic nuclides |
|---|---|
| 0.38 | $I^{126}$, $I^{128}$, $Cs^{127}$, $Cs^{129}$, $Sn^{113}$, $Xe^{127}$ |
| 0.5 | $e^+ e^-$, $Na^{24m}$, $Te^{121}$, $Cs^{132}$, $Sb^{115}$, $I^{121}$, $Cs^{128}$, $Cs^{129}$, $Co^{55}$, $In^{111m}$, $Cs^{128}$, $I^{126}$, $I^{128}$ |
| 0.7 | $Cs^{132}$, $I^{124}$, $I^{126}$, $Mn^{52}$, $In^{110}$, $Te^{119}$, $Te^{117}$ |
| 1.46 | $K^{40}$, $Na^{24}$, $Mn^{52m}$, $Mn^{52}$, $Ni^{57}$, $Mg^{24}$, $Co^{55}$ |
| 1.8 | $Al^{28}$, $Te^{119}$, $Te^{117}$, $I^{124}$, $Na^{24}$, $Mg^{26}$, $Si^{28}$ |
| 2.2 | $n$ capture, $Al^{27}$, $Na^{24}$, $Te^{117}$, $Sb^{116}$, $Ni^{57}$ |
| 2.6 | $Tl^{208}$, $Al^{27}$, $Na^{24}$ |
| 4.4 | $C^{12}$, $C^{13}$, $Na^{24}$, $Si^{29}$ |
| 5.0 | $C^{13}$, $Si^{29}$, $O^{16}$ |
| 6.0 | $O^{16}$, $Fe^{57}$, $Si^{28}$ |
| 7.6 | $Fe^{57}$, $O^{16}$, $Si^{29}$, $Al^{28}$ |

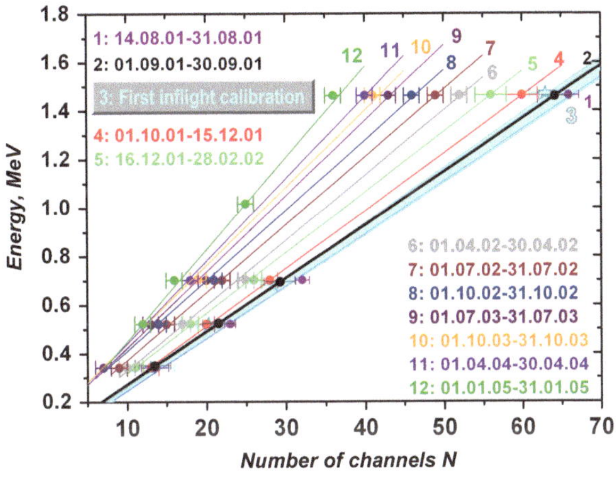


**Fig. 9** The evaluation of the AVS-F apparatus calibration with time. From [28]

spacecraft; iron was contained in the satellite orientation system; carbon and hydrogen were composed in the plastic anticoincidence shield; potassium, oxygen, silicon, nickel, and magnesium were contained in photomultipliers. The other isotopes from Table 1 were also products of interaction of cosmic rays with materials of the detector and the spacecraft. Relation between the energy $E$ of the detected particle and the number of channel $N$ in the low-energy gamma-band of the AVS-F apparatus was described by the following linear function at the 0.997 significance level:

$$E_{\text{low}\gamma}(\text{MeV}) = (0.05 \pm 0.02) + N \times (0.0219 \pm 0.0002). \tag{3}$$

Such significance level allows to use less number of lines for the subsequent calibration procedures.

Then we have studied the time-dependent changes of the calibration function using summarized monthly equatorial calibration spectra. Four background lines with average energies of 0.38 MeV, 0.5 MeV, 0.7 MeV, and 1.46 MeV were separated in each summarized spectrum and we used these lines as the base for calibration. Time dependance of calibration function is presented in Fig. 9.

The obtained curves were approximated with linear function:

$$E = A + B \times N, \tag{4}$$

The results of data analysis have shown that approximation coefficients time dependences were the following: [6, 13, 28]:

$$A = (2.1 \pm 0,9) \times 10^{-2} + t_1 \times (9.7 \pm 0.3) \times 10^{-3} - \\ -t_2 \times (3.3 \pm 0.9) \times 10^{-4} + t_3 \times (3.3 \pm 0.3) \times 10^{-6}; \tag{5}$$

$$B = (2.12 \pm 0.05) \times 10^{-2} + t_1 \times (7.5 \pm 1.1) \times 10^{-4} - \\ -t_2 \times (2.4 \pm 0.6) \times 10^{-5} + t_3 \times (3.8 \pm 0.9) \times 10^{-7}. \quad (6)$$

During 1 month the AVS-F apparatus energy threshold and conversion coefficient in the low energy gamma-band were changed for $\sim 1\,\%$ and $\sim 1.8\,\%$ correspondingly, dynamical range changed from $(0.041 \pm 0.007 \div 10.9 \pm 0.1)$ $MeV$ according to the August 2001 data up to $(0.11 \pm 0.01 \div 19.9 \pm 0.4)$ $MeV$ according to the data of January 2005 [6]. Such change of the calibration characteristics of the AVS-F apparatus in the low-energy $\gamma$-band could be associated both with the detecting element (CsI(Tl) crystal) aging, degradation of optic contact between the crystal and the photomultiplier and the changes of the electronics parameters of the spectrometric tract [13].

As it was noted, ten activation lines were separated for the background spectra of the AVS-F apparatus in the low-energy gamma-band. Six lines are also detected in the high-energy band, but it is impossible to separate them in this band because the channels width in the area of the lines detection is $\sim$ 1 MeV. Therefore well-known envelope curve with a center at 7.62 MeV [101] is used. Moreover the characteristic feature of decay of the neutral pions produced by nuclear reactions of the cosmic rays' protons was used for the calibration in this energy range [73]. As it was mentioned above, the calibration spectrum was summarized during a long time, and because the satellite was single-axis solar-oriented the flux during the period of the calibration spectrum accumulation was considered as isotropic. Correspondingly, in this case the center of the characteristic feature from the neutral pions decay was at 67.5 MeV [88].

Primary in-flight calibration of the AVS-F apparatus in the high-energy band was made during the initial period of the AVS-F apparatus operation by using the spectrum resulted from summarizing of the spectra for 2,992 equatorial orbit segments with total accumulation time of approximately 30 h (Fig. 10a, b). There were three characteristic features separated from the spectrum [73].

The first one was identified as a sum of nuclear lines produced as a result of spacecraft materials activation (as it was already mentioned above, its maximum is located at 7.62 MeV).

The nature of the second (2) and the third (3) features was determined on the basis of the spectrum summarized from 1,786 polar cap orbit segments with total accumulation time of about 18 h (Fig. 10c, d). Geomagnetic rigidity threshold in the polar caps region ($\sim$ 2 GeV) is essentially lower than in the equatorial one ($\sim$ 14.7 GeV), and the protons energy spectrum has power law shape with negative index; therefore the spectral feature caused by the protons ionization losses was more intensive in the polar cup regions accumulated spectrum than in equatorial one, i.e., the feature (2) corresponds to the relativistic protons ionization losses. Location of this feature maximum was determined as the results of relativistic protons passing through the SONG-D detector numeric simulations, but detailed discussion of this problem is outside of the frames of the present book.

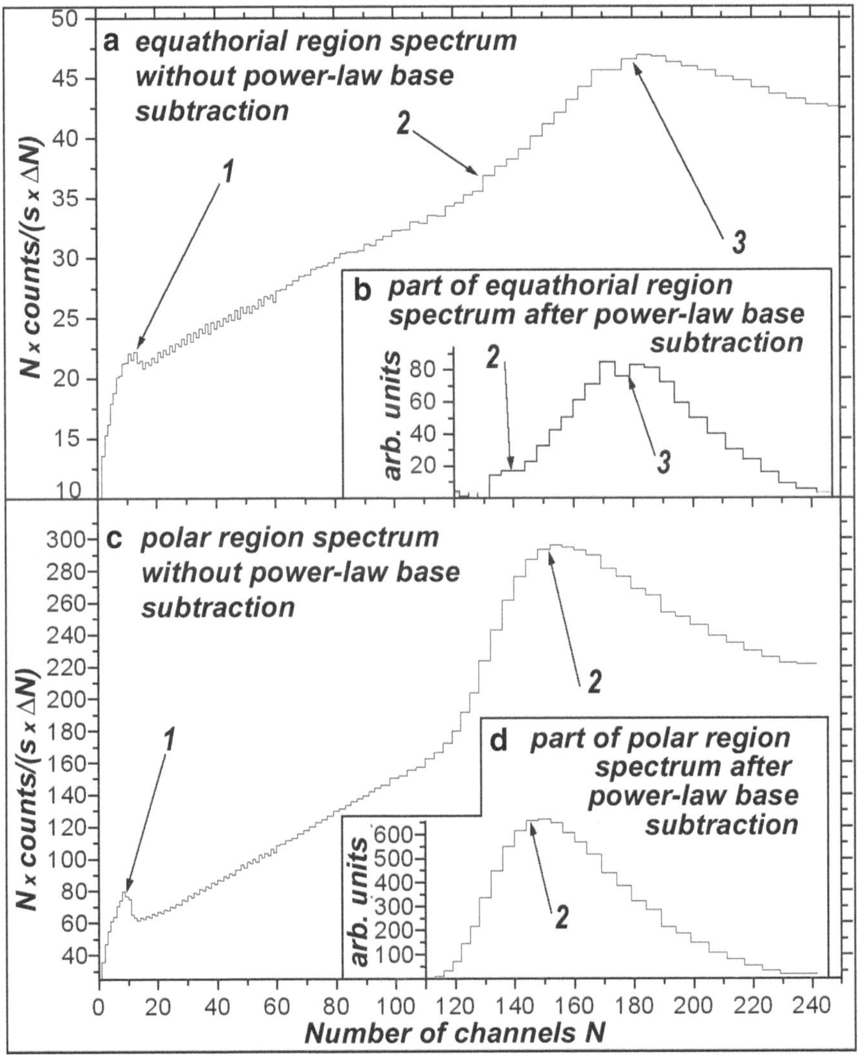


**Fig. 10** Summarized calibration energy spectra in the high-energy band in the equatorial region (**a**, **b**) and in the polar caps one (**c**, **d**) according to the AVS-F data

So the third feature was associated with registration of $\gamma$-quanta from $\pi^0$ decay (as was mentioned above, its maximum was located at 67.5 MeV). According to the results of the in-flight calibration, the AVS-F apparatus operational range in the high-energy gamma-emission registration mode was $(4.4 \pm 1.4) \div (94 \pm 3)$ MeV in November–December 2001, and dependance of the energy $E$ from the channel number $N$ was described with the following function [73]:

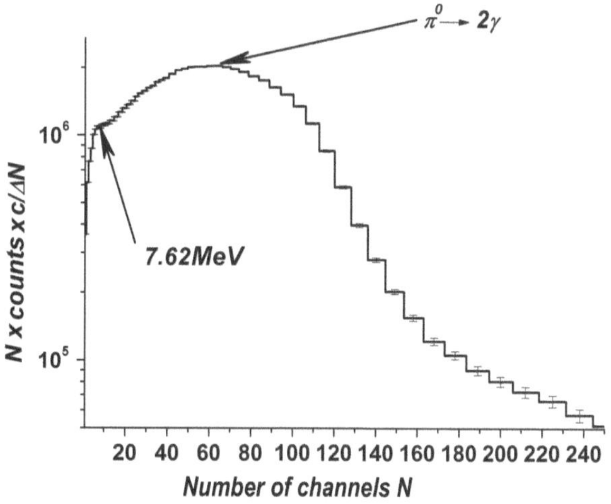


**Fig. 11** Summarized energy spectrum in the equatorial region according to the AVS-F apparatus data for January 2005, multiplied by the number of channel $N$

$$E(\text{MeV}) = (4.4 \pm 1.4) + N \times (0.36 \pm 0.01). \tag{7}$$

The AVS-F apparatus calibration characteristics in the high-energy band also varied with time due to the same effects as in the low-energy region.

The summarized equatorial region energy spectrum according to the January 2005 data is presented in Fig. 11 to illustrate the applicability of this feature for the calibration. This spectrum was multiplied by the channel number to identify spectral features more distinctly. In this figure as well as in Fig. 10a, two spectral features were separated: the first of them corresponds to the sum of activation lines (as was mentioned above, it is located at 7.62 MeV), and the second one associated with gamma-emission from $\pi^0$-mesons' decay. The analysis has shown that in January 2005 high-energy gamma-range of the AVS-F apparatus was from 2 MeV up to 260 MeV [29].

## Typical Burst Events

About 1,100 burst events with duration more than 2 s were identified due to the results of the AVS-F data analysis for the period from August 2001 till December 2005. All of these events were observed outside of the radiation belts and SAA.

Several events types were separated by the analysis of the observed burst events:

- 68 % were quasistationary equatorial precipitations—see the section "Characteristics of the Quasistationary Equatorial Precipitations";

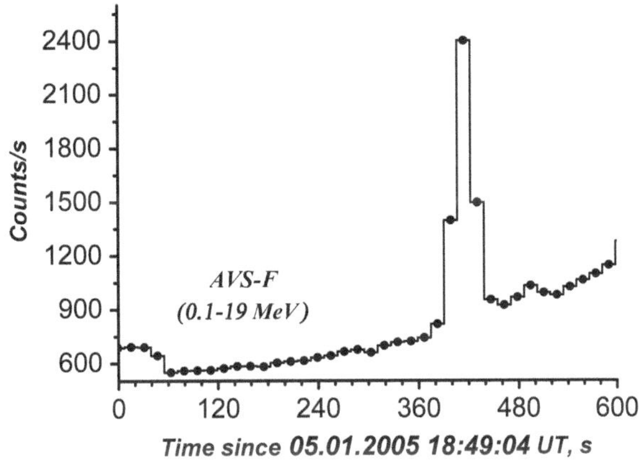


**Fig. 12** Temporal profile of the event observed in the Earth's shadow on January 5, 2005, in the low-energy $\gamma$-band of the AVS-F apparatus

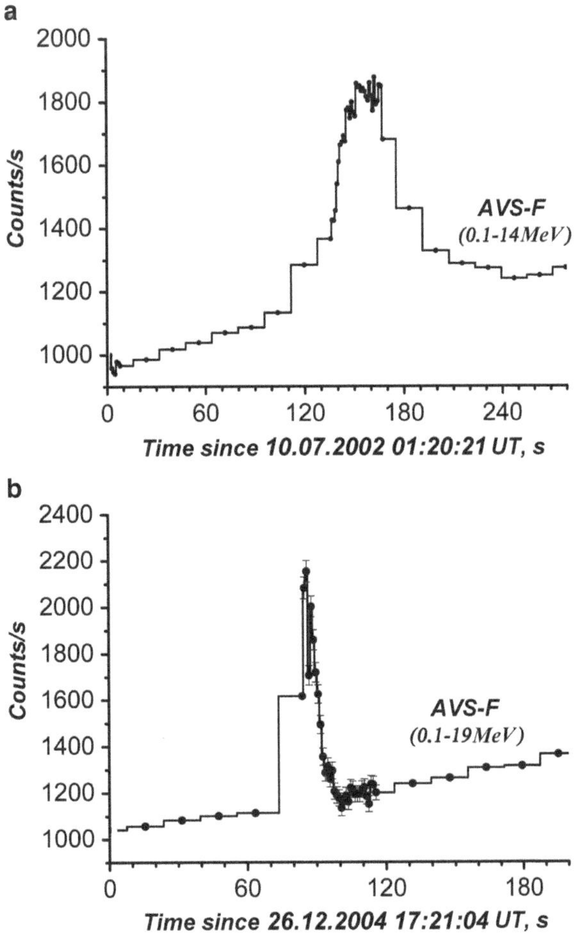


**Fig. 13** Temporal profiles of the event of July 10, 2002, in the energy band of 0.1–14 MeV (**a**) and one of December 26, 2004, in the energy band 0.1–19 MeV (**b**) observed outside the Earth's shadow

- 7 % of the detected events were associated with solar flares and were confirmed by the data of other experiments (GOES, RHESSI etc.)—see the section"Solar Flares Characteristics Analysis";
- 3 % of the identified events were gamma-ray bursts (GRB), also confirmed by the data of other experiments (SWIFT, HETE, RHESSI etc.)—see the section "Analysis of the Gamma-Ray Bursts Characteristics";
- 22 % of the detected events were needed more detailed analysis.

17 % of all the detected events (except solar flares) were observed in the Earth's shadow. Unidentified events were observed both in the Earth's shadow (Fig. 12) and out of it (Fig. 13).

## Characteristics of the Quasistationary Equatorial Precipitations

Average counts rate value of the AVS-F apparatus in the low-energy $\gamma$-band in the range of geographical latitude from $-30°$ up to $+40°$ was monotone decreased along the direction to the geomagnetic equator, and was monotone increased along the direction to the high latitudes, i.e., typical counts rate temporal (latitudinal) profile without burst events had one wide minimum in the region of geomagnetic equator. The second-degree polynomial was not enough for approximation of the minimum's width, it was necessary to use the forth- or the fifth-degree polynomial [5, 17]—see the section "AVS-F Apparatus Background Model in the $\gamma$-Bands."

During data analysis at some equatorial orbit parts for the low-energy $\gamma$-band there was found 15–30 % counts rate increases above its average value obtained by approximation of these parts with polynomials in the range of geographic latitude from $-25°$ up to $+30°$. These effects were called quasistationary equatorial precipitations [72]—see Fig. 14.

Analogous events were observed in 1997 by the LEGRI instrument on board the Spanish satellite MINISAT-01 (initial altitude of the orbit—600 km, inclination—28.5°) in the energy band of 20–100 keV [98]. In Fig. 15 an event of precipitation observed in 1997 by the matrix of the LEGRI instrument consisted of 20 CdZnTe detectors, each of 2 mm thick and 1 $sm^2$ is presented.

Integration time was 100 s, i.e., increasing of the counts rate during the time interval of 12,500–14,500 s is statistically significant ($> 3\sigma$), but it cannot be explained by the activation of the material of the detectors and satellite, because it was observed sufficiently long time after the last satellite passage of the SAA region. Unfortunately, previously these events were not studied and not marked out as separate events class; its detailed analysis started only during the AVS-F apparatus data analysis [19–22, 24].

Quasistationary precipitations were associated with the CORONAS-F satellite passages several orbit segments with counts rate increases in the low-energy band mainly in the region of geomagnetic equator.

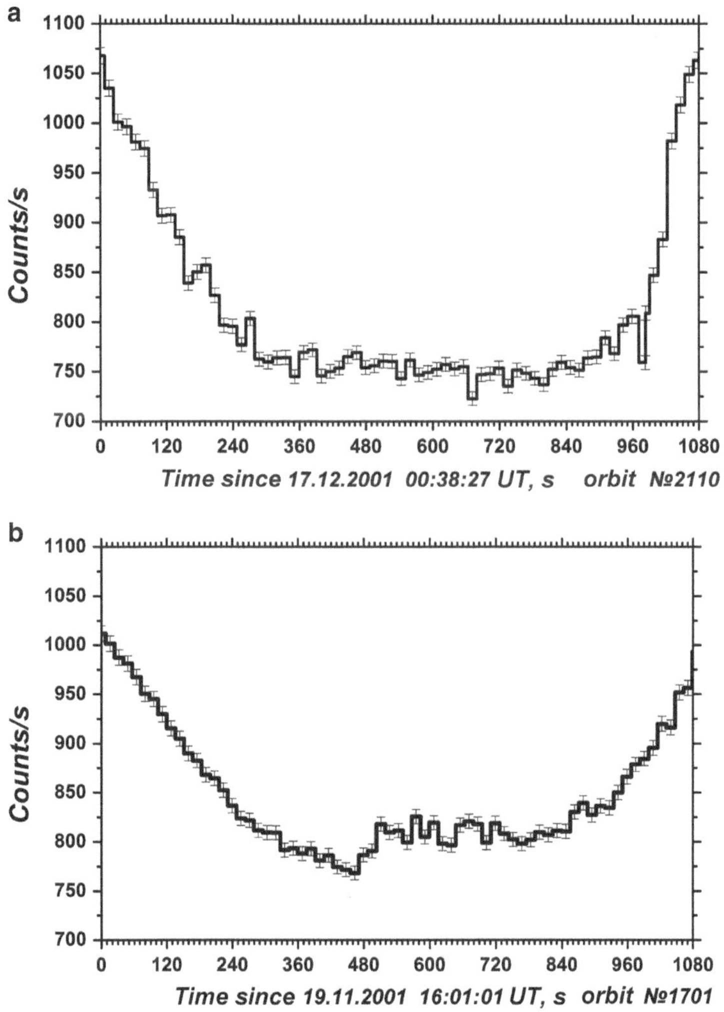


**Fig. 14** Typical temporal (latitudinal) profiles for the equatorial region: without precipitation (**a**) and with precipitation (**b**) according to the data of the AVS-F apparatus in the low-energy $\gamma$-band in the same coordinate intervals. From [20]

So far about 700 precipitations are identified during the analysis of the part of AVS-F apparatus database. Temporal (latitudinal) profiles of the most part of the precipitations ($\sim$ 44 %) are characterized with sharp counts rate rise toward South from geomagnetic equator and its slow decrease toward North. Typical temporal (latitudinal) profiles of the $\sim$ 21 % analyzed events, on the contrary, have sharp increase of the counts rate toward North and slow drop toward South from geomagnetic equator (see Fig. 16b), and $\sim$ 4 % of precipitations have essentially symmetric temporal (latitudinal) profiles (see Fig. 16c).

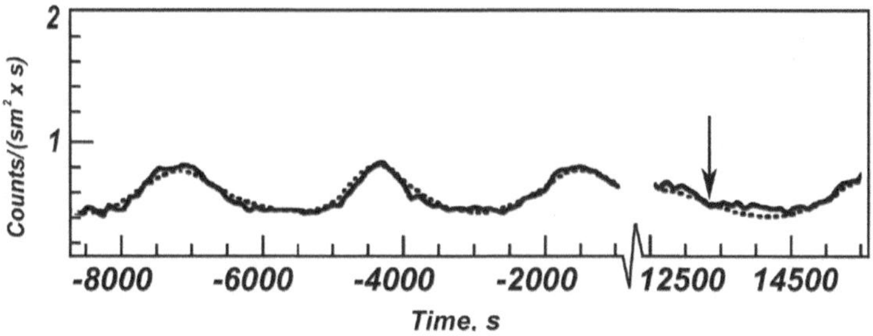


**Fig. 15** Temporal (latitudinal) profile of the hard X-ray background in the energy range of 20–100 keV according to the data of the LEGRI instrument. Excess of the counts rate over the mean model value in the interval of 12,500–14,500 s (marked by the *arrow*) is analogous to the quasistationary precipitations observed by the AVS-F apparatus (adopted from [98]). The averaged model of the LEGRI background is shown by *dashed line*, experimental data—by solid one. From [23]

These three types of precipitations were observed regardless of the spacecraft orbit altitude. Precipitations with more significant (than during the events of the I–III types) counts rate increase 1.3–1.8 times above the background value with central maximum were registered since the beginning of the 2002 (Fig. 17a). Such events were $\sim$ 4 % of the detected precipitations. Moreover several tens of faint events with symmetric profile shape were observed—see Fig. 17b. So basing on the type of the temporal (latitudinal) profiles all the quasistationary equatorial precipitations can be separated at least in four basic types [22, 23]. Temporal (latitudinal) profiles of all detected precipitations are smooth, without any statistically significant thin structure in the time scale more than 16 s.

Typical duration of the precipitations was $7 \div 10$ min for the all types events (the size of the region along geographic latitude was $20° \div 35°$) [20], but there were also short precipitations $\sim$ 3 min among the events of I and II types (their range is $\sim$ 10°) [22]—see Fig. 18. Quasistationary precipitations were observed both out of the Earth's shadow and in the Earth's shadow [19].

Such precipitations were observed at several subsequent orbit passes through the same region. Average time interval during which the quasistationary equatorial precipitations were observed in the region of 10° of geographic longitude and 30° of geographic latitude was $\sim$ 24 h, maximum one was more than 8 days [20]. Figure 19 presents the temporal profile of I type precipitation observed a day after one presented in Fig. 16a in the same coordinates range.

The forth type of precipitations and faint events with symmetric profile shape were found not long ago and its characteristics detailed analysis is still in progress. Therefore only the precipitations of I–III types features are discussed in this section.

The most part of the precipitations of I–III types were located essentially symmetrically related geomagnetic equator in the band of about 40° width, the width of the near-equator region where the precipitations of the IV type were observed was about 20° [20–22]. The counts rate map in the low-energy gamma-band

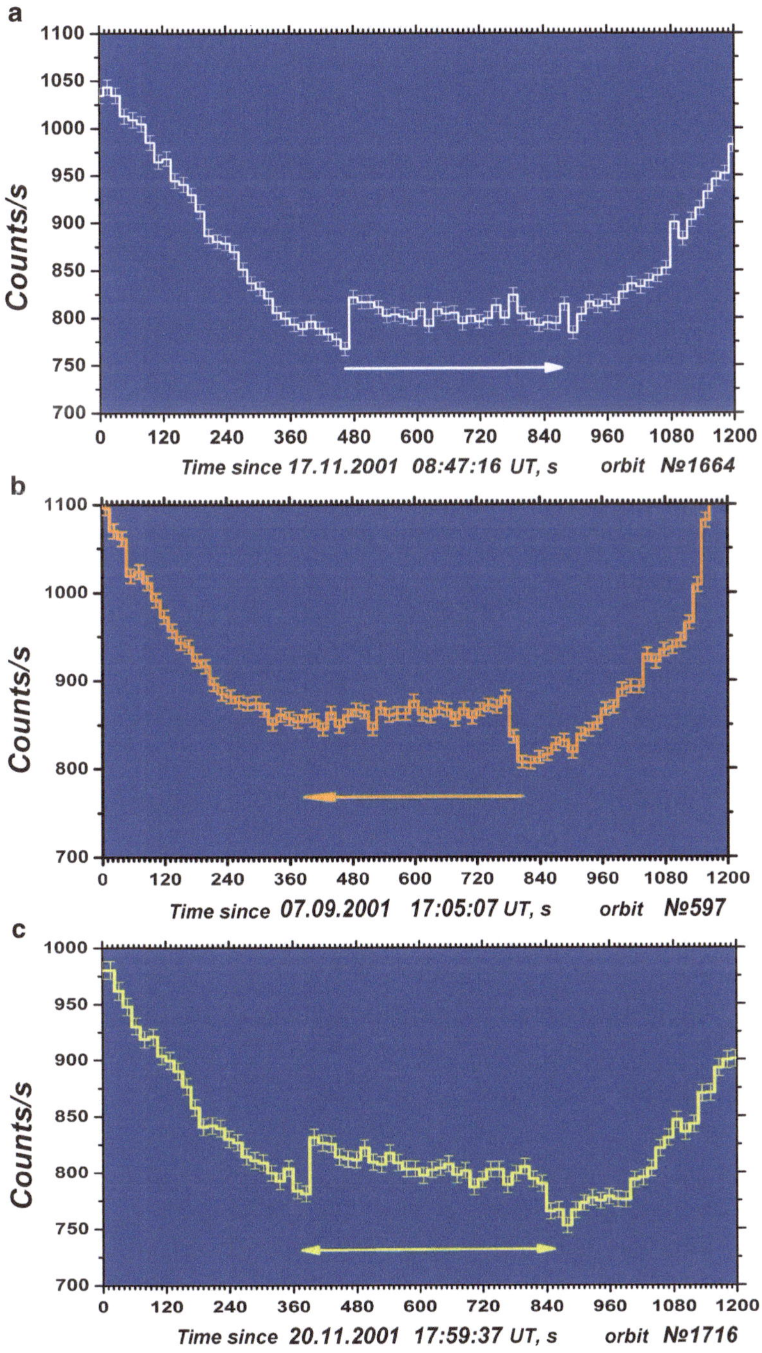


**Fig. 16** Typical temporal (latitudinal) profiles of precipitations: (**a**) type I, (**b**) type II, (**c**) type III. *Arrows* below correspond to the precipitation indication at the counts rate maps (see Figs. 20 and 21). From [24]

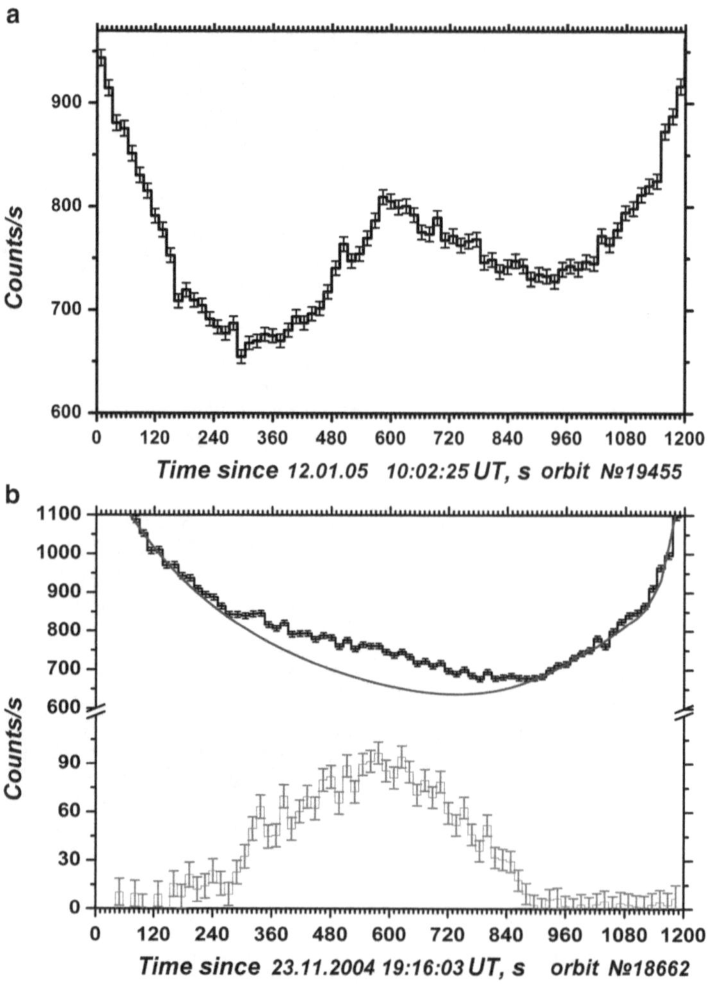


**Fig. 17** Temporal (latitudinal) profile of the IV type precipitations (**a**) and of very faint symmetric precipitations (**b**). The upper histogram in (**b**) panel presents the event without background subtraction, the *curve* shows the approximating background polynomial, and the lower histogram demonstrates the event with background subtraction. From [21]

averaged through 1.5 years of the apparatus operation with indication of the quasistationary equatorial precipitations regions is presented in Fig. 20. Projection of charged particles precipitation regions detected in different experiments on this map is presented in Fig. 21. So, the most part of the quasistationary equatorial precipitations were located in the regions of detection of electrons precipitations at the low latitudes [40] ($L \sim 1.5$—see Fig. 21) and in the region of geomagnetic equator. Precipitations of the electrons with energy $E \sim 1.3$ MeV were observed at the altitudes of 288–350 km in the regions with $L = 1.14$–$1.17$ as early as in

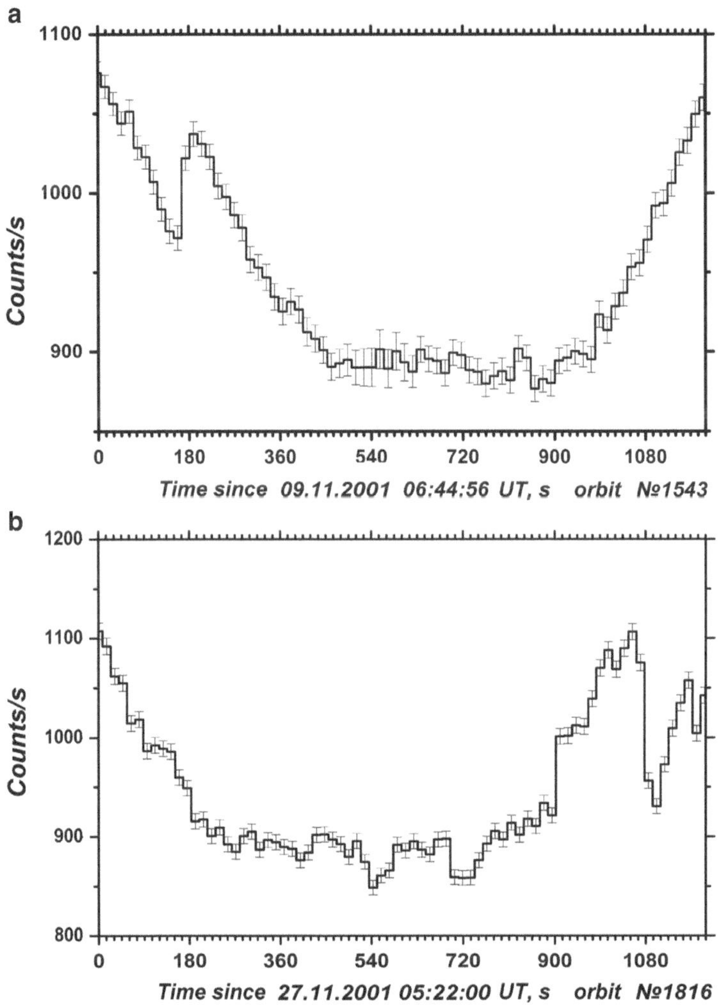


**Fig. 18** Typical temporal profiles of the short time precipitations of I (**a**) and II (**b**) types according to the AVS-F apparatus data in the energy range of 0.1–11 MeV. From [21]

1960s of the twentieth century [67]. Precipitations of the ions of hydrogen, helium, and oxygen with maximum in the region of geomagnetic equator but extended to geomagnetic latitude of $\pm 20°$ were detected by some spacecrafts, its flux intensity increases in the layer of 180–260 km, and then does not change. The detectors on board the COSMOS-900 satellite observed both day- and night-time electron precipitation at the altitude of 500 km [40].

Typical energy spectra of quasistationary equatorial precipitations are presented in Fig. 22. For all three types of precipitations the spectra are power-law [24]:

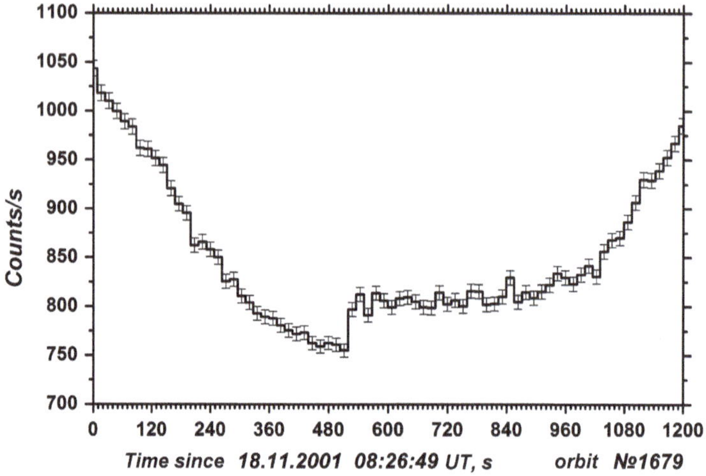


**Fig. 19** Typical temporal (latitudinal) profile of type I precipitation observed by the AVS-F apparatus in the low-energy $\gamma$-band in the same coordinate interval $\sim$ 24 h after the observation of the event presented in Fig. 16a. From [111]

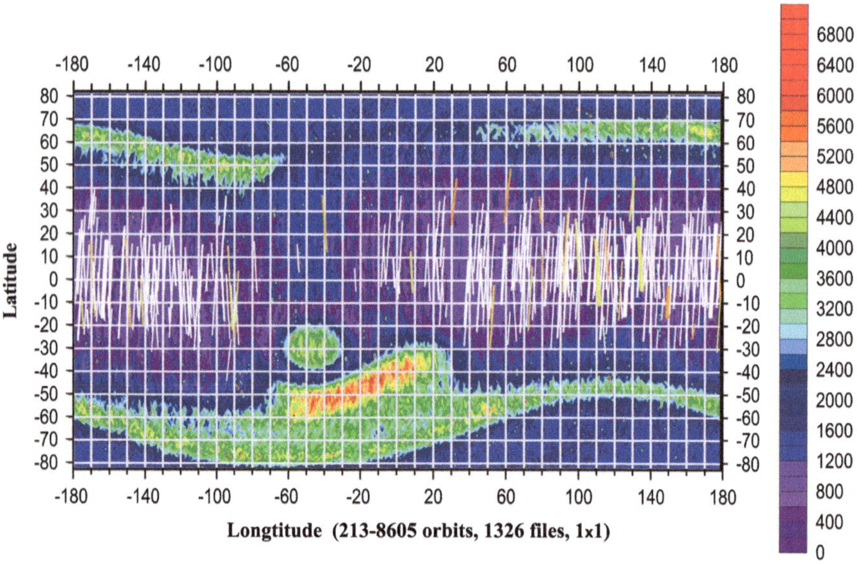


**Fig. 20** Counts rate map in the low-energy gamma-band averaged during 1.5 years of the apparatus operation. *Lines* colors depend on precipitation type (see Fig. 16). From [23]

$$I \sim E^{-\alpha}, \text{ where } \begin{cases} \alpha = 2.1 \pm 0.2, & \text{for precipitations of I and III types,} \\ \alpha = 1.8 \pm 0.1, & \text{for precipitations of II type.} \end{cases} \quad (8)$$

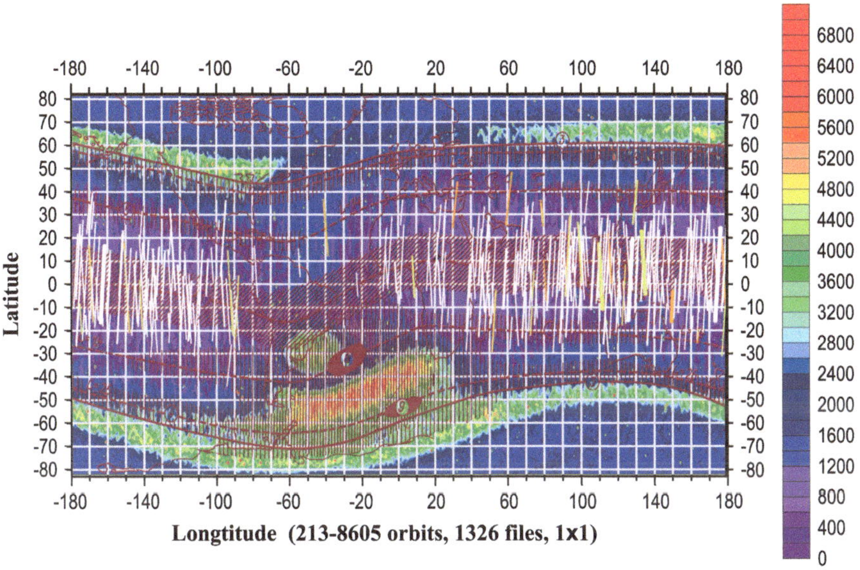


**Fig. 21** Registered during several low-orbit experiments charged particles precipitation regions projection at the map presented in Fig. 20

Shape of the spectra allows to make the conclusion that increasing of the gamma-quanta counts rate in the low-energy band of the AVS-F apparatus for some equatorial CORONAS-F satellite orbit segments was probably caused via interactions of these regions located charged particles with the material of the detector and the spacecraft.

But preliminary data analysis had shown the absence of correlation of AVS-F apparatus counts rate in the low-energy gamma-band during the quasistationary equatorial precipitations and flux of electrons (in the 0.3–0.6 MeV and 0.6–1.5 MeV bands) and protons in the range of 1–5 MeV according to the data of Cosmic Rays Monitor (Russian abbreviation is MKL) also mounted on board the CORONAS-F satellite (see description in the chapter "Scientific Set of Instruments "Solar Cosmic Rays""). Probably this effect is associated with non-sufficient sensitivity of this device (effective area of the SONG-D detector is almost ten times bigger). Only correlations of the counts rate latitudinal dependence in the regions without precipitations were observed [19, 24]. Comparison of the counts rate temporal profiles according to the AVS-F apparatus data (on the region with precipitation) and to the MKL data (at the same time interval but without any peculiarities) is presented in Fig. 23.

We have analyzed the number of the detected precipitations dependence from the geomagnetic activity level [21,22]. A series of sequential precipitations detected from September 2001 till February 2002 were selected for the analysis. Distributions of the precipitations recurrence frequency on the Kp- and Dst-indices during the studied time intervals are presented in Fig. 24. Distributions of the precipitations

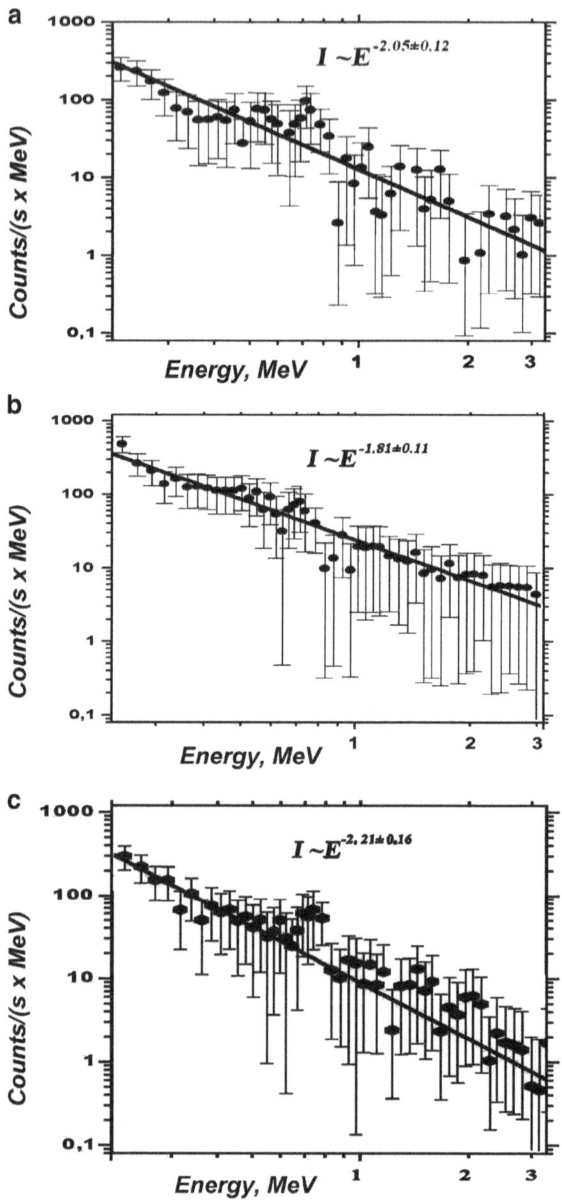


**Fig. 22** Typical energy spectra of first three types quasistationary equatorial precipitations. From [23]

amount on Kp- and Dst-indices for the events of I and II types are shown in Fig. 25a, b, and normalized amount precipitations distributions on Kp-indices for events of I and II types taking into account Kp-indices recurrence frequency are presented in Fig. 26. Corresponding subsets volumes were 251 and 118 events for the precipitations of I and II types correspondingly.

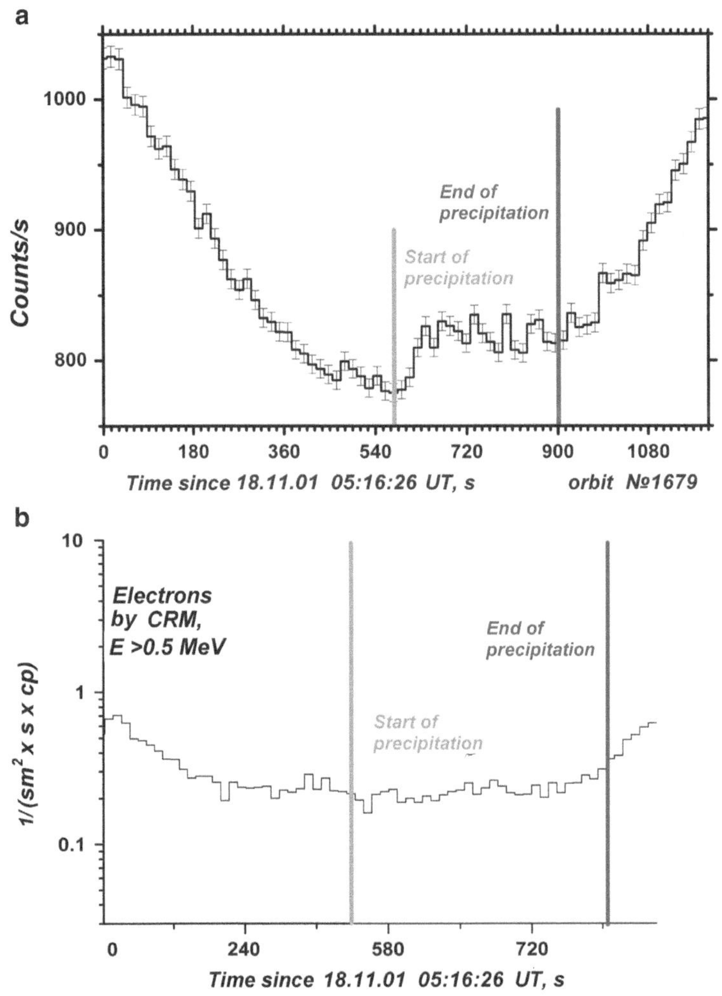


**Fig. 23** Counts rate temporal profiles in the region with precipitation according to the data of the AVS-F (**a**) and MKL (**b**)

The II type precipitations recurrence frequency increased with geomagnetic level activity increasing at the 95 % significance level [21]. Such tendency was also observed for the precipitations of I type [22, 23]—see Fig. 26. It can be associated with an effect found in the BATSE experiment, where increasing of the frequency of electrons precipitations caused by powerful UHF-transmitters (in particular, its bremsstrahlung) related to the increasing of geomagnetic activity level was observed [65]. Unfortunately, the volume of the III type precipitations subset (21 events) during the analyzed time interval is not sufficient for identification of any relation.

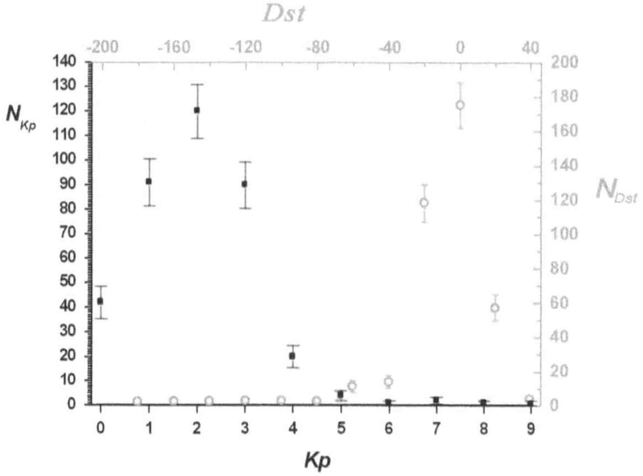


**Fig. 24** Distributions of the recurrence frequencies of Kp-indices (*black squares*, low and left axes) and of Dst-indices (*white circles*, upper and right axes) for the period from September 2001 till February 2002. From [23]

Besides, essential variations of the precipitations amount depending on the CORONAS-F satellite orbit altitude were not observed. During the period from November 14, 2001, till January 12, 2005, the orbit altitude decreased from 510 km down to 380 km. Two samples of the AVS-F apparatus data were studied: 46 precipitations observed during 9 days from November 14 till November 22, 2001 (exposure of 622,777 s, altitude ranges of 492 $\div$ 532 km), and 19 precipitations registered during 9 days from January 3 till January 12, 2005 (exposure of 318,410 s, altitudes range of 362 $\div$ 398 km). Taking into account normalizing to the exposure time these two values can be considered as coinciding in the limits of the statistic errors [19], but probably larger data samples analysis will reveal insignificant differences.

Configuration of the location area of the charged particles produced, detected by the AVS-F apparatus gamma-quanta via interactions with materials of detector and satellite, was depended on altitude: during the period from 2001 to 2004 the ratio of the AVS-F apparatus counts rate during the quasistationary equatorial precipitations to the background one counts rate decreased twice in the low-energy gamma-band (average altitude of the CORONAS-F satellite orbit decreased from 510 down to 390 km)—Figs. 27 and 16a.

## Analysis of the Terrestrial Gamma-Ray Flashes Characteristics

As it was mentioned in the section "AVS-F Apparatus Short Description," after the burst trigger the temporal profile with 1 ms time discretization was recorded in the low-energy band during 4.096 s regardless duration of the detected event. During the experiment short events with duration of several milliseconds were registered [3, 31].

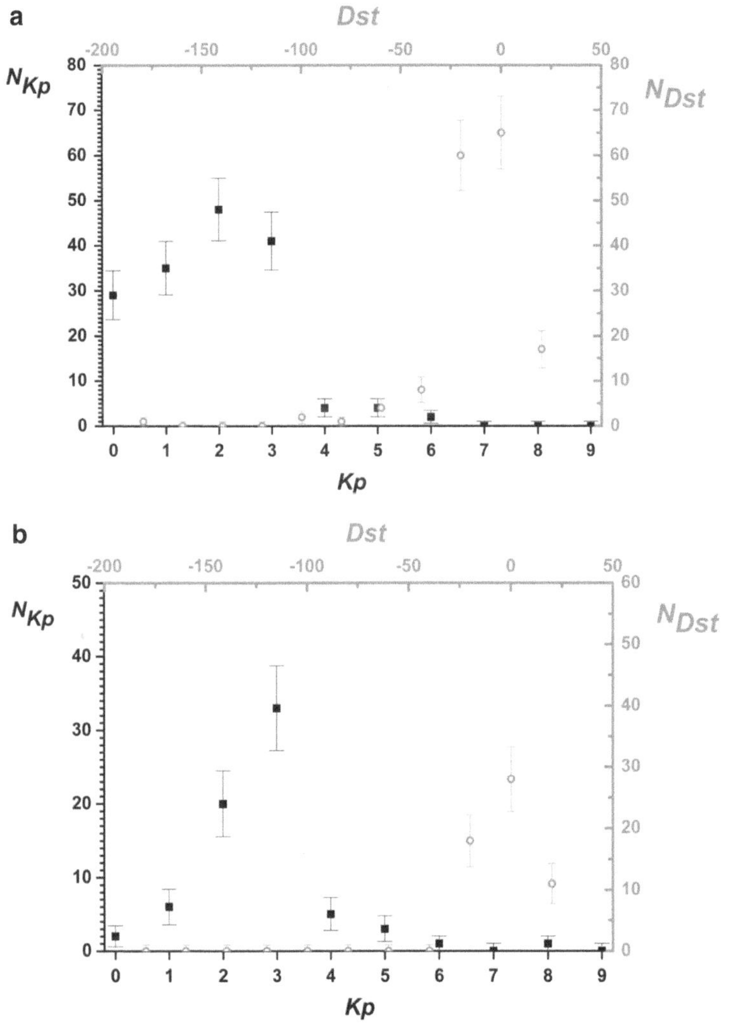


**Fig. 25** Precipitations number distributions on Kp-indices (*black squares*, low and left axes) and on Dst-indices (*white circles*, upper and right axes) for the types I (**a**) and II (**b**) events. From [23]

The observed short events can be caused by the following instrumental reasons [4]:

- fluorescence of CsI(Tl) after high-energy charged particle passing through the SONG-D detector,
- fluctuations of the number of photoelectrons or ions in the photomultiplier after high-energy charged particle passing through it;
- processes in the electronics system produced by huge energy deposition in the SONG-D detector.

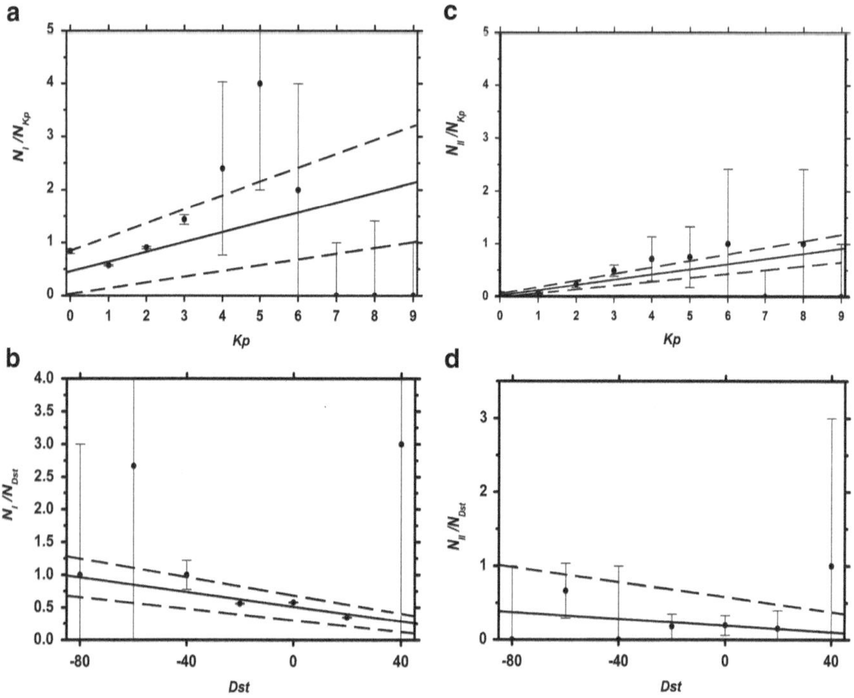


**Fig. 26** Precipitations number distributions on Kp-indices and Dst-indices for the types I (**a, b**) and II (**c, d**) events taking into account corresponding indices recurrence frequencies during the observation time. From [111]

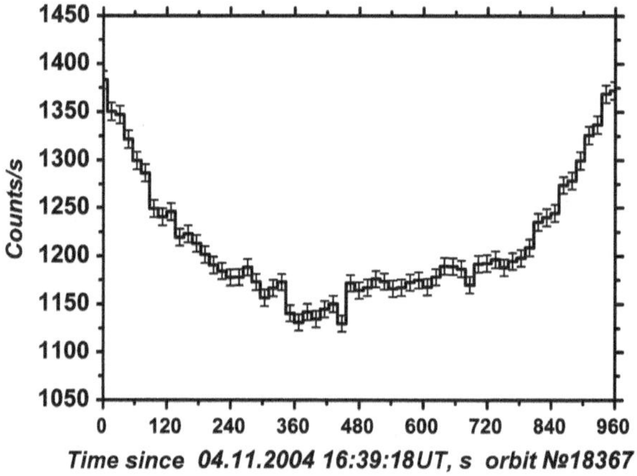


**Fig. 27** AVS-F counts rate temporal profile in the low-energy $\gamma$-band during the precipitation detected at the end of 2004

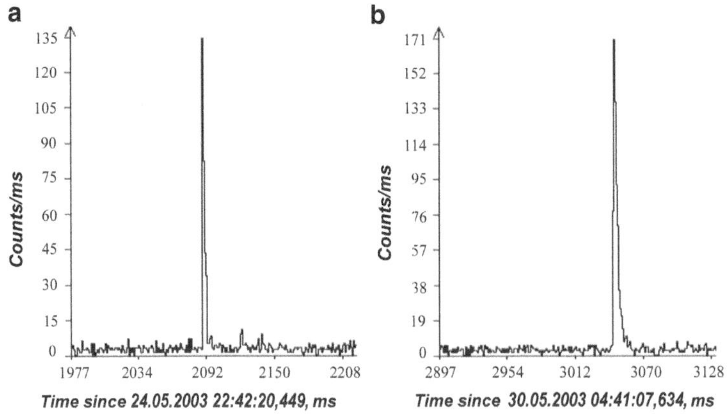


**Fig. 28** Terrestrial gamma-ray flashes typical temporal profiles according to the AVS-F data. From [31]

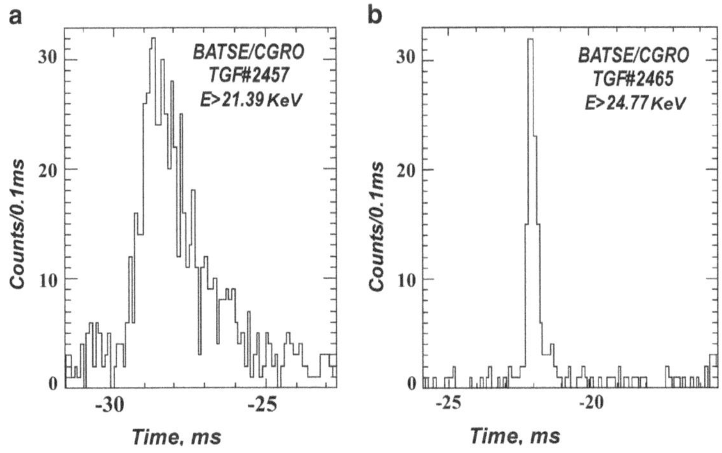


**Fig. 29** Terrestrial gamma-ray flashes typical temporal profiles according to the BATSE data. From [31]

But part of the events can be caused by the terrestrial gamma-ray flashes (TGF)—see Fig. 28, analogous to the $\gamma$-events, discovered several years ago during analysis of the BATSE data. These events were registered at the near-equatorial latitudes ($0° \div 30°$) [50]. The source of the TGF observed during the BATSE experiment was the Earth's atmosphere; moreover the flashes were observed in spite of the fact that the events coming from under the Earth's horizon had to be eliminated effectively [66].

Terrestrial gamma-ray flashes lasted for several milliseconds, their spectra were very hard, for some events gamma-emission with energy up to 17 MeV was observed by the RHESSI experiment (http://scipp.ucsc.edu/~dsmith/tgf/). Typical temporal TGFs profiles according to the BATSE instrument in the energy band of 0.02–10 MeV are presented in Fig. 29.

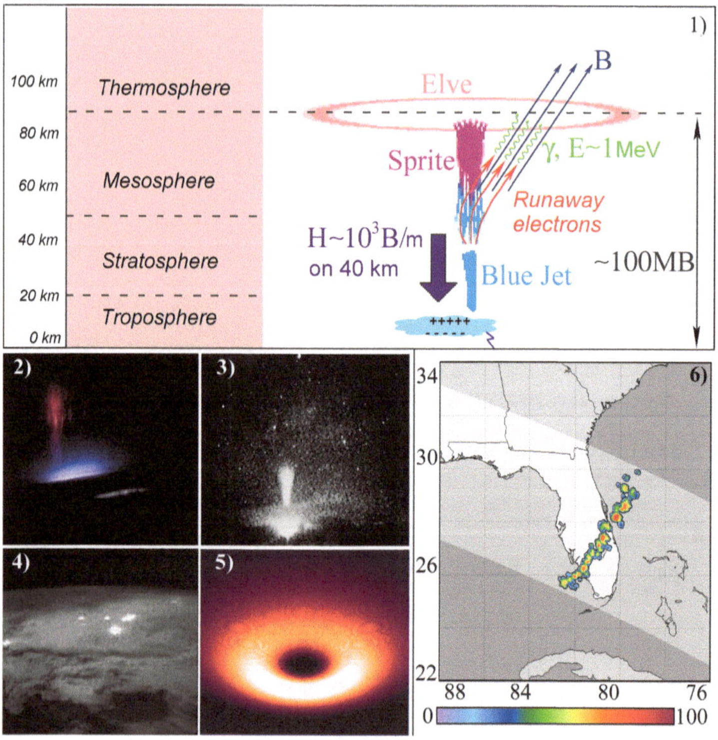


**Fig. 30** Physical phenomena associated with lightning discharge in the upper layer of the Earth's atmosphere (http://elf.gi.alaska.edu,http://umbra.nascom.nasa.gov): (1) the scheme of $\gamma$-emission generation during the lightning discharge; (2) color photo of the red sprite (RS) observed during the thunderstorm of July 4, 1994; (3) black-and-white photo of the *blue* jet (BJ) observed during the thunderstorm of July 1, 1998, above Florida; (4) black-and-white photo of light flashes during the thunderstorm of October 21, 1989, at night above Argentina, made by the camera on board space shuttle; (5) color photo of sub-ms ionospheric flash (elve) observed during the thunderstorm of July 1, 1998, above Florida; (6) distribution of light flashes intensity during the thunderstorm of March 9, 1998, above Florida according to the data of American–Japan semiconductor camera LIS (Lightning Imaging Sensor) on board the TRMM (Tropical Rainfall Measuring Mission) satellite which was launched on November 28, 1997. From [3]

It is supposed that these events were associated with thunderstorm phenomena in the upper layers of the Earth's atmosphere: they occurred during electric discharges between the upper clouds layer and the ionosphere. During this process at least four types of physical events are observed, except the lightning itself—see Fig. 30: red sprites (RS), blue jets (BJ), submillisecond ionospheric flashes (Elve), and TGF; moreover optic emission was also observed by cameras on board spacecrafts [70]. Blue jets are blue light beams (in many cases with violet component), which occur in the upper part of the charged cloud structure and directed toward the lower part of the atmosphere, and they scatter at the altitudes of $40 \div 50$ km [107]. Red sprites are optic flashes of red color, observed at the altitudes of 50–90 km directly over the thunderstorm cloud during lightning [70, 86]. Submillisecond ionospheric flashes caused by the fluorescence of nitrogen during lightning directed toward the lower part of ionosphere [70]. TGFs occur during the interaction of the atmosphere's matter and so-called runaway electrons, i.e., electrons produced during ionization of the Earth's atmosphere molecules by the cosmic rays and later accelerated during multiple scattering and interaction with the thunderstorm formation electromagnetic field. Obviously the most part of the electrons is thermalized during multiple scattering; however, the density of the electrons with energy of $E_e > 1$ MeV at the altitude of $\sim$ 10 km is $\sim$ $10^{-5}$ $cm^{-3}$ [102]. Consequently, TGFs are electrons bremsstrahlung emission directed along the electron beam; moreover average energy of gamma-quanta is $\sim$ 1 MeV, energy released during the flash—$E = 10^9$ erg, luminosity—$L = 10^{12}$ erg/s [68, 86].

TGF identification from the AVS-F apparatus database was based on temporal profiles fractal analysis, because all background fluctuations (including those causing short events resulting from instrumental reasons) in the AVS-F apparatus in the low-energy gamma-band were Poisson at the satellite orbit segments outside the radiation belts and SAA. Therefore, the time distribution of intervals between separate events contributed in the value of counts rate were exponential [31].

Fractal index of such temporal profiles was about 1.5, if it is possible to consider the counts rate mean value as a constant in the limits of the studied time interval [53, 59] (it is true for the AVS-F apparatus, because duration of each analyzed temporal profile is 4,096 ms). At the present time 47 events not associated with instrumental effects were identified from the AVS-F database [31].

Identified events are similar to TGF in duration, temporal profile's shape, and duration distribution [3, 4, 31]. Duration distribution for TGF is wider than for short events caused by instrumental effects—see Fig. 31a.

One of the events identified in theAVS-F apparatus database was detected during a powerful tropical thunderstorm Beni, and it proves its correct interpretation as TGF. This TGF was located not far from the center of the thunderstorm (Fig. 31b). Now weather conditions during the other 44 events are analyzed.

Gamma-emission during TGF is directed, width of the beam is less than one degree even taking into account scattering; therefore it is impossible to observe one and the same flash by different detectors simultaneously. But two TGFs were observed by the AVS-F apparatus on July 19, 2004, at about 21:20 in the low-energy gamma-band, coordinates of the CORONAS-F satellite during this period were close to the coordinates of the RHESSI spacecraft, which detected TGF at the same day at 21:24:01,530 (http://scipp.ucsc.edu/~dsmith/tgf/). Consequently AVS-F and RHESSI observed the processes occurred during one and the same thunderstorm formation [4]. Temporal profiles of these events according to the AVS-F apparatus data are presented in Fig. 32.

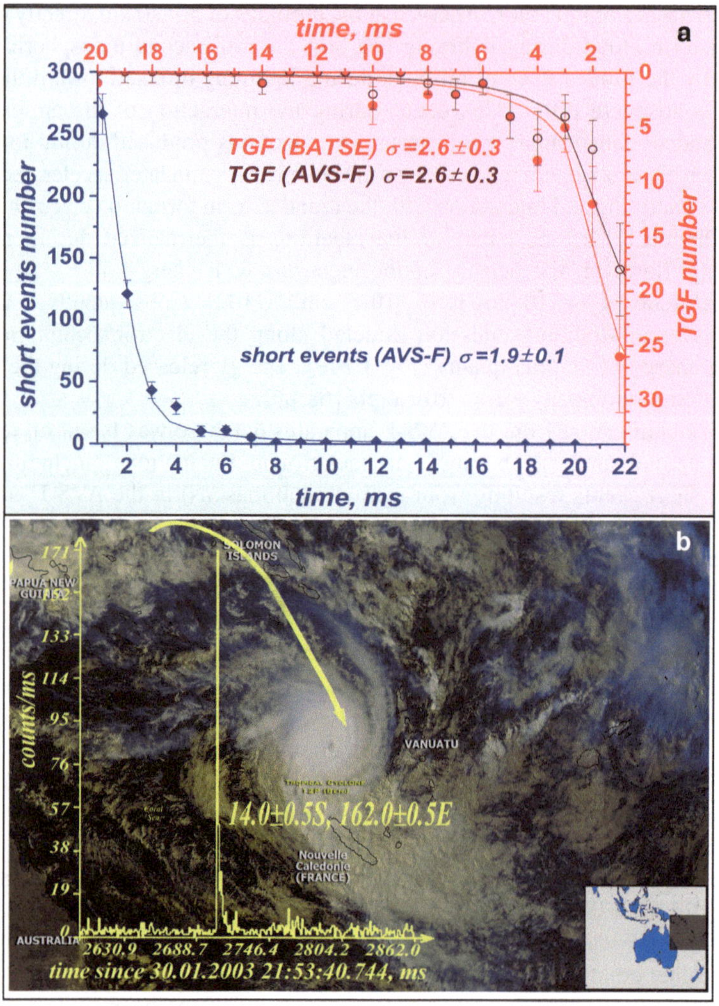


**Fig. 31** (**a**) Duration distributions for following subsets: 498 short events detected by AVS-F apparatus during January 18–20, 2003 (normalized to the sample volume), 65 TGFs registered by the BATSE instrument and 47 TGF observed by AVS-F. From [3]; (**b**) temporal profile of the TGF registered during the tropical thunderstorm Beni according to the AVS-F data. From [4]

## Analysis of the Gamma-Ray Bursts Characteristics

During the period from August 2001 till December 2005, there were over 30 gamma-ray bursts (GRB) detected by the AVS-F apparatus [25, 36].

As it was mentioned above, the minimum time for spectral information accumulation for the AVS-F was about 1 s; therefore all detected gamma-ray bursts lasted over 2 s.

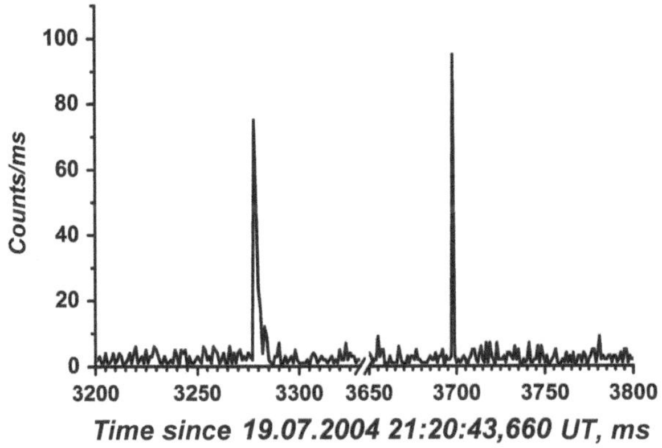


**Fig. 32** TGF observed by the data of the AVS-F apparatus on July 19, 2004, at about 21:20 UT in the low-energy gamma-band was very close in coordinates to the storm formation as well as TGF, detected by the RHESSI instrument on the same day at 21:24:01.530 UT. From [4]

Comparative characteristics of the AVS-F apparatus and other detectors of GRB [25] are operated during the same period are presented in Table 2.

Typical example of temporal profile of the gamma-ray burst detected by the AVS-F apparatus is presented in Fig. 33. The burst was observed in the equatorial region of the CORONAS-F satellite orbit. The most part of the registered by the AVS-F apparatus events were also observed by other detectors of gamma-ray bursts, for instance, by the SWIFT/BAT, RHESSI, "Konus," and HETE [36].

Gamma-ray burst GRB050126 ($z \approx 1.26$ - see http://www.mpe.mpg.de/~jcg/grbgen.html) was observed simultaneously by the AVS-F apparatus in the low-energy gamma-band and by the SWIFT/BAT [47] at 12:00:53 UT —see Fig. 34. Duration of this burst was $t_{90\text{SWIFT}} \approx 26$ s according to the SWIFT data (http://swift.gsfc.nasa.gov/docs/swift/archive/grb_table.html/), and $t_{90\text{AVS}} \approx 80$ s according to the data of the AVS-F apparatus (values of $t_{90}$ can be different depending on the sensitivity and energy band of the detector registered it). Maximum energy of $\gamma$-quanta observed during this burst by the AVS-F apparatus was $E_{\max} = 0.37 \pm 0.03$ MeV.

The burst GRB030328 ($z \approx 1{:}52$ - see http://www.mpe.mpg.de/~jcg/grbgen.html) detected at 11:20:58 UT, was observed simultaneously by the HETE and AVS-F, $t_{90\text{HETE}} \approx 100$ s (http://space.mit.edu/HETE/Bursts/), $t_{90\text{AVS}} \approx 80$ s. Its temporal profile is presented in Fig. 35. This burst is more hard than the GRB050126—the maximum energy of $\gamma$-quanta detected by the AVS-F apparatus was $E_{\max} = 0.68 \pm 0.03$ MeV.

For some GRBs emission in the high-energy $\gamma$-band of the AVS-F apparatus was also detected during the time intervals corresponding to their duration according to the data of the SWIFT, HETE, and RHESSI [25,37]. Unfortunately, redshifts are not determined for all bursts during which the high-energy $\gamma$-emission was observed by the AVS-F apparatus (http://www.mpe.mpg.de/~jcg/grbgen.html).

**Table 2** Comparative characteristics of the AVS-F apparatus and other GRB detectors

| Characteristics | AVS-F | RHESSI | SWIFT/BAT | GGS-Wind Konus |
|---|---|---|---|---|
| Effective area ($cm^2$) | $\sim 300$[a] | < 300 | 5200 | 200 |
| Field of wiev (sr) | $\sim 2\pi$ | $2\pi$ | $2\pi$ | $4\pi$ |
| Detector | CsI(Tl) | Ge | CdZnTe | NaI |
| Energy band (MeV) | 0.05–11<br>4–94<br>(in 2001)<br>0.08–22<br>2–260<br>(in 2005) | 0.05–17 | 0.015–0.15 | 0.01–10 |
| Energy resolution (FWHM) | 86 keV (13 %) for 660 keV | < 3 keV for 100 keV < 5 keV for 500 keV | 3.3 keV for 60 keV 100 keV for 660 keV | 100 keV for 660 keV |

[a] For the events detected only by the upper surface

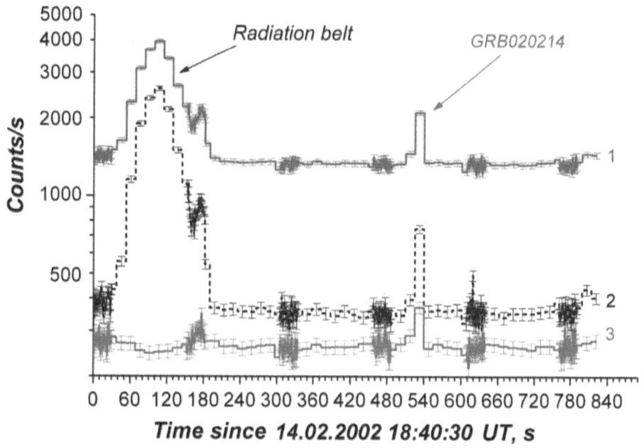


**Fig. 33** The example of gamma ray burst detected by the AVS-F apparatus and HETE-2 instruments. The first counts rate increase corresponds to the radiation belt, the second one—GRB020214. Temporal profiles of the burst are presented in three energy bands: 0.1–11 MeV (1), 0.41–11 MeV (2), and 0.563–2.34 MeV (3). From [9]

Gamma-emission during the GRB021008 (beginning at 07:00:50 UT according to the data of the RHESSI instrument) was registered in the energy band up to 7 MeV, i.e., it was detected in both $\gamma$-bands of the AVS-F apparatus. Temporal profiles of this burst by the RHESSI (http://grb.web.psi.ch/grb_list_2002.html) and AVS-F data in different energy ranges are presented in Fig. 36. Duration of this burst according to the data of the RHESSI instrument was $t_{90\mathrm{RHESSI}} \approx 13$ s in the range of 0.025–0.5 MeV, and according to the data of the AVS-F apparatus $t_{90\mathrm{AVS}} \approx 9$ s in the range of 0.1–7 MeV.

Gamma-emission of maximum energy was observed by the AVS-F apparatus during the burst GRB050525 [36], observed simultaneously by the RHESSI instrument (http://grb.web.psi.ch/grb_list_2005.html) and the AVS-F apparatus at 00:49:50 UT. Its temporal profiles in different energy ranges according to the

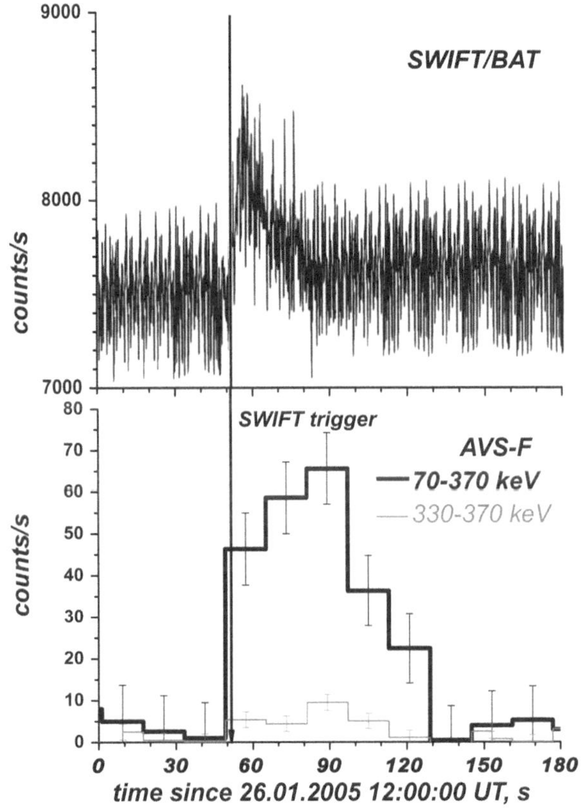


**Fig. 34** GRB050126 temporal profiles according to AVS-F and SWIFT/BAT data. From [25]

data of both experiments are presented in Fig. 37. Maximum energy of $\gamma$-quanta detected by the AVS-F apparatus for this burst was $E_{\max} = 147 \pm 3$ MeV. During the period of CORONAS-F satellite operation only AVS-F apparatus used the signal from the SONG-D detector and itself detector SONG in MKL instruments complex could provide registration of such high energy gamma-emission. Other analogous instruments were out of operation this time.

Typical energy spectra of the gamma-ray bursts according to the data of the AVS-F apparatus are presented in Fig. 38. There are no distinct spectral features in these spectra, and it is in agreement with the RHESSI instrument data in the range of 100–1500 keV [108].

## Solar Flares Characteristics Analysis

Analysis of the energy spectra and time behavior of the X-ray and gamma-emission in wide energy range during solar flares allows to obtain important information about the processes of particles acceleration in the solar flares, and

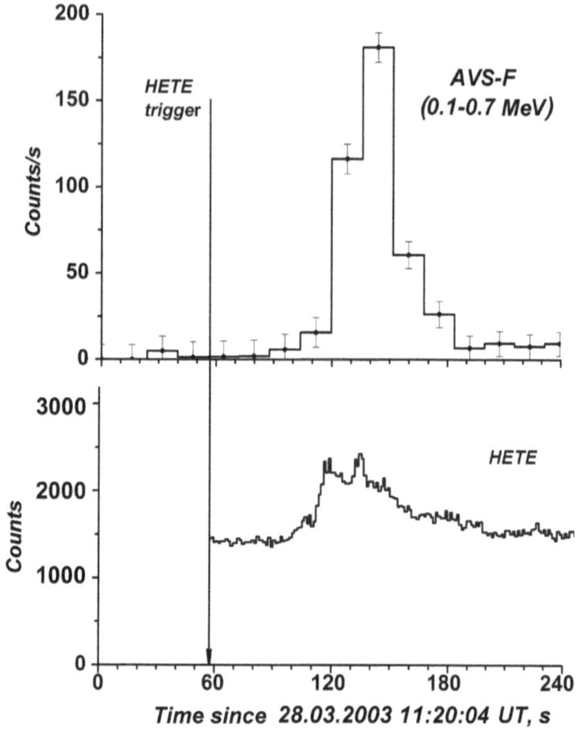


**Fig. 35** GRB030328 temporal profiles according to AVS-F and HETE data. From [25]

about the chemical composition of the accelerated particles and surrounding solar atmosphere [58, 60, 61, 81, 94, 95].

Solar flares were observed by the various detectors on board different spacecrafts (HEAO-1, HEAO-3, OSO-7, HINOTORI, CGRO, SMM, GRANAT, RHESSI, YOHKOH, SOHO, CORONAS series satellites, etc.). Taking it into account the volume of information about solar flares obtained by different instruments the observation data systematization and cataloguing become more and more important allows to simplify and promote comparison, analysis, and interpreting of the results obtained by various experiments. Now several catalogues with detailed information about the observed solar flares are published (see, for instance, [64, 82, 103]). Full catalogue of the flares from October 2002 till to the present day according to the data of the GOES, SOHO, RHESSI, YOHKOH, SDAC, LASCO, and TRACE experiments is located at http://www.lmsal.com/solarsoft/latest_events_archive.html and is updated permanently.

According to the results of the data analysis two catalogues of solar flares detected in the low- [18] and high-energy [29] ranges of the AVS-F apparatus were composed.

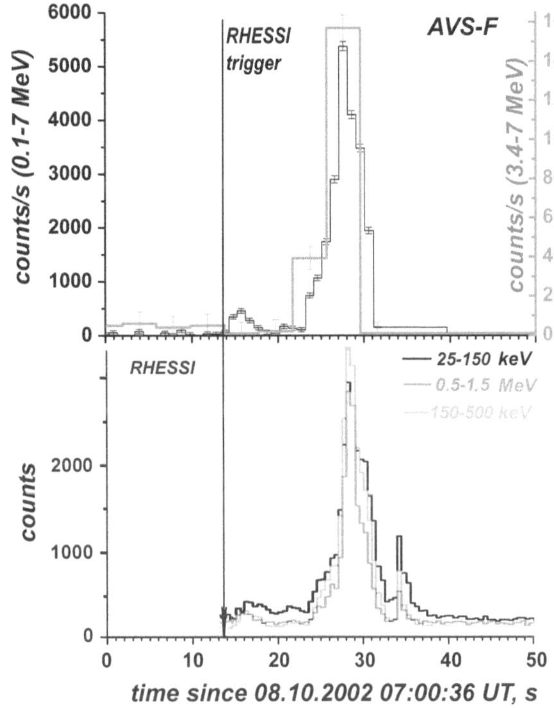


**Fig. 36** GRB021008 temporal profiles according to AVS-F and RHESSI data. From [25]

## *Catalogue of Solar Flares in the Low-Energy Gamma-Band According to the Data of the AVS-F Apparatus On board the CORONAS-F Satellite During 2001–2005*

Over 60 solar flares of class stronger than M2.0 (according to the GOES classification) were detected by the AVS-F apparatus in the low-energy gamma-band during the period from August 2001 till the beginning of December 2005. Over 30 of them were more intensive than M5.0 [11, 18].

In some cases the series of close in time powerful solar flares (with class more than M1.0) originated from the same active region were observed. For instance, in January 2005 over 200 solar flares were registered (moreover 27 of them were of M and X classes) by the detectors on board the GOES, RHESSI, SOHO, and other satellites. These flares can be grouped in the series with the active regions NOAA 10715—NOAA 10729 sources (http://www.lmsal.com/solarsoft/latest_events_archive.html). Some of the active regions produced only small flares, below C1.3 class—AR 10724, AR 10725, AR 10727, and AR 10729, while the region NOAA 10720 was a source of several flares of X class.

Six of 27 powerful flares of January series were also observed by the AVS-F apparatus [15, 32–34]. An active region NOAA 10720 produced five of them, and NOAA 10719—the other one. Moreover, during this period the AVS-F apparatus

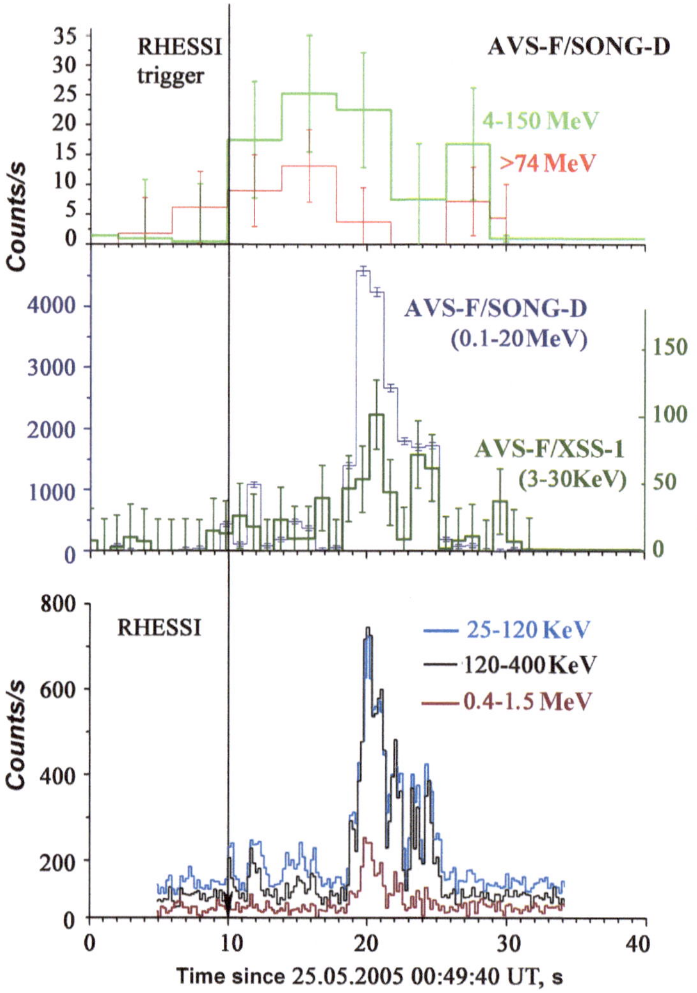


**Fig. 37** GRB050525 temporal profiles according to AVS-F and RHESSI data. From [36]

registered hard X-rays and $\gamma$-rays in the range of $E > 100$ keV during some flares of B and C classes (for more detailed discussion see the section "Hard X-ray and Gamma-Emission During Faint Solar Flares").

Another several series of the flares were observed during the period from November 3 till November 10, 2004 (http://www.lmsal.com/solarsoft/latest_events_archive.html) [109]: during this interval about 100 solar flares were registered (16 of them were of M and X classes) originated from the active regions 10689, 10691, 10693, 10696, and 10698. Moreover AR 10696 was a source of over 50 flares, 15 of which were more powerful than M1.0 class.

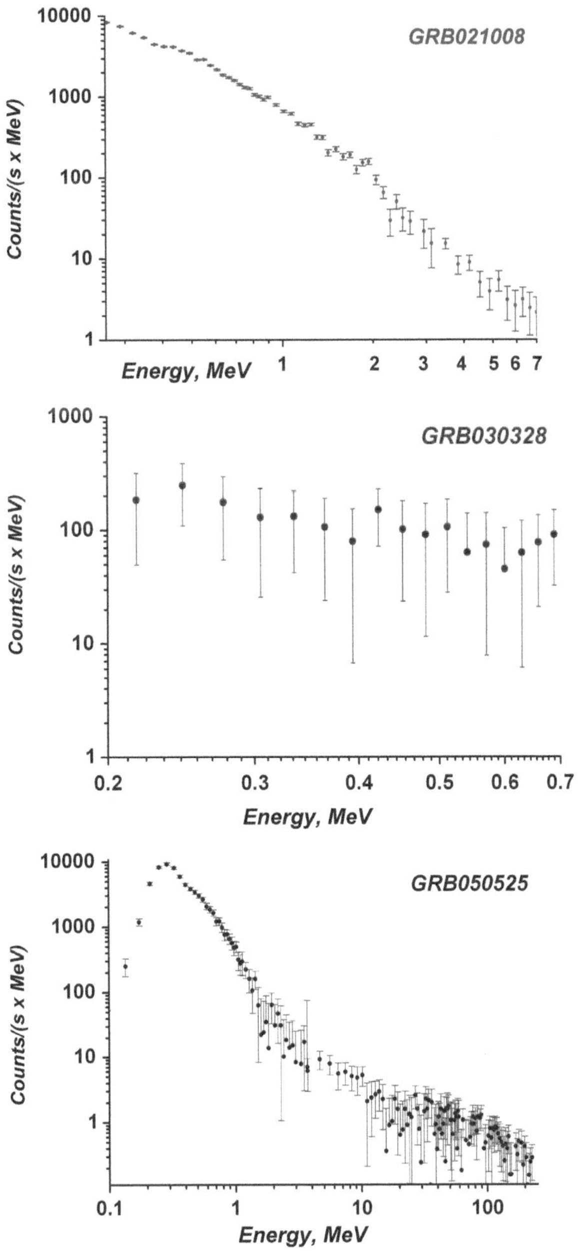


**Fig. 38** Energy spectra of the GRB021008, GRB030328 and GRB050525 according to the data of the AVS-F apparatus. From [25]

During the period from October 19 till November 15, 2003, eight active regions were detected on the Sun. They produced over 200 solar flares, and six of these active regions (10484, 10486, 10488, 10490, 10498, 10501) produced hard X-ray and $\gamma$-emission (http://www.lmsal.com/solarsoft/latest_events_archive.html).

During this period detectors on board the GOES, RHESSI and other satellites detected 60 solar flares stronger than M1.0 class. Several of these flares were also detected in the various experiments on board the CORONAS-F satellite [104], and five of them were observed by the AVS-F apparatus [11, 12, 16].

For the solar flares majority the hard X-rays and gamma-emission were observed during the rise phase of the soft X-rays according to the data of the detectors on board the GOES series satellites, for some events it was registered during its maximum [14]. Usually the gamma-emission duration during solar flares in the low-energy range of the AVS-F apparatus was essentially smaller than the duration of their X-rays emission in the ranges of 0.1–0.8 nm and 0.05–0.4 nm according to the GOES satellites data, and summarized temporal profiles in the low-energy range has simple structure with one maximum as in X-ray range [14, 18]. But for several solar flares duration of hard X-ray and $\gamma$-emission substantially coincides with the duration of the soft X-rays emission [14]. Besides, during some flares several episodes of detection of hard X-ray and $\gamma$-rays associated with the temporal profile features in the soft X-ray range (see, for instance, p. 225).

Among the flares of all the above-mentioned types there are events both with lines in energy spectrum and with smooth spectra, i.e., e-dominant.

During some events (including small flares) temporal profile's thin structure with timescale of approximately $30 \div 160$ s and $7 \div 35$ ms was observed in the AVS-F apparatus low-energy gamma-range both in the bands corresponding to flare spectral continuum and identified spectral lines.

## Solar Flares Temporal Profiles

One of the typical examples of the flares with gamma-emission observed during the rise phase of the soft X-rays emission was the flare of January 20, 2005 (X7.1) [15]. This event was the most powerful one from the series of this period (see p. 217).

It began at 06:36 UT according to the data of the detectors on board the GOES series satellite, finished at 07:26 UT, and maximum of the soft X-rays emission was observed at 07:01 UT (http://www.swpc.noaa.gov/ftpdir/warehouse/2005/2005_events.tar.gz). The source of this flare was the active region NOAA 10720. The flare of January 20, 2005, was accompanied with the coronal mass ejection (CME), and also the proton event and the GLE N 69 (Ground Level Event) were the most intensive ones beginning from 1978 and 1956, respectively—see, for instance, [43, 48, 87]. Temporal profiles of hard X-rays and $\gamma$-emission from this flare in the range of 0.1–20 MeV according to the data of the AVS-F apparatus are presented in Fig. 39.

Some of the events of November 2004 series also had similar temporal profile type (see p. 218), in particular, the flares of November 3 and 4 (of M1.6 and M5.4 classes, respectively) originated from the active region NOAA 10696. Their temporal profiles according to the data of the AVS-F and GOES-12 are presented in Fig. 40b, c. The flare of November 3 began at 03:23 UT according to the data of the detectors on board the GOES-12 satellite and ended at 03:57 UT (http://

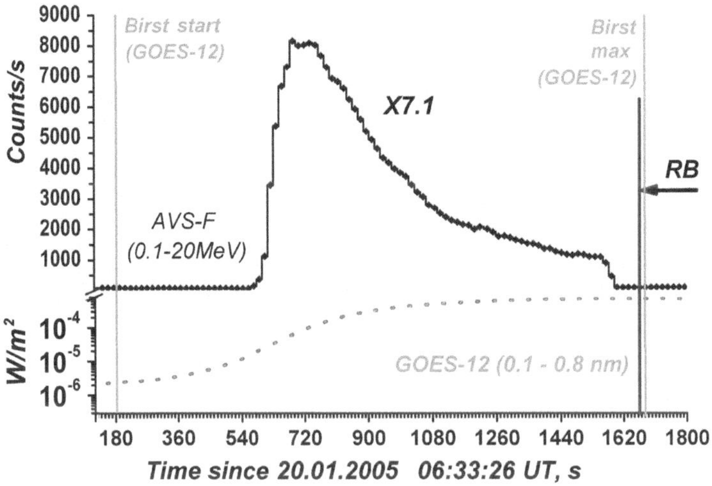


**Fig. 39** January 20, 2005 solar flare temporal profile by the data of the AVS-F apparatus (with background subtraction) in the energy range of 0.1–20 MeV and one of by the GOES-12 satellite data in the band of 0.1–0.8 nm. Time interval when the CORONAS-F satellite passed through the Earth's radiation belt is marked by the *arrow* RB. From [33]

www.swpc.noaa.gov/ftpdir/warehouse/2005/2004_events.tar.gz) [109], its emission in the range of 0.1–19 MeV was observed by the AVS-F apparatus in the polar region of the CORONAS-F orbit. The flare of November 4 was detected in the equatorial part of the CORONAS-F orbit, its soft X-rays according to the GOES-12 data was observed from 22:53 UT till 23:26 UT (http://www.swpc.noaa.gov/ftpdir/warehouse/2005/2004_events.tar.gz) [109].

Hard X-rays and gamma-emission of the flares of April 9, 2003 (of M2.5 class) and October 26, 2003 (of M7.6 class) were also registered during the X-rays rise phase of [14]—see Fig. 41a, b. The first of these flares began at 23:23 UT and ended at 23:34 UT according to the data of the GOES-10 satellite (http://www.swpc.noaa.gov/ftpdir/warehouse/2003/2003_events.tar.gz). Its source was the active region 10326.

The event of October 26, 2003, was one from the solar flares series of October–November, 2003 (see p. 219). Its source was an active region NOAA 10484 (N01W38), flare begin, maximum and end were registered in the soft X-rays band according to the data of the GOES-12 satellite at 21:34 UT, 21:40 UT, and 21:48 UT correspondingly (http://www.swpc.noaa.gov/ftpdir/warehouse/2003/2003_events.tar.gz). Unfortunately, during this flare maximum the CORONAS-F satellite passed into the radiation belt, but gamma-emission of this flare in the low-energy range of the AVS-F apparatus was recorded in the near-equator region of the satellite's orbit in the time interval from 21:36:52 UT till 21:37:56 UT [16].

The solar flare of July 17, 2002 (of M8.5 class) was a characteristic illustration of the type of solar flares, during which hard X-rays and $\gamma$-rays were observed

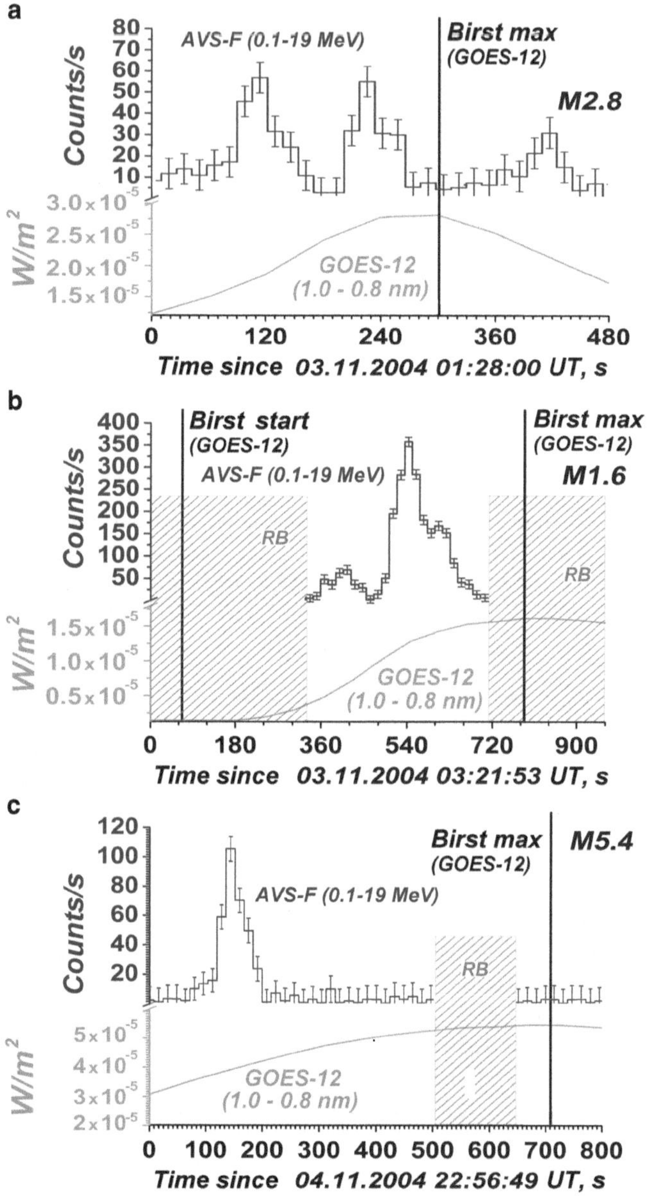


**Fig. 40** Temporal profiles of the solar flares observed on November 3–4, 2004, according to the data of the AVS-F apparatus (with background subtraction) and the GOES-12 satellite. Time intervals when the CORONAS-F satellite passed through the Earth's radiation belt are marked by hatched regions RB. From [33]

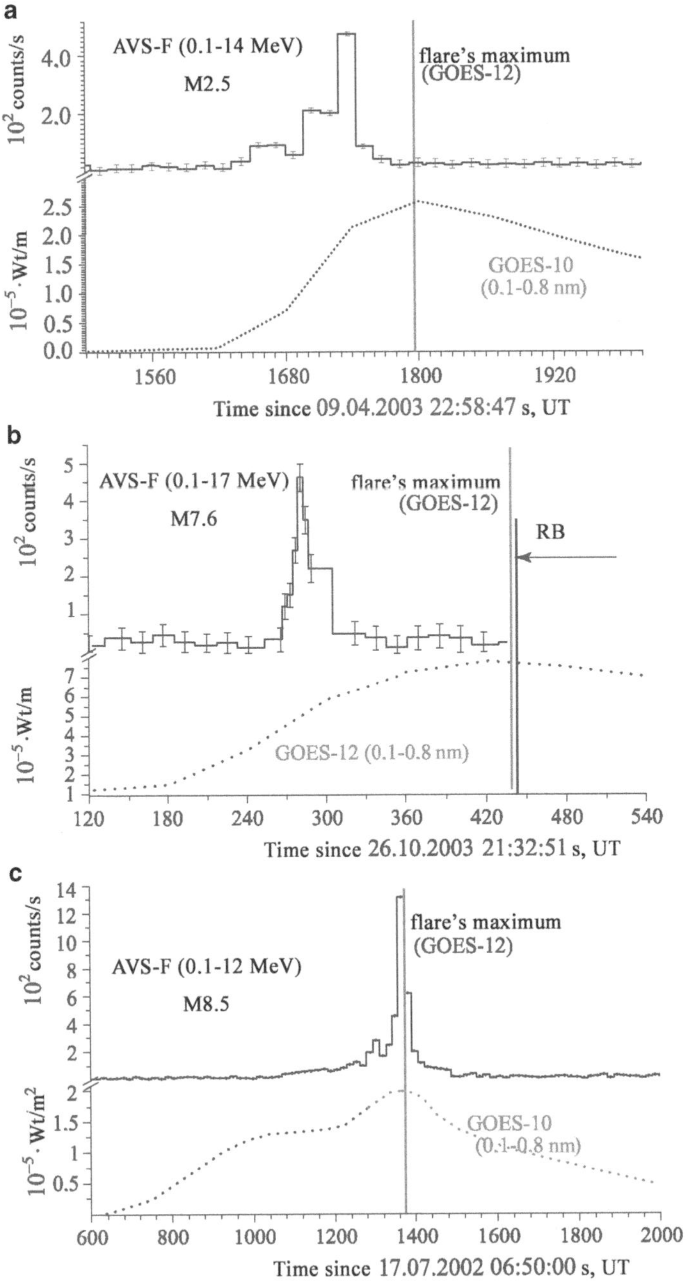


**Fig. 41** Temporal profiles of the solar flares of April 9, 2003 (**a**), from [18], October 26, 2003 (**b**), from [18], and July 17, 2002 (**c**), from [14], in the low-energy $\gamma$-band according to the data of the AVS-F apparatus (with background subtraction) and ones in soft X-ray bands according to the data of GOES series satellites. Beginning of time interval when the CORONAS-F satellite was passing through the Earth's radiation belt is marked by the *arrow* RB

during the maximum of the soft X-rays emission [14]—see Fig. 41c. According to the data of the instruments on board GOES-10 satellite this flare began at 06:58 UT, ended at 07:19 UT, and maximum of its soft X-rays emission was observed at 07:13 UT (http://www.swpc.noaa.gov/ftpdir/warehouse/2002/2002_events.tar.gz). Its source was an active region 10030. One of the series of November 2004 flares (November 3, M2.8 class) also had such type temporal profile. This flare was originated from the active region 10691, its soft X-rays emission was observed from 01:26 UT till 01:36 UT according to the data of the instruments on board GOES-12 satellite (http://www.swpc.noaa.gov/ftpdir/warehouse/2005/2004_events.tar.gz). Temporal profiles of the flare of November 3, 2004 according to the AVS-F data (with background subtraction) and to the data of the instruments on board GOES-12 satellite are presented in Fig. 40a.

The events of January 15 and 17, 2005, are also examples of this type of flares [15, 32, 33]. They were from January flares series (see p. 217), produced by an active region of NOAA 10720. The flare of January 15, 2005 (of X2.6 class) was observed during the period of 22:35–23:31 UT according to the data of the instruments on board GOES series satellites (http://www.swpc.noaa.gov/ftpdir/warehouse/2005/2004_events.tar.gz). During the maximum of the soft X-rays (23:02 UT according to the data of the GOES-12 satellite) the AVS-F apparatus registered hard X-rays and $\gamma$-emission [33]—see Fig. 42a. Solar flare of January 17, 2005 (of X3.8 class) began at 06:59 UT and ended at 10:07 UT (with maximum at 09:52 UT) according to the detectors on board the GOES series satellites data (http://www.swpc.noaa.gov/ftpdir/warehouse/2005/2004_events.tar.gz). Its hard X-rays and $\gamma$-emission were observed just after the CORONAS-F satellite leave the Earth's shadow area during maximum and decreasing of the soft X-rays emission [33]. Temporal profiles of this flare according to the data of the AVS-F apparatus (with background subtraction) in the range of 0.1–20 MeV and in the X-band according to the data of the detectors on board GOES-12 satellite are presented in Fig. 42b. During the last 9 min of the flare the CORONAS-F satellite was entered in the radiation belt, but by this time emission in the low-energy $\gamma$-band of the AVS-F apparatus already decreased to the background level [32].

Summarized temporal profiles of the flares, for which the duration of the hard X-rays and $\gamma$-emission coincides with the duration of the soft X-rays emission in the operational ranges of the AVS-F apparatus has more complicated shape, then ones in the soft X-bands according to the data of the detectors on board the GOES series satellites [12].

As an example, similar temporal profile was registered by the AVS-F apparatus during the solar flare of October 29, 2003 [84]— Fig. 43.

This flare of X10 class began at 20:38 UT, ended at 20:55 UT according to the data of the detectors on board the GOES series satellites (http://www.swpc.noaa.gov/ftpdir/warehouse/2003/2003_events.tar.gz) and was one of the series of October–November 2003 flares (see p. 219), originated from the active region NOAA 10486 (its coordinates were S15W02). Maximum of the soft X-rays of this flare was observed at 20:49 UT, but it corresponded to minimum in the energy range of 0.1–17 MeV, and maxima in this range were observed during the rise (4

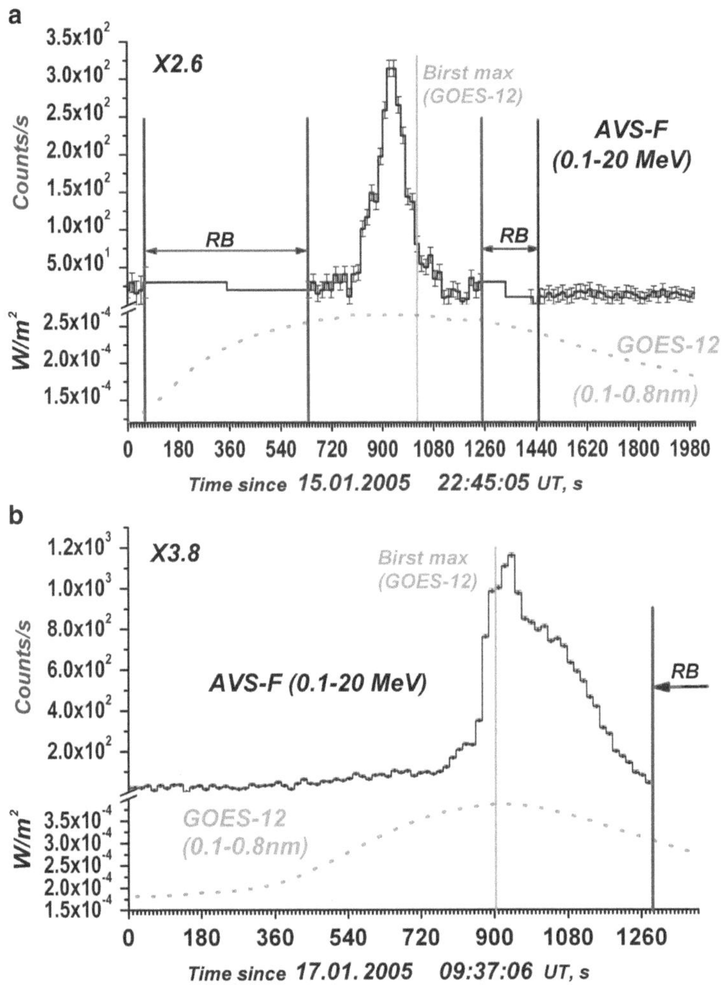


**Fig. 42** Temporal profiles for the solar flares of January 15, 2005 (**a**) and of January 17, 2005 (**b**) according to the data of the GOES-12 satellite in the band of 0.1–0.8 nm and the data of the AVS-F apparatus (with background subtraction). Time intervals when the CORONAS-F satellite was passing through the Earth's radiation belt are marked by *arrows* RB. The moment when the CORONAS-F satellite left the Earth's shadow (09:49:53 UT) corresponds to 845 s on the time axis

maxima) and decreasing (2 minima) phases of the soft X-rays emission and were not associated with the features of its temporal profile the soft X-band [78]. Thin structure with characteristic timescale of $30 \div 160$ s was observed at the summarized temporal profile of this flare in the low-energy range of the AVS-F apparatus.

Soft X-rays temporal profile of November 4, 2003 (of X18 class) solar flare has two sharp increases associated with the features of the temporal profile in the energy range of 0.1–17 MeV, as opposed to the flare of October 29, 2003 [12].

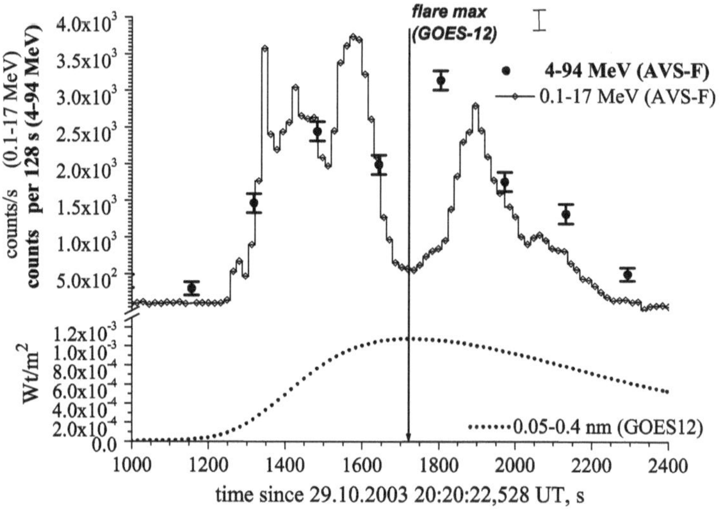


**Fig. 43** October 29, 2003, solar flare temporal profiles of AVS-F counts rate in the energy bands of 0.1–17 MeV and 4–94 MeV (with background subtraction) and one of soft X-ray flux according to the GOES-12 satellite data in the band of 0.05–0.4 nm. From [16]

The first increase was accompanied by a short impulse of gamma-emission in time interval of 19:33:01–19:34:08 UT with following decrease of the emission intensity in the range of 0.1–17 MeV down to the background level, and the second one was associated with the long episode of gamma-emission in the range of 0.1–17 MeV, lasted from 19:40:00 UT till 20:23:25 UT [16]—see Fig. 44. This flare began at 19:29 UT and ended at 20:30 UT according to the data of the detectors on board the GOES series satellites (http://www.swpc.noaa.gov/ftpdir/warehouse/2003/2003_events.tar.gz). It was the most powerful from October to November 2003 series (see p. 219). The source of the flare of November 4, 2003, was an active region NOAA 10486 (S19W83). Unfortunately, during the flare's maximum (19:44 UT according to the data of the GOES-12 satellite) the CORONAS-F satellite was passing through the radiation belt and the south polar cap [16].

If during the solar flare several episodes of hard X-rays and gamma-emission were observed, some of them could be also registered at the decreasing of the soft X-rays emission according to the data of the detectors on board the GOES series satellites, as for instance, during the flare of August 24, 2002 (of X3.1 class)—see p. 246.

## Energy Spectra of the Solar Flares

Up to six spectral lines complexes were observed during some flares, corresponding to the nuclear lines, neutron capture line (2.223 MeV), annihilation line (0.511 MeV), and $\alpha\alpha$-line—$^7$Be (0.429 MeV) + $^7$Li (0.478 MeV) [18]. Identified spectral features were observed during all time interval of hard X-rays and $\gamma$-rays

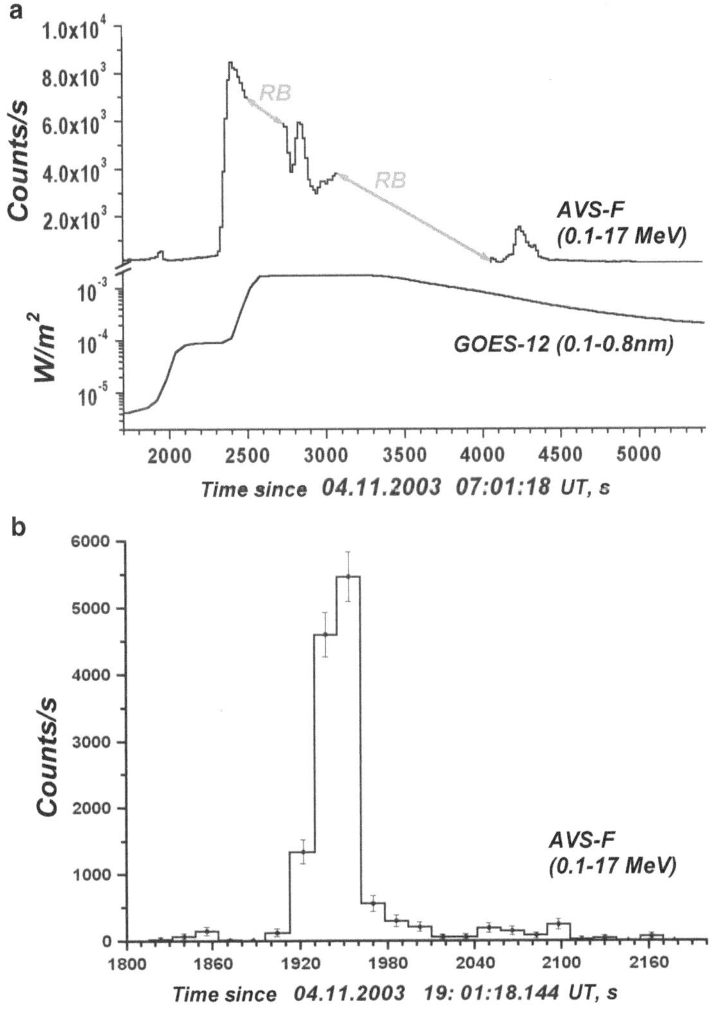


**Fig. 44** November 4, 2003 flare temporal profile of counts rate according to the AVS-F data in the low-energy gamma-band (0.1–17 MeV) without background subtraction and one of soft X-ray flux in the band of 0.05–0.4 nm by the GOES-12 satellite data (**a**) and larger scale temporal profile of the first short impulse by AVS-F data in the low-energy gamma-band (with background subtraction) (**b**). Time intervals when the CORONAS-F satellite passed through the Earth's radiation belt are marked by *arrows* RB. From [12]

registration by the AVS-F apparatus. Such lines were typical for the solar flares spectra (for instance, they were observed during October 27, 1999 [95] and July 01, 1991 [96] events).

In particular, during the October 29, 2003, solar flare five spectral lines complexes were identified in the summarized energy spectrum in the ranges of

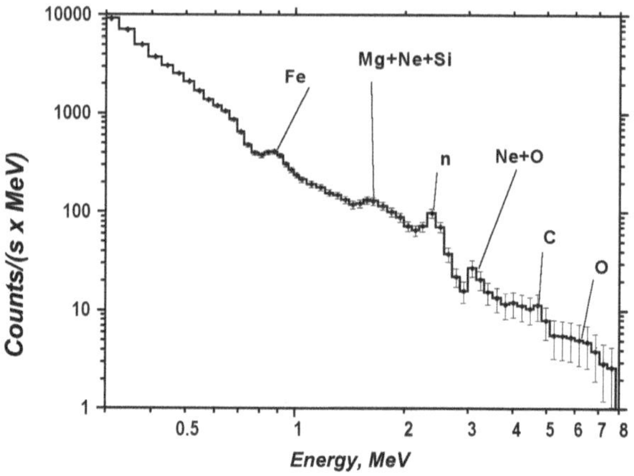


**Fig. 45** October 29, 2003, solar flare summarized energy spectrum (with background subtraction) during time interval 20:39:07–20:55:00 UT by AVS-F data. From [10]

0.81–0.94 MeV, 1.51–1.74 MeV, 2.6–3.4 MeV, 4.0–5.0 MeV, 5.3–6.9 MeV, corresponding to the following nuclear lines: $^{56}$Fe (0.847 MeV); $^{24}$Mg (1.37 MeV) + $^{20}$Ne (1.63 MeV) + $^{28}$Si (1.79 MeV); $^{20}$Ne (2.62 MeV) + $^{21}$Ne (3.18 MeV) + $^{22}$Ne (3.22 MeV) + $^{16}$O (3.2 MeV); $^{12}$C (4.44 MeV); $^{16}$O (6.13 MeV) [84]. Maximum in the range of 2.14–2.64 MeV corresponding to the neutron capture line was also revealed. October 29, 2003, solar flare summarized energy spectrum (with background subtraction) during the time interval of 20:39:07–20:55:00 UT according to the data of the AVS-F apparatus is presented in Fig. 45. Analogous spectral lines were observed during this flare according to the data of the RHESSI satellite [100]. Moreover, in the range of 0.3–0.6 MeV very faint lines (for instance, annihilation line) possibly can be observed.

Four spectral lines complexes in the ranges of 0.4–0.6 MeV ($\alpha\alpha$ + annihilation line), 1.7–2.3 MeV ($^{24}$Mg + $^{20}$Ne + $^{28}$Si + neutron capture line), 3.2–5.0 MeV ($^{22}$Ne + $^{16}$O + $^{12}$C), and 5.3–6.9 MeV ($^{16}$O) were identified in the January 20, 2005, solar flare spectrum during the time interval of 06:43:16–06:59:51 UT according to the data of the AVS-F apparatus [29, 33]—see Fig. 46.

In the spectrum of the January 17, 2005, flare also four complexes of lines at energies of 0.4–0.6 MeV ($\alpha\alpha$ + annihilation line), 0.7–0.9 MeV ($^{56}$Fe), 2.0–2.3 MeV ($^{24}$Mg + $^{20}$Ne + $^{28}$Si + neutron capture line), and 3.6–5.0 MeV ($^{12}$C) were observed during the time interval of 09:51:13–09:58:40 UT according to the data of the AVS-F apparatus [15, 32, 33] (see Fig. 47b).

During the time interval of 22:56:31–23:05:51 UT only two complexes of spectral lines were observed in the January 15, 2005, solar flare spectrum: in the range of 0.46–0.65 MeV, corresponding to the positron line, and in the band of 2.0–2.3 MeV, corresponding to the neutron capture one [32, 33] (see Fig. 48a).

Energy spectra of several solar flares observed by the AVS-F apparatus were not contained any gamma-lines, i.e., these flares were e-dominant [11, 14, 18]. Examples of such flares are presented in Fig. 48. Previously flares of this type were registered by various detectors, in particular, by the SMM/GRS [103].

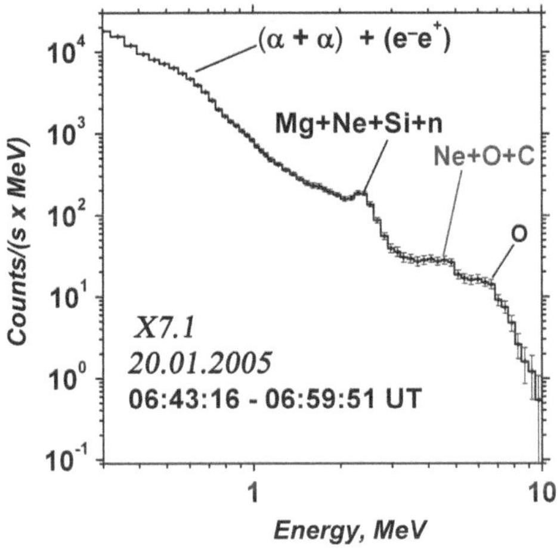


**Fig. 46** January 20, 2005, solar flare summarized energy spectrum (with background subtraction) according to the AVS-F apparatus data. From [33]

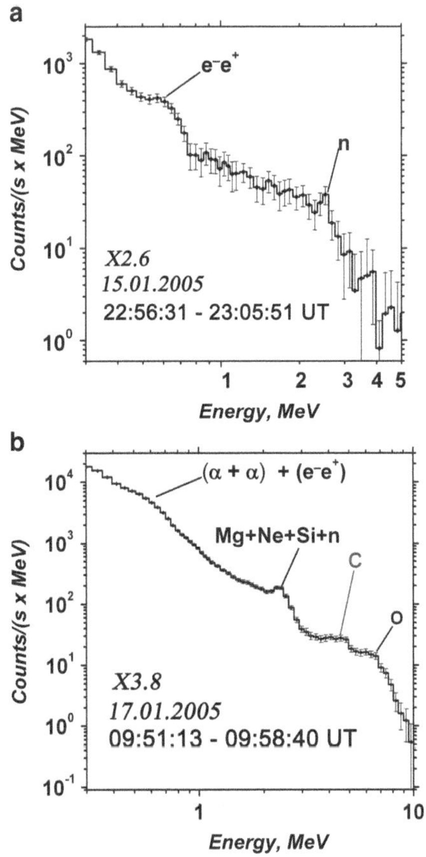


**Fig. 47** Summarized energy spectra of the solar flares of January 15, 2005 (**a**) and January 17, 2005 (**b**) according to the AVS-F apparatus data (with background subtraction). From [33]

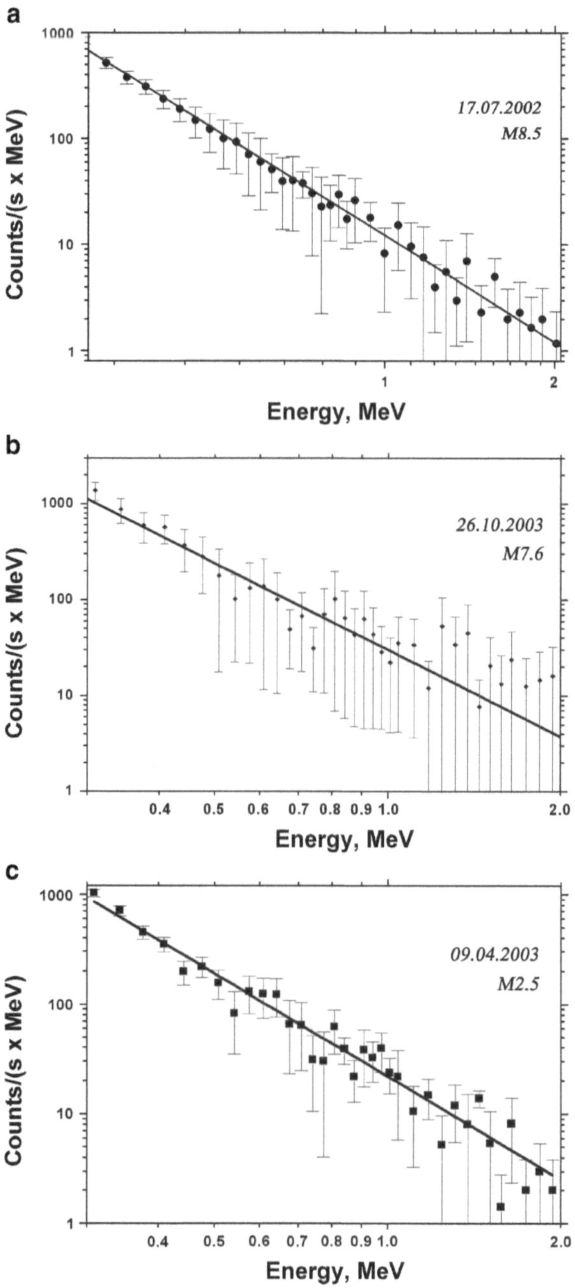


**Fig. 48** Spectra in the low-energy $\gamma$-band for the solar flares of July 17, 2002 (**a**), October 26, 2003 (**b**), and April 9, 2003 (**c**) according to the AVS-F apparatus data (with background subtraction). From [14]

**Table 3** Spectral features of the flare of October 29, 2003, and the peaks on temporal profiles in the corresponding energy ranges

| Spectral features, MeV | Center, MeV | Interpretation | Peaks in temporal profile, UT | | | | | | |
|---|---|---|---|---|---|---|---|---|---|
| 0.3–0.6 | – | Continuum | 20:42 | 20:43 | – | 20:46 | – | 20:51 | – |
| 0.81–0.94 | 0.86 | Fe | 20:42 | – | 20:45 | 20:46 | 20:48 | 20:51 | – |
| 1.51–1.74 | 1.6 | Mg+Ne+Si | 20:42 | – | 20:45 | 20:46 | – | 20:51 | 20:52 |
| 2.14–2.44 | 2,2 | n | – | – | 20:45 | 20:46 | 20:48 | 20:51 | 20:52 |
| 2.6–3.4 | 3.1 | Ne+O | – | – | – | – | – | – | 20:52 |
| 4.0–5.0 | 4.2 | C | 20:42 | 20:43 | 20:45 | 20:46 | 20:48 | 20:51 | – |
| 5.3–6.9 | 6.0 | O | – | – | 20:45 | 20:46 | 20:48 | 20:51 | – |

## Solar Flares Temporal Profiles Thin Structure

As it was already mentioned above, thin structure with characteristic timescale of $30 \div 100$ s was observed in the summarized temporal profiles of the solar flares, for which the duration of the hard X-ray and $\gamma$-emission substantially coincides with the duration of the soft X-rays—for instance, see Fig. 43. The behavior of temporal profiles of the solar flares in the energy ranges corresponding to the identified spectral features was also analyzed [16, 32–34].

Temporal profiles of the AVS-F apparatus counts rates during the October 29, 2003, solar flare in the energy ranges corresponding to the nuclear lines of the elements $^{56}$Fe, $^{24}$Mg + $^{20}$Ne + $^{28}$Si, $^{20}$Ne + $^{21}$Ne + $^{22}$Ne + $^{16}$O, $^{12}$C, $^{16}$O, neutron capture line 2.2 MeV and continuum in the energy range of 0.3–0.6 MeV, are presented in Fig. 49.

During this flare the following maxima of the gamma-emission were observed [84] (maxima characteristics are presented in Table 3): at 20:42 UT (in the ranges corresponding to the continuum and the lines of $^{56}$Fe, $^{24}$Mg + $^{20}$Ne + $^{28}$Si, $^{12}$C), at 20:43 UT (for the continuum and line of $^{12}$C), at 20:48 UT (in lines of $^{56}$Fe, n, $^{12}$C and $^{16}$O), and at 20:52 UT (in lines of $^{24}$Mg + $^{20}$Ne + $^{28}$Si, n, $^{20}$Ne + $^{21}$Ne + $^{22}$Ne + $^{16}$O). Maximum at 20:45 UT was observed in all energy bands, except ones corresponding to the continuum and the line of $^{20}$Ne + $^{21}$Ne + $^{22}$Ne + $^{16}$O. Maxima at 20:46 UT and at 20:51 UT were observed in the ranges of all lines, except the line $^{20}$Ne + $^{21}$Ne + $^{22}$Ne + $^{16}$O. Maximum at 20:49 UT corresponding to maximum emission in the soft X-band according to the data of the GOES-12 satellite was not registered in the energy band $E > 0.1$ MeV [78], but maximum at 20:48 UT corresponded to maximum of X-rays in the range of 12–25 keV observed by the RHESSI instrument (http://hesperia.gsfc.nasa.gov/hessidata/dbase/).

The October 29, 2003, solar flare emission in the low-energy gamma-band of the AVS-F apparatus was registered during approximately 1000 s, and it allowed to analyze periodograms for temporal profiles with time resolution of 16 s, including energy ranges corresponding to the identified spectral lines and continuum.

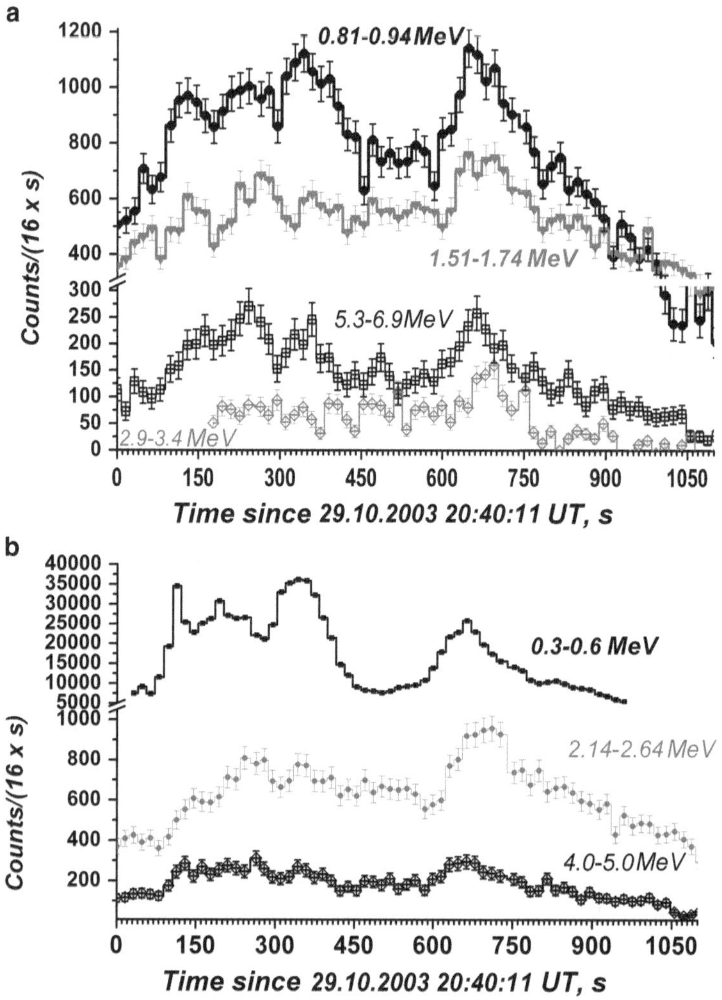


**Fig. 49** Temporal profiles of the AVS-F counts rate during the flare of October 29, 2003, in energy bands corresponding to: (**a**) nuclear lines from the elements $^{56}Fe$ (0.81–0.94 MeV), $^{24}Mg + ^{20}Ne + ^{28}Si$ (1.54–1.74 MeV), $^{20}Ne + ^{21}Ne + ^{22}Ne + ^{16}O$ (2.6–3.4 MeV), and $^{16}O$ (5.3–6.9 MeV), (**b**) continuum in the energy band of 0.3–0.6 MeV, neutron capture line 2.2 MeV, and nuclear line $^{12}C$ (4.0–5.0 MeV). From [16]

Examples of periodograms for this solar flare temporal profiles in the energy ranges corresponding to the continuum, neutron capture line, carbon and oxygen lines are presented in Fig. 50. Periodograms analysis allows to make conclusion about the presence of thin structure with timescale varied from 34 to 158 s on this solar flare temporal profiles at significance level of 99 % in various energy ranges (Table 4) [16].

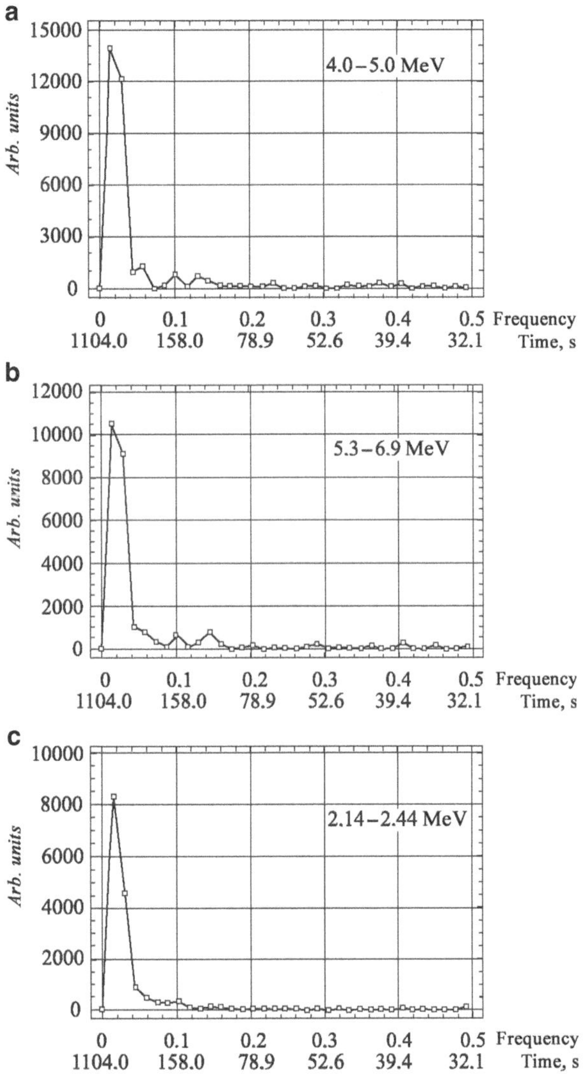


**Fig. 50** Examples of periodograms for AVS-F counts rate temporal profiles during the solar flare of October 29, 2003, in energy bands corresponding to carbon (**a**), oxygen (**b**), and neutron capture lines (**c**). From [16]

Several maxima were identified on the January 20, 2005, flare temporal profiles (see Fig. 51a). Two maxima at 06:44:36 UT and at 06:53:46 UT were coinciding in the statistic errors limits in the energy ranges of 0.15–0.30 MeV and 0.4–0.7 MeV (corresponded to the continuum and the combination of $\alpha\alpha$- and annihilation lines). One maximum at 06:46:36 UT was revealed in the ranges of 3.2–5.0 MeV and 5.3–6.9 MeV ($^{22}Ne + {}^{16}O + {}^{12}C$ and $^{16}O$), and one more at 06:47:16 UT in the band of 1.7–2.3 MeV (nuclear lines of $^{24}Mg + {}^{20}Ne + {}^{28}Si$ and neutron capture line) (see Fig. 51a). Analysis of the corresponding periodograms had shown the presence of thin structure with timescale of 33–92 s on these temporal profiles in

**Table 4** October 29, 2003, solar flare counts rate temporal profiles characteristic timescales in the various energy ranges according to AVS-F apparatus data

| Range limits, MeV | Interpretation | Temporal profiles characteristic timescales, s | | | | | | | | | | | | | | |
|---|---|---|---|---|---|---|---|---|---|---|---|---|---|---|---|---|
| 300–600 | continuum | – | – | – | 100 | – | – | – | – | – | – | – | – | – | 39 | – |
| 0.81–0.94 | Fe | 158 | – | – | – | 100 | – | – | – | – | – | – | – | 46 | – | – |
| 1.51–1.74 | Mg+Ne+Si | – | 138 | – | – | 100 | 85 | – | – | 69 | 58 | – | 48 | – | – | 34 |
| 2.14–2.44 | n | 158 | – | – | 110 | – | – | – | – | – | – | – | – | – | 39 | – |
| 2.6–3.4 | Ne+O | 158 | 138 | – | – | – | – | – | 74 | – | – | 55 | – | – | – | – |
| 4.0–5.0 | C | 158 | – | 123 | 110 | – | – | – | – | 69 | 58 | – | – | 46 | – | – |
| 5.3–6.9 | O | 158 | – | – | – | – | – | 78 | – | – | – | 55 | – | – | 39 | – |

discussed above energy ranges at the significance level of 99 % [32, 34] (see Table 5 and Fig. 51b, c).

The following general maxima were revealed in the January 17, 2005, flare temporal profiles registered in the ranges corresponding to the identified spectral features: 09:41:26 UT (0.15–0.30 MeV—continuum), 09:40:36 UT (0.4–0.7 MeV—$\alpha\alpha$ + annihilation line), 09:42:31 (0.7–0.9 MeV—$^{56}$Fe), 09:42:16 (2.0–2.3 MeV—$^{24}$Mg + $^{20}$Ne + $^{28}$Si + neutron capture line), and 09:41:54 (3.6–5.0 MeV—$^{12}$C). These temporal profiles had thin structure with time scale of 33–61 s at the significance level of 99 %, and it was proved by the analysis of corresponding periodograms—see Table 5 and Fig. 52 [32, 34].

January 15, 2005, flare temporal profile had one maximum at 23:00:19 UT in the range corresponding to the continuum, maximum at 23:00:45 UT in two other ranges (0.48–0.67 MeV—positron line and 2.0–2.3 MeV—neutron capture line). Also thin structure with time scale of 34–87 s was observed for all this flare temporal profiles at the significance level of 99 %—see Table 5 and Fig. 53 [32, 34].

Integral temporal profiles with time discretization of 1 ms in the low-energy range of the AVS-F apparatus (from one to six intervals 4.096 s long for each flare depending on its duration) were analyzed for seven solar flares. According to the results of the preliminary analysis no time regularities with duration from 2 up to 100 ms at the 99 % significance level were registered on temporal profiles of the flares of July 17, 2002 (M8.5 class), April 4, 2002 (M6.1 class), October 26, 2003 (M7.6 class), November 4, 2003 (first short episode), and September 12, 2004 (M3.2 class) [18]. A fragment of the November 4, 2003, solar flare temporal profile with time discretization of 1 ms in the energy range of 0.1–17 MeV began at 19:33:20,346 UT is presented in Fig. 54a. Periodogram for this temporal profile is shown in Fig. 54b.

As it was already mentioned in the present section, summarized temporal profile of the flare of October 29, 2003 (X10 class) had complicated structure in the energy range of 0.1–17 MeV and in the energy ranges corresponding to the features identified in this flare energy spectrum. However in all six fragments of this flare

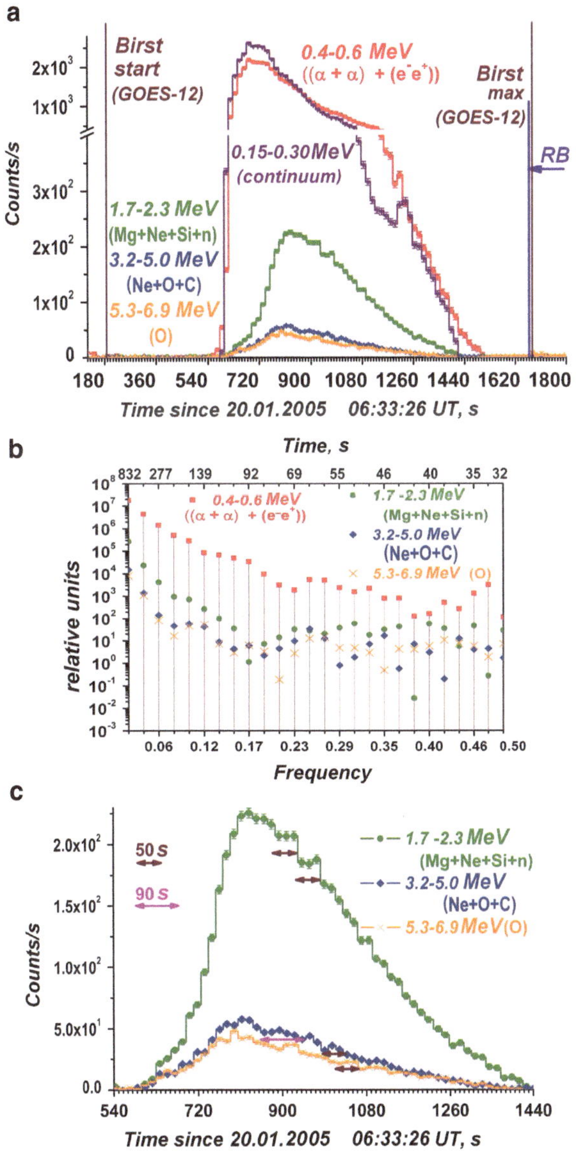


**Fig. 51** Temporal profiles of the January 20, 2005 solar flare (flare's begin was at 06:33:26.222) in the energy ranges corresponding to the separated spectral features (**a**), their periodograms (**b**), and characteristic time scales for temporal profiles of this event (**c**). Beginning of time interval when the CORONAS-F satellite passed through the Earth's radiation belt is marked by the *arrow* RB. From [32]

integral temporal profiles with time discretization of 1 ms thin structure with time scale of 2–100 ms was absent at the significance level of 99 % [16]— see Fig. 55.

Summarized temporal profile of the January 20, 2005 flare (X7.1 class) had simple structure in the energy range of 0.1–20 MeV, but thin structure with time scale of 33–92 s was observed in its temporal profiles in the energy ranges corresponding to the features identified in the energy spectrum. A fragment of this flare integral temporal profile with time discretization of 1 ms in the range of

**Table 5** Characteristic timescales of temporal profiles thin structure of the solar flares observed by the AVS-F apparatus in January 2005

| | 20.01 | | | | 15.01 | | 17.01 | | | |
|---|---|---|---|---|---|---|---|---|---|---|
| $E$, MeV | 0.4–0.7 | 2.0–2.3 | 3.2–5.0 | 5.3–6.9 | 0.4–0.7 | 2.0–2.3 | 2.0–2.3 | 3.6–5.0 | 0.7–0.9 | 0.4–0.7 |
| $\tau$,s | 83 | 69 | 92 | 92 | 61 | 87 | 61 | 64 | 61 | 61 |
| | 64 | 52 | 46 | 59 | 47 | 34 | 46 | 35 | 46 | 37 |
| | 49 | 44 | 42 | 44 | 41 | – | 33 | – | 33 | – |
| | 44 | 40 | 36 | 38 | 34 | – | – | – | – | – |
| | 38 | 35 | – | – | – | – | – | – | – | – |
| | 33 | – | – | – | – | – | – | – | – | – |

0.1–20 MeV is presented in Fig. 56a, and its periodogram in Fig. 56b. Data analysis had shown the presence of time regularities with the timescales of 7 ms, 8 ms, 22 ms, and 35 ms—see Fig. 56. The structure of these time regularities remained the same after averaging of this temporal profile over 2 s—corresponding periodogram is presented in Fig. 56a. Thus, thin structure with characteristic scales of 7–35 ms was observed on this flare integral temporal profile with time discretization of 1 ms in the energy range of 0.1–20 MeV at the confidence level of 99 % [32, 33].

## Hard X-Ray and Gamma-Emission During Faint Solar Flares

As it was mentioned above, the AVS-F apparatus registered emission with energy of $E > 50$ keV during several faint solar flashes (of B and C classes according to the GOES classification) [85].

The results of the AVS-F apparatus database analysis had shown that the existence of the component with energy $E > 50$ keV does not depend on the flare class in the soft X-range of 0.5–10 keV. Thus, for some flares of M class no statistically significant counts rate increase over the background in the range of $E > 50$ keV were detected by the AVS-F apparatus. Such flares typical example was the event of November 8, 2001, of M4.2 class (beginning at 14:59 UT, maximum at 15:35 UT, end at 16:00 UT according to the detectors on board the GOES series satellites data; http://www.swpc.noaa.gov/ftpdir/warehouse/2001/2001_events.tar.gz). However in some cases the emission in the low-energy gamma-band was observed during very faint solar flares (of B and C classes), for instance during the events of January 12, 2005, and January 14, 2005.

I According to the GOES data the solar flare of January 12, 2005, of B4.6 class was observed from 20:08 UT till 20:13 UT, with maximum of the soft X-rays emission at 20:11 UT (http://www.swpc.noaa.gov/ftpdir/warehouse/2005/2005_events.tar.gz). During all this time period the AVS-F apparatus registered significant gamma-emission increase over the background level in the low-energy gamma-range—see Fig. 57.

Moreover according to the results of the preliminary analysis thin structure with characteristic timescale of $\sim 90$ s was observed on this flare temporal profile of the

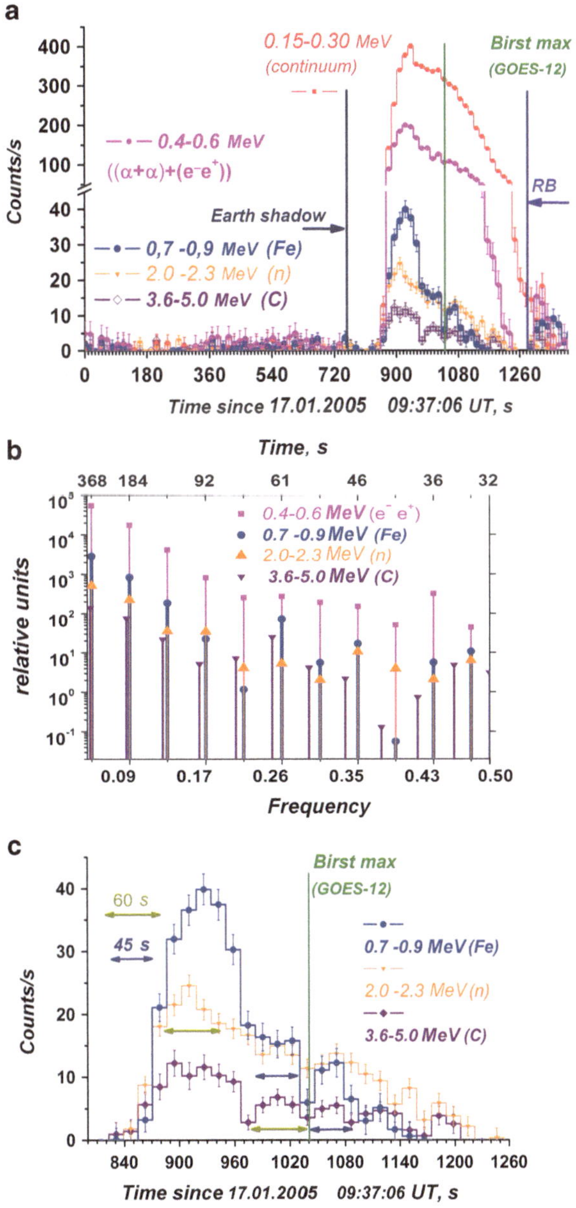


**Fig. 52** Temporal profiles for the flare of January 17, 2005 (beginning at 09:37:06.402) in the energy ranges corresponding to the separated spectral features (**a**), their periodograms (**b**), and characteristic time scales for temporal profiles of this flare (**c**). Beginning of time interval when the CORONAS-F satellite passed through the Earth's radiation belt is marked by the *arrow* RB. From [32]

AVS-F apparatus low-energy gamma-band. Maximum energy of $\gamma$-quanta detected during this flare was $E_{max} = 7.0 \pm 0.3$ MeV. Energy spectrum of the flare of January 12, 2005, is presented in Fig. 58. At the level of significance of 97 % this spectrum was approximated by power function with the index of $\alpha = -1.5 \pm 0.1$. There were no lines in energy spectrum of this flare, i.e., it was e-dominant.

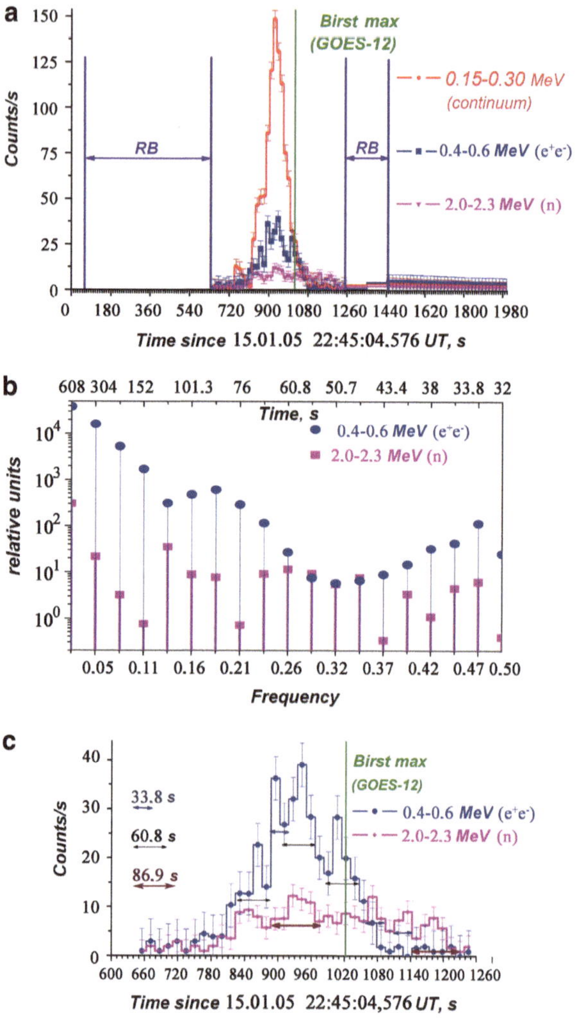


**Fig. 53** Temporal profiles of the January 15, 2005 solar flare (flare's begin was at 22:45:04.576) in the energy ranges corresponding to the separated spectral features (**a**), their periodograms (**b**), and characteristic time scales for temporal profiles of the this flare (**c**). Time intervals when the CORONAS-F satellite passed through the Earth's radiation belt are marked by the *arrows* RB. From [32]

Unfortunately, there were no data of the RHESSI satellite during this flare because of telemetry gap and spacecraft passing through the radiation belt (http://sprg.ssl.berkeley.edu/tohban/browser/?show=grth+qlpcr).

According to the data of the detectors on board the GOES series satellites C2.8 class solar flare of January 14, 2005, lasted from 17:35 UT till 17:50 UT with maximum of the soft X-ray emission at 17:42 UT (http://www.swpc.noaa.gov/ftpdir/warehouse/2005/2005_events.tar.gz). During this flare the statistically significant count rate exceeding over the background level in the low-energy range of the AVS-F apparatus was detected in time interval of 17:35–17:40 UT at the phase of the soft X-ray rise. Temporal profiles of this flare in the low-energy AVS-F

**Fig. 54** Fragment of AVS-F counts rate temporal profile in energy range of 0.1–17 MeV during the flare of November 4, 2003 with time resolution of 1 ms in the time interval from 19:33:20.346 UT till 19:33:24.443 UT (**a**) and its periodogram (**b**). From [16]

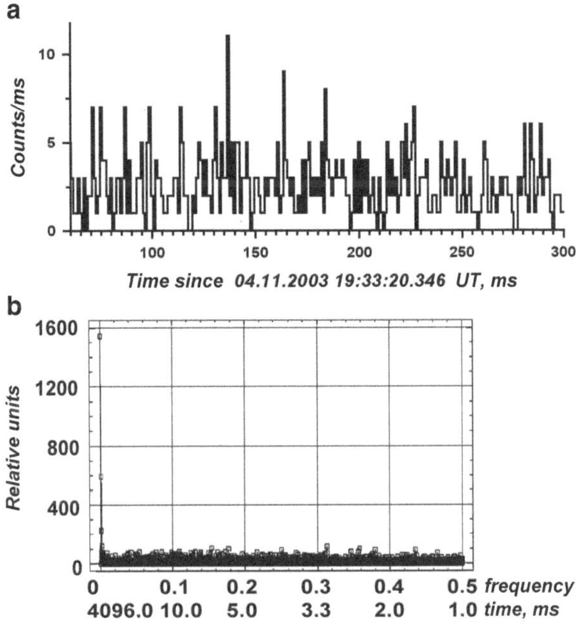


apparatus gamma-band and according to the data of GOES-12 are presented in Fig. 59. Maximum energy of $\gamma$-quanta registered during the flare of January 14, 2005, was $E_{\max} = 0.64 \pm 0.03$ MeV. Analogous to the January 12, 2005 solar flare no spectral lines were also identified in the energy spectrum of the flare of January 14, 2005—see Fig. 60. This flare was also of the e-dominant type, its energy spectrum during time interval of 17:35–17:40 UT was approximated by the power-function with the index of $\alpha = -3.1 \pm 0.2$.

Solar flare March 6, 2005, with class B1.1 was the faintest event for which emission in the energy band $E > 0.1$ MeV was registered by AVS-F data on the results of preliminary analysis [38].

Up to now the subset of 60 solar flares with classes B and C by GOES classification with the presence of statistically significant flux in energy band $E > 0.1$ MeV was studied by the data of AVS-F apparatus on board CORONAS-F satellite. The results of faint flares high-energy $\gamma$-emission analysis were discussed in detail in [38]. The shapes of temporal profiles on GOES and AVS-F/ XSS-1 data were similar at (90–99) % confidence level. Moreover, the maxima of the time derivatives of temporal profiles on GOES and AVS-F/ XSS-1 data were observed in the same time moments.

Only about 50 % of the processed faint solar flares (classes B and C on GOES classification) for which statistically significant flux in energy band $E > 0.1$ MeV was registered correspond to Neupert effect according to AVS-F data [38, 85]. Discordance of behavior of hard X-ray and $\gamma$-emission temporal profiles and time derivative of soft X-ray emission temporal profiles during faint solar flares can

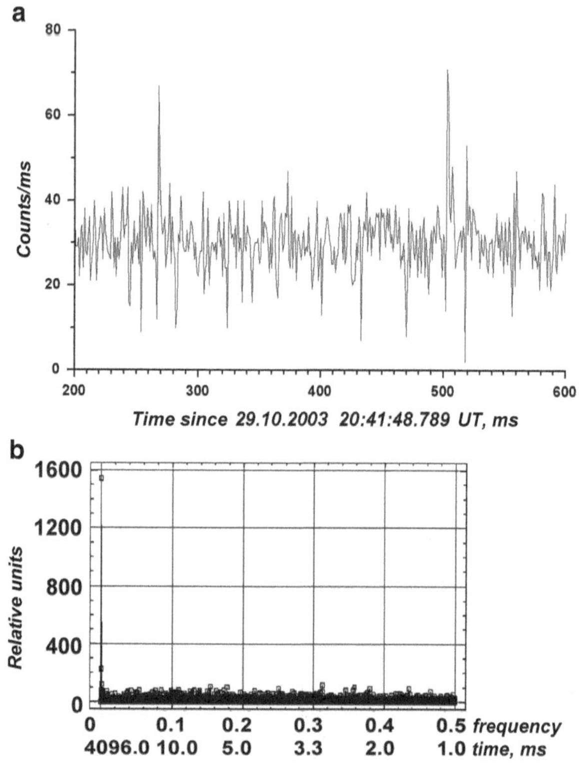


**Fig. 55** Fragment of temporal profile (**a**) of the AVS-F counts rate in the energy range of 0.1–17 MeV during the flare of October 29, 2003, with time resolution of 1 ms in the time interval from 20:41:48.789 UT till 20:41:52.886 UT and its periodogram (**b**). From [16]

indicate that in such flares occurs additional particles acceleration or dissipation of their energy, for example, due to turbulence [38].

Some thin structure with characteristic timescale 50–110 s was observed on temporal profiles of several faint events, for example on January 7 and 12, 2005 flares (classes B2.3 and B4.6 correspondingly) and on March 5, 2005 one (class B2.0) [38, 39, 85]

Energy spectra of all such solar flares had power law shape without any features by results of preliminary data analysis [38, 39].

During some faint flares high energy $\gamma$-emission was observed. For example, $\gamma$-quanta with energy up to several tens of MeV were registered during limb flare January 7, 2005 [39] and up to several MeV during disk one January 12, 2005 [38]. During these flares Neupert effect was not observed [38].

Hard X-rays emission in the range from 40 keV up to several hundreds of keV was also detected during several flares of > C7.7 class (http://smdc.sinp.msu.ru/index.py?nav=flares) in SCR experiment on board the CORONAS-F satellite (see the chapter "Scientific Set of Instruments "Solar Cosmic Rays""). The cataloging of faint solar flares (B, C classes according to the GOES classification) basing on the AVS-F apparatus data now is still in progress.

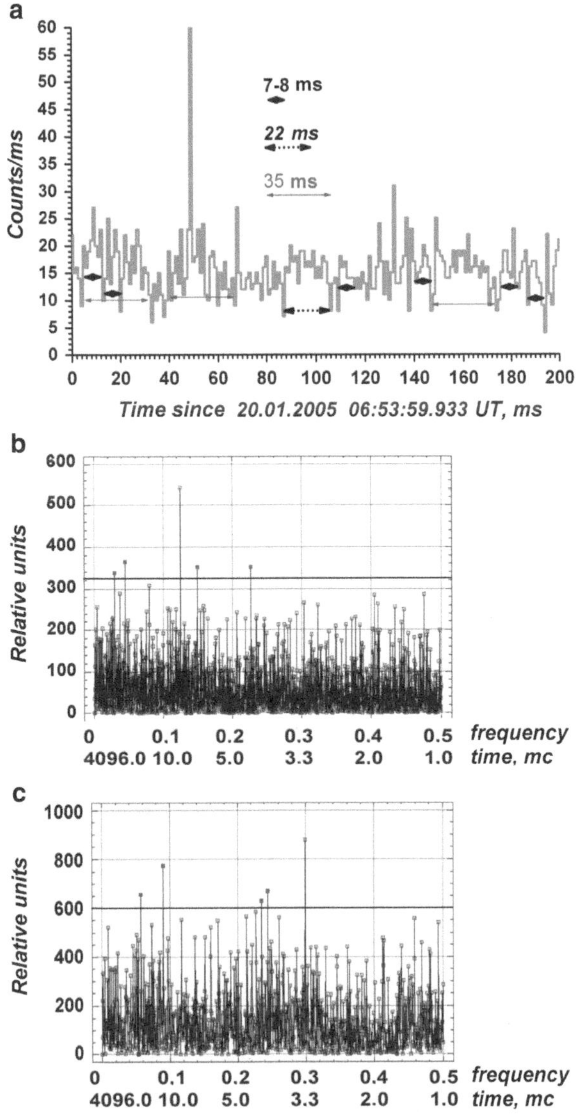


**Fig. 56** Fragment of AVS-F counts rate temporal profile of January 20, 2005, solar flare with time resolution of 1 ms in the time interval from 06:53:59.933 UT till 06:54:04.030 UT (**a**), its periodogram (**b**), and periodogram of this temporal profile averaged over 2 ms (**c**). From [32]

## *Catalogue of Solar Flares in the High-Energy Gamma-Band According to the Data of the AVS-F Apparatus On board the CORONAS-F Satellite During 2001–2005*

About one-fourth part of the solar flares registered by the AVS-F apparatus in the low-energy gamma-band during the period from August 2001 till December 2005 were accompanied by emission in the high-energy $\gamma$-band [16, 29, 32, 33].

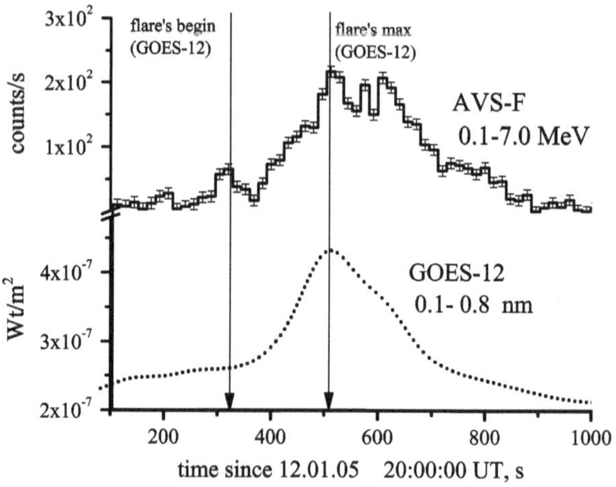


**Fig. 57** January 12, 2005, solar flare (B4.6 class) temporal profiles of the AVS-F counts rate (with background subtraction) in energy region of 0.1–7 MeV and one of the X-ray flux by the GOES-12 satellite data in the band of 0.1–0.8 nm. From [85]

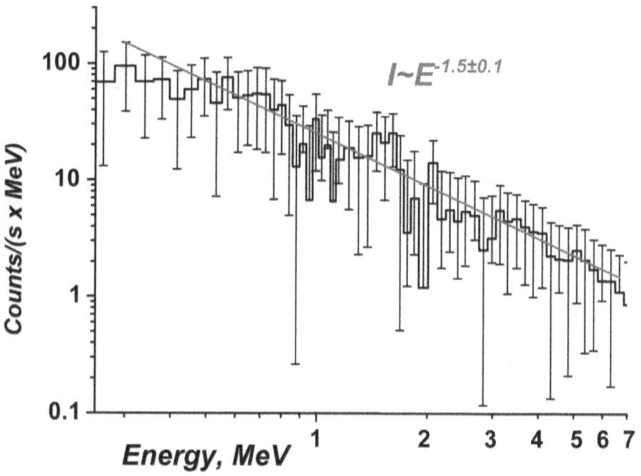


**Fig. 58** January 12, 2005 (B4.6 class) solar flare energy spectrum according to the AVS-F apparatus data (with background subtraction) in the low-energy $\gamma$-band during time interval 20:09–20:14 UT. From [85]

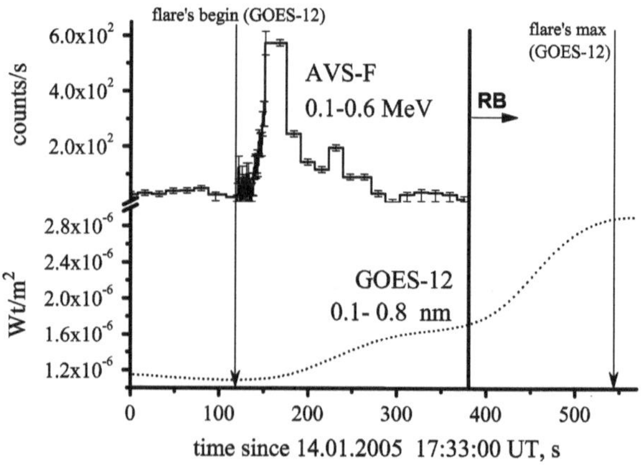


**Fig. 59** January 14, 2005 (C2.5 class) solar flare temporal profiles of the AVS-F counts rate in energy band of 0.1–0.6 MeV (with background subtraction) and one of X-ray flux by the GOES-12 satellite data in the band of 0.1–0.8 nm. From [85]

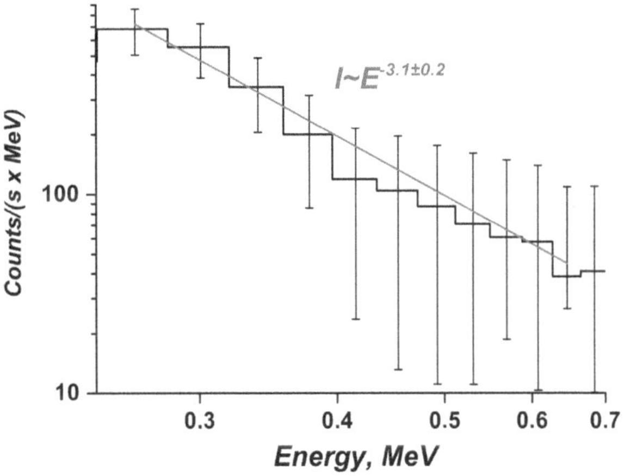


**Fig. 60** January 14, 2005 (C2.5 class) solar flare energy spectrum according to the AVS-F data in the low-energy $\gamma$-band during time interval 17:35-17-40 UT (with background subtraction). From [85]

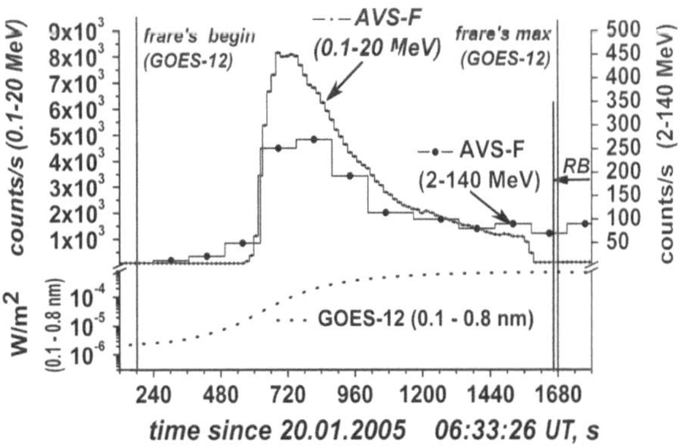


**Fig. 61** Temporal profiles of the AVS-F counts rate (with background subtraction) during the flare of January 20, 2005, in the energy range of 0.1–20 MeV and 2–140 MeV and X-ray flux in the band of 0.1–0.8 nm according to the data of the GOES-12 satellite. The moment when the CORONAS-F satellite went into the Earth's radiation belt is marked with an *arrow* RB. From [35]

The solar flare of October 29, 2003, is a typical example of this phenomenon. Duration of its $\gamma$-emission in both gamma-ranges of the AVS-F apparatus almost coincides with the duration of soft X-rays emission [16]. But as it was mentioned in the previous section the temporal profiles of the flare in both $\gamma$-bands have more complicated structure than in the soft X-ray band—see Fig. 43.

Solar flare of January 20, 2005 (X7.1 class) was also observed in both ranges of the AVS-F apparatus. Gamma-emission of this flare in both low-energy and high-energy ranges of the AVS-F apparatus was observed during the rise of the emission in the soft X-range. As it was mentioned in the previous section, it was accompanied by proton event and GLE (the most intensive [43, 48, 87, 99] for the last 27 and 49 years, respectively, although there were more powerful flares during these time intervals) and by CME. This flare gamma-emission temporal profiles (integration time of 16 s in the range of 0.1–20 MeV and 128 s in the range of 2–260 MeV) according to the data of the AVS-F had simple structure with one maximum as in the soft X-range [29, 32, 33] (Fig. 61).

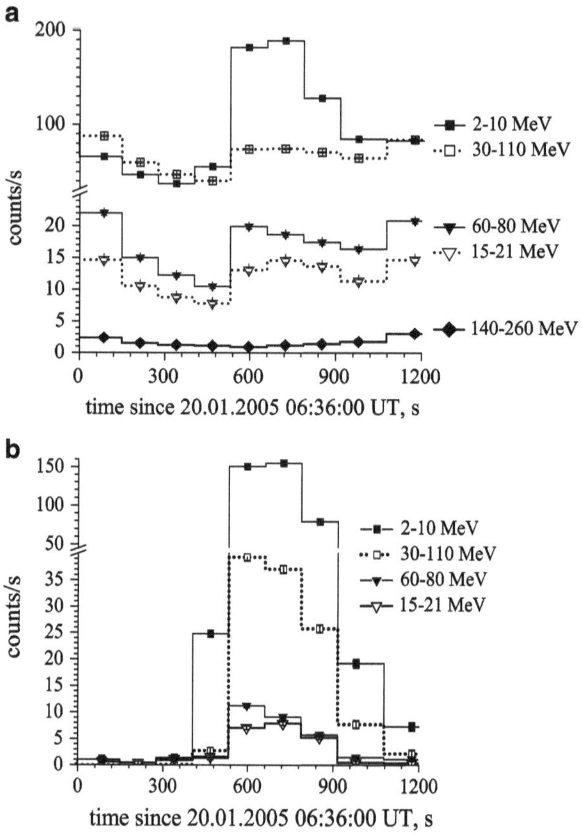


**Fig. 62** AVS-F counts rate temporal profiles during the flare of January 20, 2005, in energy bands: (**a**) 2–10 MeV, 15–21 MeV, 30–110 MeV, 60–80 MeV, and 140–260 MeV (without background subtraction); (**b**) 2–10 MeV (corresponding to registration of nuclear lines and neutron capture line), 15–21 MeV and one corresponding to gamma-rays from neutral pions decay (total range of 30–110 MeV and region of 60–80 MeV) (with background subtraction). From [26]

Hard gamma-emission temporal profiles registered by the AVS-F apparatus during this flare without background subtraction are presented in Fig. 62a. Maximum energy of $\gamma$-quanta detected during the event of January 20, 2005 was $137 \pm 4$ MeV—in the energy range of 140–260 MeV the counts rate did not exceed the background level in the statistical errors limits [29]. January 20, 2005, solar flare hard gamma-emission temporal profiles (with background subtraction) are presented in Fig. 62b. During this flare the hard gamma-emission was observed in the range corresponding to the registration of gamma-quanta from neutral pions decays [26, 29].

Summarized two-dimensional distribution of the ratio of the slow light output component to the total light output in dependence on the total energy deposition in the CsI(Tl) for this flare in time interval from 06:44:52,351 UT till 06:51:15,795 UT with background subtraction is presented in Fig. 63a. Data analysis has shown that during maximum of the flare of January 20, 2005, hard gamma-emission registration the neutrons and relativistic protons fluxes did not exceed background level in the statistical errors limits, powerful proton event, and GLE were detected later [48, 87]. According to the data of the neutron monitors "SOUTH POLE"

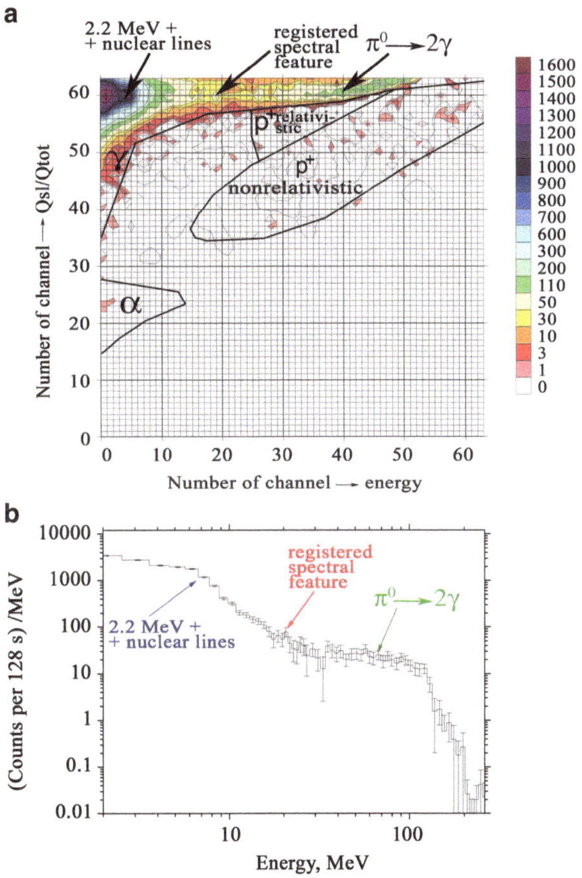


**Fig. 63** Two-dimensional distribution of the ratio of the slow light output component to the total light output dependence on the total energy deposition in the SONG-D detector (**a**) and energy spectrum (with background subtraction) in the high-energy gamma-band obtained by convolution of this distribution by the axis corresponding to the ratio of the slow light output component to the total light output (**b**) according to the data of the AVS-F apparatus for January 20, 2005, solar flare summarized during time interval 06:44:52.351–06:51:15.795 UT. From [35]

(90° S) and "Inuvik" (68.4° N 133.7° W) the counts rate rises were registered at 06:53:30 UT and at 07:05:30 UT, respectively [48, 99].

The maximum in the region of 1–6 channels of the abscissas axis corresponds to the sum of nuclear lines in the range of 2–7 MeV and to neutron capture line of 2.2 MeV (for more details see Fig. 46), and one in the region of 30–50 channels associated with registration of gamma-emission from the neutral pions decay [26]. In the region of 16–20 channels one more feature were identified. All this features are also seen in this flare energy spectrum—see Fig. 63b: wide line in the region of 2–7 MeV, corresponding to registered in the low-energy gamma-band nuclear lines and the line from neutron capture, and wide spectral feature caused by gamma-quanta from neutral pions decay [26, 29]. At the significance level of $2.5\sigma$ new non-identified spectral feature were detected in the range of 15–21 MeV [26]. But in energy spectrum registered in the maximum of high-energy gamma emission of the January 20, 2005 flare (06:47:00–06:49:08 UT) this feature was identified at the significance level of $3\sigma$.

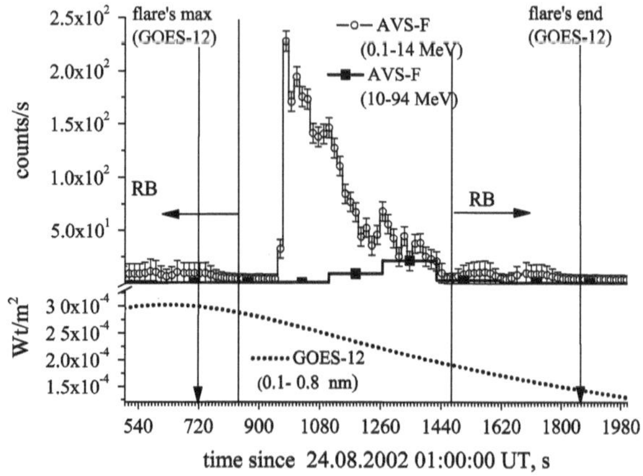


**Fig. 64** Temporal profiles of $\gamma$-emission for the flare of August 24, 2002, according to the data of the AVS-F apparatus and X-rays according to the data of the GOES-12 satellite. Time intervals when the CORONAS-F satellite was passing through the Earth's radiation belt are marked with *arrows* RB. From [29]

Spectral feature from neutral pions decay was registered earlier during several solar flares in other experiments (for instance, see [57]). The feature in the range of 15–21 MeV can be the line of 15.11 MeV ($^{12}C + ^{16}O$), or a line 20.58 MeV from radiative neutron capture by $^{3}He$, or their combination. The possibility of 15.11 MeV line detection was discussed in [51, 52], the 20.58 MeV one was only theoretically predicted in some articles (for instance, see [80, 81, 106]). Such feature in the solar flare energy spectrum was observed first time by the AVS-F apparatus [26].

The class X3.1 solar flare of August 24, 2002, was observed by the detectors on board the GOES-12 satellite in the range of 0.1–0.8 nm during the time interval of 00:49–01:31 UT, maximum of emission was registered at 01:12 UT (http://www.swpc.noaa.gov/ftpdir/warehouse/2002/2002_events.tar.gz). It originated from the active region 10069. During the flare of August 24, 2002, two episodes of gamma-emission in the AVS-F apparatus low-energy range were detected (00:57:00–01:06:16 UT and 01:15:46–01:24:02 UT) [18], but $\gamma$-emission in the range of 10–94 MeV was observed only during the second episode of its registration in the low-energy range [29] at the decreasing of emission in the soft X-band according to the data of the detectors on board GOES-12 satellite (see Fig. 64). At the end of the first episode of this flare $\gamma$-emission registration the CORONAS-F satellite went into the radiation belt, and the second episode was registered during the spacecraft passing of the polar cap. One more feature of this flare was the presence of thin structure with characteristic timescale of $\sim$ 60 s in the summarized temporal profile of the second episode of the flare in the AVS-F apparatus low-energy range [29]. It must be noted that this flare temporal profiles shapes differ essentially in the low- and high-energy ranges [18, 29].

Presence or absence of high-energy $\gamma$-emission during the solar flare do not depend on intensity of its intensity in the soft X-range [18, 29]: the flares of December 30, 2004 (M4.2) and January 20, 2005 (X7.1) were registered by the AVS-F apparatus both in the low- and the high-energy $\gamma$-ranges while the counts rate in the high-energy range did not exceed background level for a statistically significant value during the flares of January 17, 2005 (X3.8) and January 15, 2005 (X2.6), although they were observed by the AVS-F apparatus in the low-energy range [18].

## Conclusions

The whole volume of the AVS-F experiment available database is 20315 files (orbits 88–24570) contains 1 266 657 740 Bytes of information. The averaged percent of telemertry gaps was about 0.02 % per file.

In-flight calibration of the AVS-F apparatus was done both in the high-energy and in the low-energy gamma-ranges. The areas corresponding to registration of $\gamma$-quanta, relativistic protons, and secondary particles produced via neutrons interactions (including non-relativistic protons and $\alpha$-particles) were determined on two-dimensional distributions of the dependence of the ratio of the slow light output component to the total light output on the total energy deposition in the SONG-D detector. Operation of the spectrometry channel of the AVS-F apparatus was stable during all period the experiment was conducted, threshold energy and conversion coefficient were changed for $\sim$ 1 % and $\sim$ 1.8 % per month correspondingly. The obtained calibration information was used for the development of the NATALYA-2M instrument on board CORONAS-FOTON satellite [79] operated from January 31, 2009 to December 3, 2009.

Analysis of the background conditions allowed to make the AVS-F apparatus background model: in each spectral channel of $\gamma$-bands background counts rate temporal profile was approximated by the forth- or fifth-degree polynomials at the equatorial satellite's orbit segments and by the parabolic curve, linear function or by a constant in the polar regions. Such approximations were made in the concrete geomagnetic activity level (the values of Kp and Dst indices were taken into account). This approximation was used for identification of the burst events, study of their characteristics, and software development for the data analysis in the experiments on board CORONAS-PHOTON satellite [77].

At present over 1,100 events with duration more than 2 s without any coordinate relations to Earth Radiation Belts and South Atlantic Anomaly were separated on the results of preliminary analysis of AVS-F experiment database:

1. About 68 % of the identified events were associated with quasistationary equatorial precipitations—on some equatorial segments in the ranges of geographic latitude of $-25° \div +30°$ there were discovered 15–30 % counts rate increases in the low-energy gamma-band of the AVS-F apparatus over its average value obtained by approximation of these parts with polynomials.

   Four basic types of quasistationary precipitations were identified. For the precipitations with fast counts rate rise toward North and slow decrease toward South from geomagnetic equator in the latitudinal profiles (21 % of the registered precipitations—II type) the amount of the observed precipitations linearly depends on the geomagnetic activity level at 95 % significance level. For quasistationary precipitations with counts rate fast rise toward South and slow decrease toward North from geomagnetic equator (about 44 %—I type) analogous tendency was observed. The samples of the events of III type ($\sim$ 4 %, latitudinal profiles were symmetric with sharp boundaries) and of IV type (also $\sim$ 4 %, latitudinal profiles were symmetric and have sharp maximum) are still not enough for the statistical analysis.

The precipitations of I–III types were mainly located symmetrically relatively to geomagnetic equator as a band about 40° wide, width of the near-equator region where IV type precipitations were observed was about $\sim$ 20°. So, the main part of the precipitations was located in the regions of the low-latitudinal ($L \sim 1.5$) and equatorial electrons precipitations were observed. Quasistationary equatorial precipitations can be associated with charged particles precipitations in certain parts of the Earth's magnetosphere were regularly crossed by the CORONAS-F spacecraft.

2. About 7 % of the detected events were solar flares with classes stronger than M1.0 according to the GOES classification. Catalogues of solar flares detected in the low- and high-energy gamma-ranges of the AVS-F apparatus operation were made. Solar flares hard X-rays and $\gamma$-emission were mainly observed during the rise or maximum phases of the emission in the soft X-rays band according to the detectors on board the GOES series satellites data and duration of their registration is less than of the soft X-rays band. For some solar flares duration of the hard X-rays and $\gamma$-emission almost coincides with the duration of the emission in the soft X-rays band. There was also registered several flares during which the AVS-F apparatus detected more than one episode of hard X-rays and $\gamma$-emission, but in the intervals between detected episodes the count rate intensity did not exceed background level for a statistically significant value.

   According to the preliminary data analysis gamma-emission with energy over 10 MeV was registered during 12 % of the observed flares with classes stronger than M1.0. During all flares except one of August 24, 2002, the duration of $\gamma$-emission in both AVS-F apparatus gamma-ranges substantially coincides. Presence or absence of high-energy $\gamma$-emission during the solar flare does not depend on its intensity in X-range.

   In the spectra of some solar flares stronger than M1.0 in the low-energy gamma-range several spectral lines complexes were observed. Registered corresponded to $\alpha\alpha$-lines, annihilation line, nuclear lines, neutron capture line on $^1$H (2.223 MeV). In the high-energy range there were detected spectral features corresponded to the combination of the observed in the low-energy range nuclear and neutron capture on $^1$H lines (in the band of 2–7 MeV) and the one caused by the registration of gamma-quanta from neutral pions decay. In the spectrum of the January 20, 2005 solar flare the feature in the range of 15–21 MeV was detected for the first time. It can be associated with lines of 15.11 MeV ($^{12}$C + $^{16}$O) or 20.58 MeV (from neutron radiative capture on $^3$He), or with their combination.

   Energy spectra of several solar flares observed by the AVS-F apparatus were not contained any gamma-lines, i.e., these flares were e-dominant [11, 14, 18].

Previously flares of this type were registered by various detectors, in particular, by the SMM/GRS [103].

Thin structure with characteristic timescale of 30–160 s was observed at 99 % significance level on some solar flares stronger than M1.0 temporal profiles in the low-energy gamma-band in the energy ranges corresponding to the identified spectral features or whole gamma-band energy boundaries. The fragments of integral temporal profiles with time discretization of 1 ms and duration of 4.096 s in the low-energy AVS-F apparatus gamma-range for seven solar flares were analyzed (from 1 to 6 fragments for each flare). According to the results of the preliminary analysis only during the flare of January 20, 2005, thin structure with timescale from 7 ms to 35 ms was detected at 99 % confidence level in the energy range of 0.1–20 MeV.

During faint solar flares (of B and C classes according to the GOES classification) the emission in the energy band $E > 100$ keV was registered from ~60 solar flares and from several ones $\gamma$–quanta with energy up to several tens of MeV were observed [38]. Such emission was detected either at the phase of rise phase of the soft X-rays emission, or during all its duration [38, 39, 84]. The results of faint flares high-energy $\gamma$-emission analysis were discussed in detail in [38]. The shapes of temporal profiles on GOES and AVS-F/XSS-1 data were similar at (90–99) % confidence level and the maxima of the time derivatives of temporal profiles on GOES and AVS-F/XSS-1 data were observed in the same time moments. Some thin structure with characteristic timescale 50–110 s was observed on temporal profiles of several faint events [38,39]. There were no lines in these solar flares spectra, i.e., such events are e-dominant. Now a catalogue of such solar flares is composed.

3. About 3 % of the identified events were gamma-ray bursts. The most part of GRBs detected in the low-energy range of the AVS-F apparatus were confirmed by the data of the HETE, RHESSI, and SWIFT/BAT experiments, and during some bursts high-energy gamma-emission was observed: $E_{max} = 147 \pm 3$ MeV for GRB050525 [25, 36]. During the discussed time period only the SONG-D detector could provide registration of so high energy $\gamma$–emission.
4. About 22 % of the identified events are needed to be analyzed in more detail.

About 17 % of all detected events (excluding solar flares) were observed in the Earth's shadow.

Several short events with duration of 1–16 ms were registered during the experiment. They were associated with terrestrial gamma-ray flashes. These events were detected above the powerful thunderstorm formations.

Basing on the results of the data analysis during the whole period of the AVS-F apparatus operation it is possible to conclude that the AVS-F apparatus was operated in normal mode without degradation of its dynamical characteristics up to the end of the CORONAS-F operation on December 6, 2005.

## References


1. Alcaraz, J., Alpat, B., Ambrosi, G., Anderhub, H., et al.: Cosmic protons. Phys. Lett. **B490**, 27 (2000)
2. Alcaraz, J., Alvisi, D., Alpat, B., Ambrosi, G., et al.: Protons in near earth orbit. Phys. Lett. **B472**, 215 (2000)
3. Arkhangelskaja, I.V.: Application of fractal analysis for the processing of temporal profiles of non-stationary event recorded by the AVS-F instrument during the experiment onboard the spacecraft CORONAS-F. Kosmicheskaya nauka i tekhnologiya **9**(5–6), 81 (2003), Ukraine (in Russian)
4. Arkhangelskaja, I.V.: CORONAS-F/AVS-F observations of terrestrial gamma-ray flashes. In: AGU, 86(52), Fall Meet. Suppl., Abstract N AE14A-08.2005
5. Arkhangelskaja, I.V., Arkhangelsky, A.I., Kotov, Yu.D.: Model of the background of the AVS-F instrument in the low-energy gamma-range. Proceedings "Scienific session of MEPhI-2005", M.: MEPhI, vol. 7, p. 35 (2005) (in Russian)
6. Arkhangelskaya, I.V., Chervyakova, O.I., Arlhangelsky, A.I., Glyanenko, A.S., Kotov, Yu.D.: Analysis of the stable operation of sectrometry channel of the AVS-F instrument in the low-energy gamma-range. In: Proceedings of the "International Conference "CORONAS-F: Three Years of the Solar Activity Observations, 2001–2004" January 31–February 5, IZMIRAN, Troitks, p. 33. (in Russian) (2005)
7. Arkhangelsky, A.I., Glyanenko, A.S., Kotov, Yu.D., Pavlov, A.V., et al.: The AVS-F experiment of the CORONAS-F project on registration of fast-changing fluxes of cosmic and solar gamma-emission. Pribory i tekhnika eksperimenta (5), 16 (1999) (in Russian)
8. Arkhangelsky, A.I., Glyanenko, A.S., Kotov, Yu.D., Pavlov, A.V., et al.: Modernisation of the AVS-F experiment of the CORONAS-F project. In: Proceedings of the "Scientific Session MEPhI-1999", vol. 4. M.: MEPhI, p. 30 (1999) (in Russian)
9. Arkhangelskaja, I.V., Arkhangelsky, A.I., Glyanenko, A.S., Kotov, Yu.D.: In-flight calibration and the resulst of the data processing for the AVS-F instrument obtained during the experiment onboard the CORONAS-F satellite. Kosmicheskaya nauka i tekhnologiya **9**(2), 20 (2003), Ukraine. (in Russian)
10. Arkhangelskaja, I.V., Arkhangelsky, A.I., Glyanenko, A.S., Kotov, Yu.D., et al.: October 29, 2003 solar flare gamma-emission spectra investigation by SONG and AVS-F data onboard CORONAS-F satellite. In: Procedings of the "Scientific Session MEPhI-2004", M.: MEPhI, vol. 7, p. 18 (in Russian) (2004)
11. Arkhangelsky, A.I., Arkhangelskaja, I.V., Kotov, Yu.D., Glyanenko, A.S., et al.: Solar flares observed by AVS-F instrument onboard CORONAS-F satellite during 2,5 year of it's operation. In: Multi-Wavelength Investigations of Solar Activity, IAU Symposium, vol. 223, p. 441. Cambridge University Press, Cambridge (2004)
12. Arkhangelskaja, I.V., Arkhangelsky, A.I., Kotov, Yu.D., Kuznetsov, S.N.: The investigation of the spectra of solar events observed in October–November 2003. Multi-Wavelength Investigations of Solar Activity, IAU Symposium, vol. 223, p. 439. Cambridge University Press, Cambridge (2004)
13. Arkhangelskaja, I.V., Arkhangelsky, A.I., Glyanenko, A.S., Samoilenko, V.T., et al.: The stability of the AVS-F apparatus spectrometric section in the gamma-band. In: Proceedings of the 28th Russian Cosmic Ray Conference, p. 153 (2004)
14. Arkhangelskaja, I.V., Arkhangelskii, A.I., Glyanenko, A.S., Kotov, Yu.D., et al.: The solar flares observed in low energy gamma-ray band by AVS-F apparatus data onboard CORONAS-F satellite in 2001–2005 years. In: Danesy, D., Poedts, S., De Groof, A., Andries, J. (eds.) Proceedings of the $11^{th}$ European Solar Physics Meeting "The Dynamic Sun: Challenges for Theory and Observations" (ESA SP-600), p. 108.1. 11–16 September 2005, Leuven, Belgium (2005)
15. Arkhangelskaja, I.V., Arkhangelskii, A.I., Glyanenko, A.S., Kotov, Yu.D., et al.: The investigation of January 2005 solar flares gamma-emission by AVS-F apparatus data onboard

CORONAS-F satellite in 0,1–20 MeV energy band. In: Danesy, D., Poedts, S., De Groof, A., Andries, J. (eds.) Proceedings of the 11th European Solar Physics Meeting "The Dynamic Sun: Challenges for Theory and Observations" (ESA SP-600), p. 107.1. 11–16 September 2005, Leuven, Belgium (2005)
16. Arkhangelskaja, I.V., Arkhangelsky, A.I., Kotov, Yu.D., Kuznetsov, S.N., et al.: Gamma-ray radiation of solar flares in October-November 2003 according to the data obtained with the AVS-F instrument onboard the CORONAS-F satellite. Solar Syst. Res. **40**(4), 302–313 (2006)
17. Arkhangelskaja, I.V., Arkhangelsky, A.I., Kotov, Yu.D.: The AVS-F apparatus background model in the low-energy $\gamma$-band. In: Proceedings of the "International Conference "CORONAS-F: Three Years of the Solar Activity Observations, 2001–2004" January 31–February 5, IZMIRAN, Troitks, p. 28 (in Russian) (2005)
18. Arkhangelskaja, I.V., Arkhangelsky, A.I., Kotov, Yu.D., Kuznetsov, S.N., et al.: The solar flare catalog in the low-energy gamma-ray range based on the AVS-F instrument data onboard the CORONAS-F satellite in 2001–2005. Solar Syst. Res. **40**(2), 133 (2006)
19. Arkhangelskaja, I.V., Amandzholova, D.B., Arkhangelsky, A.I., Kotov, Yu.D.: The properties of the quasistationary equatorial precipitations according to the data of the AVS-F instrument onboard the CORONAS-F satellite. In: Abstracts of the "International Conference "CORONAS-F: Three Years of the Solar Activity Observations, 2001–2004" January 31–February 5, IZMIRAN, Troitsk, p. 14 (in Russian) (2005)
20. Arkhangelskaja, I.V., Amandzholova, D.B., Arkhangelsky, A.I., Kotov, Yu.D.: The stydies of quasistationary equatorial precipitaions according to the data of the AVS-F instrument. In: Procedings of the "Scientific Session MEPhI-2005", M.: MEPhI, vol. 7, p. 31 (in Russian) (2005)
21. Arkhangelskaja, I.V., Amandzholova, D.B., Arkhangelsky, A.I., Kotov, Yu.D.: The studies of dependece of quasistationary equatorial precipitations on geomagnetic activity according to the data of the AVS-F instrument. In: Procedings of the "Scientific Session MEPhI-2007", M.: MEPhI, vol. 7, p. 27 (in Russian) (2007)
22. Arkhangelskaja, I.V., Amandzholova, D.B., Arkhangelsky, A.I., Kotov, Yu.D.: Studies of the dependence of quasistationary equatorial precipitations on Kp- and Dst-indices of geomagnetic activity according to the data of the AVS-F instrument. In: Procedings of the "Scientific Session MEPhI-2006", M.: MEPhI, vol. 7, p. 17 (in Russian) (2006)
23. Arkhangelskaja, I.V., Amandzholova, D.B., Arkhangelsky, A.I., Kotov, Yu.D.: Features of quasi-stationary precipitations according to the data obtained with the AVS-F instrument onboard the CORONAS-F satellite. Solar Syst. Res. **42**(6), 536–542 (2008)
24. Amandzolova, D.B., Arkhangelskaja, I.V., Arkhangelskiy, A.I., Kotov, Yu.D., et al.: The analysis of quasistationary equatorial precipitations observation frequency dependence from the geomagnetical activity level. In: Proceedings of The $20^{th}$ European Cosmic Ray Symposium in Lisbon, Portugal (2007) http://www.lip.pt/events/2006/ecrs/proc/ecrs06-s0-189.pdf
25. Arkhangelskaja, I.V., Arkhangelsky, A.I., Kotov, Yu.D., Kuznetsov, S.N., et al.: Gamma-ray bursts recorded in 2001–2005 by the AVS-F instrument onboard the CORONAS-F satellite in the low-energy gamma-ray range. Cosmic Res. **45**(3), 261–264 (2007)
26. Arkhangelskaja, I.V., Arkhangelsky, A.I., Kotov, Yu.D., Kalmykov, P.A.: The studying of January 20, 2005 solar flare characteristics in the high energy gamma-band by AVS-F data. In: Procedings of the "Scientific Session MEPhI-2007", M.: MEPhI, vol. 7, p. 19 (in Russian) (2007)
27. Arkhangelskaja, I.V., Afonina, I.V., Arkhangelskij, A.I., Borodina, E.A., et al.: Preliminary results of database treatment from AVS-F apparatus in energy range 0,1–11 MeV. In: Proceedings of the $18^{th}$ European Cosmic Ray Symposium, Moscow, Russian (2002)
28. Arkhangelskaja, I.V., Chervyakova, O.I., Arkhangelsky, A.I., Kotov, Yu.D., et al.: The stability of the AVS-F apparatus spectrometric section in the gamma-band. In: Procedings of the "Scientific Session MEPhI-2005", M.: MEPhI, vol. 7, p. 33 (in Russian) (2005)
29. Arkhangelskaja, I.V., Arkhangelsky, A.I., Kotov, Yu.D., Glyanenko, A.S., et al.: Solar flares catalogue in the high-energy $\gamma$-band by the data of the AVS-F apparatus onboard

CORONAS-F satellite for the period of 2001–2005. In: Obridko, V.N., Zaitsev, V.V. (eds.) Proceedings of the Russian Conference "Experimental and Theoretical Investigations of Geliophysics Activity Forecasting Fundamentals", pp. 9–14, Troitsk (in Russian) (2006)
30. Arkhangelsky, A.I., Glyanenko, A.S.: The use of the MicroPC in the AVS-F apparatus onboard CORONAS-F satellite. Modern Tekhnol. Automatization (3), 58 (in Russian) (2004)
31. Arkhangelskaja, I.V.: Fractal analysis method applicability to Terrestrial Gamma-Ray Flashes separation in database of AVS-F apparatus onboard CORONAS-F satellite. In: Fullekrug, M., Mareev, E.A., Rycroft, M.J. (eds.) Proceedings of the NATO Advanced Study Institute on "Sprites, Elves and Intense Lightning Discharges", Corte, Corsica, 24–31 July 2004, Nato Science Series II: (closed), vol. 225, XVI, 399 p. Springer, 2006. p. 386
32. Arkhangelskaja, I.V., Arkhangelsky, A.I., Kotov, Yu.D., Glyanenko, A.S., et al.: January 2005 solar flares series gamma-emission temporal profiles investigation by the data of the AVS-F apparatus onboard CORONAS-F satellite. In: Obridko, V.N., Zaitsev, V.V. (eds.) Proceedings of the Russian Conference "Experimental and Theoretical Investigations of Geliophysics Activity Forecasting Fundamentals", Troitsk, pp. 15–20 (in Russian) (2006)
33. Arkhangelskaja, I.V., Arkhangelsky, A.I., Kotov, Yu.D., Kuznetsov, S.N., et al.: Gamma-ray emission from solar flares of January 2005 observed by the AVS-F apparatus onboard the CORONAS-F satellite. Solar Syst. Res. **42**(4), 351–358 (2008)
34. Arkhangelskaja, I.V., Arkhangelskij, A.I., Glyanenko, A.S., Kotov, Yu.D.: Thin structure of temporal profiles of solar flares January 15, 17 and 20, 2005 by data of AVS-F apparatus onboard CORONAS-F satellite. Adv. Space Res. **43**(4), 542–546 (2009)
35. Arkhangelskij, A.I., Arkhangelskaja, I.V., Glyanenko, A.S., Kotov, Yu.D.: AVS-F observations of gamma-ray emission during January 20, 2005 solar flare up to 140 MeV. Adv. Space Res. **43**(4), 589–593 (2009)
36. Arkhangelskaya, I.V., Arkhangelskii, A.I., Glyanenko, A.S., Kotov, Yu.D., et al.: The GRB detected in low energy gamma-ray band by AVS-F apparatus onboard CORONAS-F satellite in 2001–2005 years. In: Kleinert, H., Jantzen, R.D., Ruffini, R. (eds.) Proceedings of the Eleventh Marcel Grossmann Meeting on General Relativity, pp. 2015–2018. World Scientific, Singapore (2008)
37. Arkhangelskaya, I.V., Arkhangelskii, A.I., Kotov, Yu.D. Kuznetsov S.N., et al.: Gamma-ray bursts observed by AVS-F apparatus onboard CORONAS-F satellite in the low-energy gamma-band during the period 2001–2005. 29 Russian Cosmic Ray Conference, Proceedings, pp. 2015–2018, GA11 (2006)
38. Arkhangelskaja, I., Arkhangelskiy, A., Kotov, Y., et al.: The emission in the region E > 0:1 MeV during disk and limb faint solar flares. Adv. Space Res. **51**, 1996 (2013)
39. Arkhangelskaja, I.V., Kostina, M.S., Arkhangelsky, A.I., et al.: Faint solar flares with hard X-ray and gamma emission observed by AVS-F onboard CORONAS-F satellite. In: Proceedings of the 37th COSPAR Scientific Assembly, Canada, p. 121 (2008)
40. Avakyan, S.V., Vdovin, V.I., Pustornakov, V.F.: Ionizing and penetrating radiaiton in the near-Earth space. SPb, Gosgidromet. (in Russian) (1994)
41. Barashenkov, V.S., Toneev, V.D.: Interactions of high-energy particles and atomic nuclei with nuclei. M.: Atomizdat, 1972. (in Russian)
42. Belousova, I.V., Bogovalov, S.V., Glyanenko, A.S., Klepikov, V.Yu. et al.: Background conditions in the range from 30 keV to 5 MeV in the orbit of CORONAS satellites. J. Moscow Phys. Soc. (6), 415 (1996)
43. Bieber, J.W., Clem, J., Evenson, P., et al.: Largest GLE in half a century: neutron monitor observations of the January 20, 2005 event. In: Acharya, B.S., Gupta, S., Jagadeesan, P., et al. (eds.) Proceedings of the 29th International Cosmic Ray Conference. 3–10 August 2005, Pune, India, vol. 1, p. 237. Tata Institute of Fundamental research, Mumbai (2005)
44. Bogovalov, S.V., Glyanenko, A.S., Grigoriev, A.I., Zhuravlev, V.I., et al.: Persectives of the studies of fast-changing X-ray and gamma-emission of the Sun by the CORONAS-I experiment. In: Proceedings of the III International Meeting CORONAS-I, Kaluga-88, L.: LPhTI, p. 130 (1989) (in Russian)

45. Bogomolov, A.V., Bogomolov, V.V., Denisov, Yu.I., Kudryavtsev, M.I., et al.: Characteristics of the components of the background gamma-emission and neutrons on the "Mir" station orbits. Space Res. **38**(4), 355 (2000)
46. Bogomolov, A.V., Britvich, G.I., Myagkova, I.N., Ryumin, S.P.: Identification of neutrons from the background of gamma-quanta during registration by CsI(Tl) detectors. PTE (1), 13 (in Russian) (1996)
47. Butler, N.R., Kocevski, D., Bloom, J.S., et al.: A complete catalog of swift GRB spectra and durations: demise of a physical origin for pre-swift high-energy correlations. Appl. J. **671**(1), 656 (2007)
48. Cane, H.V., Erickson, W.C., Kaiser, M.L., Raymond, J.C., et al.: Why did the January 20 2005 GLE Have Such a Rapid Onset? Fall 2005 AGU Abstract N SH21A-05
49. Charalambous, P.M., Dean, A.J., Lewis, R.A., Dipper, N.A.: The background noise in space born low-energy gamma-ray telescopes. Nucl. Instr. Meth **A283**, 533 (1985)
50. Cohen, M.B., Inan, U.S., Fishman, G.L.: Terrestrial gamma-ray flashes observed on BATSE/CGRO and ELF/VLF radio atmospherics. In: American Geophysical Union, Fall Meeting, abstract N AE33A-0944 (2005)
51. Crannell, C.J., Crannell, H., Ramaty, R.: Solar gamma Rays above 8 MeV. Astrophys. J. **229** 762 (1979)
52. Crannell, C.J., Lang, F.L.: Nuclear gamma-ray ratios as spectral diagnostics for protons accelerated in solar flares. Proceedings of the Workshop on Nuclear Spectroscopy of Astrophysical Sourses, held in Washington, DC, December 14–16, p. 18 (1987)
53. Crownover, R.M.: Introduction to Fractals and Chaos. Jones and Bartlett Publ., London (1995)
54. Dean, A.J., Lei, F., Knight, P.J.: The space radiation environment and background noise in astronomical Îş-ray telescopes. J. Br. Interplanetary Soc. 53(3–4), 91 (2000)
55. Dean, A.J., Lei, F., Knight, P.J.: The space radiation environment and background noise in astronomical $\gamma$-ray telescopes. J. Br. Interplanetary Soc. **53(3–4)**, 91 (2000)
56. Devicheva, E.A., Dobrovolsky, G.F., Kovalevskaya, M.A., et al.: Complex study of the Sun and solar-Earth physics. In: Proceedings of the III International Meeting CORONAS-I, Kaluga-88, L.: LPhTI, p. 99 (1989) (in Russian)
57. Dunphy, P.P., Chupp, E.L., Bertsch, D., Schneid, E., et al.: Neutrons and pion-decay gamma-rays from the solar flare of 1991 June 11. Bull. Am. Astron. Soc. **28**, 857 (1996)
58. Forrest, D.J., Chupp, E.L.: Simultaneous acceleration of electrons and ions in solar flares. Nature **305**, 291 (1983)
59. Fractals in physics: In: Proceedings of the VI International Symposium on Fractals in Physics. M: Nauka (1988)
60. Gan, W.Q.: Spectral evolution of energetic protons in solar flares. Astrophys. J. **496**, 992 (1998)
61. Gan, W.Q.: Solar gamma-ray spectroscopy and abundance of elements. Chin. J. Astron. Astrophys. **26**, 255 (2002)
62. Gehrels, P.M.: Instrumental background in gamma-ray spectrometers flown in low Earth orbit. Nucl. Instr. Meth. **A313**, 513 (1992)
63. Glyanenko, A.S., Arkhangelsky, A.I.: Analysis of the operation of the controller based on octagon systems-4020 plate in the AVS-F experiment onboard the CORONAS-F satellite during 3.5 years of operation. In: Abstracts of the "International Conference "CORONAS-F: Three Years of the Solar Activity Observations, 2001–2004" January 31–February 5, IZMIRAN, Troitks, p. 32 (2005) (in Russian)
64. Hartman, R.C., Bertsch, D.L., Bloom, S.D., Chen, A.W., et al.: The third EGRET catalog of high-energy gamma-ray sources. Astrophys. J. Suppl. **123**, 79 (1999)
65. Horack, J.M., Fishman, G.J., Meegan, C.A., Wilson, R.B., et al.: BATSE observations of bremsstrahlung from electron precipitation events. In: Gamma-ray bursts, Proceedings of the Workshop, University of Alabama, Huntsville, 16–18 Oct 1991, p. 373 (1991)
66. Horack, J.M., Koshut, T.M., Mallozzi, R.S., Storey, S.D., et al.: Implications of the BATSE data for a helocentric origin of gamma-ray bursts. Astropys. J. **429**, 319 (1994)

67. Imhof, W.L., Smith, R.V.: Observation of nearly monoenergetic high-energy electrons in the inner radiation belt. Phys. Rev. Lett. **14**(22), 886 (1965)
68. Inan, U.S., Reising, S.C., Fishman, G.J., Horack, J.M.: Geophys. Res. Lett. **23**(9), 1017 (1996)
69. Kalmykov, P.A., Arkhangelskaja, I.V., Arkhangelsky, A.I., Kotov, Yu.D., et al.: Modelling of the response of the detector of the AVS-F instrument of the CORONAS-F scientific equipment complex for the fluxes of space protons in the equatorial region of the Earth magnetosphere. Proceedings of the "Scientific Session of MEPhI-2007", M.: MEPhI, vol. 7, p. 21 (in Russian) (2007)
70. Koshak, W.J., Krider, E.P.J.: Geophys. Res. **94**, 1165 (1999)
71. Kotov, Yu.D., Belousova, I.V., Glyanenko, A.S., Bogovalov, S.V., et al.: Background Conditions aboard the KORONAS-I satellite as measured by the AVS instrument. Cosmic Res. **37**(2), 42 (1999)
72. Kotov, Yu.D., Arkhangelskaja, I.V., Arkhangelsky, A.I., Glyanenko, A.S., et al.: Preliminary results of the processing of the data of the AVS-F instrument in the energy range of 0.1–11 MeV. Izvestiya RAN. Seriya phyzicheskaya, **66**(11), 1666 (2002) (in Russian)
73. Kotov, Yu.D., Arkhangelskya, I.V., Arkhangelky, A.I., Glyanenko, A.S., et al.: Preliminary results of the processing of the AVS-F instrument data in the high-energy range. In: Proceedings "Scientific session MEPhI-2003", M.: MEPhI, vol. 7, p. 20 (in Russian) (2003)
74. Kuznetsov, V.D.: The results of study of the sun and solar-terrestrial relations with the CORONAS-F satellite: a review. Solar Syst. Res. **39**(6), 433–441 (2005)
75. Kuznetsov, V.D., Charikov, Yu.E., Kotov, Yu.D., Kuznetsov, S.N., et al.: A review of the solar results from CORONAS-F satellite. In: Multi-Wavelength Investigations of Solar Activity, IAU Symposium, vol. 223, p. 357. Cambridge University Press, Cambridge (2004)
76. Kuznetsov, V.D., Sobelman, I.I., Zitnik, I.A., Kotov, Yu.D., et al.: Results of solar observations on-board the "CORONAS-F" satellite. In: Proceedings of the 35th COSPAR Scientific Assembly, Paris, France, p. 812, 18–25 July 2004
77. Kotov, Yu.D. The features of the high energy emission generation during solar flares and possibilities of CORONAS-PHOTON satellite project of its investigation. In: 28 Russian Cosmic Ray Conference, Electronic Proceedings, 7–11 June 2004, M.: MEPhI, p. 2112 (2004)
78. Kotov, Yu.D., Arkhangelskaya, I.V., Arkhangelsky, A.I., Glyanenko, A.S., et al.: Study of gamma rays of solar flare of October 29, 2003 according to data of AVS-F and SONG instruments onboard CORONAS-F satellite. Bull Russian Acad Sci. Phys. **69**(6), 859 (2005)
79. Kotov, Yu.D. and COR-PHOT TEAM.: Satellite project "CORONAS-PHOTON" for study of solar hard radiation. In: Proceedings of the 35th COSPAR Scientific Assembly, p. 1283, 18–25 July 2004, Paris, France
80. Kuzhevskij, B.M. Nuclear processes in the solar athmosphere and solar cosmic emission M. Energoatomizdat (1985) (in Russian)
81. Kuzhevskij, B.M., Gan, W.Q., Miroshnichenko, L.I.: The role of nuclei-nuclei interactions in the production of gamma-ray lines in solar flares. Chin. J. Astron. Astrophys. **5**(3), 295 (2005)
82. Kuznetsov, S.N., Kudela, K., Myagkova, I.N., Yushkov, B.Yu.: X-ray and gamma-emission solar flare catalogue obtained by SONG onboard CORONAS-F satellite. In: Proceedings ISCS 2003 Symposium, "Solar Variability as an Input to the Earth's Environment", Tatranska Lomnica, Slovakia, 23–28 June 2003 (ESA SP-535, September 2003), p. 683
83. Kuznetsov, S.N., Bogomolov, A.V., Gordeev, Yu.P., Gotseluk, Yu.V., et al.: Preliminary results of SKL experiment onboard CORONAS-I satellite. Izvestiya RAS. Phys. **59**(4), 2 (in Russian) (1995)
84. Kotov, Yu.D., Arkhangelskaja, I.V., Arkhangelsky, A.I., Glyanenko, A.S., et al.: Study of gamma rays of solar flare of October 29, 2003 according to data of AVS-F and SONG instruments onboard the CORONAS-F satellite. Bull. Ross. Akad. Nauk: Fiz. **69**(6), 859–862 (2005)
85. Kostina, M.S., Arkhangelskaja, I.V., Arkhangelsky, A.I., Kotov, Yu.D., et al.: The investigation of the Neupert effect in faint solar flares on AVS-F data. In: Proceedings of the 31th International Cosmic Ray Conference. 7–15 July 2009, Lodz, Poland, pp.1551.1–1551.4. Published on CDROM

86. Lehtinen, N.G., Walt, M., Inan, U.S.: Geophys. Res. Lett. **23**(19), 2645 (1996)
87. Mewaldt, R.A., Looper, M.D., Cohen, C.M., Mason, G.M., et al.: Space Weather Implications of the 20 January 2005 solar energetic particle event. American Geophysical Union, Spring Meeting 2005, N SH32A-05 (2005)
88. Murzin, V.S.: Introduction to the Cosmic Rays Physics. Atomizdat, Moscow (1979)
89. Nemets, O.F., Gofman, Yu.V.: Nuclear Physics Reference Book. "Naukova dumka", Kiev (1975)
90. Pankov, V.M., Prokhin, V.L., Shkurkin, Yu.G., Glyanenko, A.S., et al.: X-ray semiconductor (CdTe) spectrometer for solar flare and preflare studies. Radiophys. Quant. Electron. **39**(11–12), 1002–1005 (1996)
91. Pankov, V.M., Prokhin, V.L., Khavenson, N.G.: The XSS-1 X-ray CdTe spectrometer onboard the CORONAS-F satellite. Solar Syst. Res. **40**(4), 314–318 (2006)
92. Porras, E., Sanchez, F., Reglero, V., Cordier, B., et al.: Production rate of proton-induced isotopes in different materials. Nucl. Instrum. Methods Phys. Res. B **160**, 73 (2002)
93. Physics values reference book., // M.: Energoatomizdat (1991)
94. Ramaty, R., Murphy, R.J.: Nuclear processes and accelerated particles in solar flares. Space. Sci. Rev. **45**, 213 (1987)
95. Ramaty, R., Mandzhavidze, N.: Gamma rays from solar flares. In: Proceedings of the IAU Symposium. Highly Energetics Physical Processes and Mechanisms for Emission from Astrophysical Plasmas, Montana State University, Bozeman, 6–10 July 1999, vol. 195, p. 12 (2000)
96. Ramaty, R., Mandzhavidze, N., Barat, C., Lockwood, J.A., et al.: The Giant 1991 June 1 flare: evidence for gamma-ray production in the corona and accelerated heavy ion abundance enhancements from gamma-ray spectroscopy. Astrophys. J. **479**, 458 (1997)
97. Rubin, B.C., Lei, F., Fishman, G.J., Finger, M.H., et al.: A model of the gamma-ray background on the BATSE experiment. Astron. Astrophys. Suppl. **120**, 687 (1996)
98. Sanchez, F., Ballestros, F., Robert, A., Reglero, V., et al.: Background in low Earth orbits measured by LEGRI telescope—short and long term variability. Nucl. Instr. Meth. **B155**, 160 (1999)
99. Share, G.H., Murphy, R.J., Smith, D.M., Shih, A.Y., et al.: RHESSI Observations of the 2005 January 20 solar flare. In: Proceedings of the SHINE 2006 Workshop, p. 39. Zermatt Resort, Midway, Utah, DC. July 31–August 4 (2006)
100. Shih, A., Smith, D., Lin, R., Schwartz, R., et al.: Temporal variability of gamma-ray lines from the X-class solar flares of October–November 2003. $35^{th}$ COSPAR Scientific Assembly, p. 3221. Held 18–25 July 2004, in Paris, France (2005)
101. Thompson, D.J., Bertch, D.L., Fichtel, C.E., Hartman, R.C., et al.: Calibration of the energetic gamma-ray experiment telescope (EGRET) for the Compton gamma-ray observatory. AJSS **86**, 629 (1993)
102. Unan, U.S.: Lighting effects at high altitudes: sprites,elves, and terrestrial gamma ray flashes. C.R. Physique 2002 (3), 1411 (2002)
103. Vestrand, W.T., Forrest, D.J., Rieger, E., et al.: The solar maximum mission atlas of gamma-ray flares. Astrophys. J. Suppl. **120**, 409 (1999)
104. Veselovsky, I.S., Panasyuk, M.I., Avdyushin, S.I., Bazilevskaya, G.A., et al.: Solar and heliospheric phenomena in October–November 2003: causes and effects. Cosmic Res. **42**(5), 435–488 (2004)
105. Vishnevsky, O.V., Glyanenko, A.S., Pavlov, A.V., Pankov, V.M., et al.: Kalibration of the semiconductor detector XSS-1 of the CORONAS-F project. In: Proceedings of the "Sceintific session of MEPh-2001", M.: MEPhI, vol. 7, p. 37 (2001) (in Russian)
106. Voronchev, V.T., Kukulin, V.I., Kuzhevskij, B.M.: Nucl. Instrum. Methods Phys. Res. A **525**, 626 (2004)
107. Wescott, E.M., Sentman, D.D., Heavner, M.J., Hampton, D.L., et al.: Geophys. Res. Lett. **23**(16), 2153 (1996)
108. Wigger, C., Hajdas, W., Zehnder, A., et al.: Proceedings of the Conference on Swift and GRBs: Unveiling the Relativistic Universe, San Servolo, Venice, 5–9 June 2006, Shpringer (2006)

109. Yermolaev, Yu.I., Zelenyi, L.M., Kuznetsov, V.D., Chertok, I.M., et al.: Magnetic storm of November, 2004: Solar, heliospheric, and magnetospheric disturbances. J. Atmos. Solar-Terrest. Phys. **70**(2–4), 334–341 (2008)
110. http://swift.gsfc.nasa.gov/docs/swift/archive/grb_table.html/
111. Arkhangelskaja, I.V., Amandzholova, D.B., Arkhangelsky, A.I., Kotov, Yu.D.: In: Procedings of the "Scientific session MEPhI-2005", M.: MEPhI. vol. 7, p. 31 (2005) (in Russian)

# Variability of Extreme Ultraviolet Fluxes at Various Timescales as Measured On board the CORONAS Space Mission (SUFR-SP-K and VUSS-L Experiments)

**A.A. Nusinov, T.V. Kazachevskaya, V.V. Katyushina, P.M. Svidsky, and D.A. Gonyukh**

## Introduction

The launch of the CORONAS satellites was of fundamental importance to the development of the national space weather service. The flights of the CORONAS automatic solar observatory gave us a clear idea of the space division of the service and its particular elements, as well as of the possibility to realize it by the effort of the national research institutions.

The extreme ultraviolet (EUV) radiation is the most important factor of solar activity, which determines to a great extent the energy of the solar–terrestrial coupling, the state of the Earth's upper atmosphere and ionosphere, and the space weather as a whole. EUV measurements in various spectral bands have become a common practice as part of the monitoring of solar activity along with the patrol observations of the traditional indices, such as the solar radio emission and Wolf numbers. Routine EUV observations have been carried out for more than three decades, since 1976 when they were started on board the AE-E satellite. These observations are not an easy task, because they require sophisticated spectral equipment able to operate for a long time on board a spacecraft. On the other hand, it is well known (e.g., see [7,8,35]) that the radiation fluxes in different EUV ranges are closely related. This allows us, in accordance with the idea suggested in [7], to obtain adequate information on the overall EUV spectrum from the measurements taken in some isolated reference lines (e.g., see [3]). The commonly used reference lines are the brightest chromospheric lines, such as HeI 58.4 nm or HeII 30.4 nm. As was shown in [9, 26], the $L_\alpha$ flux ($\lambda = 121.6$ nm) can be used as a solar

A.A. Nusinov (✉) · T.V. Kazachevskaya · V.V. Katyushina · P.M. Svidsky
Fedorov Institute of Applied Geophysics, Roshydromet, Moscow, Russia
e-mail: nusinov@mail.ru

D.A. Gonyukh
Central Design Bureau of Hydrometeorological Equipment, Roshydromet, Moscow, Russia

V. Kuznetsov (ed.), *The Coronas-F Space Mission*, Astrophysics and Space Science Library 400, DOI 10.1007/978-3-642-39268-9_8,

activity index for determining radiation fluxes in the far ultraviolet wavelength range ($\lambda = 120$–$250$ nm).

At the same time, the study of correlation between the $L_\alpha$ line intensity and the fluxes in individual EUV lines or spectral bands ($\lambda \leqslant 120$ nm, EUV range) would allow extending this conclusion also to the short-wave spectral range. This approach offers a possibility of monitoring EUV radiation with the aid of relatively simple photometers in the $L_\alpha$ line (e.g., CORONAS-F/VUSS-L) or the total flux photometers at the wavelengths smaller than 130 nm (SUFR). The absolute calibration of the SUFR device carried out with the use of an electronic synchrotron [1] makes it possible to fill the gaps in the EUV observation series for the period 1999–2003 with CORONAS-F data.

Below, we discuss the main results of the CORONAS-I and CORONAS-F measurements relevant to the following problems:

- patrol observations and acquisition of data on the absolute values of solar radiation fluxes at wavelengths shorter than 130 nm in the epochs of minimum and maximum of solar activity;
- establishing a correlation between the radiation fluxes in the $L_\alpha$ line and in various EUV ranges on the basis of CORONAS-F measurements and determining the fundamental characteristics of flux variation in cycle 23;
- observation and study of solar flares;
- satellite observations of solar eclipses in the EUV spectral range.

## The SUFR and VUSS-L Devices for Measuring Solar EUV Fluxes On board the CORONAS Spacecrafts

### *A Brief Description of the Equipment*

Instruments of two types were mounted on board the CORONAS satellites for regular measurements of the intensity of solar ionizing EUV radiation at wavelengths $\lambda < 130$ nm and in the hydrogen $L_\alpha$ line (121.6 nm). The instruments were jointly developed at the Institute of Applied Geophysics and at the Central Design Office of Hydrometeorological Instrument Production of the Russian Federal Service for Hydrometeorology and Environmental Monitoring (Roshydromet). The CORONAS-I satellite was equipped with a solar ultraviolet radiometer (SUFR-Sp-K) that performed measurements at wavelengths $\lambda \leqslant 130$ nm and in the hydrogen $L_\alpha$ line ($\lambda = 121.6$ nm). The SUFR radiometer is based on the thermoluminescent method of registration. Instruments of this type have been installed on several satellites and are described in detail in [11, 12, 14–16, 30]. CORONAS-F was equipped with a similar radiometer. For spectral measurements in the 30.4 nm range, a vacuum ultraviolet solar spectrometer (VUSS) based on the principle of collisional photoelectronic spectrometry [23] was tested on board the CORONAS-I satellite.

Spectral measurements on board the CORONAS-F satellite in the wavelength range including the hydrogen $L_\alpha$ line (121.6 nm) were made with a vacuum ultraviolet Lyman-alpha solar spectrometer (VUSS-L) [28].

The design of the instrument is based on the principle of a simple spectrophotometer, in which the radiation in a narrow spectral band is separated by filters and is transformed to electronic signal carrying information about the radiation intensity. A "solar-blind" photodiode is used as a photo detector. The sensitivity to long-wave radiation is limited because the cathode is covered with Cu J whose photoemission decreases drastically at the wavelengths $\lambda > 130$ nm [40]. The sensitivity band is bounded from below because the input window made of $Mg F_2$ does not transmit radiation at $\lambda < 110$ nm [22]. The separation of the spectral band is the result of superposition of the sensitivity function of the photocathode and the transmission band of the input filter made of $Mg F_2$. As shown by the estimates of the instrument response to the solar radiation [38], the main part of the signal is associated with the narrow band in the vicinity of the $L_\alpha$ line. A broad spectral band at 180 nm also contributes significantly to the signal. A detailed analysis demonstrates that the flux variations in the $L_\alpha$ line are responsible for variations in the VUSS signal. Therefore, the VUSS instrument can be used to patrol the radiation in this line.

The practical application of the instrument shows that the input windows of photo detectors, which are made of magnesium fluoride and lithium fluoride, degrade considerably under the action of UV radiation. This leads to variations in the instrument sensitivity that must be taken into account when processing the measurements. To do this, we need to know the degree of degradation. The problem was solved in the following way. In addition to the main photo detector, we introduced a second, similar one (called reference detector) and a source of UV radiation—a hydrogen LGV lamp (referred to as a reference lamp, RL). The instrument is designed so that the solar radiation does not fall on the reference photo detector, while the emission from the reference lamp hits both detectors at equal angles. Thus, the window of the reference photo detector does not degrade by the action of the solar UV flux. So, if the reference lamp is switched on in the absence of solar radiation (during the night part of the spacecraft's orbit), the signals of the main and reference photo detectors can be compared to find the extent of degradation of the input window of the former and to take into account the variations in the device sensitivity when processing the data. The calibration of the photo detectors is usually performed along the shadowed part of the satellite orbit when the calibration-on command is issued.

It is also necessary to take into account that the instrument's orientation with respect to the Sun changes during the flight. For this purpose, the instrument is supplied with an optical orientation sensor whose signal is proportional to the cosine of the angle between the normal to the sensor plane and direction to the Sun. The information on stability of the electric regimes of the photo detectors and reference lamp is also transmitted to the Earth. Thus, the operation of the VUSS-L spectrometer is based on quite a complicated algorithm determined by the control program. The possibility of in-flight calibration is particularly important when long-term trends in the behavior of solar UV radiation are studied.

The instruments under discussion are constructed as monoblocks separately mounted on the CORONAS-F board panel. The telemetry information from the SUFR and VUSS-L instruments was acquired on board the satellite, partially processed at IZMIRAN, and then transferred to the Institute of Applied Geophysics for subsequent processing and analysis.

**Analysis of On board Operation of the SUFR Sp-K Instrument and Data Processing Method.** In the first months of operation on board the satellite, some characteristics of the instrument changed, apparently, under the influence of the ambient conditions (a similar effect had been observed earlier, e.g., on the Interball and Electro satellites, as well as on some other home and foreign spacecraft). One could note an abrupt fall of transmission of the $Mg F_2$ crystalline filter and a gradual decrease of signals from the detectors without filter. To refine calibrations during the flight, the SUFR Sp-K radiometer is supplied with special luminescent standards, which allow necessary corrections to be made. A small systematic decrease of the standard signal and a significant drop of signal from the detectors without filter were recorded since August 2001 when the measurements had started. The signal from long-persistent luminophores was used to correct everyday data in the same way as it was done on the Interball tail probe [15, 16].

**Analysis of On board Operation of the VUSS-L Instrument and Data Processing Method.** Measurements were taken all over the CORONAS-F operation period. The main results were obtained during the daytime part of the orbit, though the sensitivity of the instrument was sufficient to detect the emission variability on the night side and near the terminator, as well as the enhancement of UV radiation during solar flares. A great many flares of various intensities were recorded during the mission, from the minor flare of X-ray class C9 to the most intensive flares of importance X5.3. The record of a minor flare (M1.9) observed on 21 January 2003 is shown by way of example in Fig. 1. The left panel represents the VUSS-L signal. The null signal corresponds to the periods when the satellite was on the night side of the orbit; a short 4 V pulse corresponds to the calibration signal. The part of orbit corresponding to the flare is marked by a circle. The right panel illustrates the isolated and smoothed VUSS signal for the time interval under discussion. One can see that the signal amplification during the flare is insignificant, less than 1 %. The transformer used to digitize the analog signal in the data acquisition and processing system allowed us to reliably identify even weaker flares (down to importance C9) and to study in detail the variability of the flare-generated EUV fluxes.

There were periods (from 15 August 2001 to 2005) when the VUSS-L signal displayed irregular behavior due to unstable sensitivity of the device. Let us explain this considering by way of example the data processing for a few months. The solid line in Fig. 2 represents the telemetry signal for the period from March 19 to June 21, 2003. The abscissa shows the time in modified Julian days (MJD); the ordinate is the VUSS signal in Volts less the background. One can see that, between March 19 and June 21, 2003, the signal, first, changes from 5.8 to 2.1 V and then (from May 17

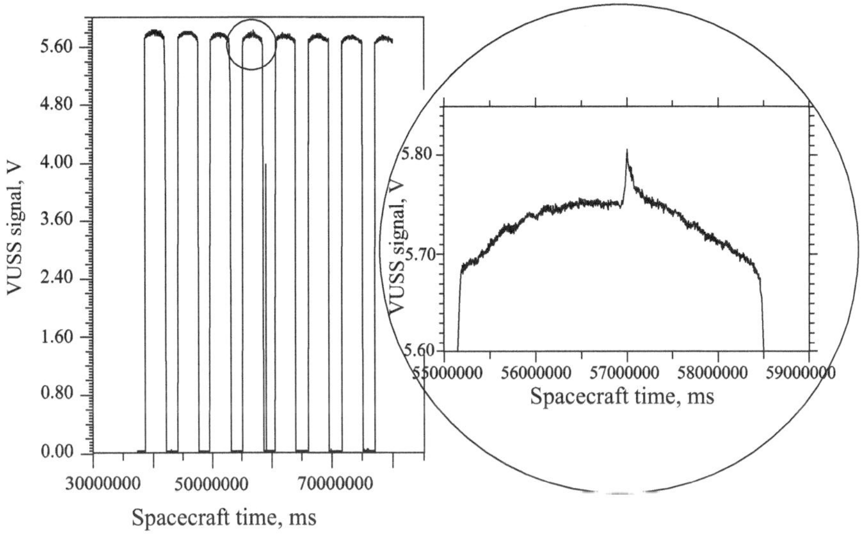


**Fig. 1** VUSS-L signal variations during the telemetry session and during the flare of 21.01.2003 (X-ray class M1.9). From [30]

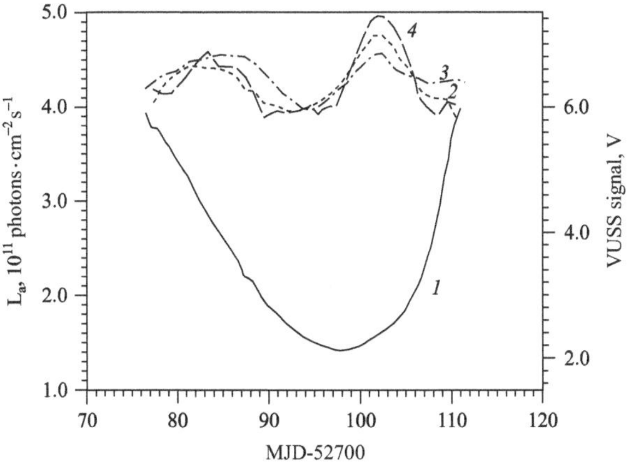


**Fig. 2** VUSS-L signal for the period from March 19 to June 21, 2003, $L_\alpha$ emission flux: 1, VUSS-L signal; 2, reconstructed signal for the same period; 3, SORCE (Solar Radiation and Climate Experiment) measurements; and 4, the flux value inferred from the ionospheric index

to June 21), increases up to 6.0. Such a scale of EUV variations is inconsistent with the present-day notions and indicates that the sensitivity of the instrument might have changed. The problem arises of how useful heliophysical information can be gained by reconstructing the signal distorted because of reduced sensitivity of the device. A special method for reconstructing the $L_\alpha$ line intensity in the periods of a significant loss of sensitivity of VUSS-L was developed and tested.

**Processing Method.** The VUSS signal measured over a lapse of time $V(t)$ is the result of combination of the solar radiation flux with the instrument sensitivity on a given day. We can write down

$$V(t) = S(t) \cdot k(t), \tag{1}$$

where $V(t)$ is the measured signal; $S(t)$ is the solar radiation flux, and $k(t)$ is the sensitivity of the device.

In turn, the solar radiation flux can be represented as

$$S(t) = f(t) + s(t), \tag{2}$$

where $f(t)$ accounts for slow variations of the emission over an activity cycle and $s(t)$ for day-to-day variations.

It is natural to suggest that the device sensitivity $k(t)$ changes slowly over a short time interval, too. Therefore, if we approximate the signal variation $V(t)$ by a smooth function (a polynomial, exponent, etc.) and, then, exclude such approximation due to the change of sensitivity, we shall obtain the useful signal generated by solar radiation.

The above equations were used to calculate the radiation flux in the H $L_\alpha$ line. The function $f(t)$ approximating the trend of solar activity for the period under consideration (exponent or polynomial) was selected on the basis of data on H $L_\alpha$ flux variations inferred from the ionospheric index [25]. It must be emphasized that only the type of the function is used in this case. The absolute calibration was made using the mean $L_\alpha$ intensities inferred also from the ionospheric index [25].

Figure 2 illustrates the VUSS-L signal 1 before and 2 after processing by the method described above for the selected time interval from March 19 to June 21, 2003. The polynomials of the 4th and 5th degree were used in Eq. (2) as the approximating functions $f(t)$. The same figure represents 4 the emission flux in the H $L_\alpha$ line as inferred from the ionospheric index and 3 as measured directly in the SORCE/SOLSTICE experiment [39]. One can see that the time behavior of the $L_\alpha$ flux revealed as a result of processing of the VUSS-L data agrees both with the direct SORCE measurements and with the $L_\alpha$ intensity values inferred from the ionospheric index.

**Processing Results.** The processing method described above was applied to all data obtained with varying sensitivity. They are shown in Fig. 3. The signals over an isolated time interval were normalized to the mean value for the given interval. Then, the approximating function $f(t)$ in Eq. (2) was selected individually.

Figure 3 shows the signal obtained as a result of preliminary processing of the VUSS-L telemetry (dotted line) and the reconstructed $L_\alpha$ flux for the period 2001–2003 (solid line) as a function of time in MJD.

The calibration of space-borne equipment used to measure the $L_\alpha$ emission intensity and its time variations is an extremely important and complicated task.

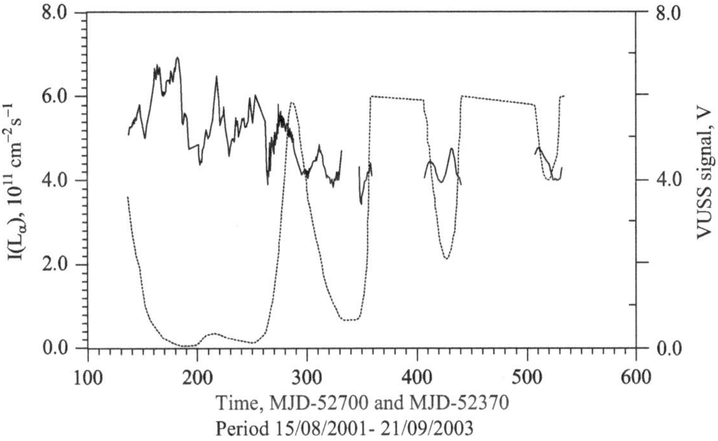


**Fig. 3** The VUSS-L signal for the period 15.08.2001 to 21.09.2003 (*dotted line*) and the reconstructed $L_\alpha$ flux (*solid line*)

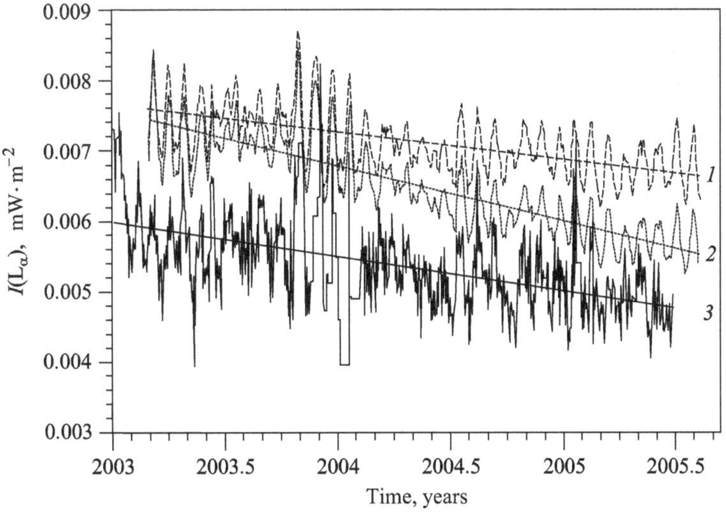


**Fig. 4** Comparison of the calculated and measured $L_\alpha$ intensities for the period from March 2003 to September 2005: 1, SORCE/SOLSTICE measurements; 2, SORCE/XPS measurements; and 3, intensities calculated from the ionospheric index [25]

In Fig. 4, the $L_\alpha$ intensity variations calculated from the ionospheric index are juxtaposed with direct measurements taken with the XPS (XUV Photometer System) and SOLSTICE instruments on board the SORCE space mission. The abscissa shows the time (MJD); the ordinate shows the $L_\alpha$ emission in units $10^{11}$ ($cm^{-2}\,s^{-1}$). One can see that the $L_\alpha$ intensities and their time variation calculated from the ionospheric index agree with the direct SORCE measurements.

The difference does not exceed a few percent of the measured value. Figure 4 reveals also a noticeable difference between simultaneous readings of the XPS and SOLSTICE instruments installed on one satellite. Since at the early stage of the SORCE mission, the readings of both instruments coincided within 1–1.5 %, this discrepancy suggests a change of sensitivity of the equipment. An additional opportunity to calibrate the satellite data obtained with different instruments is associated with calculating $L_\alpha$ intensities from the ionospheric index. In view of the aforesaid, the VUSS-L measurements were calibrated using the ionospheric index.

Thus, before the SORCE measurements were started, we had obtained reliable $L_\alpha$ data from the CORONAS-F satellite filling thereby the gaps that existed since the UARS [36] and Electro [15] missions.

## Variability of the Solar UV Radiation with the Activity Level as Inferred from the CORONAS-I and CORONAS-F Measurements

### *Patrol Observations of the Solar EUV Radiation*

A primary task of the experiment was to make patrol measurements of the solar EUV flux at the wavelength of 130 nm and in the H $L_\alpha$ line (121.6 nm). The CORONAS-I and CORONAS-F observations were carried out in different phases of the activity cycle; therefore, the flux values obtained and their variations were different.

**CORONAS-I measurements** were taken with the SUFR radiometer from March to June 1994, i.e., in the period close to the activity minimum. The 10-cm solar radio flux ($F_{10.7}$) was ranging from 80 to 100, while the flux at $\lambda < 130$ nm changed insignificantly and was, on the average, $I\ (\lambda < 130) = 7.5\,\mathrm{erg\,cm^{-2}\,s^{-1}}$. The H $L_\alpha$ line intensity was $5.5\text{–}6.1\,\mathrm{erg\,cm^{-2}\,s^{-1}}$.

**CORONAS-F measurements** were taken in the epoch of maximum of the activity cycle. The 10-cm flux was ranging from 140 to 280. The emission intensity in the hydrogen line $I(L_\alpha)$ was $6.8\text{–}8.2\,\mathrm{erg\,cm^{-2}\,s^{-1}}$, and the EUV flux at $\lambda < 130$ nm was, on the average, $I\ (\lambda < 130) = 11\text{–}13\,\mathrm{erg\,cm^{-2}\,s^{-1}}$.

## Variability of the Solar EUV Radiation with the Activity Level as Inferred from CORONAS-F and SOHO Observations

1. The CORONAS-F observation period was characterized by significant X-ray and EUV variations, including major solar flares. We have performed a joint analysis of EUV variations recorded with the CORONAS-F/SUFR-Sp-K equipment [15] and the intensity of the $\lambda = 26\text{–}34$ nm and $\lambda = 0.1\text{–}50$ nm emissions measured with the SOHO equipment for EUV monitoring (SEM). To study daily

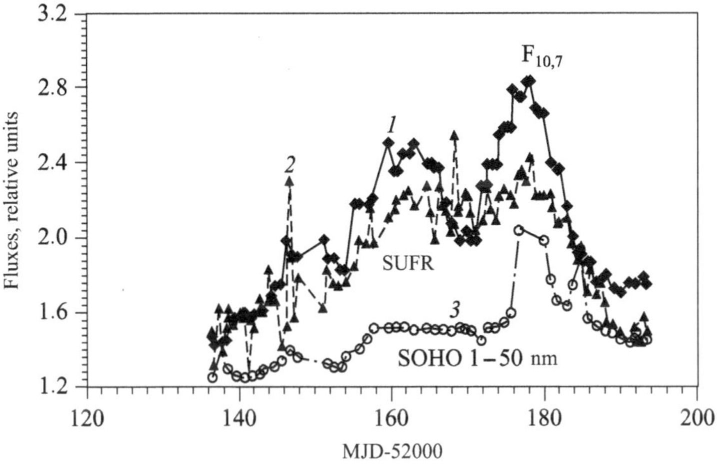


**Fig. 5** Variations in the solar radiation flux as inferred from satellite measurements: 1, radio emission index $F_{10.7}$; 2, CORONAS-F/SUFR—(0.1–130 nm); 3, SOHO/SEM (0.1–50 nm). The data are normalized to the beginning of CORONAS-F measurements (15.08.2001). From [17]

variations in the solar radiation flux, we compared the daily mean data of the CORONAS-F/SUFR measurements in the wavelength range $\lambda < 130$ nm and the SOHO daily mean data reduced to 1 a.u. Figure 5 illustrates flux variations at the wavelengths $\lambda < 130$ nm (SUFR data) and $\lambda = 0.1$–50 nm (SOHO data). The EUV flux (in relative units normalized to the beginning of measurements) measured with the SUFR (CORONAS-F) and SEM (SOHO) facilities and the 10-cm radio flux ($F_{10.7}$) are plotted against the time in MJD. Daily variations of the values under consideration have similar character, but differ significantly in amplitude: variations of the 10-cm radio flux reach 90 % of the measured value, while variations in the EUV spectrum make up 50 % for $\lambda < 130$ nm and ~55 % for $\lambda = 0.1$–50 nm (SOHO)

In Fig. 6, the data of the CORONAS-F/SUFR experiment are compared with the SOHO measurements at 0.1–50 nm. The straight lines show the r.m.s. spread range $\pm 1\sigma$ of the measured values. The points for September 25–27, 2001 corresponding to flares fall outside this range. The data in Fig. 6 are normalized to unity for the beginning of measurements. As seen from Fig. 5, this is when the values were smallest for the given observation period. Figure 6 demonstrates a good correlation between the SOHO/SEM and SUFR measurements so that the radiation flux in one range (0.1–50 or 0.1–130 nm) can be determined from the data obtained in the other with an error no more than 10 %. This supports the concept of the solar radiation patrol based on SUFR measurements in the range of 0.1–130 nm.

2. The calibration characteristics of the thermoluminescent detector [11] used in the SUFR experiment allowed us to estimate the $L_\alpha$ flux from the total 0.1–130 nm flux data obtained with a detector without filter. The result is

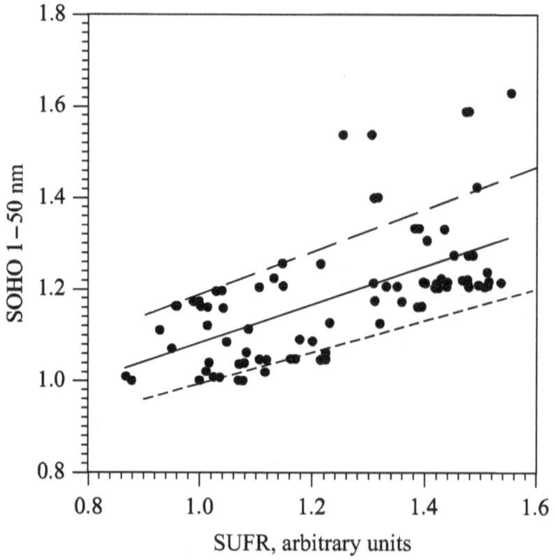


**Fig. 6** Correlation between the emission fluxes measured with the CORONAS-F/SUFR (0.1–130 nm) and SOHO/SEM (0.1–50 nm) instruments. From [17]

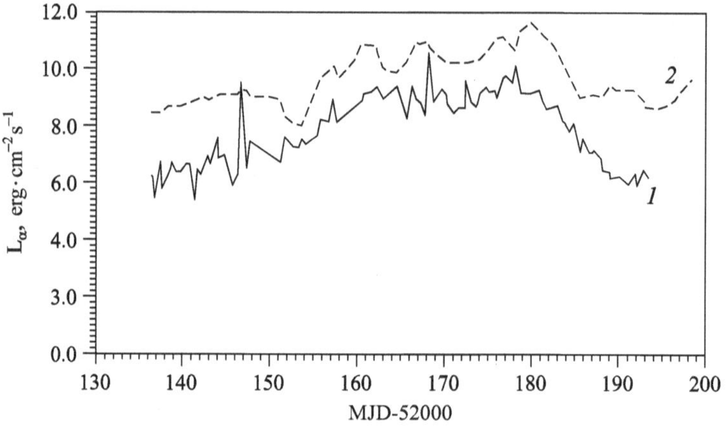


**Fig. 7** Variations of the emission intensity in the $L_\alpha$ line as inferred 1 from the CORONAS-F measurements and 2 from the ionospheric activity index $I$ $L_\alpha$. From [30]

shown with a solid line in Fig. 7. The dashed line represents for comparison the daily mean $L_\alpha$ flux estimated from the $E$-layer critical frequencies on the basis of the ionospheric index [25] and corresponding to the calibration of the UARS/SOLSTICE instrument [37, 44, 45]. One can see that the daily variations of the emission intensity obtained from the CORONAS-F data and ionospheric measurements agree fairly well, but the calibrations of the SUFR and UARS SOLSTICE devices differ.

**Fig. 8** Comparison of the $L_\alpha$ measurements (UARS data) and EUV measurements in the range of 26–34 nm (SOHO data)

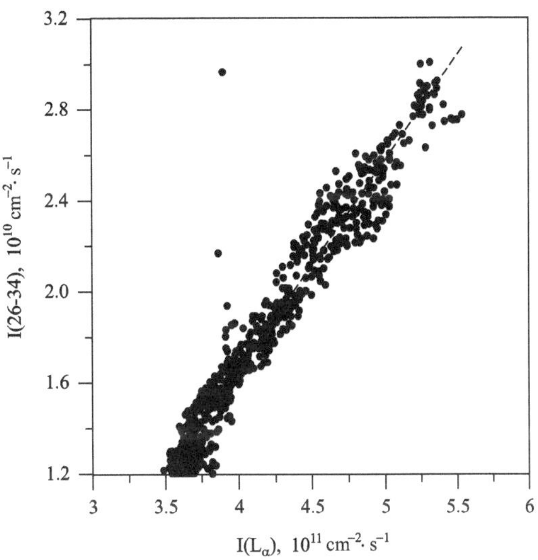


## The Study of Relationship Between the Radiation Fluxes in the $L_\alpha$ Line and in the 26–34 and 0.1–50 nm Spectral Bands

In view of the possibility to use $L_\alpha$ measurements for EUV patrol, it would be interesting to compare the $L_\alpha$ data with simultaneous measurements in various EUV ranges taken on other satellites.

The comparison was made with UARS measurements in the $L_\alpha$ line and SOHO EUV measurements in the spectral bands of 26–34 and 0.1–50 nm for the period from January 1996 to December 1999.

A high correlation amounting to 0.974 was shown to exist between the Lyman-alpha radiation flux and the fluxes at 26–34 and 0.1–50 nm. This confirms that observations in the $L_\alpha$ line can be used for monitoring the total spectrum of the solar short-wave radiation.

The relationship between the fluxes in the aforementioned spectral bands is represented in Fig. 8. Large deviations from the regression line (dashed line) correspond to the flare events. A similar relation between the fluxes in the range of 0.1–130 nm and the $L_\alpha$ flux is described above for the SUFR device (see Fig. 6). The ratio of the standard deviation to the mean reconstructed flux is ~10 % for the total flux measurements with SUFR and somewhat smaller (about 8 %) for the $L_\alpha$ measurements. This attests to the advantage of using $L_\alpha$ patrol measurements for monitoring the total EUV spectrum.

**Table 1** Satellite measurements in the H $L_\alpha$ line in cycle 23

| Period of measurements | SC | Experiment | References |
|---|---|---|---|
| 1991–1999 | UARS | SOLSTICE | Rottman [37], Woods [45] |
| 1994 | CORONAS-I | SUFR | Kazachevskaya et al. [15] |
| 1995–1996 | INTERBALL Tail probe | SUFR | Kazachevskaya et al. [15] |
| 1995–1998 | ELECTRO | SUFR | Kazachevskaya et al. [15] |
| 2001–2003 | CORONAS-F | VUSS-L | Nusinov et al. [33] |
| 2003, continued | SORCE | SOLSTICE, XPS | Rottman [39] |

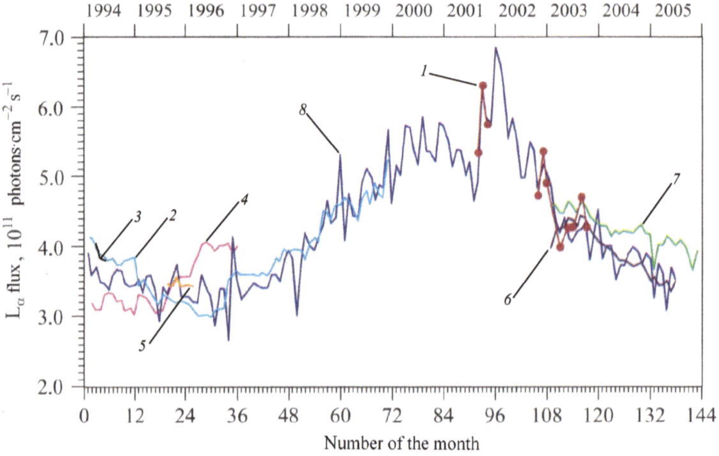


**Fig. 9** Measurements of H $L_\alpha$ radiation flux ($\lambda$ = 121.6 nm) during 23rd solar activity cycle from 1994 to 2005 according to direct satellites measurements and ionospheric data: VUSS Coronas-F (1), UARS SOLSTICE (2), "Electro" (3), CORONAS-I (4), Interball (5), SORCE XPS (6), SORCE SOLSTICE (7), and ionospheric data (8)

## The H $L_\alpha$ Hydrogen Line in the Activity Cycle 23 as Measured On board the CORONAS and Other Space Missions

Measurements on board the CORONAS satellites provided us with information on the emission intensity in the H $L_\alpha$ line in the 23rd activity cycle. Data from various satellites are available for the period from 1994 to 2005. Though the instruments and calibrations differ, yet all data plotted in one figure give an idea of intensity variations in the H $L_\alpha$ line throughout the cycle. CORONAS measurements complement and fill gaps in the foreign data.

The observation periods, satellites, and instruments used to measure the H $L_\alpha$ emission, as well as the corresponding references are summarized in Table 1.

Figure 9 represents the monthly mean emission intensities in the H $L_\alpha$ line (cm$^{-2}$ s$^{-1}$) measured directly during 1994–2005 on the satellites listed in Table 1.

The data from different satellites are denoted with different symbols (see the figure caption). The solid line represents the monthly mean H $L_\alpha$ intensities derived from ionospheric data by the Nusinov's method [25]. Direct measurements cover almost an entire cycle with gaps from November 1999 (UARS) to August 2001 (the beginning of CORONAS-F measurements). Unfortunately, measurements could not be taken at the peak of activity in 2002 for technical reasons. However, the gaps in direct observations were filled with information on the $L_\alpha$ intensities obtained from ionospheric data. It must be noted that the procedure of recovery of $L_\alpha$ intensities from ionospheric data based initially on UARS measurements (1999) was never corrected since, yet it ensured a good agreement with the SORCE measurements in 2003.

## The Scale of Day-to-Day Variations of EUV Intensity over an Activity Cycle

The measurements performed make it possible to compare the scale of daily and 27-day variations in the ultraviolet, X-ray, and radio wavelength ranges. The results of intensity measurements in the EUV range (SUFR Sp-K) and H $L_\alpha$ line (VUSS-L) are represented in Figs. 5 and 10.

The figures demonstrate solar activity variations due to the 27-day rotation of the Sun and the evolution of particular active regions. They represent the index $F_{10.7}$ and the emission intensities in the H $L_\alpha$ line ($\lambda = 121.6$ nm) both measured on CORONAS-F and derived from ionospheric data, as well as the SOHO/SEM measurements in the range $\lambda = 26$–$34$ nm.

Solar activity variations were characterized, as usual, by variations in the 10-cm radio flux, which changed from 147 to 280 units, i.e., by 90 %, during the operation period of CORONAS-F. Let us consider the scale of daily and 27-day variations as it was done in [17, 29].

The daily (day-to-day) variations measured with the CORONAS-F/SUFR instrument in the wavelength range $\lambda < 130$ nm and in the H $L_\alpha$ line were ~5 % of the measured value, and 27-day variations amounted to ~50 %.

The daily variations measured in the period under discussion with the SOHO/SME device at the wavelengths $\lambda = 0.1$–$50$ and $\lambda = 26$–$34$ nm were, respectively, 1.8 and 1.5 %. The 27-day variations measured in the same period were more impressive: the intensity increased approximately by a factor of 2.9 in both wavelength ranges. In this study, the emission from major solar flares was left out of consideration as it was in the analysis of the SUFR data.

According to GOES measurements in the wavelength range $\lambda = 0.1 - 0.8$ nm, the daily mean variation (for the days without flares) amounts to 53 % of the measured value, i.e., is quite significant. The variation with the activity level over the period of measurements is 10–30 times larger.

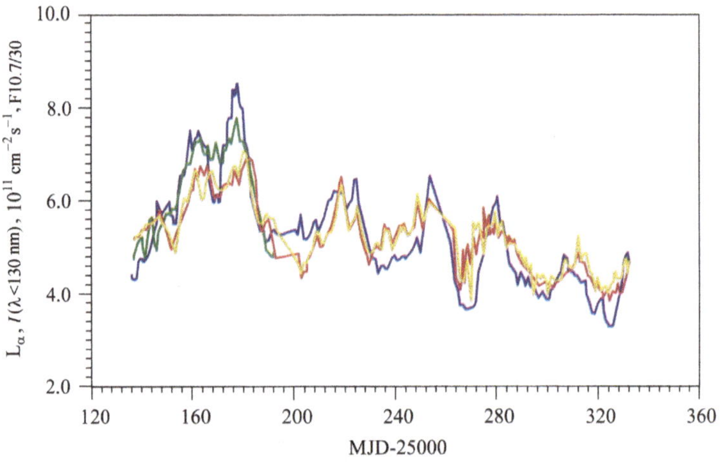


**Fig. 10** Variations of the solar radiation measurements averaged for a day during the period from August 2001 to December 2002: SUFR-Coronas-F ($\lambda < 130$ nm)—*green line*, H $L_\alpha$ ($\lambda = 121.6$ nm) VUSS-CORONAS-F—*red*, H $L_\alpha$ (of ionospheric data)—the *yellow*, and the index of the radio emission $F_{10.7}$—*blue line*

Intensity variation in the H $L_\alpha$ line ($\lambda = 121.6$ nm) for the period under discussion can also be estimated from the ionospheric index I L [25]. The mean daily (day-to-day) variation was ~7 % of the measured value (~1.8 % as obtained from smoothed data), and the variation with the activity level for the same period was ~30–40 %.

As seen from the values given above, the flux variations both from day to day and with the activity level decrease as the wavelength increases. Our measurements have corroborated this well-known conclusion and have revealed a good agreement between the H $L_\alpha$ intensity variations derived from direct SUFR measurements and inferred from the ionospheric index.

## The Study of Short-Wave Ultraviolet and X-Ray Radiation from Solar Flares in 2001–2003

The study of EUV spectrum is an essential part of investigation of evolution processes in solar flares. The radiation in this range of the spectrum is generated in the upper chromosphere and in the transition layer and characterizes the conditions therein in the period of the flare evolution. EUV observations contribute essentially to our knowledge of the flare evolution mechanisms; however, unlike X-ray observations, they are scarce.

Long-term EUV measurements with the CORONAS-F SUFR and VUSS instruments [29, 30] allow us to make up for this deficiency. It should be noted that these instruments measure the emission of the Sun as a star and do not provide information on the flare processes in the particular active regions (AR). The particular active regions were studied using the high-resolution imaging equipment designed at the Lebedev Physical Institute [46] and installed on board the CORONAS satellites.

An important task is obtaining information on the scale of variations of the total EUV flux during solar flares, since the intensity and spectrum of EUV radiation control the conditions in the Earth ionosphere and upper atmosphere and are necessary for modeling and forecasting the parameters of these media. Below, we shall discuss observations of flares of various intensities (X-ray class C, M, and X), as well as the general conclusions drawn from the CORONAS-F EUV data.

**Flare Observations in 2001–2003.** During the period 2001–2003, the CORONAS-F/VUSS-L and SUFR experiments recorded flares of various intensities. Unfortunately, the specific observation conditions on board a satellite entering the Earth shadow made it impossible to record all events. The low-intensity flares were sometimes lost against the flux fluctuations and self-noise of the equipment. Therefore, only X-ray flares of importance not lower than M1 and, occasionally (under the particularly favorable conditions), of importance C could be recorded. This restricted the number of flares used for the analysis. Some observation data and preliminary results are described in the paper by Nusinov and Kazachevskaya [28]. The data on solar flares obtained with the VUSS-L and SUFR instruments on board the CORONAS-F mission were analyzed together with the SOHO/SEM EUV measurements at the wavelengths of 26–34 and 0.1–50 nm and the GOES X-ray measurements in the range of 0.1–0.8 nm. We used 15-min SOHO/SEM data and 1-min data from GOES.

*The Major Flare of 25 August 2001* One of the most intensive flares recorded with the SUFR and VUSS instruments was the flare of X-ray class X5.3 (optical importance 3B), which occurred on August 25, 2001. Figure 11 illustrates the VUSS measurements in the H $L_\alpha$ line taken during this flare, as well as the time variation of X-ray emission in the wavelength range of 0.1–0.8 nm derived from GOES-8 measurements (hereinafter denoted $X$) and EUV variations in the 26–34 nm range as measured with the SOHO/SEM spectrometer.

The gap in the VUSS data between the 40th and 50th minutes is due to the loss of information between the successive communication sessions. It is interesting to note that the maxima of different radiation fluxes occur at different times: the principal X-ray maximum in the range of 0.1–0.8 nm is observed by ~13 min later than the EUV maximum. The variation scales differ, too. The VUSS signal varies by ~2.7 %, while the variation of X-rays is many factors of ten. The VUSS signal depends not only on the H $L_\alpha$ flux but also on the emission in a broad (~40 nm) spectral band in the vicinity of 180 nm. The contribution of the latter to the total signal is about 70 % [30]. The emission of the Sun as a star in this range changes little; therefore, we can attribute the signal variations during a flare to variations of the H $L_\alpha$ intensity alone, and the intensity estimates obtained at such times must be

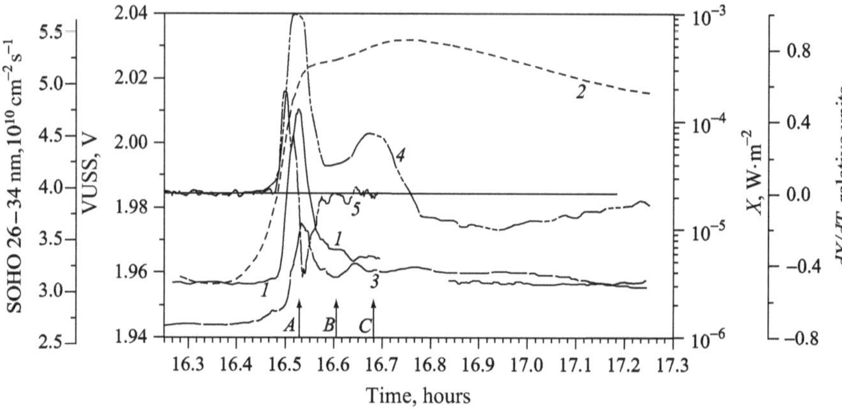


**Fig. 11** Solar flux variations in the X-ray (0.1–0.8 nm) and UV (120 and 26–34 nm) ranges during the flare of 25.08.2001: 1, $V$ (VUSS signal); 2, $X$ (GOES, 0.1–0.8 nm); 3, SOHO, 26–34 nm; 4, $dX/dt$; and 5, $dV/dt$. From [29]

increased threefold. This yields an increase of the H $L_\alpha$ flux during the flare under discussion by ~8–10 %. This value agrees fairly well with the estimates by other authors [22] discussed above and with our SUFR measurements of solar flares. It is important that the flare duration in the UV range (a few minutes) is much shorter than in the X-ray range. These properties are typical of most flares observed and will be summarized below. A detailed analysis of time variations of the emission fluxes suggests that they correspond to superposition of two or even three successive flares. This is obvious from the behavior of the time derivatives for the X-ray (line 4) and VUSS (line 5) signals. In Fig. 11, the probable flare moments denoted with arrows $A$, $B$, and $C$ correspond to the transition of the VUSS signal derivative through zero. They are clearly seen also in the 26–34 nm band (line 3). As a result of superposition, the rise rate of the GOES signal in the X-ray range changes, while the additional maxima in the VUSS signal are less pronounced. For flare $A$, the flux maximum at 26–34 nm lags behind $L_\alpha$ by ~1 min and the X-ray maximum, by ~4 min. For flare $B$, the delay times are, respectively, ~2 and ~8 min.

The emission flux from the Sun as a star in the X-ray ($\lambda$ < 10–12 nm) and ultraviolet ($\lambda$ < 130 nm) wavelength ranges, as well as in the $L_\alpha$ line (121.6 nm) was estimated from the SUFR measurements. The X-ray variations inferred from the GOES and SUFR data are virtually synchronous, and the UV radiation measured with open detectors (without filter) appears even earlier than X-rays. The open detectors revealed an increase of the energy flux at the wavelengths $\lambda$ < 130 nm (i.e., soft X-rays and EUV) by ~20–30 %. Thus, in the flare under discussion, the radiation fluxes in different wavelength ranges behave in essentially different ways. The NOAA catalogues treat this flare as an isolated one. However, the analysis of its behavior in the EUV range suggests that it is, most likely, a complex event, i.e., a superposition of a few successive flares in one active region. Similar events were

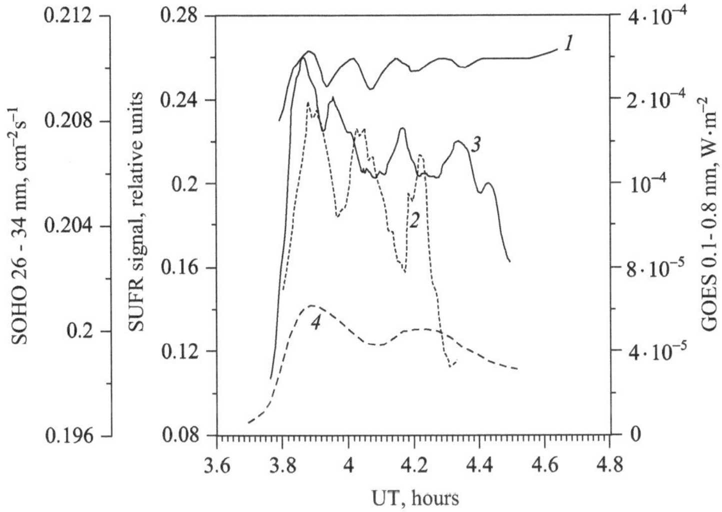


**Fig. 12** The class M5.6 X-ray flare recorded on September 16, 2001, 03:39–04:18 UT. The SUFR measurements started after the satellite had left the Earth shadow. 1: SUFR signal $\lambda < 130$ nm; 2: SUFR signal $\lambda < 12$ nm; 3: SOHO 26–34 nm; and 4: $X$ (GOES 0.1–0.8 nm). From [29]

observed with the SUFR device. Let us consider them using a flare of importance M as an example.

*The Flare of September 16, 2001* Like the flare of August 25, the other flares also display a few maxima in the EUV range and 1–2 X-ray maxima. Figure 12 shows by way of example the CORONAS-F/SUFR measurements obtained during the flare of September 16, 2001 (X-ray class M5.6, 03:39–04:18 UT) and the GOES measurements in the range of 0.1–0.8 nm. The abscissa is the time in fractions of an hour (the fine graduations are minutes). The ordinate shows the radiation intensity in relative units of the detector signals: without filter—$\lambda < 130$ nm (line 1) and with filter—$\lambda < 12$ nm (line 2). The figure represents also the time variation of the emission in the range $\lambda = 0.1$–$0.8$ nm as inferred from GOES measurements (line 4). In this flare, the detectors without filter recorded an increase of the signal at $\lambda < 130$ nm by $\sim$7 %. The detectors with filter for the range $\lambda < 10$–$12$ nm (aluminum and Mylar film) displayed an increase by a factor of $\sim$1.5. Again, we have a complex time structure of the signal from EUV radiation recorded with the SUFR equipment. While the GOES/$X$ detector reveals two maxima (the flare is mentioned in the catalogues as an isolated event), the SUFR records display four bursts in the channel $\lambda < 130$ nm and three bursts in soft X-rays $\lambda < 12$ nm. A lot of successive peaks are also seen in other EUV ranges. In particular, line 3 represents SOHO measurements in the range of 26–34 nm. One can readily see four successive peaks, all preceding the X-ray maxima detected with SUFR and $X$.

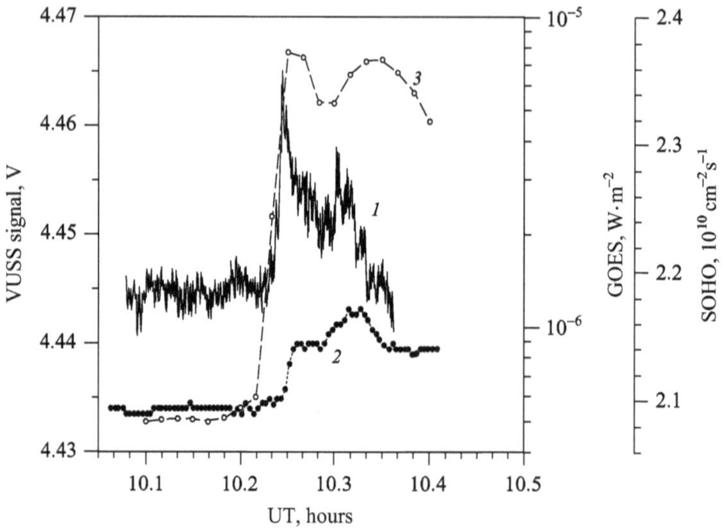


**Fig. 13** The flare of 29 March 2003, 10:11–10:25 UT, X-ray class C7.2: 1, VUSS signal; 2, SOHO 26–34 nm; 3, (GOES 0.1–0.8 nm). From [29]

*The Flare of March 29, 2003* Multiple bursts in a single flare are clearly noticeable in one of the weakest flares ever observed—a short-lasting flare of importance C7.2 that occurred on March 29, 2003 at 10:11 UT (Fig. 13). The VUSS signal (line 1) shows that the flare is also complex and consists of a number of successive individual flares. As in other cases, the UV maxima precede the X-ray ones (line 3). Line 2 corresponds to the 26–34 nm flux measured on SOHO. It is interesting to note that the first maximum on this line is virtually absent; the second maximum precedes the corresponding X-ray maximum and lags behind the maximum in the VUSS signal.

*Observations of the Flare-Generated EUV Radiation On board the SOHO Satellite.* The SOHO/SEM measurements of the solar ultraviolet radiation in the wavelength ranges of 26–34 and 0.1–50 nm give us an opportunity to check and refine our earlier conclusion on the basis of extensive experimental material. The SOHO/SEM data are available on the web site http://www.usc.edu/dept/space_science/sem_first.htm. We analyzed the total of 27 flares of different importance and X-ray class recorded in the period 2001–2003. For some flares, the data are available from both SOHO and CORONAS-F missions and for the others, from SOHO only. It is shown in Figs. 9–11 that the EUV spectrum has two or more maxima unlike the X-ray range where a single maximum is present. A similar pattern of flare evolution was observed also on the SOHO mission: of the 27 flares under examination, 15 had two maxima. These characterize two phases in the flare evolution. Note that the main variations in the VUSS-L signal coincide with the first, impulsive phase of the flare as shown in Figs. 11–13.

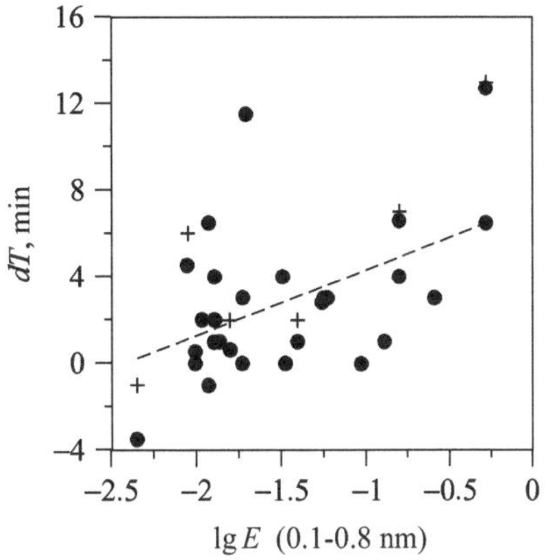


**Fig. 14** The delay $dT$ of the X-ray maximum (GOES, 0.1–0.8 nm) relative to the EUV maximum (VUSS and SOHO/SEM) as a function of X-ray flux in the maximum phase of the flare: *filled circle*—SOHO measurements in the range of 26–34 nm; *plus symbol*—VUSS measurements in the vicinity of 120 nm. From [29]

## The Time of the Flare Maximum Phase in the EUV and X-Ray Spectral Ranges

Our studies have shown that in most of the flares observed, the maximum phase in EUV precedes the X-ray maximum; the ultraviolet emission follows 2–13 min ahead of X-rays [28]. A similar result was obtained from the SOHO data on 27 solar flares observed in the same period.

Figure 14 represents the time difference (in minutes) between the EUV and X-ray maxima of a flare as a function of X-ray flux in the maximum phase.

The abscissa is the flux intensity in the 0.1–0.8 nm range at the flare maximum measured on GOES-8 (in logarithmic scale). The ordinate shows the delay time between the X-ray and EUV maxima as inferred from the VUSS and SOHO/SEM data. The VUSS and SOHO/SEM measurements in the range of 26–34 nm are seen to agree: in most cases, the maximum observed in the EUV range precedes the X-ray maximum, the delay time increasing with the increase of the flare intensity. Although the spread of points for minor flares is rather large, and the number of major flares observed is insufficient to plot reliable regression curves, the tendency is obvious.

## Radiation Energy in the Ultraviolet and X-Ray Spectral Bands at the Flare Maximum as Inferred from the CORONAS-F, SOHO, and GOES Satellite Measurements

A basic characteristic of solar flares is the amount of energy released both at the maximum and in the course of a flare. In Fig. 15, the flare energy in the EUV and X-ray spectral bands is represented on a log–log scale. The abscissa is the energy (erg $cm^{-2}$ $s^{-1}$) of the 0.1–0.8 nm X-ray flux at the flare maximum;

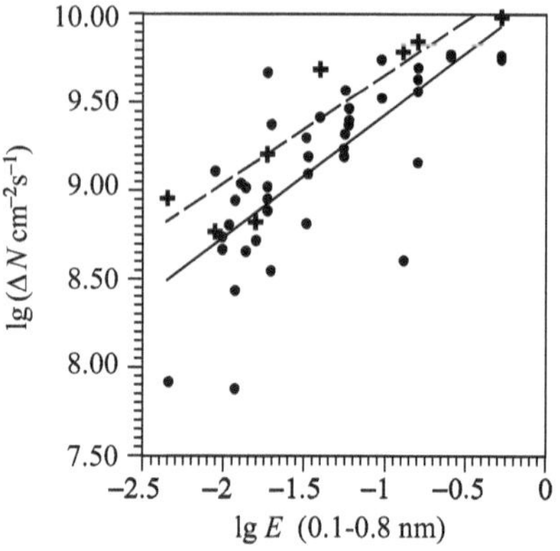


**Fig. 15** Relationship between the amplitudes of solar flares in the X-ray (0.1–0.8 nm) and ultraviolet wavelength ranges: *filled circle*—SOHO measurements in the range of 26–34 nm; *plus symbol*—VUSS measurements in the vicinity of 120 nm. From [29]

the ordinate shows the increase of EUV radiation in the vicinity of the H $L_\alpha$ spectral line (CORONAS-F/VUSS data) and in the range of 26–34 nm (SOHO/SEM data). The H $L_\alpha$ line intensity under undisturbed conditions was borrowed from the work by Nusinov [25] and was recalculated to fit the flare measurements obtained. Straight lines can be drawn through the observation points as shown in Fig. 15; i.e., the flare-generated fluxes are related by a power-law dependence within nearly two orders of variation of the X-ray emission. Such dependence is characteristic of the flare-generated fluxes at different wavelengths of the X-ray spectrum [27]. It is explained by peculiarities of the radiation spectrum of plasma with the changing temperature and emission measure. Although in the case under discussion, the emission originates in different regions and at different heights, the type of relationship between the radiation fluxes is conserved. The same relationship was derived earlier from the flare measurements at $\lambda < 130$ nm taken with a similar device (SUFR) installed on board the Prognoz-7 satellite [13].

## Discussion of Results

As shown above, the majority of the flare events are complex phenomena resulting from superposition of several, nearly simultaneous flares. In studying these events, the measurements of EUV radiation provide a much more comprehensive variation pattern than do the soft X-ray measurements.

In the next section, we shall discuss observation data similar to ours on the delay of the maximum X-ray flux in the range of 0.1–0.8 nm relative to harder emissions (50–150 keV, 150–500 keV, 0.5–1.3 MeV, and 4–7 MeV ranges) obtained with the CORONAS-F/SONG instrument [18]. A delay of the maximum of soft X-rays relative to harder components has been reported since the beginning of multiwave measurements (e.g., see the paper by Den and Somov [4,41] and references therein). On the other hand, the flare-associated hard X-ray and EUV emissions vary almost synchronously, which suggests their generation in the same regions.

A similar conclusion on the time profile of major flares was drawn by Lee et al. [20], who compared X-ray observations of the importance X1.1 flare of March 22, 2000 in the range of 0.1–0.8 nm with the measurements of hard X-rays and radio emission in the range of 1.6–18.0 HHz. The delay time obtained was ~4 min. As noted in [34], a short-term enhancement of hard X-rays in several energy ranges from 25 to 230 keV recorded by the Yohkoh mission during the flare of importance X5.6 on 6 April 2001 preceded by ~6 min the radiation in the range of 0.1–0.8 nm measured on GOES. A good time correlation was found to exist between the ultraviolet radiation from the flare of 25 August 2001 recorded on SOHO in the Lyman continuum and the $H_{\alpha}$ emission [21]. A delay of the soft X-ray maximum (GOES, 0.1–0.8 nm) relative to the hard X-ray flux (53–93 keV) was revealed from Yohkoh observations [2]. It is to be noted that, unlike the soft component, the flare-generated hard X-ray emission has a complex structure consisting of a number of successive pulses. Warren and Warshall [43] revealed a good spatial agreement between the sources of hard X-ray pulses recorded on Yohkoh and the EUV sources observed with the TRACE on board telescope. These sources are located at the feet of the coronal loops heated almost simultaneously. The delay suggests that the heating of the flare loops begins at their feet in the chromosphere or at the bottom of the transition layer where the emission at the wavelengths of about 100 nm originates and, then, it spreads up to the top where the main part of the X-ray flux is emitted. This hypothesis is supported by the fact that the time it takes for the soft X-rays to reach their peak value increases with the energy released at the flare maximum.

The main conclusions we have arrived at are as follows:

1. The observation of solar flares on board the CORONAS-F mission in the EUV wavelength range revealed some particularities in their time variation that suggest nearly simultaneous eruption of several flares in one active region. This effect is usually blurred in soft X-rays, but is clearly pronounced in the hard X-ray and EUV ranges (e.g., at 26–34 nm).
2. The data analyzed corroborate the earlier estimates of the increase of EUV fluxes during a flare (no more than ~10 % in the H $L_{\alpha}$ line even in the most intensive flares), as well as the probable scenario of flare evolution from the lower layers of the solar atmosphere to the corona.

## Ultraviolet, Hard X-Ray, and Gamma-Ray Emission of Solar Flares Recorded with the VUSS-L and SONG Instruments in 2001–2003

Measurements of the solar electromagnetic emission in several spectral bands expand significantly the scope of information on the properties of plasma in the flare region and its heating mechanisms. For example, routine X-ray measurements in two adjacent spectral ranges on board the GOES satellites carried out within the

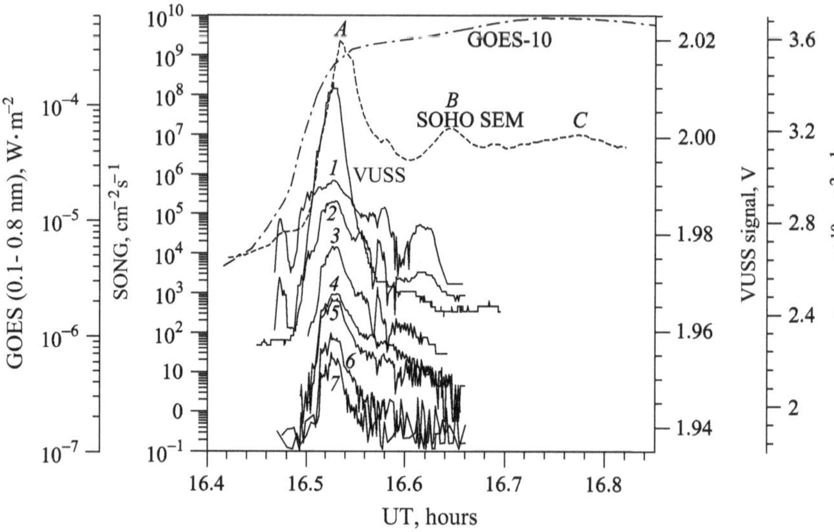


**Fig. 16** Electromagnetic radiation fluxes from the class X5.3 flare of 25 August 2001 measured in different energy ranges with the CORONAS-F/VUSS (*heavy solid line*) and SONG (*thin solid lines*) instruments, as well as the SOHO/SEM (26–34 nm, *dashed line*) and GOES-10 (0.1–0.8 nm, *dash-dotted line*) measurements. The energy of quanta in different SONG channels are numbered as follows: 1, 50–150 keV; 2, 150–500 keV; 3, 0.5–1.3 MeV; 4, 1.3–4 MeV; 5, 4–7 MeV; 6, 7–15 MeV; and 7, 15–26 MeV. From [32]

frames of the solar patrol program provide data on the temperature and emission measure of the flare plasma and their variation in the course of a flare [42]. Above, we analyzed simultaneous measurements on board the CORONAS-F, SOHO, and GOES-10 satellites [29]. Now, we shall consider the first results of a joint analysis of simultaneous measurements taken with the CORONAS-F/VUSS (about ~120 nm) and SONG instruments during several major flares in 2001–2003 [18, 32].

**The Results of Measurements.** The data on solar flares from the CORONAS-F/VUSS-L experiment were analyzed together with the SONG measurements of hard X-rays and gamma-rays with the energies from 50 keV to 200 MeV [18], SOHO/SEM EUV measurements in the wavelength range of 26–34 nm, and GOES X-ray measurements in the range of 0.1–0.8 nm. The results of the analysis for the flare of 25 August 2001 are represented in Fig. 16 (cf. Fig. 11). One can see that the hard X-ray emission occurs and reaches its maximum virtually simultaneously with the EUV emission in the H $L_\alpha$ line. An additional peak at about 16:58 UT (hours and fractions of hour) detected in three lower-energy channels of the SONG instrument was due to the passage of the satellite through the Earth's outer radiation belt where the lower-energy SONG channels were measuring, besides the flare radiation, the bremstrahlung emission of the outer belt electrons. The ultraviolet emission in the range of 26–34 nm reached its maximum approximately half a minute later. Note that we mean here the first maximum (denoted as *A* in Fig. 16). As shown above,

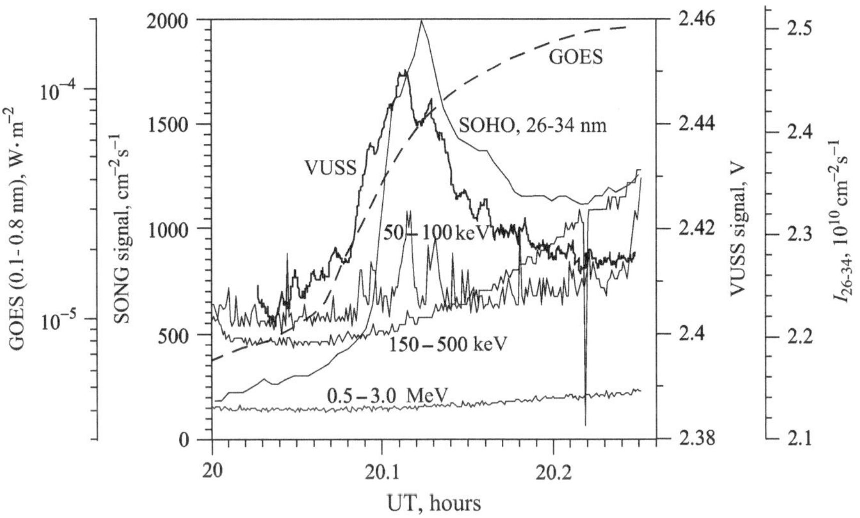


**Fig. 17** CORONAS-F/VUSS and SONG measurements of the X1.8 flare on 11 June 2003 and simultaneous measurements in the wavelength range of 26–34 nm on the SOHO satellite and in the range of 0.1–0.8 nm on the GOES geostationary satellite. The energy intervals for each of the SONG channels are indicated next to the corresponding lines. From [32]

the flare under examination might be a superposition of three nearly simultaneous events (see also [29]).

Synchronization of flux variations in the $L_\alpha$ line and in hard X-rays was also observed in other flares, e.g., in the X1.3 flare of 27 May 2003 and X1.8 flare of 11 June 2003 (Fig. 17). In the latter, the flux variations are not only observed in the energy range of 50–150 keV but also noticeable in the range of 150–500 keV.

The examples cited above allow us to conclude that the maximum energy of electron beams responsible for the X-ray bremstrahlung of solar flares can reach 500 keV and, in some particularly intensive flares, it can even exceed tens of MeV.

It is to be noted that in all flares under examination, the soft X-ray emission reached its maximum significantly (a few minutes) later than the UV and hard X-ray emissions whose maxima virtually coincide in time. This effect was discussed earlier by Nusinov and Kazachevskaya [28] who attributed it to propagation of the flare energy from the lower to the upper layers in the solar atmosphere. The study of radiation fluxes measured simultaneously in different spectral bands makes it possible to estimate the energy transfer rate. As seen from Figs. 16 and 17, the delay of the 26–34 nm radiation relative to the H $L_\alpha$ flux is about 0.3–0.5 min. The H $L_\alpha$ line forms at a height of ~2,000 km, and the He II line (30.4 nm), which contributes primarily to the 26–34 nm flux, forms at about 2,500 km (e.g., see [6, p. 252]). Hence, the heat propagation velocity is ~15–25 km/s. This is much less than the value of ~80–100 km/s, which can be derived from the typical delay of soft X-rays with respect to the UV and hard X-ray emissions if we assume the source of the former to be at $\sim$50,000 km and the mean delay time for major flares of class X to be $\sim$10 min.

The principal conclusions are as follows.

1. The variations in ultraviolet radiation and hard X-rays recorded during solar flares on board the CORONAS-F mission have been analyzed jointly. It is shown that, in the flare impulsive phase, the UV radiation in the H $L_\alpha$ line varies synchronously with hard X-rays and forestalls significantly the maximum in soft X-rays. Such behavior corresponds to the concept of successive heating of higher and higher regions in the solar atmosphere and the transfer of energy from the lower layers to the coronal area of the flare.
2. It is shown that in the flare impulsive phase, the energy of electron beams may amount to 500 keV, and the energy transfer rate may be as high as tens of kilometers per second.

## Observations of the Annular Solar Eclipse of 31 May 2003 in the Far Ultraviolet Spectrum

On 31 May 2003, the CORONAS-F satellite crossed the path of the annular solar eclipse. This was accompanied by the change of signals of the on board equipment. The passage through the eclipse zone was recorded by the VUSS-L and DIFOS instruments. The eclipse is also seen clearly on X-ray images from the SPIRIT telescope.

The annular solar eclipse of 31 May 2003 could be observed in the North hemisphere. It began early in the morning in northern Scotland and swept quickly across Iceland and portions of Greenland. At Reykjavik, the eclipse began at 03:08 UT, when the Sun was under the horizon, and ended at 05:01 UT. The maximum phase was recorded at 04:03 UT. A partial eclipse could be observed from most of the Europe (except Spain and Portugal) and the Middle East as well as from central and northern Asia.

CORONAS-F crossed the zone of the annular solar eclipse of 31 May 2003 three times. During these passages, visible radiation and EUV radiation in the vicinity of the hydrogen $L_\alpha$ line (121.6 nm) were measured with the VUSS-L instrument. The DIFOS photometer measured radiations in the infrared, visible, and near ultraviolet spectral ranges [19], and the SPIRIT telescope measured X-ray emission [46].

The behavior of the VUSS signals and the signals from the additional optical detector EOS in these periods is illustrated in Fig. 18. One can see several simultaneous consecutive falls in the signals of both instruments. The rectangular falls correspond to the satellite passage to the night side of the Earth and the triangle ones to the moments it was traversing the eclipse zone. The signal from the EOS detector operating in the visible range is represented as a mere illustration of the eclipse effect. During the first passage across the eclipse zone, the instrument signal

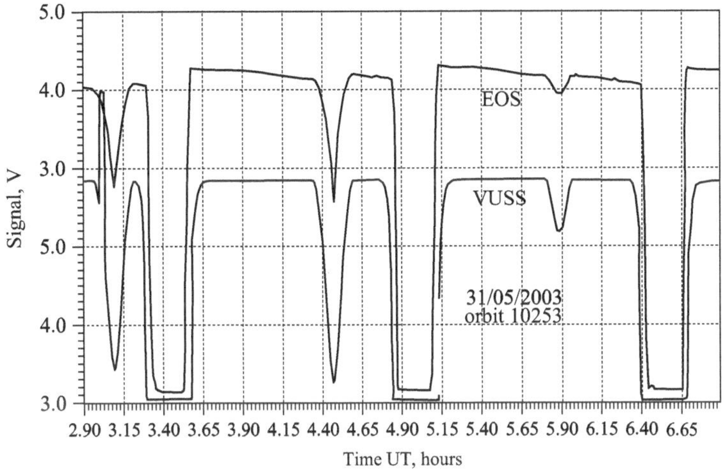


**Fig. 18** Time variations of the VUSS and EOS signals during the triple passage across the eclipse zone. From [10]

was superposed by the calibration signal. Therefore, the data for this period were excluded from consideration.

The multichannel DIFOS photometer on board the CORONAS-F spacecraft [19] detected intensity variations of the solar emission in six spectral ranges.

**Specific Features of Eclipse Observations from a Spacecraft.** In [31], Nusinov et al. present an eclipse map adopted from [5]. The lines on the map show the segments of the CORONAS-F orbits corresponding to the periods when the eclipse effect was recorded by the VUSS instrument. Note that this map can only be used as a reference, because eclipse observations from a rapidly moving celestial body, such as an artificial Earth satellite are specific. In this case, not only does the unocculted part of the Sun observed from the satellite change its size smoothly (as for a fixed observer), but it also changes its position on the Sun rapidly. Therefore, the ephemeredes of the Sun, the Moon, and the observer (satellite) must be calculated.

The corresponding calculations were performed by Prof. K.V. Kuimov at the Sternberg Astronomical Institute (see [31]). They comprise the right ascension and declination, the angular radii of the Sun and the Moon, the angle between the Sun and the Moon, the distance to the Moon, and, finally, the geometric parameter $X$, which is the area fraction of the solar disk not occulted by the Moon: $X = 1 - s/(\pi r_s^2)$.

The calculations show that the annular phase of the eclipse corresponding to $X = 0.11$ could be observed at the heights of the CORONAS-F orbit. Outside the eclipse zone, $X$ is, naturally, equal to 1.

**Table 2** Normalized VUSS signal and geometrical factor X during three consequative CORONAS-F orbits

| Orbit number/time, UTC (h:min:s) | 1/ 03:05:39 | 2/ 04:28:20 | 3/ 05:53:22 |
|---|---|---|---|
| VUSS | 0.15 | 0.09 | 0.76 |
| $X$ | 0.20 | 0.14 | 0.74 |

**Results of Observations with the VUSS Instrument.** In order to compare the VUSS measurements in the eclipse zone with variations of the geometric parameter $X$, the readings of the instrument were normalized by dividing them by the maximum value measured outside the eclipse.

The moments when the maximum effect was observed in each orbit (the VUSS signal and the $X$ parameter were minimal) are listed in Table 2.

As inferred from the table, the satellite did not enter the annular zone though it was passing rather close during the second flyby. The same conclusion follows from the map (see [31]) and from the value $X = 0.14$, which is close to $X = 0,11$ (see above).

Note that the minimal normalized VUSS signal was smaller than the minimal $X$ parameter during the first two satellite flights across the eclipse zone. This is quite surprising and can be explained either by the presence of isolated bright elements distributed over the solar disk or by darkening to the limb of the solar disk at the wavelength $\lambda_{\text{eff}}$, which should be considered to effectively describe the sensitivity band of the instrument. Since the instrument was designed to measure radiation variations in the hydrogen $L_\alpha$ line (121.6 nm), one should expect brightening rather than darkening to the limb (e.g., see [24]). Therefore, we are bound to speak of the effective wavelength, which is probably longer than 121.6 nm.

As for the individual bright elements distributed over the solar disk, one could expect to see them due to a high solar activity on that day and on the subsequent days. Three active regions (365, 368, and 373) were present on the visible side of the Sun. One of them demonstrated violent flare activity. For example, an M-class X-ray flare occurred in that region shortly before the eclipse; the flare was accompanied by a coronal mass ejection and by the arrival of high-energy proton fluxes at the Earth's orbit.

Let us estimate what luminosity contrast between the active region and the entire solar disk is necessary to obtain the observed relation.

Let $I_0$ be the full-disk flux and $dI$ be the flux from the active region. Then, we obtain that, for the signal minimum in the second flyby, the ratio of the flux from the maximum occulted area to that from the full disk with allowance for the active region flux is $I_0 = X/(I_0 + dI) = 0.09$ (see the table). Since the parameter $X = 0.14$, the value of $dI/I_0$ is equal to 0.5, i.e., exceeds manifold the previous estimates [24].

The calculations were made at 1-s time steps. Since the VUSS measurements were taken at shorter intervals, the data processing was carried out by assigning the $n$th second the value obtained by averaging the instrument readings over the time interval $n - 0.5\,\text{s} < \text{UT} \leqslant n + 0.5\,\text{s}$.

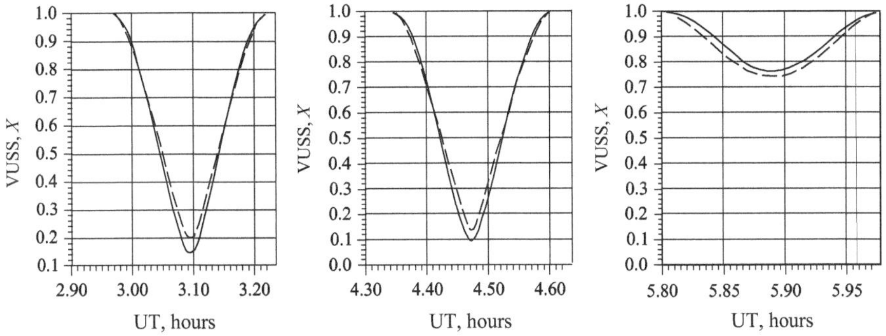


**Fig. 19** Time variations of the normalized VUSS signal and the $X$ parameter during the triple flyby of the CORONAS-F spacecraft through the eclipse zone. The *solid curves* are the measurements averaged over 1-s intervals; the *dashed curves* show the $X$ parameter calculated for each second. From [10]

The normalized VUSS signal and the $X$ parameter are compared in detail in Fig. 19, which illustrates the results of all three flights of the CORONAS-F spacecraft through the eclipse zone. The solid curves represent the measurements averaged over 1-s intervals; the dashed curves show the $X$ parameter calculated for each second.

One can see that in the first two flybys, when the maximum eclipse effect is observed, the $X$ parameter exceeds the normalized VUSS signal. In the third flyby, the relation is opposite. Note, however, that in the first two flybys, the normalized VUSS signal is also higher than $X$, when the latter exceeds 0.65. Since the minimum value of $X$ during the third flyby is 0.74, the relation between the values under consideration does not contradict the result obtained in the first two flybys.

This is seen better in Fig. 20, which represents the ratio of the VUSS signal to the $X$ parameter as a function of $X$ for all three flights past the eclipse zone. One can see that the dependences for all three flybys merge into a single curve and the above-mentioned ratio becomes equal to unity at $X \approx 0.63$.

The excess of the normalized VUSS signal over the $X$ parameter (VUSS/X $> 1$) at $X > 0.63$ can be attributed to the effect of active regions, while the inverse relation of these magnitudes at $X < 0.63$ seems to be due to the darkening to the limb.

## Comparison with the SPIRIT Telescope Data

The effect of nonuniformity of the brightness distribution over the solar disk can also be revealed from the eclipse data obtained with the SPIRIT telescope on board the CORONAS-F mission [46]. The ratio of the normalized signal to $X$ is above unity in all channels. This agrees with the observed asymmetry of brightness distribution

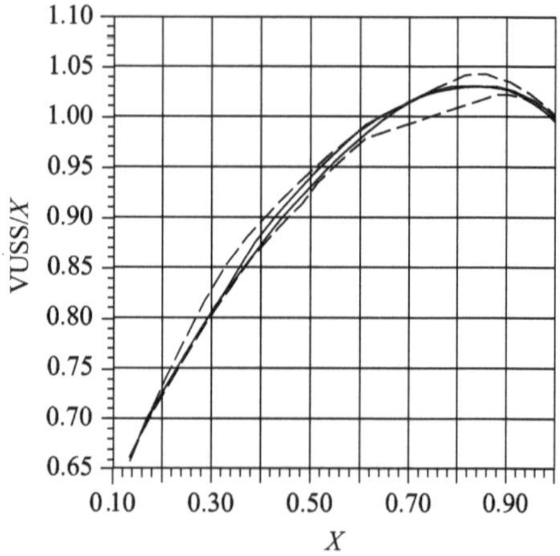


**Fig. 20** The ratio of the normalized VUSS signal to the $X$ parameter as a function of $X$ for all three flights past the eclipse zone. From [31]

over the solar disk associated with active regions in the central part and at the western limb of the disk, as well as with coronal holes at the center [31]. In the $L_\alpha$ line, the asymmetry is much weaker; therefore, the ratio of the normalized flux to $X$ is close to unity (except for the last value).

The main conclusions from the eclipse observations are as follows:

1. The solar eclipse was observed in various spectral ranges from the EUV to infrared spectrum. It was demonstrated that one and the same solar eclipse can be observed many times during successive flybys of the satellite and, thus, the informative value of the eclipse data increases significantly.
2. The character of variation of the radiation fluxes in different eclipse phases was shown to depend on the radiation wavelength and to differ from geometric variations calculated with allowance for the mutual position of the Sun, the Moon, and the satellite in each of the consecutive eclipse flybys. These differences are due to at least two factors: (1) different limb darkening/brightening functions for different wavelengths and (2) different contribution and contrast of active regions. In what follows, we are going to find the limb darkening functions that agree with observations and to determine the contribution of active regions.

## Conclusion

The telemetry data from the CORONAS-F/VUSS-L and SUFR-Sp-K instruments, after having been processed, provided us with information on the solar EUV radiation, its characteristics, and variation. The instruments were designed jointly by the Institute of Applied Geophysics and the Central Design Office of Hydrometeorological Instrument Production of the Russian Federal Service for Hydrometeorology and Environmental Monitoring (Roshydromet).

During 4.5 years of operation of the mission, a vast experimental database was built up. The scientific data obtained can be divided into several groups.

The scope of the cycle and day-to-day variations of ultraviolet radiation was determined depending on the level of solar activity (CORONAS-I and CORONAS-F measurements).

The data on radiation intensity in the H $L_\alpha$ line based on direct satellite measurements were obtained for the period of 1994–2005 (full cycle 23). The measurements made on board the CORONAS mission satellites filled the gaps in the series of data from foreign missions.

The short-wave ultraviolet and X-ray emissions of solar flares were studied for the period 2001–2003. Amplitude characteristics of EUV radiation were obtained, and their relationship with X-ray variations and the fluxes at about 30 nm was established. Power dependence was found to relate the soft X-ray and EUV intensities. This allows us to use X-ray patrol data to estimate variations in the flare-generated UV flux. The results obtained are important for interpreting variations that occur in the ionosphere during solar flares.

The time characteristics of solar flares were studied in the ultraviolet, hard X-ray, and gamma-ray spectra in the period 2001–2003 using the VUSS-L and SONG data. The relative delay times were obtained for the flare radiation fluxes at different wavelengths. This study changed our views of the physics of solar flares and revealed a new phenomenon, namely, that the flare effects at different heights in the ionosphere arise at different (though close) instants of time, the delay of the soft X-ray flux increasing with the increase of the flare intensity. This result ought to be taken into account when simulating the ionospheric effects of solar flares.

Eclipse observations in the EUV spectral range were carried out, including the unique observation of the annular solar eclipse of 31 May 2003. These observations will allow us to obtain a comprehensive pattern of EUV brightness distribution over the solar disk.

A wealth of experimental data on solar EUV radiation obtained on board the CORONAS satellites is of particular value to further advances in the field of heliophysics. Besides, the data collected will be useful in developing the space weather monitoring service and sophisticating the observation methods.

## References


1. Anevsky, S.I., Vernyi, A.E., Gonyukh, D.A., et al.: Calibration of the absolute spectral sensitivity of a solar UV radiometer with the use of synchrotron radiation. Nucl. Instrum. Methods Phys. Res. **A308**, 165–168 (1991)
2. Asai, A., Yokoyama, T., Shimojo, M., Masuda, S., Shibata, K.: Flare ribbon expansion and energy release rate. Astrophys. J. **611**(1), 557–567 (2004)
3. Bruevich, E.A., Nusinov, A.A.: Short-wave radiation spectrum for aeronomic calculations at different levels of solar activity. Geomagn. Aeron. **24**(4), 581–585 (1984) (in Russian)
4. Den, O.G., Somov, B.V.: Dissipation of the magnetic field in a high-temperature plasma as a mechanism of energy release in solar flares. Astron. Zh. **66**(2), 294–306 (1989) (in Russian)

5. Espenak, F., Anderson, J.: Annular and total solar eclipsis of 2003. NASA 2003 Eclipse Bull. NASA/tp-2002-211618, 12 (2003). http//ssd.jpl.nasa.gov/
6. Gibson, E.G.: The Quiet Sun, 356 pp. NASA Spec. Pull, NASA-SP-303, (1973)
7. Hinteregger, H.E.: The extreme ultraviolet solar spectrum and its variation during a solar cycle. Ann. Geophys. **26**, 547 (1970)
8. Hinteregger, H.E., Fukui, K., Gilson, B.R.: Observational, reference and model data on solar EUV, from measurements on AE-E. Geophys. Res. Lett. **8**(11), 1147–1150 (1981)
9. Katyushina, V.V., Nusinov, A.A.: The use of data on the solar radiation flux in the $L_\alpha$ line for the control of FUV spectrum. Geomagn. Aeron. **33**(5), 167–170 (1993) (in Russian)
10. Katyushina, V.V., Kazachevskaya, T.V., Nusinov, A.A.: In: Stepanov, A.V., Benevolenskaya, E.E., Kosovichev, A.G. (eds.) Multi-wavelength Investigations of Solar Activity. Proceedings IAU Symposium No. 223. 2004 International Astronomical Union, St. Petersburg (2004). doi:10.1017/S1743921304006246
11. Kazachevskaya, T.V., Kvater, G.S., Mishchenko, E.D., Selantyev, V.V.: Registration of vacuum ultraviolet radiation with the thermoluminescent phosphorus. Opt. Spectrosc. **40**, 410–411 (1976) (in Russian)
12. Kazachevskaya, T.V., Ivanov-Kholodny, G.S., Gonyukh, D.A.: Estimation of the magnitude of the solar short-wave radiation flux based on the AES measurements in 1978–1979. Geomagn. Aeron. **25**, 995–997 (1985) (in Russian)
13. Kazachevskaya, T.V., Bruevich, E.A., Ivanov-Kholodny, G.S.: Satellite observations of chromospheric solar flares in the EUV spectral range. Solnechnye Dannye (3), 68–72 (1986) (in Russian)
14. Kazachevskaya, T.V., Avdyushin, S.I., Gonyukh, D.A., et al.: Solar flux and spectrum measurements in the EUV spectral region on board the KORONAS-I satellite. Radiophys. Quantum Electron. **39**(11–12), 1011–1014 (1996)
15. Kazachevskaya, T.V., Nusinov, A.A., Gonyukh, D.A.: Measurements of the solar short-wave ultraviolet flux on board the ≪Electro≫ and ≪Interball≫ satellites in 1995–1996. Izv. RAN, Ser. Fiz. **62**(6), 237–239 (1998) (in Russian)
16. Kazachevskaya, T.V., Nusinov, A.A., Svidsky, P.M., et al.: Measurement of short-wave solar ionizing radiation flux on the satellite INTERBALL-1 (tail probe). Cosmic Res. **36**(3), 284–286 (1998) (in Russian)
17. Kazachevskaya, T.V., Nusinov, A.A., Katyushina, V.V., Gonyukh, D.A.: Variations in solar shortwave radiation in the solar activity cycle as measured by the CORONAS satellites. Sol. Syst. Res. **42**(4), 359 (2008)
18. Kuznetsov, S.N., Kudela, K., Myagkova, I.N., Yushkov, B.Yu.: X-ray and gamma-, emission Solar flare catalogue obtained by SONG on board CORONAS-F satellite. In: ISCS Symposium, ≪Solar Variability as an Input to the Earth's Environment≫, Tatranská Lomnica, Slovakia, 8 June 2003. ESA SP-535, September 2003, pp. 683–685
19. Lebedev, N.I., Kuznetsov, V.D., Oraevsky, V.N., Staude, J., Kostyk, R.I.: DIFOS helioseismic experiment on board the CORONAS-F mission. Astron. Rep. **48**(10), 871 (2004)
20. Lee, J., Gallagher, P., Gary, D., Nita, G., Choe, G., Bong, S.-C., Yun, H.S.: H$\alpha$, extreme-ultraviolet, and microwave observations of the March 22 solar flare and spontaneous magnetic reconnection. Astrophys. J. **585**(1), 524–535 (2003)
21. Lemaire, P., Gouttebroze, P., Vial, J.-C., Curdt, W., Schuhle, U., Wilhelm, K.: Flare observation of the Sun as a star by SUMER/SOHO in the hydrogen Lyman continuum. Astron. Astrophys. **418**, 737–742 (2004)
22. Makarova, E.A., Kharitonov, A.V., Kazachevskaya, T.V.: The Solar Radiation Flux, 176 p. Nauka, Moscow (1991) (in Russian)
23. Mishchenko, E.D., Abdullaev, A.M., Kazachevskaya, T.V., Kvater, G.S.: Vaccum ultraviolet solar spectrometer based on the priniciple collisional photoelectronic spectrometry. Opt. Spectrosc. **44**(6), 1076–1080 (1978) (in Russian)
24. Nusinov, A.A.: The brightness variation of ultraviolet sources at different wavelengths as a function of their position on the Sun. Solnechnye Dannye (1), 94–97 (1993) (in Russian)
25. Nusinov, A.A: The ionosphere as a natural detector for studying long-term variations in the solar geoeffective fluxes. Geomagn. Aeron. **44**(6), 718–725 (2004)

26. Nusinov, A.A., Katyushina, V.V.: Lyman alpha line intensity as a Solar activity index. Sol. Phys. **152**(1), 201–206 (1994)
27. Nusinov, A.A., Chulankin, D.I.: Soft X-ray variations during solar flares. Geomagn. Aeron. **37**(1), 9 (1997)
28. Nusinov, A.A., Kazachevskaya, T.V.: The variations ultraviolet radiation during large solar flares as observed on board the CORONAS-F spacecraft. Geomagn. Aeron. **45**(3), 375–379 (2005)
29. Nusinov, A.A., Kazachevskaya, T.V.: Extreme ultraviolet and X-ray emission of solar flares as observed from the CORONAS-F spacecraft in 2001–2003. Sol. Syst. Res. **40**(2), 111–116 (2006)
30. Nusinov, A.A., Kazachevskaya, T.V., Katyushina, V.V., Svidsky, P.M., et al.: Measurement of extreme ultraviolet solar radiation in different wavelength intervals onboard the CORONAS satellites: instruments and main results. Sol. Syst. Res. **39**(6), 470–478 (2005)
31. Nusinov, A.A., Katyushina, V.V., Kazachevskaya, T.V., Kuimov, K.V., Bugaenko, O.I., Slemzin, V.A., Lebedev, N.I.: An annular eclipse on May 31, 2003, observed in extreme ultraviolet. Sol. Syst. Res. **40**(4), 348 (2006)
32. Nusinov, A.A., Kazachevskaya, T.V., Myagkova, I.N., Kuznetsov, S.N., Yushkov, B.Yu.: Ultraviolet, hard X-ray, and gamma-ray emission of solar flares recorded by VUSS-L and SONG instruments in 2001–2003. Sol. Syst. Res. **40**(4), 282–285 (2006)
33. Nusinov, A.A., Kazachevskaya, T.V., Katyushina, V.V.: A flux of EUV emission measured onboard the ≪CORONAS≫ artificial satellites during minimum of 23-rd cycle of solar activity. Adv. Space Res. **37**, 246–252 (2006)
34. Qiu, J., Lee, J., Gray, D.E.: Impulsive and gradual nonthermal emissions in an X-class flare. Astrophys. J. **603**(1 Pt. 1), 335–347 (2004)
35. Rawer, K., Tai, H.S.: Correlated variations of solar EUV line emissions. Adv. Space Res. **37**(2), 234–237 (2006)
36. Rottman, G.J.: Observations of solar UV and EUV variability. Adv. Space Res. **8**(7), 53–66 (1988)
37. Rottman, G.J., Thomas, N., Woods, T.N., Thomas, P.: Sparn, solar-stellar irradiance comparison experiment 1: 1. instrument; design and operation. J. Geophys. Res. **98**, 10667–10677 (1993)
38. Rottman, G.J., Woods, T.N., White, O.R., London, J.: Irradiance observations from the UARS SOLSTICE experiment. In: Pap, J.M., Frohlich, C., Hudson, H.S., Solanki, S. (eds.) The Sun as a Variable Star, pp. 73–80. Cambridge University Press, Cambridge (1994)
39. Rottman, G.J., Woods, T.N., McClintock, W.: SORCE solar UV irradiance results. Adv. Space Res. **37**(2), 201–208 (2006)
40. Sommer, A.H.: Photoemissive Materials: Preparation, Properties and Uses, 268 pp. Wiley, New York (1969)
41. Somov, V.V., Oreshina, I.V., Lyubimov, G.P.: Topological model of a major solar flare. Astron. Rep. **48**(3), 246 (2004)
42. Thomas, R.J., Starr, R., Crannell, C.J.: Expressions to determine temperatures and emission measures for solar X-ray events from GOES measurements. Sol. Phys. **95**, 323–329 (1985)
43. Warren, H.P., Warshall, A.D.: Ultraviolet flare ribbon brightenings and the onset of hard X-ray emission. Astrophys. J. **560**(1 Pt. 2), L87–L90 (2002)
44. Woods, T.N., Thomas, N., Rottman, G.J., Ucker, G.J.: Solar-stellar irradiance comparison experiment 1: 2. Instrument; calibrations. J. Geophys. Res. **98**, 10679–10694 (1993)
45. Woods, T.N., Tobiska, W.K., Rottman, G.J., Worden, J.R.: Improved solar Lyman alpha irradiance modeling from 1947 to 1999 based on UARS observations. J. Geophys. Res. **105**, 27195–27215 (2000)
46. Zhitnik, I.A., Kuzin, S.V., Sobelman, O.I., et al.: Main results of the SPIRIT experiment onboard the CORONAS-F satellite. Sol. Syst. Res. **39**(6), 442–452 (2005)

# Scientific Set of Instruments "Solar Cosmic Rays"

**S.N. Kuznetsov[1], A.V. Bogomolov, V.I. Galkin, Yu.I. Denisov[2], A.N. Podorolsky, S.P. Ryumin, K. Kudela, and J. Rojko**

Complex analysis of the generation of high-energy neutral radiation (gamma-quants and neutrons) during the solar flares on board the SMM satellite has shown that this study provides important information about the generation of the solar energetic particles.

The development of a complex program of the studies of solar activity and its influence on the near-Earth space began in the USSR in the middle of 1980s of the twentieth century. It resulted in the development of a number of satellites of CORONAS series. A set of scientific instruments SCR (Solar Cosmic Rays) was developed by the scientists of SINP MSU and IEP SAS in order to study relations between the radiation conditions in the near-Earth space and solar activity. This instrument was installed on board the satellites CORONAS-I and CORONAS-F launched to the orbit on March 2, 1994 and July 30, 2001, respectively. It consisted of the following three instruments.

SONG (SOlar Neutrons and Gammas) instrument is intended for registration of X-rays and gamma-emission within the energy range of 30 keV to 200 MeV, and neutrons with energies over 20 MeV. Besides, this device could measure the flux and the spectrum of electrons within the range of 2–108 MeV and proton flux with energies > 75 MeV.

---

[1]Deceased

[2]Deceased

S.N. Kuznetsov · A.V. Bogomolov (✉) · V.I. Galkin · Yu.I. Denisov · A.N. Podorolsky · S.P. Ryumin
D.V. Skobeltsyn Institute of Nuclear Physics, M.V. Lomonosov Moscow State University (SINP MSU), Moscow, Russia
e-mail: aabboogg@rambler.ru

K. Kudela · J. Rojko
Institute of Experimental Physics, Slovak Academy of Sciences (IEP SAS), Kosice, Slovakia

V. Kuznetsov (ed.), *The Coronas-F Space Mission*, Astrophysics and Space Science Library 400, DOI 10.1007/978-3-642-39268-9_9,

MKL (Russian abbreviation for Monitor of Cosmic Rays) instrument is intended for registration of electron flux within energy range of 0.3–12 MeV and proton flux with energies from 1 MeV up to ~90 MeV.

CCR-3 (Composition of Cosmic Radiation) instrument measures the densities of the fluxes of nuclei within the charge range 1–10 and energy range from few up to dozens MeV per nucleon: $1.5 < E < 19$ MeV/nucleon for the nuclei $^{1}H$ and $^{4}He$; $3.7 < E < 46$ MeV/nucleon for the nuclei up to $^{20}Ne$.

The SONG instrument consists of two blocks: detecting unit SONG-DM and electronics unit SONG-EM. Information recorded by the SONG-DM unit is processed and output to telemetry by means of the SONG-EM unit.

The SONG-DM device (solar neutrons, gamma-radiation) is intended for registration of neutral and charged particles within the following energy ranges:

- fluxes and energy release spectra of neutrons with energies of >20 MeV;
- fluxes and spectra of $\gamma$ with energies of 0.03–200 MeV;
- fluxes and spectra of electrons with energies of 11–108 MeV;
- fluxes of protons with energies of 200–300 MeV;
- total fluxes of electrons with energies of >55 MeV and protons with energies of >70 MeV.

The instrument is intended for registration of neutron flux up to $\sim 100\,\mathrm{cm^{-2}\,s^{-1}}$, $\gamma$—up to $\sim 400\,\mathrm{cm^{-2}\,s^{-1}}$, electrons—up to $\sim 400\,\mathrm{cm^{-2}\,s^{-1}\,sr^{-1}}$, protons—up to $\sim 600\,\mathrm{cm^{-2}\,s^{-1}}$.

Overall size and mounting support: $350 \times 350 \times 560\,\mathrm{mm^3}$; $(167 \pm 0.2)$ mm + $(167 \pm 0.2)$ mm.

Mass of the instrument does not exceed 50 kg.

Power consumption of the instrument under standard conditions and power supply of $27^{+7}_{-3}$ V does not exceed 8 W.

The instrument is located outside of the object. $X$-axis is directed to the Sun with an accuracy of $\pm 2°$.

The SONG-DM instrument consists of radiation detectors, sampling and analyzing circuits, pulse devices, voltage converter for the detectors, and electronics supplying inside one block. Appearance of the instrument is presented in Fig. 1.

The SONG-DM instrument is operated combined with the SONG-EM manufactured by Slovakian colleagues.

**The Method of Particles Registration.** The structure of the SONG-DM detector is presented in Fig. 2.

In some way the method of particles registration by the SONG instrument is analogous to that used by the instruments GRS [3], OSSE [4], and GRS [5], mounted on board the satellites SMM, CGRO, and "Yohkoh", respectively. In this instruments gammas were registered in non-organic crystals of the matter with big $Z$. In the instruments on board the satellites SMM and CGRO they used NaJ-crystals, and CsJ-crystals behind them were used in the anticoincidence mode for the analysis of the detailed spectra. CsJ-crystals were used for the neutrons' registration. BGO crystals with big effective $Z$ are more suitable for registration of gammas with energies of ~2 MeV. In the SONG instrument CsJ-crystal is used for gammas

**Fig. 1** Photo of the SONG-DM instrument mounted on the support stand

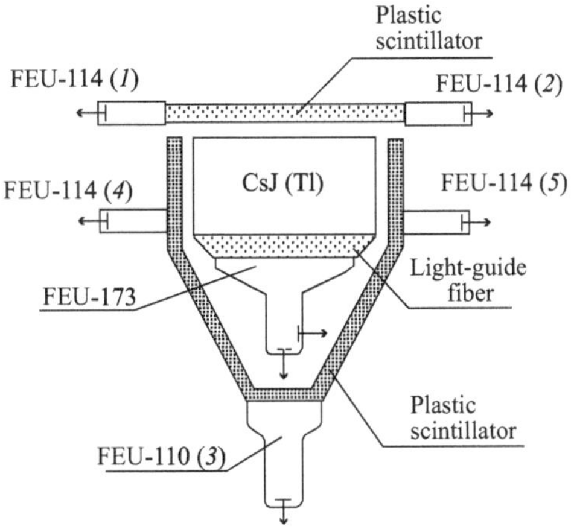


**Fig. 2** The structure of the SONG-DM instrument

registration: it has rather smaller energy resolution comparing with NaJ, but impulse shape depends on specific ionization of the particle [1, 2], therefore it is possible to differ impulses produced by gammas and neutrons.

The particles are registered by scintillation counter which consists of CsJ(Tl) (diameter of 200 mm and height of 100 mm) scanned by a photomultiplier FEU-173. In order to protect this scintillation detector from the charged particles it is covered with an active anticoincidence protection consisted of two parts. The lower part is a truncated cone and cylinder in optic contact, this part is manufactured from plastic scintillator with 20 mm thick walls and protects the basic counter from below and laterally. The truncated cone is scanned by a photomultiplier FEU-110 from below (3 in Fig. 2); the cylinder is scanned by two photomultipliers FEU-114 (4, 5) laterally from two diametric points. The upper protection part is a disc also produced from plastic scintillator 10 mm thick; it is scanned by two photomultipliers FEU-114

(1, 2) from two diametric points. This scintillator protects the basic detector from above and, besides, along with the basic detector forms a telescope for electrons registration. The total protection of the basic counter is not less than 1.3 g/cm$^2$, of the upper anticoincidence disc—0.05 g/cm$^2$.

After amplification the impulses from the anodes of the anticoincidence photomultipliers come to discriminators with threshold energy of 1 MeV and then through OR-circuit to gate circuit with the basic detector's impulses which are registered by the discriminator with minimum threshold (0.03 MeV). Coincidence with the impulses of the basic detector essentially increases maximum instrument's counting rate. Counting rate of the anticoincidence cone, anticoincidence cylinder, and anticoincidence discs along with counting rates of the impulse at the output of the gain circuit (general inhibit) are sent to telemetry.

If detected event is produced by a neutral particle, i.e., the response of the basic detector is not accompanied by a signal of the anticoincidence protection, neutrons and gammas separation circuit connected to the last dynode of the basic detector's photomultiplier analyzes the impulse shape in CsJ(Tl) crystal. Intensities of the fast ($I_F$) and slow ($I_S$) components of light flash luminescence in CsJ(Tl) are compared in the circuit. Luminescence constants for these components are 0.5–0.7 and 7 ms, respectively. Value of the ration $I_S/I_F$ depends on specific ionization of the detected particles and varies from 1.0 for electrons with energies of 0.66 MeV to 0.35 for $\alpha$-particles with energies of $\sim$5 MeV. In the instrument the neutrons are detected by charged particles produced by non-elastic interactions with Cs and J nuclei, mainly, protons, deuterons, $\alpha$-particles and heavier nuclei, or by recoil nuclei produced by elastic scattering of the neutrons at the same nuclei, i.e., by the particles with high specific ionization. Specific ionization of electrons and positrons produced by interactions of gammas with CsJ at the energies within the operating range of the instrument is close to minimum. Separation circuit $n \leftrightarrow \gamma$ produces a signal "neutron" ("$n$") at $I_S/I_F < 0.9$, which corresponds to the protons with energies of <100 MeV. At the amplitudes of 0.1–3 MeV (in terms of electron equivalent) the separation circuit $n \leftrightarrow \gamma$ produces a signal "n1" (C1 parameter) and at the amplitudes of 3–60 MeV—a signal "n2" (C2 parameter). The first area corresponds to the registration of neutrons with energies of up to 1,000 MeV by recoil nuclei (Cs and J nuclei at which neutrons scattered elastically), the second—to the registration of the neutrons by nuclear reactions. Due to Coulomb barrier energy spectra of the protons produced by the reactions with Cs and J nuclei are limited with energy of $\sim$4 MeV, $\alpha$-particles—$\sim$10 MeV and more higher energies for the nuclei with $Z > 2$. Due to nonlinear character of light output of CsJ(Tl)-crystal amplitudes of the signals produced by the protons with energies of 4 MeV are the same as those produced by the electrons with energies of $\sim$3 MeV. Amplitudes of the light impulses for the recoil nuclei in the operating volume of the instrument are tenfold smaller than for the electrons of the same energy; during elastic interactions the neutrons deliver to Cs and J nuclei not more than 1/30 part of their energy.

Simultaneous registration of the neutrons by nuclear reactions and by recoil nuclei duplicates the effective area of the detector. In-flight operation has shown that recoil nuclei are excited and de-energized by gammas.

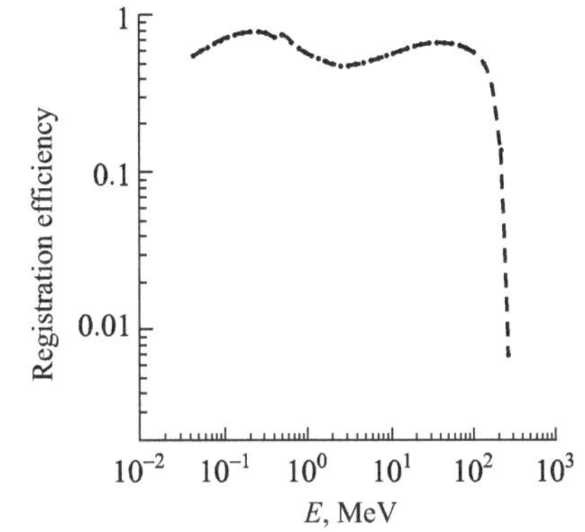


**Fig. 3** Relation of gammas registration efficiency for SONG-DM instrument from energy

Energy release spectrum of recoil nuclei is measured in the instrument by two-channel amplitude analyzer, and energy release spectrum of nuclear reactions products—by means of six-channel analyzer. In the case of the basic detector's response with no signal of anticoincidence protection both "n1" and "n2" signal is regarded as detection of gamma. Partly such events are caused by neutrons but as a rule their flux is essentially smaller than gamma-flux. Spectrum of such events within the range of 0.1–200 MeV is measured in the instrument by means of ten-channel amplitude analyzer, all neutral particles with energy release over 200 MeV are also registered by a separate channel. Besides, in order to register gamma-lines this spectrum within the range of 0.3–20 MeV is measured by 238-channel logarithmic analyzer. All signals of the basic detector apart from those sending to the separation circuit $n \leftrightarrow \gamma$ are taken from the plate load of the photomultiplier FEU-173.

Simultaneous presence of a signal of disc plastic scintillator with amplitude of 1–6 MeV, coinciding signal of the basic detector with amplitude of 3–100 MeV and no signal of the low part of the anticoincidence protection are the sign of registration of electrons with energies of 11–108 MeV. Spectrum of such electrons is measured in the instrument by means of six-channel amplitude analyzer.

Detected particle is regarded as a proton with energy from 200 to 300 MeV if its energy release in the basic detector exceeds 200 MeV.

Cosmic rays particles—electrons with energies over 50 MeV and protons with energies over 70 MeV—are recorded as events with energy release over 50 MeV in CsJ.

Maximum effective area of the instrument for neutron registration by nuclear reactions is 34 $cm^2$. Maximum effective area for gamma-registration is $\sim$280 $cm^2$. Efficiency of gamma counting depends on energy. Gammas registration efficiency versus their energy is presented in Fig. 3.

Effective area for registration of total flux of electrons and protons with energies over 55 and 70 MeV, respectively, is equal to $\sim$170 $cm^2$. Geometry factor for registration of electrons is 520 $cm^2$ sr.

The most important elements of the instrument's structure—low-voltage converter and high-voltage converter feeding the photomultiplier FEU-173, separation circuit $n \leftrightarrow \gamma$, and 238-channel amplitude analyzer—are duplicated. Operation is switched to the fallback circuits by means of radio command.

Besides the listed signals the SONG-DM instrument also receives information about the operated variant (basic or fallback) of the circuit of neutron identification and 240-channel analyzer (C6-parameter). Simultaneously with measurements made by the SONG-DM instrument information about the flux and spectra of gammas with energies of 0.2–80 MeV and neutrons with energy release of 5–80 MeV is also processed in the AVS-F instrument; and it allows to provide more detailed information on the detected particles and to increase the measurements reliability.

Signals from the plate load of the basic photomultiplier are sent from the SONG-DM instrument directly to the AVS-F instrument and impulses of parameters "total inhibit" (C7), "protons of 200 MeV" (P2), "protons of >70 MeV, electrons of >55 MeV" (P1), and "SONG flare" are also transmitted to the AVS-F instrument through the SONG-EM instrument. In order to improve dynamic ranges of amplitude registration signals are sent to the AVS-F instrument both from the output of U2-amplifier through the connector Sh 151 and directly from the plate load of FEU-173.

The MKL instrument is intended for registration of charged particles within the following energy ranges:

- spectra and fluxes of protons within the energy range of 1–90 MeV;
- spectra and fluxes of electrons within the energy range of 0.5–12 MeV;
- $\alpha$-particles flux with energies of 100–140 MeV;
- electron flux with energy of $\geqslant$1.0 MeV;
- proton flux with energies of >10 MeV;
- total flux of electrons with energies of $\geqslant$1.6 MeV and protons with energies of $\geqslant$23 MeV;
- total flux of electrons with energies of $\geqslant$1.6 MeV and protons with energies of $\geqslant$90 MeV.

The instrument is designed for the registration of the flux up to $2 \cdot 10^5\,\mathrm{cm^{-2}\,s^{-1}\,sr^{-1}}$. Information from the instrument is transmitted to the information storage and processing system (ISPS).

Overall size of the instrument is $200 \times 250 \times 250 \pm 5$ mm, its mass does not exceed 4.6 kg.

Power consumption of the instrument under standard conditions and power supply of 27 V does not exceed 4.0 W. Appearance of the instrument is presented in Fig. 4.

The instrument consists of one block with radiation detectors, pulse generators, gate circuits, and voltage converters for the detectors and electronics along with device for information output to telemetry inside it. Two detector units of the instrument are arranged on the upper panel of the case. Inside the case there are devices for amplification and generation of impulses, circuits for information

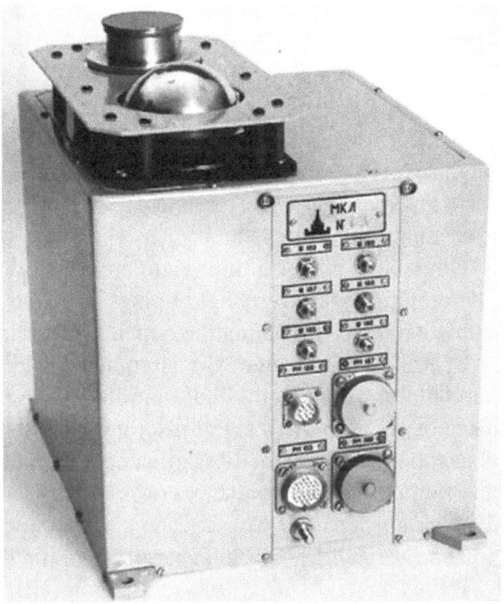


**Fig. 4** Photo of the MKL instrument

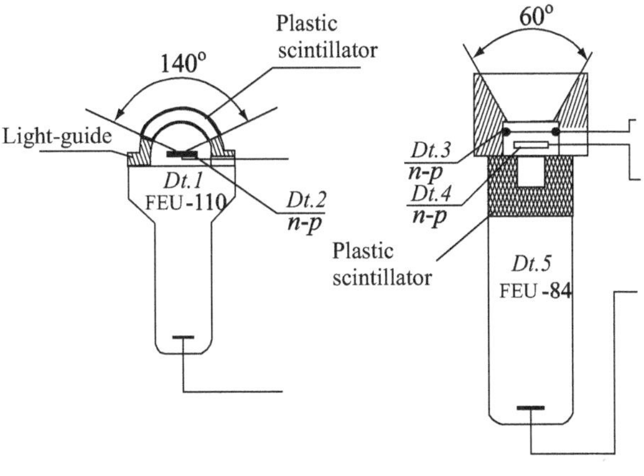


**Fig. 5** Schematic arrangement of the MKL instrument

transmission to telemetry, voltage converter for the detectors and electronics, photomultipliers.

Detector part of the instrument consists of two telescopic systems. Functional diagram of the instrument is presented in Fig. 5. Proton flux with energies within the range of 23–90 MeV, $\alpha$-particles with energies of 100–140 MeV along with flux of electrons with energy $E_e > 1.6$ MeV, and protons with energy $E_p > 90$ MeV are measured by means of telescopic system which is presented in the upper left part of Fig. 5.

Scintillation detector *Dt.1* is a part of spherical layer of plastic scintillator based on polystyrene inside the area enclosed the cone with a peak in the sphere's center and span angle of 140°. Layer's thickness is ~4.0 mm, its radius is 40 mm. A semiconductor detector *Dt.2* 12 mm in diameter and 2 mm thick is located in the center of the sphere.

Specific ionization of the particles for their movement in the considered system is used for particles' identification and determination of their energy. Amplitude of impulse produces by the charged particle ($\Delta E$) is measured by scintillator detector; for further analysis the particles with energy release in semiconductor detector exceeding double ionization of a relativistic single-charged particle are selected. Moving in the matter of the scintillator detector the charged particle ejects protons. Photomultiplier converts light flashes into electric impulses. Light comes to photocathode of FEU-110 along the optical fiber produced of organic glass.

Scintillator is covered with aluminum protection 0.5 mm thick. Signals from the photomultiplier scanning the scintillator detector and signals from semiconductor detector come to the integral discriminators through the gain and pulse generating detectors. Each integral discriminator is built in a logical component in such a way that formed channel produces a response only to the particles which energy belongs to a certain interval. Consequently proton spectrum getting through the detector system *Dt.1* and *Dt.2* within energy ranges of 23–90 MeV is determined.

Hemispherical layer of scintillator *Dt.1* is used as single detector in order to record electrons with $E_e = 0.5$–1.0 MeV, protons with energies of $E_p = 14$–26 MeV and protons with energies of $E_p \geqslant 10$ MeV. Geometry factor of the detector is ~30 cm$^2$ sr. Discrimination threshold for the protons with energies of $E_p = 14$–26 MeV is chosen to be equal ~10 MeV.

Spectra of electron flux with energies of $E_e = 0.5$–12 MeV and proton flux with energies of $E_p = 1$–5 MeV are measured by means of telescopic system consisting of three detectors (the right part of Fig. 5).

*Dt.3* is a thin *n-p*-detector of ~10 mm diameter and 50 mkm thickness; *Dt.4* is a thick *n-p*-detector of 12 mm diameter and 2,000 mkm thickness; *Dt.5* is a jointed scintillation detector of Phoswich type. It consists of CsJ crystal of 10 mm diameter and height. CsJ has a higher density comparing with Si-detectors and allows to detect electrons with higher energy. In order to protect it from the particles penetrating laterally and from below detector is covered with a 10-mm thick layer of plastic scintillator. Both scintillators are scanned by one photomultiplier FEU-84. The events in the crystal and plastic scintillator (light flashes during the charged particles' passing) are separated by means of electronic separation circuit according to the difference in de-excitation time. Events happened in plastic scintillator prohibit the further signal processing. In order to protect the telescope from light it is covered with a 10 mkm thick aluminum foil from above.

Input aperture of telescope is enclosed by aluminum collimator with span angle of ~60°. Geometry factor of the system is equal to ~0.5 cm$^2$ sr. Effective geometry factors for the particles recorded by the MKL instrument are presented in Table 1.

After impulse gain, generation, and separation, the signals from the photomultiplier and semiconductor detectors come to the discriminator. Discriminator threshold is adjusted in such a way that discriminator produces a response only when a particle of a certain energy is passing. Signals from discriminator come to the input

**Table 1** Characteristics of the particles recorded by the MKL instrument

| No. | Parameter (MeV) | Effective geometry factor ($cm^2$ sr) |
|---|---|---|
| 1 | $E_e > 0.5$; $E_p > 9$ | 30 |
| 2 | $E_e > 1$, $E_p > 9$ | 30 |
| 3 | p 14–26 | 30 |
| 4 | p 23–26 | 2 |
| 5 | p 26–50 | 2 |
| 6 | p 50–90 | 2 |
| 7 | $E_e > 1.6$; $E_p > 90$ | 2 |
| 8 | $\alpha$ 23–33 | 2 |
| 9 | $E_e > 1.6$; $E_p > 23$ | 2.45 |
| 10 | p 1–5 | 0.5 |
| 11 | e 0.3–0.6 | 0.425 |
| 12 | e 0.6–1.5 | 0.44 |
| 13 | e 1.5–3 | 0.39 |
| 14 | e 3–6 | 0.43 |
| 15 | e 6–12 | 0.47 |

of coincidence circuits and then output parameters of the instrument in the form of counting rates of certain particles within the corresponding energy ranges are sent to the source of numeric arrays. Information from the MKL instrument is recorded by telemetry and sent to the Earth by means of ISPS by means of numeric arrays output from the MKL instrument to the SONG-EM block of the SCR instrument and then to the ISPS system.

The CCR-3 (composition of cosmic rays) instrument is intended for spectrometry measurements of the flux of the cosmic rays nuclear component. It records density of the nuclei flux within the charges range of 1–10 and with energies from few up to tens MeV/nucleon: $1.5 < E < 19$ MeV/nucleon for the nuclei $^1H$ and $^4He$; $3.7 < E < 46$ MeV/nucleon for the nuclei $^{20}Ne$.

Overall size of the instrument is equal to 200 × 208 × 270 $mm^3$, mass—5.0 kg.

The CCR-3 instrument consists of one block; its appearance is presented in Fig. 6.

The instrument consists of a telescope, circuits of analogous and digital processing of signals, communication circuit for on board radio-technical system and DIOGENESS-PRAM block, power supply for the detector and electronics.

The telescope includes four silicon detectors 25(50), 100, 2,000, and 2,000 mkm thick. Input angle determined by collimator and distance between two first detectors is equal to 50°.

Measurements are based on determination of energy release by three first detectors. Amplitudes of the impulses of nuclei stopped in the second and third detectors and entered the telescopes within the given space angle are analyzed.

In order to provide operation under control at big loads (up to $10^5$ Hz) short impulses generators on the shorted delay circuits are placed after charge-sensitive amplifier. Then impulses pass through the devices of sample and storage (DSS) which act as impulse linear transmitter and lengthener.

**Fig. 6** Photo of the CCR-3 instrument

Fast discriminators and coincidence circuits produce signals which control opening of DSS which store the amplitudes of input signals till the end of amplitude-digital converting.

Control signals $p$, $\alpha$, $z$, conforming passing of the nuclei with $z \geqslant 1$, $z \geqslant 2$, $z \geqslant 3$, are produced by means of current summation unit which sums up impulses of the three first detectors with corresponding coefficients: $I_1 + k_2 I_2 + k_3 I_3$.

For the detectors 25, 100, and 2,000 mkm thick $k_2 = 0.3$ and $k_3 = 0.1$.

Information from the CCR-3 instrument is sent to ISPS through the DIOGENESS-PRAM device. Indications of the analogous-digital converter are transmitted to the DIOGENESS-PRAM (PRAM) instrument through the lock words controlled from this block. Parameters counting rates (counting parameters) $p2$, $p3$, $\alpha 2$, $\alpha 3$, $z2$, and $z3$ are determined by means of gages $2^8$, called over by the PRAM block. Information is sent to PRAM block in the form of parallel code through one digital 8-bit channel—data bus. Counting channels, analyzer's channels, and state counters are connected to this data but in a certain order according to the information obtained by the CCR-3 instrument and a program contained in the PRAM block.

State counter includes preliminary information about the nucleus detected by the analyzer's channels and about the proton intensity. Information recorded in the state counter determines poll mode and the instrument's operation procedure.

The instrument is mounted outside the object under the blanket. Telescope's axis is directed toward the antisolar point.

The CCR-3 is a monitoring instrument with a good charge resolution. Its resolution can be demonstrated by the biaxial diagram of analyzers indications.

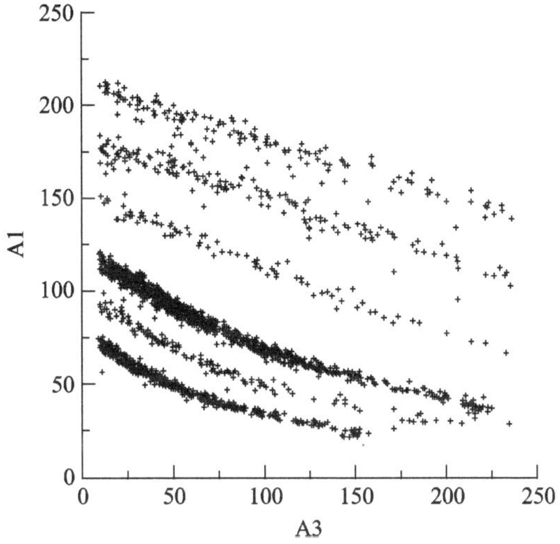


**Fig. 7** Diagram of registration of different elements by the CCR-3 instrument in flight

Figure 7 presents the data of A1–A3 analyzers during several hours of the flare of November 22, 2001.

In the diagram it is easy to separate tracks of C, N, O, Ne, Mg, and Si nuclei.

The CCR instrument provided information from the moment of the launching in 2001 till March 2004. During this period information on the SCR nuclei flux observed in the near-Earth space after a number of big flares on September 24, 2001, November 4 and 22, 2001, April 22, 2002, and a series of flares in November, 2003 was obtained.

Obtained information includes data on nuclei intensity flux (protons, Helium nuclei, and nuclei with $Z \geqslant 6$) within two energy ranges, energy spectra for the fluxes of separate nuclei from C up to Si, and relative abundance of separate nuclei in SCR flux.

## References

1. Bogomolov, A.V., Britvich, G.I., Myagkova, I.N., Ryumin, S.P.: Identification of neutrons against the gamma-ray background by CsI(Tl) crystal detectors. Instrum. Exp. Tech. Vol. 39, No. 1, 13–17 (1996)
2. Bogomolov, A.V., Myagkova, I.N.: Registration of neutrons and gammas in laboratory conditions by means of separation basing on the shape of scintillation impulse in the CsJ(Tl) crystal. Moscow Univ. Bull. Phys. Astron. (2), 61–63 (1997) (in Russian)
3. Forrest, D.J., Chupp, E.L., Ryan, J.M., et al.: The Gamma ray spectrometer for the solar maximum mission. Sol. Phys. **65**, 15–23 (1980)
4. Murphy, R.J., Share, G.H., Grove, J.E., et al.: Vestruns OSSE observations of the 4 June 1991 solar flare. In: Ryan, J.M., Vestrand, W.T. (eds.) High-Energy Solar Phenomena—A New Era of Spacecraft Measurements, Waterville Valley, 1993, pp. 15–20
5. Ogawara, Y.: Japanese satellite program on the solar research for the coming solar maximum. 1. Observations of high-energy electromagnetic radiations. Space Sci. Rev. **51**, 175–184 (1989)

# Protons Acceleration in Solar Flares: The Results of the Analysis of Gamma-emission and Neutrons Recorded by the SONG Instrument Onboard the CORONAS-F Satellite

**S.N. Kuznetsov[1], V.G. Kurt, B.Yu. Yushkov, I.N. Myagkova, V.I. Galkin, and K. Kudela**

## Introduction

Observation of gamma-ray emission with energy of 0.3–3000 MeV is the most informative method to study the particle acceleration processes and therefore to understand better the process of the flare energy release itself. This emission arises as a result of interactions of high-energy particles with the matter of the solar atmosphere. It does not undergo significant changes leaving high-density regions and comes to the Earth's orbit keeping information about the processes acting in the acceleration region. A review of energetic ions in the solar atmosphere deduced from observations of neutrons and gamma-rays can be found in [1, 2]. Low energy ions manifest nowise themselves in the solar atmosphere until their energy reaches $\sim$ 10 MeV/nucleon. Information concerning fluxes and spectra of ions of these low energies can be obtained only by the direct measurements of particles fluxes in the space. If ions are accelerated up to higher energies, then they start to interact with dense matter of the solar atmosphere producing neutral emissions.

Narrow gamma-line emission appears as a result of nuclear reactions of the protons, $\alpha$-particles, and heavy nuclei accelerated up to the energy of 10–30 MeV/nucleon, with the protons, helium nuclei, and other heavy nuclei of the solar atmosphere. The energy of the photons of these gamma-lines lies in the range of 0.15–17 MeV. The most intensive lines appear in the de-excitation

[1]Deceased

S.N. Kuznetsov · V.G. Kurt · B.Yu. Yushkov (✉) · I.N. Myagkova · V.I. Galkin
D. V. Skobeltsyn Institute of Nuclear Physics, M. V. Lomonosov Moscow State University (SINP MSU)
e-mail: clef@srd.sinp.msu.ru

K. Kudela
Institute of Experimental Physics, Slovak Academy of Sciences (IEP SAS), Kosice, Slovakia

V. Kuznetsov (ed.), *The Coronas-F Space Mission*, Astrophysics and Space Science Library 400, DOI 10.1007/978-3-642-39268-9_10,

process of the nuclei $^{12}C$ (4.438 MeV) and $^{16}O$ (6.129 MeV). Protons with energy of $\sim 10-30$ MeV are the most effective for the excitation of these nuclei. When the proton energy grows up then the cross section of these reactions falls down considerably. Life time of the excitation states does not exceed $10^{-12}$ s, therefore the lines are emitted without visual delay and so they are called *direct lines*. So far there are 17 such relatively narrow lines [3].

Neutrons as well as direct gamma-lines appear as a result of the nuclear reaction of the accelerated ions with the matter of the solar atmosphere. The most important reaction for the neutron generation is interaction of protons with the nuclei $^4He$. Energy threshold of this reaction is 30 MeV [4]. The "fate" of the neutrons generated during these reactions can be different. The neutrons which stayed at the Sun can be captured by nuclei before they decay. Capture reaction $^3He + n \rightarrow {}^3H + p$ goes without photons emission while the reaction $H + n \rightarrow {}^2H + \gamma$ generates photons with energy of $E_\gamma = 2.223$ MeV. This line lags behind with regard to the direct gamma-lines at approximately 80 s because the neutron must be decelerated to the thermal energy before it will be captured by proton.

High-energy neutrons (several hundreds MeV) are generated during the interaction of high-energy protons and $\alpha$-particles with the matter of the solar atmosphere. Some of these neutrons can leave the Sun. Part of them decay in flight according to the reaction $n \rightarrow p + e^- + \gamma$ (life time of free neutron is $\approx 15$ min). This time is comparable to the propagation time of high-energy neutrons from the Sun to Earth. The first solar neutron signal at the Earth's orbit was detected by Chupp et al. [5] with Gamma Ray Spectrometer (GRS) on Solar Maximum Mission (SMM) during the solar flare on June 21, 1980. Second time GRS recorded a solar neutron signal after the June 3, 1982 flare [6]. Besides, some neutron events were separated in the time series of well-situated (in a sub-solar point) high-mountain neutron monitors [7–11]. Neutron measurements provide one more informative channel for the studies of the particle acceleration up to high energies.

Protons with energies $> 30$–50 MeV do not manifest themselves in the solar atmosphere until their energy exceeds the threshold of $\pi$-meson generation—about 300 MeV. High-energy photons arise as a result of either the neutral pion ($\pi^0$) decay or the charged pion ($\pi^\pm$) decay accompanied by generation of electrons and positrons which in turn interact with solar matter and create high-energy bremsstrahlung [12–15]. The $\pi^0$ (rest mass $E_0 = 135$ MeV) decay produces gamma-emission with a maximum in the energy range of 60–100 MeV that is formed due to Doppler broadening. If the energy of protons is significantly higher than pion generation threshold, continuous spectrum of gamma-emission has the same shape as the spectrum of the parent protons. Maximum energy of the photon spectra is comparable with maximum values of proton energy.

Primary accelerated electrons generate bremsstrahlung with a maximum energy equal to the energy of these electrons. In particular the strong sub-peaks of bremsstrahlung up to 100–300 MeV were observed from several solar flares [16–19]. Thus the total gamma-ray spectrum at energies above 10 MeV is a combination of a power-law bremsstrahlung spectrum and a pion-decay emission spectrum. To distinguish the pion-decay feature in the gamma-emission spectrum, a detector with an appropriate sensitivity and spectral resolution is necessary. Moreover, the intensity of the pion-decay emission must exceed the bremsstrahlung background in a range of 60–200 MeV.

For the first time such hard gamma-emission with photon energy > 60 MeV was observed with SMM/GRS during the flares of June 21, 1980 [5] and June 3, 1982 [20]. In the latter case this emission was attributed to the $\pi^0$ decay. Later the spectra with reliably separated pion-decay line and high-energy tail (> 60 MeV) were observed during several powerful flares near the maximum of the 22nd solar activity cycle (December 16, 1988, March 6, 1989, May 24, 1990, March 22, 1991, June 11 and 15, 1991) in experiments with SMM [21], GRANAT [22, 23], CGRO [24], and GAMMA [16, 25]. A brief review of these observations can be found in [26, 27]. The latest observation of the pion-produced gamma emission was performed by *Fermi* with Large Area Telescope during the modest solar flare on June 12, 2010 [28].

The SONG instrument onboard CORONAS-F satellite was the only experiment which was designed for the measurement of high-energy emissions and it operated during the 23rd solar activity cycle [29–31]. It was based on a CsI(Tl) crystal with 20 cm diameter and 10 cm height [19, 32]. The detailed instrument description can be found in the chapter "Scientific Set of Instruments ≪Solar Cosmic Rays≫" of this book. SONG had the appropriate sensitivity, spectral, and time resolution to resolve the above task.

The CORONAS-F orbit parameters did not allow to detect solar neutral emissions during satellite location in the Earth's shadow and in the radiation belts. Due to this reason and due to telemetry failures and information lacunas several powerful flares were missed or recorded only partially. In particular several major flares were missed which occurred in an active region NOAA 10486 during the famous period of high solar activity from October 23 till November 4, 2003 [33] and in the active region NOAA 10720 during January 2005. However, the SONG instrument detected hard X-ray (HXR) and gamma-ray emission from five flares of >X1 importance in October–November 2003. During the active period of January 2005 two powerful flares—January 15 and January 20—were observed. The last one seems to be the most experimentally ensured event during the whole history of the solar flares observations [34].

During the operation of the CORONAS-F SONG has recorded four events in which the $\pi^0$-decay line was reliably separated. Observation conditions during these flares allowed to measure temporal profile of high-energy emission during the impulsive phase of these flares and to determine the onset time of the pion-decay emission for certain.

Table 1 is a catalogue of solar flares with high-energy gamma-emission observations. Before CORONAS-F operation only nine events with gamma-emission > 60 MeV were detected and only five neutron events were measured onboard the satellites. Dataset of these events is short even in comparison with the set of measurements of gamma-continuum within the energy ranges > 300 keV (there are about 700–800 such events, including those recorded by the SONG instrument). Especially it is infinitely small in comparison with the database of soft X-rays (SXR)

**Table 1** Solar flares with gamma-ray energy $E_\gamma > 60$ MeV observed onboard satellites

| Flare date | Flare coordinates | Experiment | References |
|---|---|---|---|
| 21.06.1980 | N20W90 | SMM/GRS | [5] |
| 03.06.1982 | S09E72 | SMM/GRS | [20] |
| 16.12.1988 | N27E33 | SMM/GRS | [21] |
| 06.03.1989 | N35E69 | SMM/GRS | [21] |
| 24.05.1990 | N36W76 | GRANAT/PHEBUS | [22] |
| 22.03.1991 | S26E28 | GRANAT/PHEBUS | [23] |
| 26.03.1991 | S09W23 | GAMMA/GAMMA-1 | [16] |
| 11.06.1991 | N31W17 | CGRO/EGRET | [24] |
| 15.06.1991 | N33W69 | GAMMA/GAMMA-1 | [25] |
| 25.08.2001 | S17E34 | CORONAS-F/SONG | [29–31] |
| 28.10.2003 | S16E08 | CORONAS-F/SONG | [18,19,29–31,35] |
| 04.11.2003 | S19W83 | CORONAS-F/SONG | [29–31] |
| 20.01.2005 | N14W61 | CORONAS-F/SONG | [18,29–31,34,36] |
| 12.06.2010 | N23W43 | *Fermi*/LAT | [28] |

flares, which consists of 60,000 events. So it is easy to see the importance for the high-energy solar physics of the measurements performed by SONG. Measurements of high-energy gamma-emission and neutron fluxes are very rare, so each event must be scrutinized. In particular comparison of temporal behavior and spectra of high-energy gamma-emission with other flare manifestations can clarify the acceleration processes and dynamics of high-energy electrons and protons in the solar atmosphere.

## Experimental Data

The time dependence and evolution of gamma-spectrum were analyzed in detail in order to determine the moment of the electrons and protons acceleration during the flare process. These measurements provided an opportunity to follow up and to draw out a fine structure of the particle acceleration episodes.

**Solar flare of October 28, 2003.** The polar cap is the region where on one hand the solar energetic particles can penetrate freely into the Earth's magnetosphere, and on the other hand the radiation belt particles do not make any contribution into the detector count rate. So, the time interval when the CORONAS-F crossed the polar cap, i.e., the region of $L$-shell values > 12 was suitable for the observation of both neutral emissions and charged particles coming after the flare. The information on the local conditions and the count rates of some SONG channels during the time interval 07:30–12:00 (hereinafter the Universal Time is used) on October 28, 2003, are presented in Fig. 1.

Interplanetary magnetic field (IMF) and solar wind velocity $V_{sw}$ and density $N_{sw}$ were almost constant during the time interval of 08:00–12:00. The value of $B_z$-component of IMF which influences on the magnetosphere conditions did not exceed 5 nT during the time interval of 10:00–12:00. Geomagnetic activity

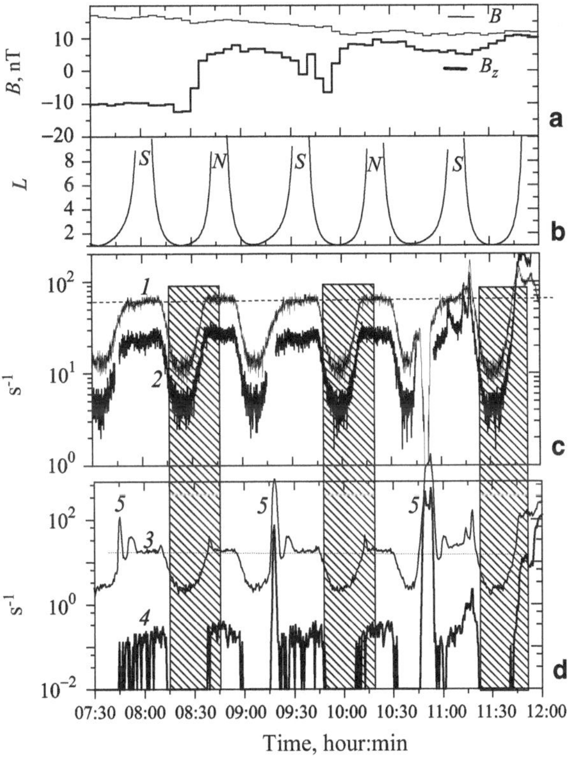


**Fig. 1** The overview of the SONG measurements during the event of October 28, 2003: *a*—Interplanetary Magnetic Field and its $B_z$- component; *b*—*L* parameter for the CORONAS-F orbit, *N* and *S* relate to the crossing of the northern and southern polar caps by the satellite, respectively; *c*—counting rates of the protons with energy of $E_p$ = 300–400 MeV (*1*) and gamma-emission at the energy of $E_\gamma$ = 60–90 MeV (*2*); *d*—counting rates of the electrons with energy of $E_e$ = 14–18.5 MeV (*3*) and neutrons with energy release of 40–60 MeV (*4*). Time intervals when the satellite was in the shadow are dashed. South–Atlantic Anomaly region is indicated with number *5* (taken from [29]). From [19]

indices varied weakly. So we can conclude that during this period there were no variations which could result in some short cosmic rays variations, geomagnetic field disturbances, or changes of cut-off rigidity for the charged particle penetration comparing with other crossings of the polar cap by the satellite. It ensure us that during the observations of the neutral and charged emission from the flare of October 28 the previous state of the solar activity did not influence on the contribution of charged particles into the counting rate of all channels comparing with the previous orbits. According to the logic of the electronic schemes of the instrument the count rate of the upper gamma channels was a sum of photons and neutrons events. However, count rate of gamma-channels was always ∼ 100 times higher than count rate of the neutron channels (see Fig. 1).

Net count rates in all detector channels were determined in this flare and hereafter after subtraction of the smoothed background count rate measured in the previous orbits under similar orbital and geomagnetic conditions [19]. The procedure of the background subtraction during the flare of October 28 is illustrated in Fig. 2 taken from [19].

Count rates of the SONG channels during the flare of October 28, 2003 measured with 4-second time resolution are presented in Fig. 3. Comparison of the time profiles of different channels shows that during the flare development several acceleration episodes were realized. We divided the temporal behavior of this flare

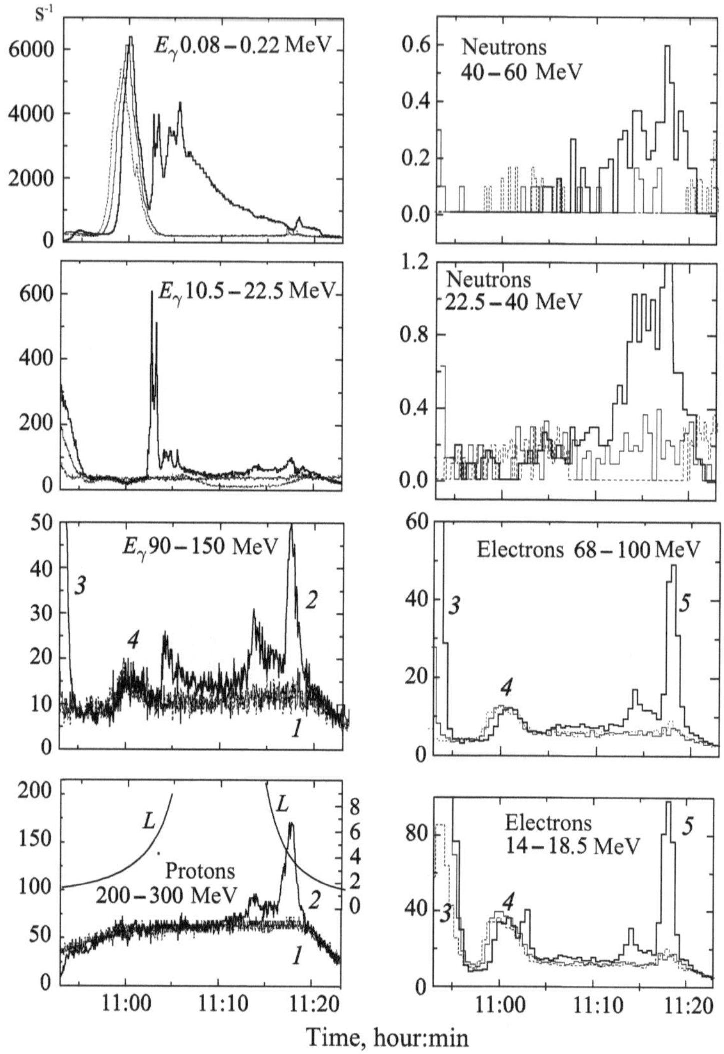


**Fig. 2** Comparison of the measurements done by the SONG instrument during the solar flare of October 28, 2003 with the measurements in the previous orbits. *Thin dashed line*, *dotted line* and number *1* indicate background measurements at the previous orbits, time-shifted to the moment of the flare. *Thick lines* and number *2* relate to the measurements during the flare. South-Atlantic Anomaly is indicated with number *3*, external radiation belt—-with *4*. *5* relates to the increasing in electron channels resulted from the spurious count of high-energy protons (taken from [19]). From [19]

in two of the most pronounced intervals. First episode was very impulsive one (it lasted from 11:02:30 to 11:03:51). The upper channel with a visible increase was 60–90 MeV. The second acceleration episode started at 11:03:51 and it was

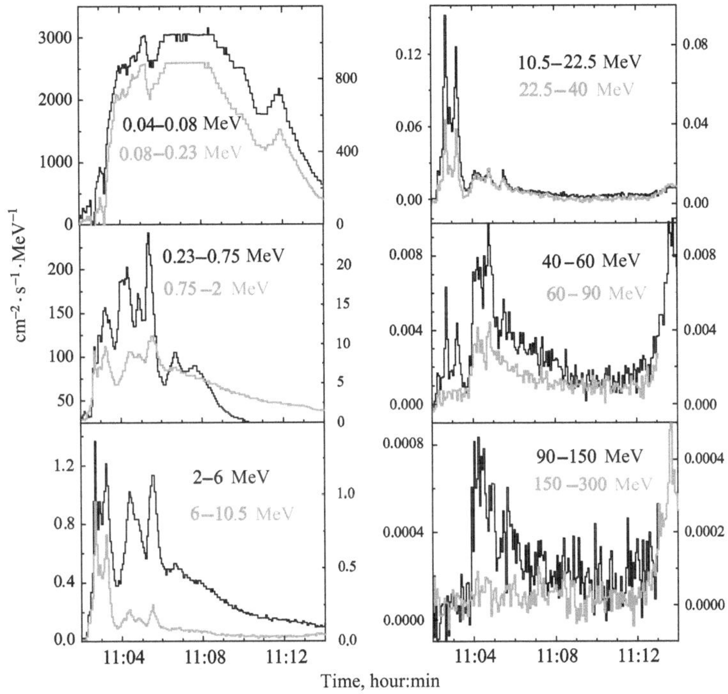


**Fig. 3** Time profiles of the gamma-emission count rates of recorded by SONG during the flare of October 28, 2003 (taken from [19]). From [19]

observed till the moment when high-energy protons arrived at CORONAS-F. From 11:12:30 protons started to contaminate gamma and neutron channels. Beginning from 11:03:51 photons with energy of >60 MeV appeared. They were also recorded by the highest channel of the SONG instrument, 150–300 MeV. This high-energy emission was lasting at least during 530 s and varied slightly.

Figure 4 presents data of SONG obtained by the amplitude analyzer in two time intervals of 30 s duration. The first one was chosen before the flare, and the second one during the flare maximum. The line seen between 80 and 115 channels of the second amplitude spectrum corresponds to the energy of 2.223 MeV line created through the neutron capture reaction. Time dependence of this line obtained by the amplitude analyzer together with time profiles of wide channels SONG in the energy interval 0.75–6 MeV is presented in Fig. 5. This figure presents also the time behavior of this line measured by INTEGRAL/SPI [37]. The 2.223 MeV line was also detected by the RHESSI after the satellite out of the shadow at 11:06 and it was observed till 11:30 [38].

We remind that the neutron capture line is delayed at 80 s comparing with the appearance of the prompt narrow gamma lines. Figure 5 shows that intensity of this line began to increase a bit later than the beginning of the first acceleration episode. The intensity of this line suddenly increased at 11:04:30, i.e., a little later the onset

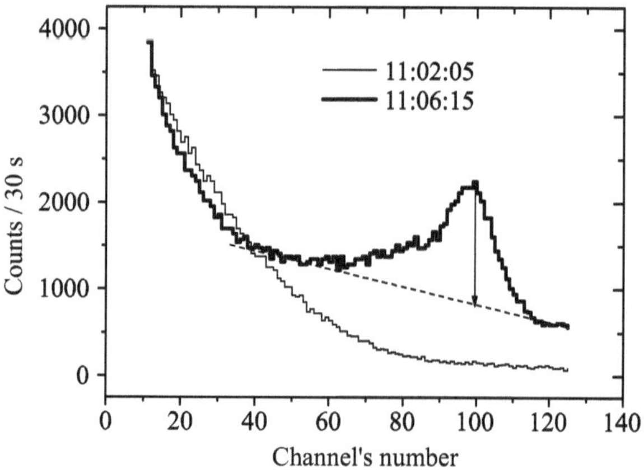


**Fig. 4** Data of the SONG amplitude analyzer during the flare of October 28, 2003. *Dashed line* shows the evaluation of the background continuum spectrum. *Vertical arrow* at the peak of full absorption relates to the line of 2.223 MeV

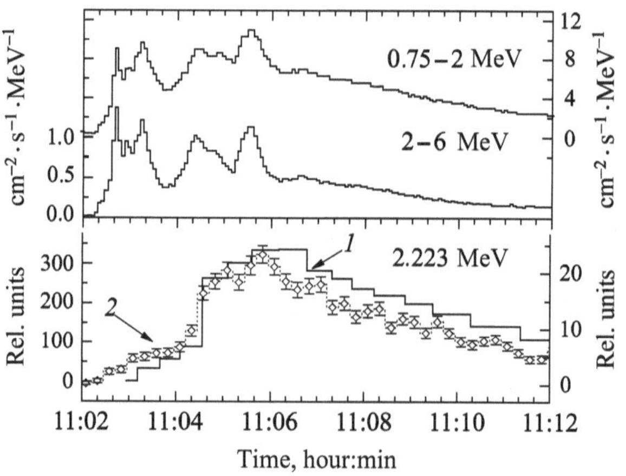


**Fig. 5** Time profiles of the gamma-emission observed during the flare of October 28, 2003. *Upper panel*: the SONG measurements in broad channels. *Lower panel*: time profiles of gamma-line 2.223 MeV observed by the SONG amplitude analyzer (curve *1*) and by SPI/INTEGRAL [37] (points *2*)

time of the second episode. Presence of nuclear lines during the first episode proves the protons acceleration up to 30 MeV. The sudden increase of both 2.223 MeV line and prompt narrow gamma lines [37] show that the number of interacting protons of these energies increased in the second episode at 11:04.

To restore the incident gamma-emission spectra we fitted the deposited count-rate spectra for selected time intervals with simulated deposited energy spectra prepared using the GEANT3.21 code. This procedure is described in [19, 39]. The restored spectra accumulated in three time intervals are presented in Fig. 6. It should be noted that two first SONG channels (0.04–0.08 and 0.08–0.23 MeV) were overloaded. Therefore we did not use count rate of these channels in the incident spectra restoration. We presented also the HXR fluxes in the energy range of 15–100 keV measured by the SPR-N patrol detector onboard the CORONAS-F [40] because the SPR-N detector has never been saturated.

A significant photon flux of the energy up to 90 MeV was observed in the first interval, but at the higher energy fluxes did not exceed the value of

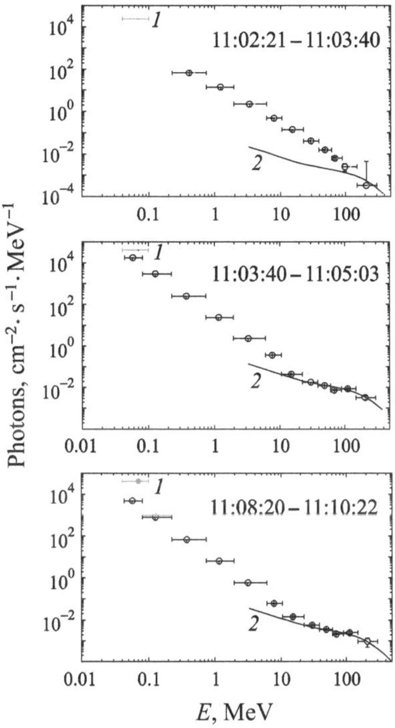


**Fig. 6** Gamma-emission spectra during the flare of October 28, 2003, according to the SONG measurements. Points *1* indicate HXR flux, measured by the SPR-N instrument [40]. The line *2* marks gamma-emission spectrum related to the pion-decay

$10^{-3}$ cm$^{-2}$ s$^{-1}$ MeV$^{-1}$. The pion-decay component could not be separated over the bremsstrahlung background in this time interval. It means that the $\pi^0$-decay process was weak or absent at all. Therefore the number of protons with energies $> 300$ MeV was too small to produce an observed emission. The bremsstrahlung produced by electrons with energies of 60–90 MeV dominated in high-energy emission during this time.

The intensity of gamma-emission at energies above 60 MeV increased drastically at 11:03:51 (see Fig. 3). The restored spectrum demonstrates an appearance of the feature produced by neutral-pion decay (see Fig. 4, middle panel). Observation of this feature allowed us to determine the onset time of high-energy proton acceleration. This feature is also seen in the third spectrum. Further intensity of pion-decay emission decreased gradually. The detailed study of SONG measurement of HXR and gamma-ray emission and neutrons fluxes during this flare can be found in [19].

*Solar flare of November 4, 2003* was partly a behind-the-limb one. SXR detectors onboard the GOES satellites were saturated within the 19:43–19:58 interval. However, Kiplinger and Garcia [41] have attributed importance $> X28$ to this flare with maximum at 19:47. In addition to a very powerful SXR burst, radio emission up to 405 GHz frequency was observed [42].

The flare began when the CORONAS-F crossed the outer radiation belt. A procedure of the background subtraction from the SONG channel count rates is shown in Fig. 7. It is seen that CORONAS-F came out the radiation belt region at 19:42:14

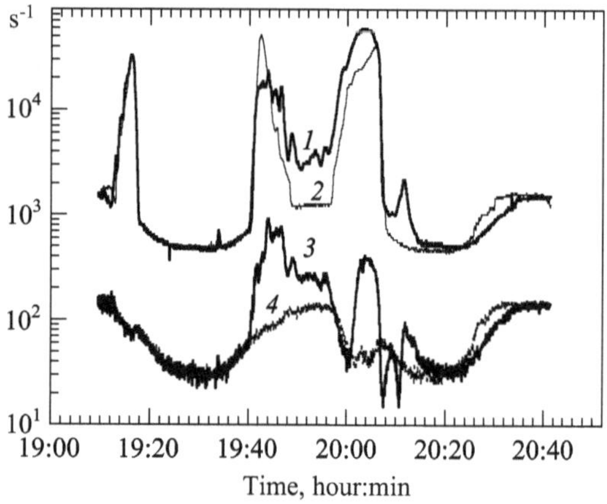


**Fig. 7** The SONG instrument's data during the solar flare of November 4, 2003 compared to the previous orbit. Numbers *1* and *2* relate to the channel of 0.23–0.75 MeV, numbers *3* and *4*—to the channel of 2–6 MeV. *Thick lines* and numbers *1, 3* are related to the measurements during the flare, *thin lines* and numbers 2, 4—to the background data, shifted in time to the moment of the flare

and observed high-energy gamma rays from this flare till the arrival of the intense flux of solar neutrons at 19:50. Beginning from this moment we could not reliably separate the detector signals with energies > 20 MeV caused by photons from the signals caused by neutrons. We compared our data with the measurements of the 25–150 keV HXR emission with the Ulysses spacecraft located at ~114° westward from the Earth–Sun line at a distance of 5.28 AU from the Sun and measured X-ray emission from 19:33 [43]. The maximum of the Ulysses count rate was observed at approximately 19:44 taking into account the distance to Ulysses. This comparison had verified that our measurements really cover a time interval of the flare main energy release [30, 31].

Time profiles of the emissions measured by the SONG instrument during this flare are presented in Fig. 8, and the flare spectrum accumulated over the time interval 19:43:00–19:45:50 is presented in Fig. 9. The feature indicated the presence of the photon fluxes from $\pi^0$—decay is clearly seen.

It should be noted that no GLE associated with this flare was observed. Solar proton event (SPE) was not also something outstanding: proton fluxes measured by GOES were ~350 $cm^{-2} \cdot s^{-1} \cdot sr^{-1}$ at the energy > 10 MeV and 1.3 $cm^{-2} \cdot s^{-1} \cdot sr^{-1}$ at the energies > 100 MeV, although this flare was one of the most powerful events over 30 years of observations. Probably the flare location near the western limb as well as coronal mass ejections (CME) and shock wave propagation through the interplanetary space prevented the bulk of the particles to get near Earth.

**Flare of January 20, 2005.** Throughout the history of flare studies this flare was apparently the event with the best set of various experimental data (see, e.g., [34, 39, 44]). Radio data and RHESSI measurements indicated that intense acceleration of electrons up to energies 100 keV and protons up to energies 10–30 MeV began at approximately 06:42 [34]. The increase of > 40 keV HXR and gamma-rays in the SONG data was distinguished over the background count rates after 06:43:30 when the CORONAS-F satellite moved in the equatorial region (see Fig. 10). SONG

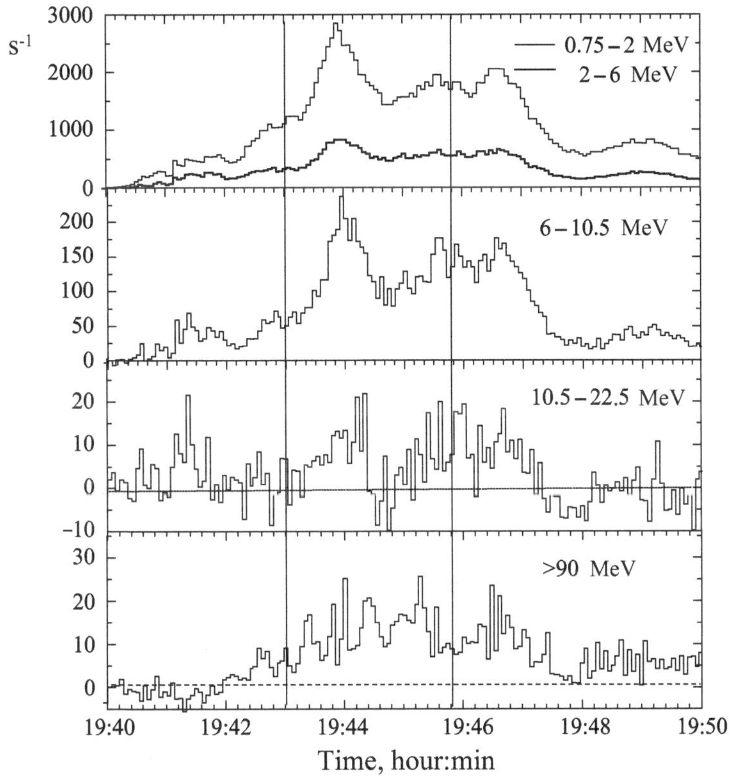


**Fig. 8** Time profiles of gamma-emission counting rate measured by SONG during the flare of November 4, 2003. The interval for which gamma-emission spectrum was calculated is separated by *vertical lines* (taken from [30]). From [30]

began to measure high-energy protons at 06:54. As a result separation of gamma-events from the total detector count rate became unclear from this moment.

Figure 11 presents net time profiles of gamma emission recorded by SONG during this flare. Two episodes of energy release are seen in the time profiles similar to those observed during the flare of October 28, 2003. Small but statistically significant photon fluxes with energy $> 60$ MeV were measured during the first episode. Considerable increase of high-energy emission intensity started at 06:45:30 and reached its maximum at 06:47:00.

Figure 12 presents time profile of the neutron-capture line obtained with the SONG amplitude analyzer along with the time profile of this line measured by RHESSI [45]. The increase of this line intensity began a bit later than the beginning of the first acceleration episode and that neutron generation. It increased suddenly during the second episode a bit later than 06:45:30 when high-energy emission intensity began also to rise.

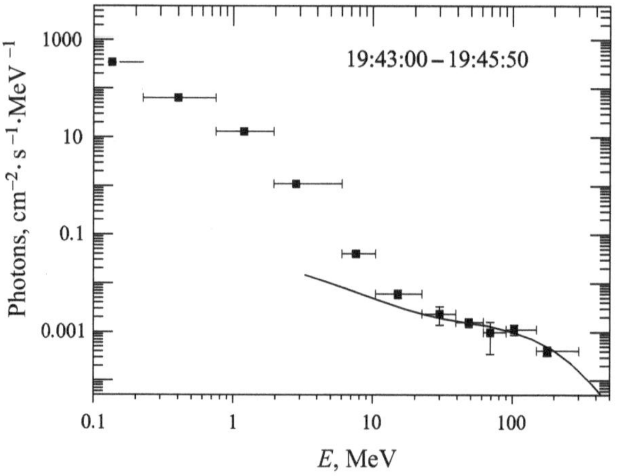


**Fig. 9** Gamma-emission spectrum during the flare of November 4, 2003 according to the SONG measurements. *Solid line* presents the spectrum of gamma-emission related to pion-decay

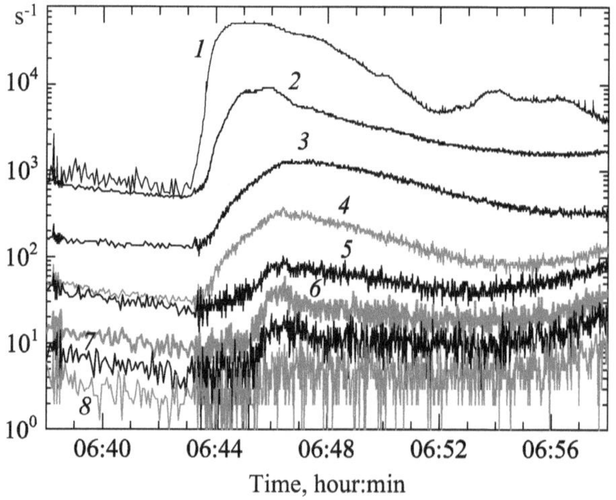


**Fig. 10** Time dependence of the SONG instrument counting rate on January 20, 2005 during the satellite's crossing of the equatorial region. The curve *1* presents the channel of 43–80 keV, the curve *2*—230–780 keV, *3*—0.78–2 MeV, *4*—2–6.2 MeV, *5*—6.2–11 MeV, *6*—23–40 MeV, *7*—62–93 MeV, *8*—93–155 MeV

Gamma-emission spectra (see Fig. 13) show that the gamma-emission fluxes were rather intensive and the broad plateau in the spectrum created by $\pi^0$-decay can be considered as reliable even in the first time interval (flux within the interval of 90–150 MeV was $\approx 3 \cdot 10^{-4}$ photons $\cdot$ cm$^{-2}\cdot$ s$^{-1}\cdot$ MeV$^{-1}$). Beginning from 06:45:30 gamma-emission intensity with energy > 100 MeV significantly increase. This plateau remained well observable till 06:54, when high-energy protons started to contaminate the count rates of SONG channels.

Observation of the broad line caused by $\pi^0$-decay provides information about the appearance time of high-energy protons in the solar atmosphere [46]. This time in turn allowed us to compare time interval of proton acceleration with the time of GLE onset and to calculate escaping time of the high energy particles from the Sun. The distance between Earth and the Sun was 0.984 AU on January 20, 2005,

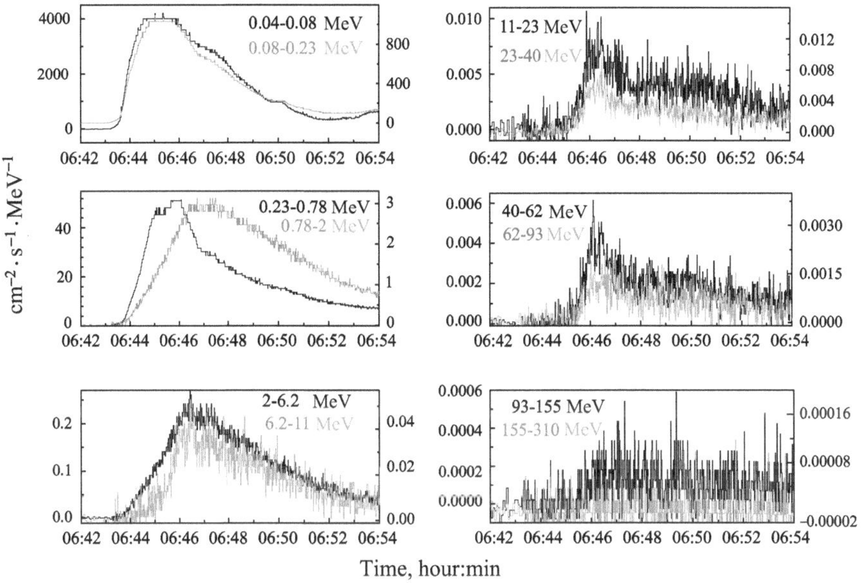


**Fig. 11** Time profiles of gamma-emission counting rate, measured by SONG during the flare of January 20, 2005

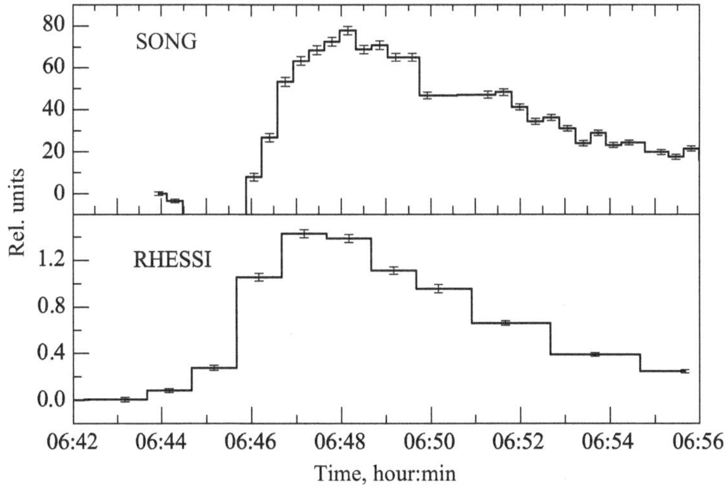


**Fig. 12** Time profile of gamma-emission flux within the line of 2.223 MeV measured with SONG and RHESSI [45] during the flare of January 20, 2005. From [45]

so photon time of flight was $t = 490$ s. Therefore, the solar time of the beginning and the time of the maximum of gamma-emission generation can be determined as 06:37:27 and 06:38:50, respectively. These moments were the onset and the

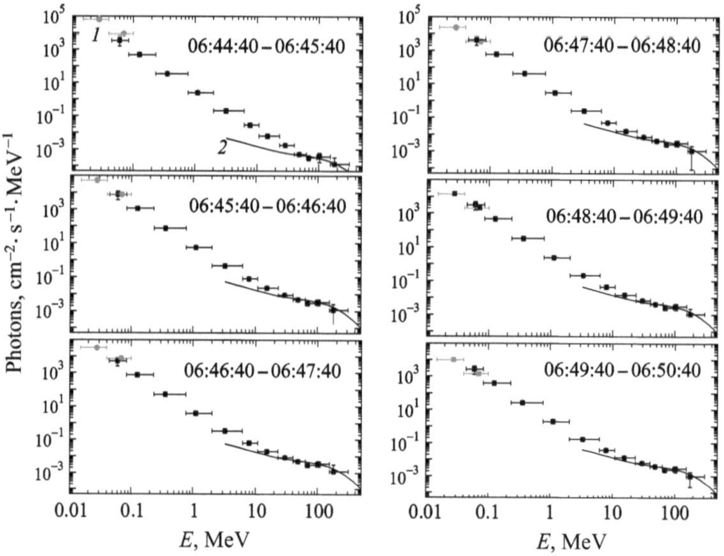


**Fig. 13** Gamma-emission spectra during the flare of January 20, 2005 according to the SONG measurements. *Dots 1* indicate HXR flux, measured by the SPR-N instrument [40], *line 2*—gamma-emission spectrum related to the pion-decay. From [40]

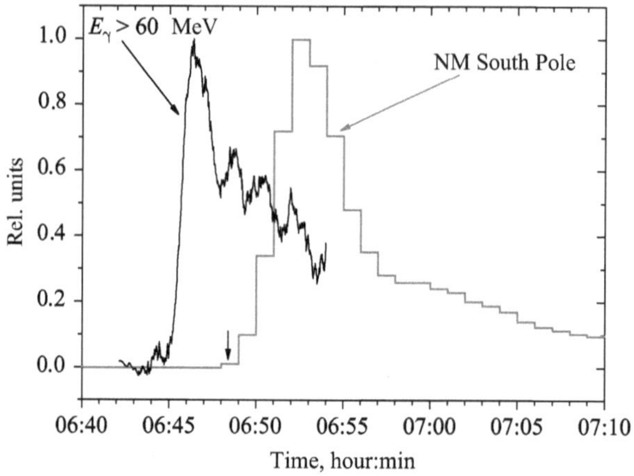


**Fig. 14** Count rate of the South Pole neutron monitor and high-energy gamma-emission counting rate measured with SONG during the flare of January 20, 2005. *Vertical arrow* indicates the moment of the NM's counting rate increase

efficiency maximum of proton acceleration. These moments can be compared with the GLE onset time that can be found from the neutron monitor network data. Figure 14 presents the count rate of the high-latitude South Pole NM. The 4 % increase of count rate was recorded in the time interval 06:48–06:49. The time delay of GLE onset relative to the detection of $\pi^0$–decay emission between 06:45:30 and 06:47:00 is $\sim 120 - 150$ s.

The threshold energy of particles detected by high-latitude NM is $\sim$450 MeV due to an atmospheric absorption, but the effective particle energy exceeds the 1 GeV (see, e.g., [47]) The latter energy corresponds to the particle velocity no less than the $v = 0.875\,c$ ($c$ is the speed of light). The IMF field lines are in the form of Archimedean spirals. Their parameters (length and heliolongitude of field line connecting the Sun and Earth) are usually determined by values of solar wind speed averaged over few days prior to the flare. The mean free path of individual particles may be less than the length of the nominal values of IMF line [48], but not less than the distance from the Sun to Earth in this epoch. If the particles propagate with low scattering or without, it then their trajectory lengths are close to the IMF line length. The propagation time of protons with energy of 1 GeV assuming the path length between 1 and 1.3 AU is equal to 581–755 s, respectively.

The GLE onset was determined as 06:48–06:49. This means that protons escaped the Sun not later than 06:38–06:39 which is in a good agreement with time of their appearance in the corona determined from the gamma-emission observation as 06:37:27–06:38:50. The obtained result reliably proves that the first high-energy protons detected at Earth left the Sun just after their acceleration in the solar corona [18]. Thus, this result makes impossible an assumption that the first protons were accelerated later, for example, at the shock front in the upper solar corona.

## Solar Neutrons Recorded by the SONG Instrument

Neutrons are expected to appear in any flare, in which the pion-decay emission is observed, and vice versa, due to their common origin. In fact high-energy neutrons were observed after three considered flares except the January 20, 2005 flare, although the observation conditions were favorable for neutrons detection by SONG during the latter flare. No neutrons were observed on during this flare by ground-level detectors as well. The absence of neutrons is difficult to understand in view of the fact that the pion-decay emission was quite intense.

During the flare of August 25, 2001, the increase of the counting rate in the SONG neutron channels was observed from 16:35 until 16:39 (see Fig. 15). Note that the minimal value of neutron energy measured by the SONG detector is 20 MeV. This threshold is caused by the energy dependence of cross section of neutrons interaction with Cs and J nuclei with production of secondary charged particles. Efficiency of SONG neutron detection is maximal within the energy range of 50–200 MeV [49]. The upper limit of the neutron detection is $\approx$ 800 MeV, because there is a high probability of the production of secondary charged particles with energies high enough to escape from the crystal into the anticoincidence shield. As a result, the effective area of the SONG instrument decreases by a factor of 3 compared to its maximum value for neutron energy 600 MeV and virtually approaches zero for neutron energy of 800 MeV.

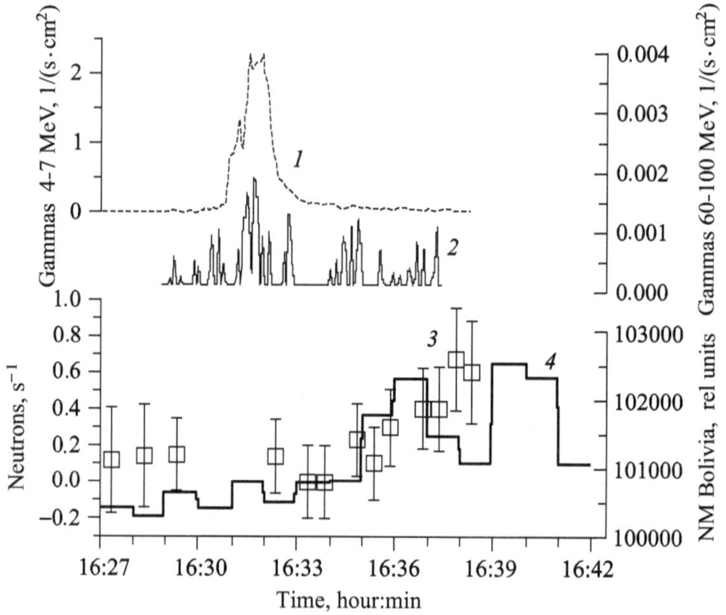


**Fig. 15** Time profiles of the total increase in the neutron channels of the SONG instrument (15–26, 26–40, 40–60, and 60–100 MeV) (*3*), the gamma-emission fluxes in the channels of 4–7 MeV(1) and 60–100 MeV (2), and the increase of the counting rate on the Chacaltaya neutron monitor (*4*) during the solar flare on August 25, 2001 (taken from [29]). From [40]

Since neutrons must be generated at the same time when $\pi^0$-decay maximum was observed (at 16:30:20 $\pm$ 20 s), and taking into account that light propagation time from the Sun to Earth is about 500 s we can estimate the energy of the neutrons as 200–400 MeV. The end of our measurements was at 16:39 due to the lack of telemetry data. Therefore the lower limit of 200 MeV is rather formal. Comparing the time profiles of the increase measured by SONG and the Chacaltaya NM, we see that both of them occurred almost simultaneously. The Chacaltaya ground-based NM is located at an altitude of $H = 5200$ m above the sea level, its coordinates are 16.2° S, and 292° E, and the geomagnetic cutoff rigidity is $R = 13.1$ GV. Sun zenith angle was 26.5° during the flare of August 25, 2001. Thus all the necessary conditions required for observation of solar neutrons were performed. The statistical excess of the NM count rate during the period of 16:34–16:39 was estimated to be $4.7\sigma$ [8]. No GLE and SPE were observed since the flare occurred at a longitude of 34° E. Therefore it is most natural to explain the increase of the NM count rate just by the arrival of the flare generated neutrons. Direct measurements of SONG can say with confidence that the high-altitude equatorial NM in Bolivia recorded actually solar neutrons.

The second time solar neutrons were recorded by SONG after the flare of October 28, 2003. The counts rates of SONG neutron channels during this flare are presented in Fig. 16. We estimate the neutron energy as $\sim$500 MeV considering that neutrons

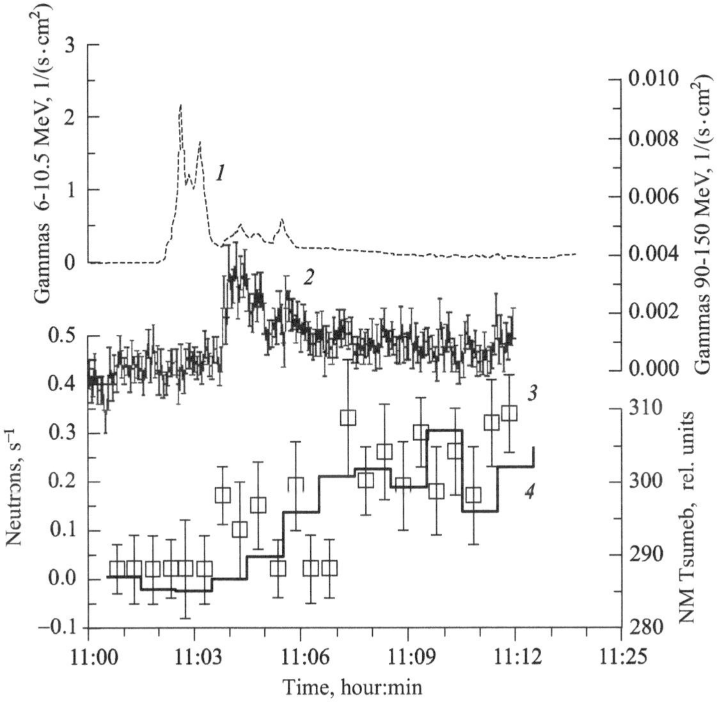


**Fig. 16** Time profiles of the total increase in the neutron channels of the SONG instrument (*3*), the gamma-emission fluxes in the channels of 6–10.5 MeV (*1*) and 90–150 MeV (*2*), and the increase of the counting rate on the Tsumeb neutron monitor (*4*) during the solar flare on October 28, 2003 (taken from [29]). From [40]

were created simultaneously with the neutral pions. The Tsumeb NM (19.2° S, 7.6° E, $H = 1240$ m, $R = 9.29$ GV) recorded an enhancement due to solar neutrons estimated as $3.4 \pm 0.3\,\%$ [47, 50]. Comparison of the Tsumeb NM data with the SONG measurements showed that the increase of this NM count rate began 2 or 3 min earlier than the increase observed by the SONG instrument. The energy of the accelerated protons in this event exceeded several GeV [51]. Consequently, the neutrons which came to Earth could also have fairly high energies. Of course, higher energy neutrons will reach Earth earlier than lower-energy ones. This is a reason why the beginning of the count rate increase measured by SONG was slightly delayed compared to the Tsumeb NM data. The bulk of charged particles accelerated during this flare came to the Earth later, after 11:12 that was shown by both SONG data (200–300 MeV proton channel) and ground-based neutron monitors [50, 51].

The third neutron event recorded by the SONG during the solar flare of November 4, 2003, is shown in Fig. 17 along with the data of the Mexico NM located at 19.33° N, 99.18° W, $H = 2274$ m, $R = 9.53$ GV. The increases of the count rate of the Mexico and Haleakala NMs on November 4, 2003, were interpreted as arrival of solar neutrons [10]. The neutron energy was estimated to

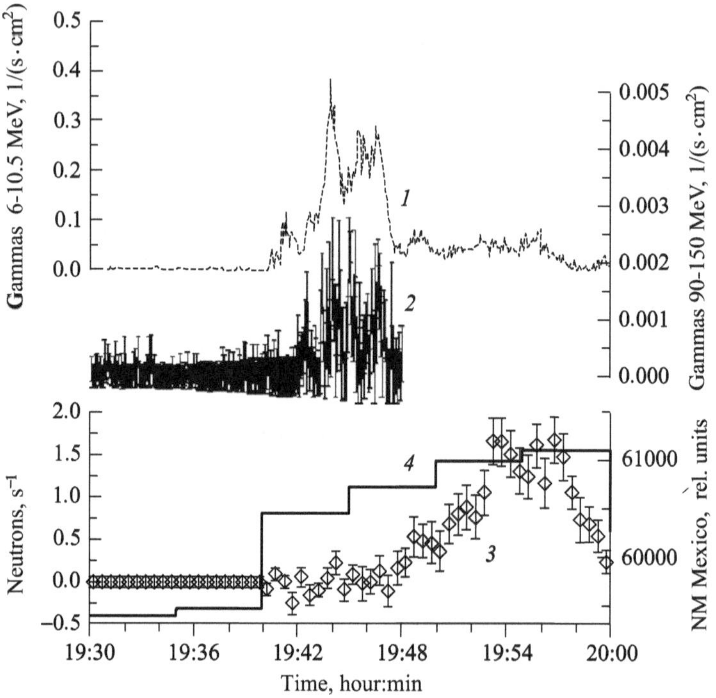


**Fig. 17** Time profiles of the total increase in the neutron channels of the SONG instrument (*3*), the gamma-emission fluxes (*1*, *2*), and the increase of the counting rate of the Mexico neutron monitor (*4*) during the solar flare on November 4, 2003. The legend is the same as in Fig. 16 (taken from [29]). From [40]

be 90–200 MeV by the method described above. It should be noted that during this event the neutron flux was high enough, so increases in the counts rate of SONG channels with different values of energy deposition were recorded in different moments. The increase in the lower energy channels count rate in which the contribution of less energetic particles was larger was recorded somewhat later than in the higher-energy channels. Figure 17 shows that during this flare the earlier increase of the count rate of the neutron monitor compared with SONG was seen even more clearly than in the October 28, 2003 event.

## Conclusions

In this chapter we presented data on gamma-ray emission and the neutron fluxes produced by four powerful solar flares, namely: August 25, 2001, October 28, 2003, November 4, 2003, and January 20, 2005 recorded by the SONG instrument. We have found:

1. The flares on October 28, 2003, November 4, 2003, and January 20, 2005 developed according to the same scenario: these events were similar in their durations and exhibited two essentially different acceleration episodes.
2. During the first episode of 1–3 min duration electrons were accelerated to the ultra-relativistic energies ~10–60 MeV. Protons were perhaps accelerated up to 10–30 MeV. The latter conclusion is proved by observations of neutron-capture gamma line (SONG, RHESSI, and INTEGRAL) and prompt gamma-lines (INTEGRAL and RHESSI). Protons with energy above the threshold of the pion production (~ 300 MeV) could be accelerated in the first episode. However, their number was small. Indeed, the spectral feature caused by the $\pi^0$-decay was not separated above bremsstrahlung in the October 28, 2003 flare. The upper limit of pion-produced fluxes was found as $\leq 1 \cdot 10^3$ phot $\cdot$ cm$^{-2}\cdot$ s$^{-1}\cdot$ MeV$^{-1}$ at $E = 100$ MeV in the other three flares. The combination of these results indicates that the spectrum shape of the accelerated protons was an exponential or power law with spectral index $S > 4 - 4.5$ and with rollover at high energies, i.e., it was quite soft.
3. The efficiency of protons acceleration and possibly the hardness of proton spectrum increased abruptly. This conclusion follows from the sharp increase of intensity of both the narrow gamma-lines and the pion-decay emission. Rise time of the photon flux at energies > 60 MeV does not exceed 10 s.
4. Duration of high-energy gamma emission caused by the presence of protons with energy above 300 MeV in the flare volume exceeds 10 min.
5. The revealing time of the $\pi^0$–decay emission excess above the primary electron bremsstrahlung determinates the onset time of high-energy proton appearance in the solar atmosphere. Knowledge of this moment allowed us to compare the time of protons acceleration at the Sun with time of the GLE onset and to calculate the escaping time of the first particles which come to Earth. It was shown that the protons responsible for the generation of neutral pions in the solar atmosphere and the protons recorded by the NMs at the GLE beginning belong to one and the same population of the accelerated particles [18, 31, 34, 46].
6. The SONG instrument recorded solar neutron fluxes immediately after the flares on August 25, 2001, October 28, 2003, and November 4, 2003. Observation of high-energy neutrons is an additional evidence of efficient particle acceleration up to energies of several hundred MeV on the Sun during impulsive phase of these flares.

Observations of high-energy gamma-emission with energy over 20 MeV and high-energy neutrons performed by CORONAS/SONG during 2001–2005 were unique for the 23rd solar activity cycle. These data complement significantly the results of earlier measurements of high-energy gamma emission and neutron fluxes provided by the SMM/GRS, CGRO/EGRET, GRANAT/PHEBUS, and GAMMA/GAMMA1. The characteristics of high-energy neutral radiation of solar flares recorded by SONG are generally similar to the characteristics of the radiation observed earlier during similar large and compact flares. It allows us to suppose that temporal development of these events is common. It became clear that the models which are in reasonable agreement with measurements at lower energies do not explain the essential properties of high-energy particle acceleration processes which are working in solar flares.

This work was supported in part by the Russian Federation for Basic Research (projects No 05-02-17487 and 09-02-01145) and a "University of Russia" Grant (UR.02.02.509/05-1). K.Kudela wishes to acknowledge grant agency VEGA, project 2/0040/13 for support.

## Recent Results Obtained After Publication of the Russian Edition

After the completion of the Russian version of this book, we continued our research and obtained some new results.

1. We have advanced the routine of the incident spectrum restoration from the SONG count rate spectrum. This routine described in [19, 31, 39] permits to separate best-fitted fluxes of bremsstrahlung continuum and the pion-decay gamma emission. The restored spectra of the flares under consideration are presented in [19, 30, 31]. We have also obtained the combined spectrum of the August 25, 2001 flare in the energy range of 0.01–500 MeV (see Fig. 18). We used data on thermal SXR (GOES), electron bremsstrahlung and gamma-lines (SPR-N, SONG and Yohkoh) as well as the pion-decay gamma emission (SONG) [31]. Time profiles and restored spectra of gamma emission from the flare of August 25, 2001, observed by SONG are presented in [30, 31].
2. We compared the magnetic-flux change rate during the October 28, 2003 flare [52] with time profiles of gamma-ray emissions with different energies and origin (see Fig. 19). The count rate in the SONG channel of 6–10.5 MeV was caused mainly by high-energy electron bremsstrahlung. The narrow gamma-line emission observed by SPI/INTEGRAL [53] was produced by the interactions of protons with energies 10–30 MeV. The SONG count rate above 90 MeV was attributed to the pion-decay gamma emission which in turn is generated by protons with energies $> 300$ MeV. The magnetic-flux change rate is usually considered as a manifestation of flare magnetic reconnection. A strong temporal correlation between the magnetic-flux change rate and the observed pion-decay emission is seen. So the most efficient acceleration of high-energy protons ($\geq 300$ MeV) coincides in time best of all with the highest change rate of the magnetic flux during the impulsive phase of this flare [19].
3. Comparison of the gamma-ray fluxes in certain energy ranges (the neutron-capture gamma-line 2.2 MeV, $F_{2.2}$, the nuclear line emission in the range 4–7 MeV, $F_{4-7}$, and the pion-decay gamma-ray emission at 100 MeV, $F_{\pi^0}$) makes it possible to estimate the energy spectrum shape of protons responsible for the production of these emissions (e.g., [12–14, 24]). We have calculated the fluences of the pion-decay component using the best-fit spectral parameters of the

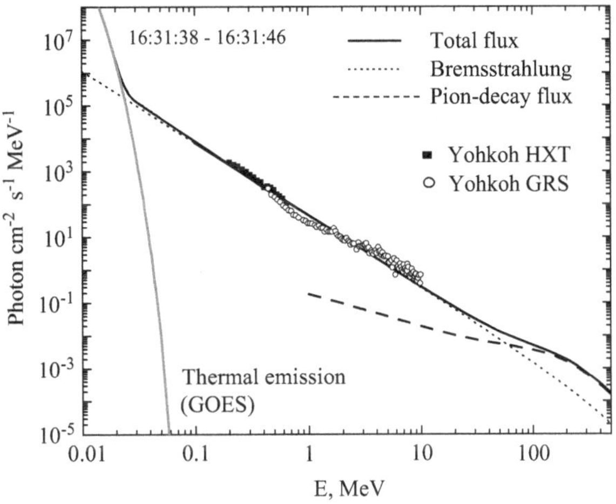


**Fig. 18** Incident spectrum of electromagnetic emissions from the August 25, 2001 solar flare (taken from [31]). From [31]

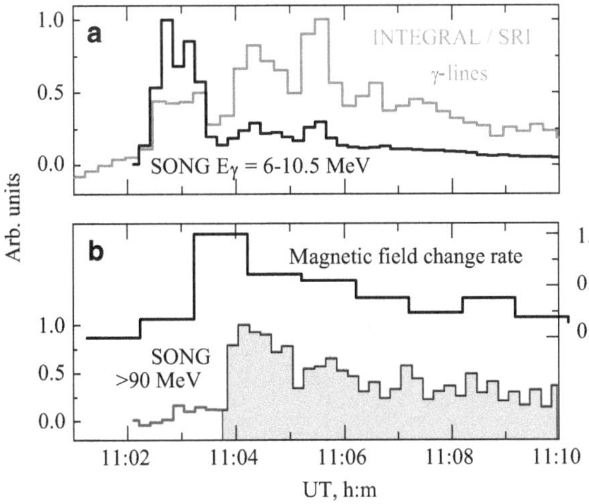


**Fig. 19** Gamma-ray emission and the magnetic-flux change rate (panel (**b**), *black curve*, right Y-axis) during the October 28, 2003 flare. *Gray color* in panel (**b**) marks the time interval of undoubted observation of the pion-decay gamma-emission

gamma-ray emission measured by SONG near the maximum of the October 28, 2003 flare [19]. The value of $F_{4-7}$ was taken to be equal to the sum of the 4.4 and 6.1 MeV lines measured by SPI/INTEGRAL [53]. The spectrum index in the proton energy range 30–300 MeV found from the $F_{\pi^0}/F_{4-7}$ ratio was $S \sim 3 \pm 0.4$. The power-law spectrum in the proton energy range 10–30 MeV was found to be $3 < S < 4$ from the $F_{2.2}/F_{4-7}$ ratio in the same time interval [53]. In the same way, it was found $S \approx 3.4$ for the extended phase after 11:06 from RHESSI data [38]. Thus, all these estimations are consistent with a power law spectrum with an index $S = 3 - 3.5$ in the total 10–300 MeV range. A spectral index of the solar particles during this event was found to be $\sim 3.5$ in the energy range 0.5–10 GeV from NMs data [51]. A similar value for the proton spectrum was obtained from CORONAS-F data using the effect of

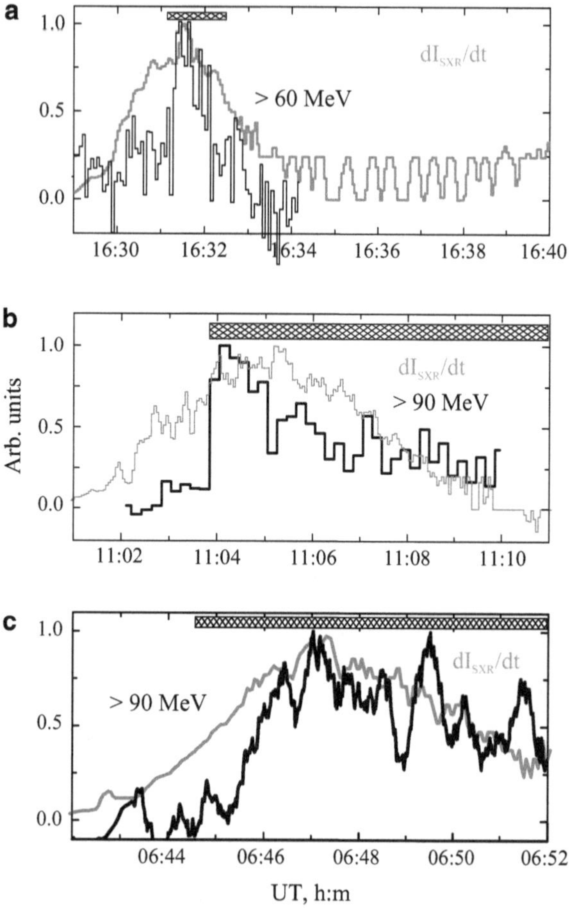


**Fig. 20** Comparison of the derivative of SXR, $dI_{SXR}/dt$, calculated from data of GOES monitors (*gray curves*) and high-energy gamma emission observed with SONG (*black curves*). Panel (**a**) presents data on the August 25, 2001 solar flare; (**b**)—October 28, 2003; and (**c**)—January 20, 2005. *Shaded rectangles* denote the intervals of undoubted observation of the pion-decay gamma-emission

geomagnetic cutoff [54]. We believe that close agreement between these values and the spectral index derived from the gamma-ray measurements reflects a common origin of the accelerated and escaping protons. Obtained values of the proton spectrum index combined with gamma-ray fluences permitted to estimate the total number of interacting particles with energies $> 30$ MeV during this flare [19].

4. We compared time profiles of various flare electromagnetic emissions as well as the time profile of the derivative of SXR, $dI_{SXR}/dt$, calculated from data of GOES monitors. We found that the time interval of maximum intensity of the pion-decay gamma emission coincides best with the time interval of maximum positive value of $dI_{SXR}/dt$ during the flares on August 25, 2001, October 28, 2003, and January 20, 2005 (see Fig. 20). This means that the proton acceleration to sub-relativistic energies occurs with maximum efficiency close to the time of the maximum flare energy release [31].

5. We compared the observational time of appearance of the pion-decay gamma emission with the onset time of high-energy protons detected at 1 AU during several events. We found that first particles were observed with a short (< 5 min) delay [55]. To determine the onset time of these particle we applied the method of superposed epoch to data of several NMs [31]. This delay is consistent with the minimal time of particle Sun–Earth propagation. This means that accelerated particles interacting at the Sun and producing the pion-decay gamma emission and first particles arriving to Earth belong to one and the same population of the accelerated particles [31, 55].

## References


1. Vilmer, N., MacKinnon, A.L., Hurford, G.J.: Properties of energetic ions in the solar atmosphere from gamma-ray and neutron observations. Space Sci. Rev. **159**, 167 (2011)
2. Miroshnichenko, L.I., Gan, W.Q.: Particle acceleration and gamma rays in solar flares: Recent observations and new modeling. Adv. Space Res. **50**, 736 (2012)
3. Share, G.H., Murphy, R.J.: Accelerated and ambient He abundances from gamma-ray line measurements of flares. Astrophys. J. **508**, 876 (1998)
4. Ramaty, R., Kozlovsky, B., Lingenfelter, R.E.: Solar gamma rays. Space Sci. Rev. **18**, 341 (1975)
5. Chupp, E.L., Forrest, D.J., Ryan, J.M. et al.: A direct observation of solar neutrons following the 01:18 UT flare on 1980 June 21. Astrophys. J. **263**, L95 (1982).
6. Chupp, E.L., Debrunner, H., Flückiger, E. et al.: Solar neutron emissivity during the large flare on 1982 June 3. Astrophys. J. **318**, 913 (1987)
7. Debrunner, H. et al.: The solar cosmic ray neutron event on June 3, 1982. Proceedings of the 18th ICRC, vol. 4, p. 75, Bangalore (1983)
8. Watanabe, K. et al.: Solar neutron event in association with a large solar flare on August 25, 2001. Proceedings of the 28th ICRC, p. 3179, Tsukuba (2003)
9. Watanabe, K., Muraki, Y., Matsubara, Y. et al.: Solar neutron events associated with large solar flares in solar cycle. Fall Meeting 2005. AGU. SH13A-0292.
10. Watanabe, K., Gros, M., Stoker, P.H. et al.: Solar neutron events of 2003 October–November. Astrophys. J. **636**, 1135 (2006)
11. Sako, T., Watanabe, K., Muraki, Y. et al.: Long-lived solar neutron emission in comparison with electron-produced radiation in the 2005 September 7 solar flare. Astrophys. J. **661**, L69 (2006)
12. Ramaty, R., Murphy, R.J.: Nuclear processes and accelerated particles in solar flares. Space Sci. Rev. **45**, 213 (1987)
13. Murphy, R.J., Dermer, C.D., Ramaty, R.: High-energy processes in solar flares. Astrophys. J. Suppl. **63**, 721 (1987)
14. Mandzhavidze, N., Ramaty, R.: Gamma rays from pion decay: evidence for long-term trapping of particles in solar flares. Astrophys. J. **396**, L111 (1992)
15. Lingenfelter, R.: Solar flare neutrons and gamma rays. High Energy Solar Phenomena / Eds. Ryan J.M., Vestrand W. T. AIP, NY AIP. 1994, P. 77
16. Akimov, V.V., Leikov, N.G., Kurt, V.G., Chertok, I.M.: The GAMMA-1 data on the March 26, 1991 solar flare // ibid, P. 130
17. Rieger, E.: Solar flares: high-energy radiation and particles. Sol. Phys. **121**, 323 (1989)
18. Kuznetsov, S.N., Kurt, V.G., Yushkov, B.Yu., Kudela, K.: CORONAS-F satellite data on the delay between the proton acceleration on the Sun and their detection at 1 AU. Proceedings of 30th ICRC, vol. 1, p. 121, Merida (2008)

19. Kuznetsov, S.N., Kurt, V.G., Yushkov, B.Yu., Kudela, K., Galkin, V.I.: Gamma-Ray and High-Energy-Neutron Measurements on CORONAS-F during the Solar Flare of 28 October 2003. Sol. Phys. **268**, 175 (2011)
20. Forrest, D.J., Vestrand, W.T., Chupp, E.L. et al.: Very energetic gamma-rays from the June 3, 1982 solar flare. Adv. Space Res. **6**, 115 (1986)
21. Alexander, D., Dunphy, P.P., MacKinnon, A.L.: High energy gamma-ray emission from solar flares: constraining the accelerated proton spectrum. Sol. Phys. **151**, 147 (1994)
22. Talon, R., Trottet, G., Vilmer, N. et al.: Hard X-ray and ?-ray observations of solar flares with the Phebus experiment. Sol. Phys. **147**, 137 (1993)
23. Terekhov, O.V., Sunyaev, R.A., Tkachenko, A.Yu. et al.: Deuterium synthesis during the solar flare of March 22, 1991 (Granat data). Astron. Lett. **22**, 143 (1996)
24. Dunphy, P.P., Chupp, E.L., Bertsch, D.L. et al.: Gamma-rays and neutrons as a probe of flare proton spectra: the solar flare of 11 June 1991. Sol. Phys. **187**, 45 (1999)
25. Akimov, V.V., Ambroz, P., Belov, A.V. et al.: Evidence for prolonged acceleration based on a detailed analysis of the long-duration solar gamma-ray flare of June 15, 1991. Sol. Phys. **166**, 107 (1996)
26. Chupp, E.L., Trottet, G., Dunphy, P.P., Rieger, E.: What we know and what we do not know about high energy neutral emissions from solar flares (a challenge for future missions). Proceedings of 28th ICRC, p. 3171, Tsukuba (2003)
27. Chupp, E.L., Ryan, J.M.: High energy neutron and pion-decay gamma-ray emissions from solar flares. Res. Astron. Astrophys. **9**, 11 (2009)
28. Ackermann, M., Ajello, M., Allafort, A. et al. FERMI detection of ?-ray emission from the M2 soft X-ray flare on 2010 June 12. Astrophys. J. **745**, 144 (2012)
29. Kuznetsov, S.N., Kurt, V.G., Myagkova, I.N. et al.: Gamma-ray emission and neutrons from solar flares recorded by the SONG instrument in 2001–2004. Sol. Syst. Res. **40**, 104 (2006)
30. Kurt, V.G., Yushkov, B.Yu., Kudela, K., Galkin, V.I.: High-energy gamma radiation of solar flares as an indicator of acceleration or energetic protons. Cosmic Res. **48**, 70 (2010)
31. Kurt, V.G., Kudela, K., Yushkov, B.Yu., Galkin, V.I.: On the Onset Time of Several SPE/GLE Events: Indications from High-Energy Gamma-Ray and Neutron Measurements by CORONAS-F. Adv. Astron. **2013**. Paper 690921 (2013)
32. Kuznetsov, S.N., Bogomolov, A.V., Gordeev, Yu.P. et al.: Preliminary results of the experiment with the SCR scientific set of instruments onboard the CORONAS-I satellite. Bull. Russ. Acad. Sci. Phys. **59**(4), 2 (1995)
33. Veselovsky, I.S., Panasyuk, M.I., Avdyusin, S.I. et al.: Solar and Heliospheric Phenomena in October-November 2003: Causes and Effects. Cosmic Res. **42**, 438 (2004)
34. Grechnev, V.V., Kurt, V.G., Chertok, I.M. et al.: An extreme solar event of 20 January 2005: properties of the flare and origin of energetic particles. Sol. Phys. **252**, 149 (2008)
35. Kuznetsov, S.N., Kurt, V.G., Yushkov, B.Yu. et al.: 28 October 2003 flare: high-energy gamma emission, type II radio emission and solar particle observations. Int. J. Mod. Phys. A. **20**, 6705 (2005).
36. Kuznetsov, S.N., Kurt, V.G., Yushkov, B.Yu. et al.: Proton acceleration during 20 January 2005 solar flare: CORONAS-F observations of high-energy gamma emission and GLE. Contrib. Astron. Obs. Skalnate Pleso **36**, 85 (2006)
37. Gros, M., Tatischeff, V., Kiener, J. et al.: INTEGRAL/SPI observation of the 2003 October 28 solar flare // Proc. 5th INTEGRAL Workshop / Eds. Schonfelder V., Lichti G., Winkler C. 2004. ESA SP-552, P. 669
38. Share, G.H., Murphy, R.J., Smith, D.M. et al.: RHESSI $e^+ - e^-$ annihilation radiation observations: implications for conditions in the flaring solar chromosphere. Astrophys. J. **615**, L169 (2004)
39. Masson, S., Klein, K.-L., Butikofer, R., Flückiger, E., Kurt, V., Yushkov, B., Kruker, S.: Acceleration of Relativistic Protons During the 20 January 2005 Flare and CME. Sol. Phys. 257, 305 (2009)

40. Zhitnik, I.A., Logachev, Yu.I., Bogomolov, A.V. et al.: Polarization, temporal, and spectral parameters of solar flare hard X-rays as measured by the SPR-N instrument onboard the CORONAS-F satellite. Sol. Syst. Res. **40**, 93 (2006)
41. Kiplinger, A., Garcia, H.: Soft X-ray parameters of the great flares of active region 486. Bull. Am. Astron. Soc. **36**, 739 (2004)
42. Kaufmann, P., Raulin, J.-P., Giménez de Castro C.G. et al.: A new solar burst spectral component emitting only in the terahertz range. Astrophys. J. **603**, L121 (2004)
43. Kane, S.R., McTiernan, J.M., Hurley, K.: Multispacecraft observations of the hard X-ray emission from the giant solar flare on 2003 November 4. Astron. Astrophys. **433**, 1133 (2005)
44. Bouratzis, C., Preka-Papadema, P., Hillaris, A. et al.: Radio observations of the 20 January 2005 X-class flare. Sol. Phys. **267**, 343 (2010)
45. Share, G.H. et al.: RHESSI spectroscopic observations of the 2005 January 20 solar flare. SPD Meeting. American Astron. Soc. 2006, P. 37
46. Kuznetsov, S.N., Kurt, V.G., Yushkov, B.Yu., Kudela, K.: Determination of instants of the particle acceleration up to relativistic energies in solar flares according to the SONG measurements onboard the CORONAS-F satellite. Bull. Russ. Acad. Sci. Phys. **70**, 1665 (2006)
47. Plainaki, C., Belov, A., Eroshenko, E., Kurt, V. et al.: Unexpected burst of solar activity recorded by neutron monitors during October November 2003. Adv. Space Res. **35**, 691 (2005)
48. Pei, C., Jokipii, J.R., Giacalone, J.: Effect of a random magnetic field on the onset times of solar particle events. Astrophys. J. **641**, 1222 (2009)
49. Panasyuk, M.I. et al.: Modeling of Neutron Fluxes on Near-Earth Spacecrafts from Experimental Results of SINP. Proceedings of RADECS 2000 Workshop, P. 9. Université Catholique de Louvain, Belgium (2001)
50. Bieber, J.W., Clem, J., Evenson, P. et al.: Relativistic solar neutrons and protons on 28 October 2003. Geophys. Res. Lett. **32**, L03S02 (2005)
51. Miroshnichenko, L.I., Klein, K.-L., Trottet, G. et al.: Relativistic nucleon and electron production in the 2003 October 28 solar event. J. Geophys. Res. **110**, A09S08 (2005)
52. Miklenic, C.H., Veronig, A.M., Vršnak, B.: Magnetic flux change rates and nonthermal flare emission. Astron. Astrophys. **499**, 893 (2009)
53. Kiener, J., Gros, M., Tatischeff, V., Weidenspointner, G.: Properties of the energetic particle distributions during the October 28, 2003 solar flare from INTEGRAL/SPI observations. Astron. Astrophys. **445**, 725 (2006)
54. Kuznetsov, S.N., Yushkov, B.Yu, Kudela, K.: Measurement of the spectrum of relativistic protons from solar flares on October 28 and November 2, 2003 onboard the CORONAS-F satellite. Cosmic Res. 45, 373 (2007)
55. Kurt, V.G., Yushkov, B.Yu., Belov, A.V.: On the ground level enhancement beginning. Astron. Lett. **36**, 520 (2010)

# Dynamics of the Relativistic Electrons Flux of the Earth Outer Radiation Belt Based on the MKL Instrument Data

**S.N. Kuznetsov[1], I.N. Myagkova, E.A. Muravieva, B.Yu. Yushkov, L.I. Starostin[2], Yu.I. Denisov[3], and K. Kudela**

The results of continuous monitoring of the intensity variations of relativistic electron flux (1.5–3 MeV and 3–6 MeV) in the outer radiation belt of the Earth which was done by the MKL instrument on board the CORONAS-F satellite are presented. The main purpose of the work is analysis of the dynamics of relativistic electron flux in the outer radiation belt of the Earth in particular long increasing of their intensity 10 and more days after magnetic storms. The analysis is based on the experimental data of 2002 when essentially long magnetically quiet periods were observed along with magnetic storms of different intensity. The presented materials show that all magnetic storms of both high intensity ($Dst \approx -200$ nT) and middle and even low intensity resulted in sharp decrease of intensity of electron flux with energy of 1.5–6 MeV. Then during several days (on the phase of the magnetic storm recovery) the intensity of the relativistic electron flux was recovered. As a rule on the 7–10th day from the beginning of the storm significant (for an order of magnitude and over) increase of the relativistic electron flux began at $L$ =3–4.5, and it was observed during 15–20 days, and sometimes even longer.

---

[1]Deceased

[2]Deceased

[3]Deceased

S.N. Kuznetsov · I.N. Myagkova (✉) · E.A. Muravieva · B.Yu. Yushkov · L.I. Starostin · Yu.I. Denisov
D.V. Skobeltsyn Institute of Nuclear Physics, M.V. Lomonosov Moscow State University (SINP MSU), Vorobyevy Gory, Moscow 119991, Russia
e-mail: irina@srd.sinp.msu.ru

K. Kudela
Institute of Experimental Physics, Slovak Academy of Sciences (IEP SAS), Kosice, Slovakia

V. Kuznetsov (ed.), *The Coronas-F Space Mission*, Astrophysics and Space Science Library 400, DOI 10.1007/978-3-642-39268-9_11, 

## Introduction

It is known that the structure of the Earth's magnetosphere and, consequently, of its radiation belts is determined by the interaction of the magnetosphere with solar wind (SW) and Interplanetary Magnetic Field (IMF) (for instance, see book [13] and its references). Coronal mass ejections (CME) produced on the Sun and characterized by high velocity and matter's density coming to the Earth's orbit significantly influence on the Earth's radiation belts conditions. The studies have shown that if CME passes near the Earth its magnetosphere decreases sharply, the region of closed L-shells [4] also decreases, the night plasma layer comes nearer to the Earth and its current and magnetic field in the tail of the magnetosphere increase. The Earth radiation belts particles which before the storm were at the outer shells are thrown out of the magnetosphere. These processes carry differently depending on parallel or antiparallel direction of the CME magnetic field to the geomagnetic field direction. So, under otherwise equal conditions, the currents which cause increasing of *Dst*-variation amplitude are more powerful at negative (with southern orientation) *Bz*-component of IMF. Along with CME both quazi-stationary and sporadic SW fluxes also influence significantly on the Earth's magnetosphere. Quazi-stationary high-speed fluxes which cause recurrent geomagnetic disturbances of middle and low intensity are related to the coronal holes at the Sun, and sporadic high-speed fluxes are currently considered to be related to CMEs and solar flares. SW's geo-effectivity, i.e. effectiveness of SW's energy transmission to the Earth's magnetosphere also depends on the IMF's orientation and maximum at negative, southern orientation and at high values of *Bz*.

Experimental studies of the electron flux variations in the Earth outer radiation belt were made from the end of 1960s by different scientific groups [1, 2, 14–16]. Comparison of the variations of the outer radiation belt at different altitudes has shown [3, 16] that these variations are of similar character, but differ in value. The question about the source (or sources) of powerful and long increases of relativistic electron fluxes in the Earth radiation belts is still open despite researches of many scientific groups. In [9] the data obtained during the experiment on board the SAMPEX satellite (orbit's altitude of 600 km, inclination of 82°) was studied. The authors have shown that there is a type of correlation (at the level not more than 40 %) for the relativistic electron flux of the outer radiation belt and SW's velocity with 2-days delaying (at $L = 10$) and 4-days delaying (at $L = 6.6$). Basing on the analysis results the authors conclude that the electrons of the solar wind must not be considered as the main source of the relativistic electrons of the Earth radiation belts.

Dynamics of the Earth radiation belts electron fluxes with energy of 0.3–3 MeV at the altitude of 400–500 km during the most powerful magnetic storms of 2001–2005 is described in detail in papers [6,7,12,17] and in the chapter "Dynamics of the Earth Radiation Belts During the Strong Magnetic Storms" of the present book [8]. The main purpose of the present chapter is analysis of the long increases of relativistic electron flux in the outer Earth radiation belt which were observed a few

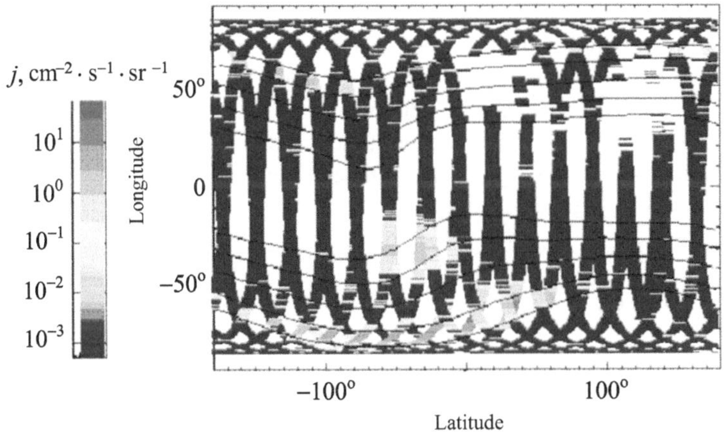


**Fig. 1** Geographical distribution of relativistic electron flux ($E_e$ = 1.5–3 MeV), recorded on board the CORONAS-F satellite on July 31, 2002

days after the end of magnetic storms and search of their relations with variations of SW's, IMF's, and geomagnetic field's parameters.

## Experiment

The MKL instrument (Russian abbreviation for monitor of cosmic rays) on board the CORONAS-F satellite was intended for the measurements of the flux and spectra of the protons and electrons within the energy ranges of 1–90 MeV and 0.3–12 MeV, respectively [5]. The flux and the spectra of relativistic electrons were measured by means of a semiconductor telescope, a part of the MKL instrument. The telescope consisted of two semiconductor detectors 0.05 mm and 2.0 mm thick and CsJ crystal 1.0 cm thick surrounded with anticoincidence plastic scintillator 0.5 cm thick. The electrons were recorded within 5 energy ranges (MeV): 0.3–0.6, 0.6–1.5, 1.5–3, 3–6, and 6–12. Telescope's aperture was $\sim 23°$, it was oriented in antisolar direction. Detailed description of the MKL instrument is presented in the chapter "Scientific Set of Instruments "Solar Cosmic Rays"" of the present book.

During the period studied in the present paper (from January till December 2002) altitude of the orbit of the CORONAS-F satellite decreased from 500 to 450 km. Figure 1 presents an example of geographical distribution of relativistic electron flux ($E_e$ = 1.5–3 MeV), recorded by the MKL instrument on June 31, 2002. Solid lines indicate the equal values of $L$-parameter (1.3; 1.6; 2.2; 3.5; 5.5).

The orbit of the CORONAS-F satellite did not allow to determine maximum intensity of the captured particles in the region of South-Atlantic Anomaly for one orbit pass; therefore, the analysis was based on several consequent passes and it resulted in a sufficient rough time resolution—1 day.

As seen in Fig. 1 the data for the direct and reverse orbit passes differs. It proves the anisotropy of the studied particles. In July 2002 direct orbit passes corresponded to the morning of magnetic local time ($\sim$ 9 h MLT), and reverse—to the evening ($\sim$ 21 h MLT). For further analysis the authors used data for morning passes crossed Brasilian Anomaly (the region of minimum intensity of the Earth magnetic field B) where as it is seen from Fig. 1 relativistic electrons intensity in the outer radiation belt at the studied altitudes is maximum.

## Data Analysis

The upper panels of Figs. 2–5 present dynamics of relativistic electron flux within the energy range of 1.5–3 MeV in DOY-coordinates (Day Of Year)–$L$ (the color indicates the intensity of particles' flux) during the period of January–December 2002 at the altitude of 500–450 km. The middle panel of Fig. 5 also presents variations of the electrons of the higher energy (3–6 MeV). Three lower panels of Figs. 2–5 present time profiles of $Dst$-variation, SW speed, and $Bz$-component of IMF (top down).

Each of the presented pictures will be analyzed in detail. Figure 2 shows that there were neither strong magnetic storms ($Dst < -100$ nT) nor SCR events accompanied with a flux of solar electrons with energy of $> 1,5$ MeV during the period from January till March 2002. During this period high-speed SW flux reached the Earth's orbit four times and caused several weak recurrent magnetic storms. Nevertheless these storms with relatively low amplitude happened at negative $Bz$, and resulted in sharp decreasing of relativistic electron flux of the outer RB ($L = 3$–4.5). It is especially obvious for the storms of February 6 (47 DOY, $Dst = -80$ nT) and March 1 (100 DOY, $Dst = -70$ nT). It should be noted that 10 days after the beginning of the storm of February 6 a noticeable increase of electron flux with energy of 1.5–3 MeV was observed at $L = 3.5$–4.5. The flux increased approximately for an order of magnitude comparing with the conditions of the outer belt before the storm (on February 1–5). Figure 2 also proves that this increase lasted for 10–12 days till the above-mentioned storm of March 1. The increase after the storm of March 1 was significantly lower. The following sharp decrease of electron flux intensity during the period of March 19–21 (78–81 DOY) is quite unusual because it started after a very small storm with $Dst$ amplitude less than $-50$ nT, caused by coming of SW flux with speed of $\sim$ 500 km/s to the Earth.

During the following three months (see Fig. 3) the series of storms happened from April 17 till April 21 (maximum value of $Dst = -150$ nT, minimum $Bz \sim -18$ nT) became the most significant disturbance of geomagnetic situation. Detailed analysis of electrons' variations in the outer ERB for this period, exhaustion of the relativistic electron belt and its reasons was included in paper [10]. Presented there intensity profiles for different $L$ show that after April 21 electron flux maximum with energy of 1.5–3 MeV increases for not less than an order of magnitude comparing with the condition before the storm and moves a little to the Earth—from c $L = 3.7$–3.8 to $L = 3.1$–3.2.

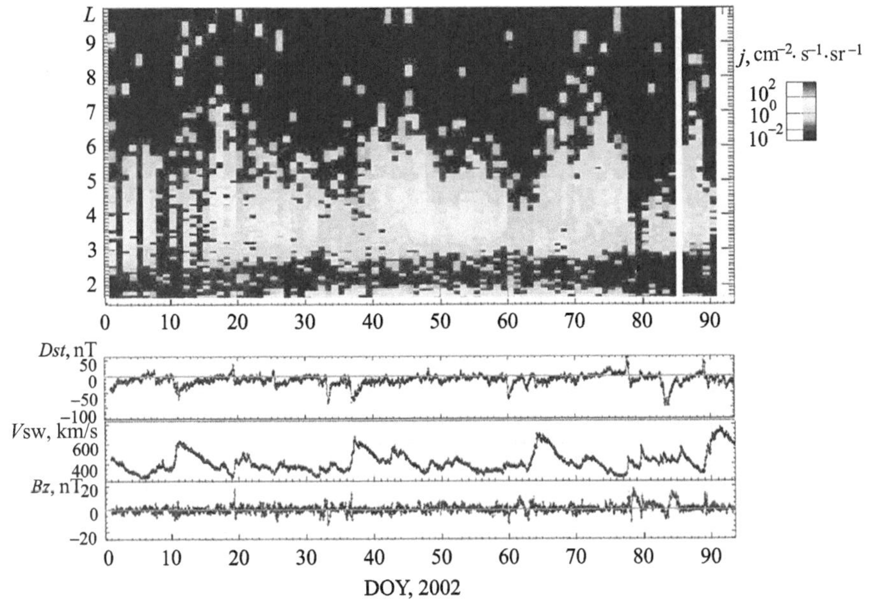


**Fig. 2** Dynamics of relativistic ($E_e$ = 1.5–3 MeV) electrons of the outer RB during January–March 2002 (*upper panel*), time dependence of *Dst*-variation, SW speed and *Bz*-component of IMF (*top down*)

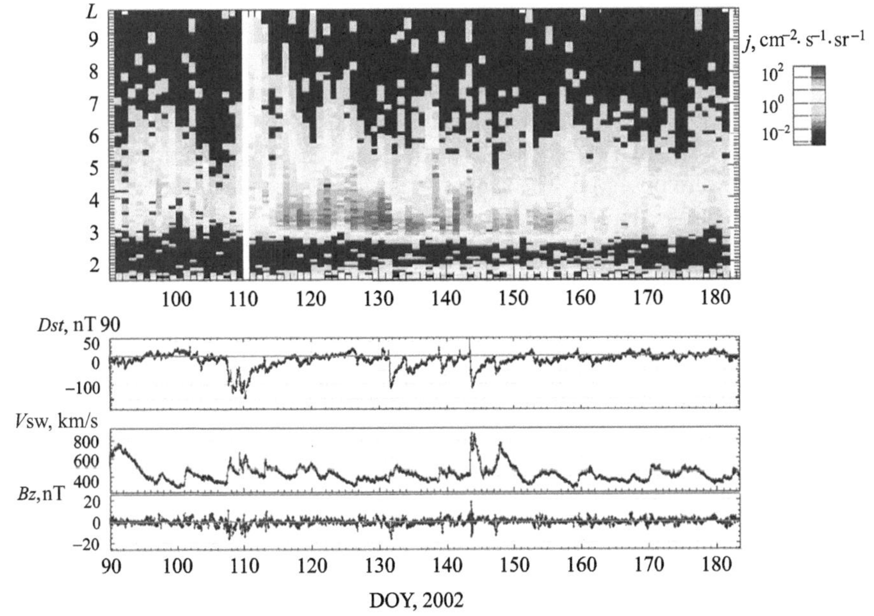


**Fig. 3** The same as in Fig. 2, for the period of April–June 2002

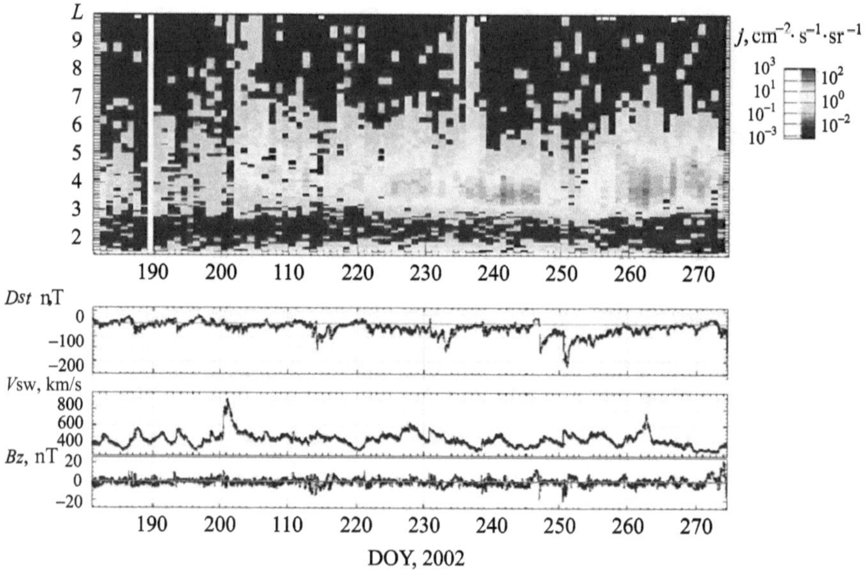


**Fig. 4** The same as in Fig. 2, for the period of July–October 2002

From Fig. 3 it is also concluded that on April 21, 2002 SCR electron flux also increased—electrons with energy of 1.5–3 MeV were observed not only at the outer RB but also at the higher $L$-shells, i.e. in the polar caps. The question of energetic SCR electrons as an additional source of relativistic electrons in the outer RB still remains in abeyance. But the fact that SCR are not the only source of the relativistic electron flux intensity in the outer RB is proved by the analysis of electron flux with energies of 1.5–3 MeV, for instance, during the period of February or June 2002, when there were no SCR electrons but the increases of relativistic electrons in the outer RB were observed. The same conclusion was made during the studies of relativistic electrons dynamics variations during the first half of 2005 [11], when during the period of March–April intensity of the outer RB electrons increased without SCR electron flux. In general, comparing Figs. 2 and 3 it is clearly seen that variations of electron flux intensity for energies of 1.5–3 MeV in the outer RB was significantly stronger during the period of April–May 2002 than during the period of January–March 2002. It should be noted that the increase started on April 22 (on April 21 the region of the outer RB was still filled with SCR electron fluxes) was lasted during approximately 20 days till May 12 when a new decrease of electron flux intensity (energy of 1.5–3 MeV) was caused by a storm with $Dst = -105$ nT (at negative $Bz \approx -19$ nT).

The next significant decrease of electron intensity was caused by the storm on May 24 ($Dst = -110$ nT), which in turn became a result of coming of a high-speed SW flux (with maximum speed of over 850 km/s) to the Earth. Maximum intensity of the recovered relativistic electron belt was almost for an order of magnitude

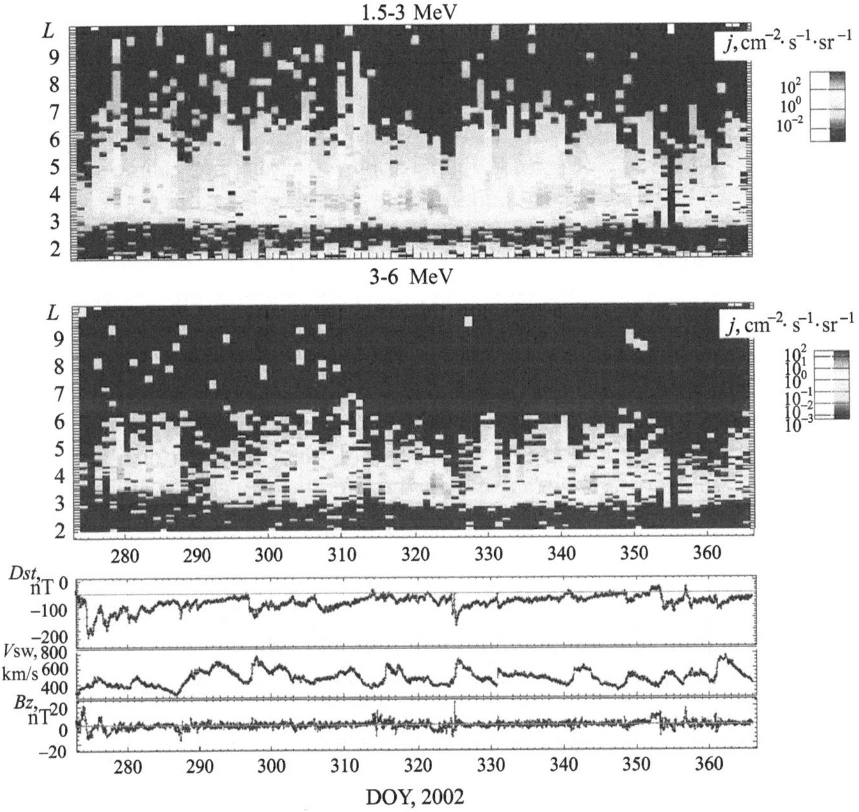


**Fig. 5** The same as in Fig. 2 and variations of electron flux (energy 3–6 MeV), for the period of October–December 2002

lower than intensity detected after the previous recovery and kept approximately till the beginning of the second decade of June (see Fig. 3). Then maximum of the relativistic electron outer belt became much more wider moving to the bigger $L$, and its intensity decreased for approximately an order of magnitude. Such type of dynamics of the outer RB of the Earth allows to suppose a significant role of diffusion processes during this flare- and magnetically calm period.

Step-by-step decrease of the intensity of electron flux of 1.5–3 MeV in the outer ERB lasted till June 20 (see Fig. 4). That day after the flare of X3.3 class in soft X-ray emission (according to the GOES satellite) SCR electrons of 1.5–3 MeV energy were detected in the polar caps. Later electron fluxes primarily appeared in the region of $L = 5$–6 (approximately 205–207 DOY), and then spread to $L = 3$–4.5. During the next two months at the main phases of the storms with *Dst* varied from $-50$ up to $-150$ nT electrons' intensity decreased at $L = 3$–3.5 many times, but there were only three sufficiently long intensity increases. The first, the smallest of them, happened from 224 till 230 DOY. It's interesting that previous to this increase intensity decrease was caused by very small storm with $Dst = -50$ nT. The second increase began approximately on 235 DOY (August 23), happened at sufficiently narrow interval of $L$—from 3.5 up to 4.5 and lasted till 247 DOY (September 4), when two consequent storms with an interval of 4 days caused at first significant intensity decrease and then practically full disappearance of relativistic electron belt. The third, the most powerful and long (approximately 20 days) increase had a wide maximum at $L$ (from 3 up to 5.5) was observed 8 days after the powerful storm of September 8 ($Dst = -180$ nT).

During the period of October–December 2002 the most powerful storm happened in the very beginning of this period, but just after it several storms with smaller intensity happened. It could cause depression of the following electron intensity increase.

After the storms of October 24 (297 DOY, $Dst = -100$ nT) and November 21 (325 DOY, $Dst = -145$ nT) related to the coming of high-speed SW flux intensity decrease and subsequent long increase of relativistic electron flux were also observed. Figure 5 presents intensity variations for the electrons of 3–6 MeV along with variations of relativistic electron flux for 1.5–3 MeV. It can be seen that not only intensity decreases but also long increases for the fluxes of particles of different energy happen almost simultaneously. In general, the character of the variations—depletion during the main phase of the storm, relatively fast (during several days) recovery and then increase for an order of magnitude and over which lasts for 10–20 days, and sometimes till the following storm which causes a new depletion—kept the same for the period of October–December 2002.

The studies of the mechanisms of relativistic electron flux increases in the outer ERB which happen several days after magnetic storm of different intensity related to coming both CMEs and high-speed SW fluxes to the Earth are interesting not only as a fundamental problem but also as an applied aspect. The reason of it is that electrons of several MeV energy can cause serious damage to electronic circuits of the instruments on board spacecrafts with an orbit crossing the outer radiation belt of the Earth.

## Conclusions

The following conclusions were made from the analysis of the dynamics of relativistic electron flux at the altitude of 450–500 km, recorded during the experiment on board the CORONAS-F satellite:

1. At the main phase of almost each magnetic storm regardless of its amplitude and reasons (coming to the Earth of CME or high-speed SW flux) simultaneously with decreasing of the size of magnetosphere and outer radiation belt a sharp decrease of relativistic electrons' intensity in the outer ERB, i.e. substantial disappearance of relativistic electron belt is observed.

2. Variations of relativistic electron flux in the outer ERB develop according to the following scenario. After depletion of the relativistic electron belt during the main phase of the storm relatively fast (during several days) recovery of their intensity happens, then it increases for an order of magnitude and over, which can last for several weeks, sometimes till the next storm, which causes a new belt's depletion. It should also be noted that significant increase of relativistic electron flux intensity in the outer ERB (up to 10–50 times comparing to the conditions before the storm) also takes place after relatively small geomagnetic disturbances ($Dst \approx -50$–100 nT).
3. Analysis of possible relations between relativistic electron flux variations in the outer ERB and the parameters of IMF and SW has shown a tendency to the increasing of the outer ERB relativistic electron flux 7–10 days after high-speed SW coming (with different time shift depending on $L$ and electrons' energy).
4. There was no significant correlation between the outer ERB relativistic electron flux and SCR high-energy electron flux.

K.Kudela wishes to acknowledge the grant agency VEGA, project 2/0040/13, for support.

## References


1. Baker, D.N., et al.: Recurrent geomagnetic storms and relativistic electron enhancements in the outer magnetosphere: ISTP coordinated measurements. J. Geophys. Res. **102**(A7), 14141–14148 (1997)
2. Blake, J.B., Kolasinski, W.A., Fillius, R.W., Mullen, E.G.: Injection of electrons and protons with energies of tens of MeV into $L < 3$ on March 24, 1991. Geophys. Res. Lett. **19**, 821–833 (1992)
3. Emelyanenko, S.P., Kuznetsov, S.N., Stolpovsky, V.G.: The outer radiation belt during the strong magnetic storm. Cosmic Res. **16**(4), 529 (1978) (in Russian)
4. Kuznetsov, S.N., Suvorova, A.V., Dmitriev, A.V.: Shape and size of the magnetopause associated with parameters of the interplanetary medium. Geomagnetism Aeronomy **38**(6), 697 (1998)
5. Kuznetsov, S.N., Kudela, K., Ryumin, S.P., Gotselyuk, Yu.V.: CORONAS-F satellite—tasks for study of particle acceleration. Adv. Space Res. **30**, 1857–1860 (2002)
6. Kuznetsov, S.N., Yushkov, B.Yu., Kudela, K., Myagkova, I.N., et al.: Dynamics of the Earth's radiation belts during the magnetic storm of November 6$^{th}$, 2001. Adv. Space Res. **36**(10), 1997–2002 (2005)
7. Kuznetsov, S.N., Myagkova, I.N., Yushkov, B.Yu., et al.: Dynamics of the Earth Radiation Belts During Strong Magnetic Storms Based on CORONAS-F Data. Sol. Syst. Res. **41**(4), 338 (2007)
8. Kuznetsov, S.N., Denisov, Yu.I., Lazutin, L.L., et al.: Dynamics of the radiation belts during the Strong magnetic storms. The present book. Chap. 12
9. Li, X., Baker, D.N., Temerin, M., Cayton, T.E., et al.: Multisatellite observations of the outer zone electron variation during the November 3–4, 1993, magnetic storm. J. Geophys. Res. **102A**, 14123–14140 (1997)
10. Myagkova, I.N, Kuznetsov, S.N., Yushkov, B.Yu., et al.: Dynamics of the Earth's radiation belts during time period April 14–24, 2002: Experimental data. In: The Inner Magnetosphere: Physics and Modeling. Geophysical Monograph, vol. 155, pp. 127–134. AGU, Washington (2005)

11. Myagkova, I.N., Kuznetsov, S.N., Panasyuk, M.I. et al.: Solar flares, solar energetic particle events and their influence on near-Earth environment in May 2005 as observed by CORONAS-F and Universitetskiy-Tatiana spacecrafts. Sun Geosphere. **1**(2), 32–36 (2006)
12. Panasyuk, M.I., Kuznetsov, S.N., Lazutin, L.L., et al.: Magnetic storms in October 2003. Cosmic Res. **42**(5), 489 (2004)
13. Tverskoy, B.A.: The basis of theoretical space physics (2004) (in Russian)
14. Vakulov, P.V., Kovrygina, L.M., Mineev, Yu.V., Tverskaya, L.V.: Dynamics of the outer belt of high-energy electrons during the fair magnetic storm. Geomagnetism Aeronomy **15**(6), 1028 (1975)
15. West, A.I., Buck, R.M., Davidson, G.T.: The dynamics of energetic electrons in the Earth's outer radiation belt during 1968 as observed by the Lawrence Livermore National Laboratory's spectrometer on OGO-5. J. Geophys. Res. **86**, 2111–2138 (1981)
16. Williams, D.J., Arens, J.F., Lanzerotti, L.J.: Observations of trapped electrons at low and high altitudes. J. Geophys. Res. **73**, 5673–5696 (1968)
17. Yermolaev, Yu.I., Zelenyi, L.M., Zastenker, G.N., et al.: A Year Later: Solar, heliospheric, and magnetospheric disturbances in November 2004. Geomagnetism Aeronomy **45**(6), 681 (2005)

# Dynamics of the Earth Radiation Belts During the Strong Magnetic Storms

S.N. Kuznetsov[1], Yu.I. Denisov[2], L.L. Lazutin, I.N. Myagkova, E.A. Muravieva, B.Yu. Yushkov, K. Kudela, R. Bucik, and M. Slivka

The results of the experimental studies of the variations of electron flux of energy of 0.3–6 MeV at the outer radiation belt of the Earth during several strongest magnetic storms of the period of August 2001–May 2005 according to the experiment onboard the CORONAS-F satellite are reported. Experimental data recorded during the magnetic storms of 06.11.2001 ($Dst = -257$ nT), 24.11.2001 ($Dst = -221$ nT), 29–30.10.2003 ($Dst = -400$ nT) and 15.05.2005 ($Dst = -260$ nT) are analyzed. It is shown that all these storms caused sharp decreases of the intensity of the ERB electron flux for energy of several hundreds keV and substantial disappearance of the outer ERB for the electrons with energies of $>$ 1.5 MeV. At the recovery phase of the storm the recovered outer ERB was observed much nearer to the Earth, near $L =$ 2.5–3, where before the storm it was observed a slot between the inner and the outer radiation belts. The decrease of the electron flux intensity is associated with a sharp decrease of magnetosphere's size during the main phase of the storm. For all studied cases variations of the ERB were sufficiently long (not less than several days).

[1]Deceased

[2]Deceased

S.N. Kuznetsov · Yu.I. Denisov · L.L. Lazutin (✉) · I.N. Myagkova · E.A. Muravieva · B.Yu. Yushkov
D.V. Skobeltsyn Institute of Nuclear Physics M.V. Lomonosov Moscow State University (SINP MSU), Vorobyevy Gory, Moscow 119991, Russia
e-mail: lll@srd.sinp.msu.ru

K. Kudela · R. Bucik · M. Slivka
Institute of Experimental Physics, Slovak Academy of Sciences (IEP SAS), Kosice, Slovakia

V. Kuznetsov (ed.), *The Coronas-F Space Mission*, Astrophysics and Space Science Library 400, DOI 10.1007/978-3-642-39268-9_12,

## Introduction

According to previous results (see, for instance, [2, 6, 9–11]) there are significant variations of the electron flux at the outer Earth's radiation belt (ERB) both during magnetic storms and during quiet periods. Comparative analysis of the electron flux variations at the outer ERB at high and low altitudes [2, 11] has shown that the observed variations are similar in good manner, but different in magnitude. In this chapter dynamics of electron belt for energy of 0.3–3 MeV during a number of powerful magnetic storms observed during 2001–2005 is studied basing on the experimental data of the CORONAS-F satellite.

## Characteristics of the Magnetic Storms

All magnetic storms in review of the present paper were caused by coronal mass ejections (CMEs) related to the solar flares. Simultaneously these flares produced solar energetic particles. Table 1 presents the time of appearance and linear speed of CMEs, caused concerned storms, and the moments of the sudden commence (SC) of the magnetic disturbances. It's impossible to say when the third and fifth CME came to the Earth, because the SC was not observed.

We'll review the variations of the parameters of the Earth's magnetosphere, solar wind (SW), interplanetary magnetic field (IMF), and geomagnetic indices for all the events in detail. Figure 1a,b present the following information for the storms of November 6 and 24, 2001, consequently (top down): the upper panel—the values of the distance up to the subsolar point of the magnetopause $X(0)$ (thick line) and ratio of the distance up to the magnetopause at an angle of 90° to the direction to the Sun $R(90)$ to the distance up to the subsolar point $R(90)/X(0)$ (thin line with dots), according to the method described in [3]; the second panel—SW's speed $V$ (thick line) and its density $N$ (thin line with dots); the third panel—$Bz$-component of IMF (thick line) and full value of IMF $B$ (thin line with dots); the lower panel—the variations of geomagnetic index *Hsym* (thick line) and *AE* (thin line) indices.

Figure 1a shows that the decrease *Hsym*, i.e. the main phase of the storm began directly after the sudden commence of the storm of November 6, 2001. Most probably it is associated with negative $Bz$-component of IMF of the front part of CME. At 3 o'clock $Bz$ value reached its minimum and began to increase, *Hsym* also reached its minimum and kept at this level till $\sim$ 8 o'clock, although $Bz$-value became positive approximately at 05:00, and from 01:53 till $\sim$ 08:00 SW's speed and density kept almost constant. Detailed analysis of this storm and ERB's dynamics for this period is presented in paper [5].

Apparently the magnetosphere and high-speed SW flux with increased density are kept in dynamical balance. Analysis of variations of the size and the shape of the magnetosphere are of particular interest. At positive $Bz$ the magnetic field near $X(0)$ is redoubled comparing to dipole's one, and $R(90) \sim 1.44X(0)$. At negative $Bz$ magnetic field near $X(0)$ is close to dipole's one because of reconnection and $R(90)$ can be equal to $\sim 1.9X(0)$. Taking into consideration decrease of $X(0)$, we can

**Table 1** Characteristics of CMEs and SCs

| Date and time of CME's appearance, hour:min (UT) | CME's speed, km/s | Date and time of registration of the sudden commence of the magnetic storm, hour:min |
|---|---|---|
| 04.11.2001, 16:20 | 1810 | 06.11.2001, 01:53 |
| 22.11.2001, 20:36 | 1443 | 24.11.2001, 05:56 |
| 22.11.2001, 23:30 | 1437 | 24.11.2001, – |
| 28.11.2003, 11:30 | 2459 | 29.11.2003, 06:11 |
| 29.11.2003, 20:54 | 2029 | – |
| 13.05.2005, 17:12 | 1689 | 15.05.2005, 02:38 |

conclude that relative cross-section of the magnetosphere keeps constant, and the magnetic flux transferred to the tail of the magnetosphere increases. As a rule, under these conditions magnetic substorms are observed and $AE$-index increases. Figure 1a also shows that from 08:00 till 13:00 at positive $Bz$ SW's pressure decreased, and $Hsym$ increased. Then, from 13:00 till 16:30 at low SW's density $Bz$'s value was equal to $-30$ nT, and it resulted in a certain oncoming of $X(0)$ to the Earth at increase of cross-section of the magnetosphere. Change of $Bz$ at 16:30 resulted in moving of subsolar point of the magnetopause away.

The data on the IMF and the Earth's magnetospheric disturbances during the storm of November 24, 2001 are presented in Fig. 1b. After the sudden arrival of magnetic distubrances at 05:56 the strongly pronounced first phase of the storm was observed during approximately 50 min. The main phase of the storm with minimum $Hsym = -130$ nT lasted till 08:00. The second main phase of the magnetic storm developed from 11:00 till 13:00. It is supposed that it was associated with coming of the second CME. After the end of the most powerful sub-storm for the studied period at 14:30 the recovery phase started and magnetosphere's size increased up to its size before the magnetic storm.

In a similar vein we'll study the magnetic storms of October 29–30, 2003 happened during the solar extreme events. Figure 2 shows that on 29.10.2003 at 06:11 CME with a speed of $\sim$ 2000 km/s came to the Earth and caused sudden commence of the magnetic storm. Almost immediately the main phase of the storm (maximum in amplitude value of $Hsym$ was approximately $-305$ nT) started. It lasted for less than an hour because of a short period of negative $Bz$ existence (see upper panel of Fig. 3 (thick line)). Then the recovery phase with fluctuations $Bz$ and $Hsym$ lasted till 13:00 (thick line at the lower panel of Fig. 3). Thin lines with points at the upper and lower panels of Fig. 3 indicate variations of the interplanetary magnetic field $B$ and $AE$-index, relatively. Then from 13:00 till 24:00 of October 29, 2003 the main phase of the second storm happened at negative $Bz$ and SW speed decreasing from approximately 08:00. The next high-speed SW flux came to the Earth at $\sim$ 17:00 (see Fig. 2), and the third powerful storm began. Maximum of the main phase of the third storm was observed at 23:00 of November 30, 2003. By the time SW's speed dramatically decreased again. Detailed analysis of the dynamics of IMF's and SW's parameters as well as of geomagnetic indices and the outer radiation belt electron flux during the storms of November 2001 and October–November 2003 made by the authors of this chapter is presented in [4].

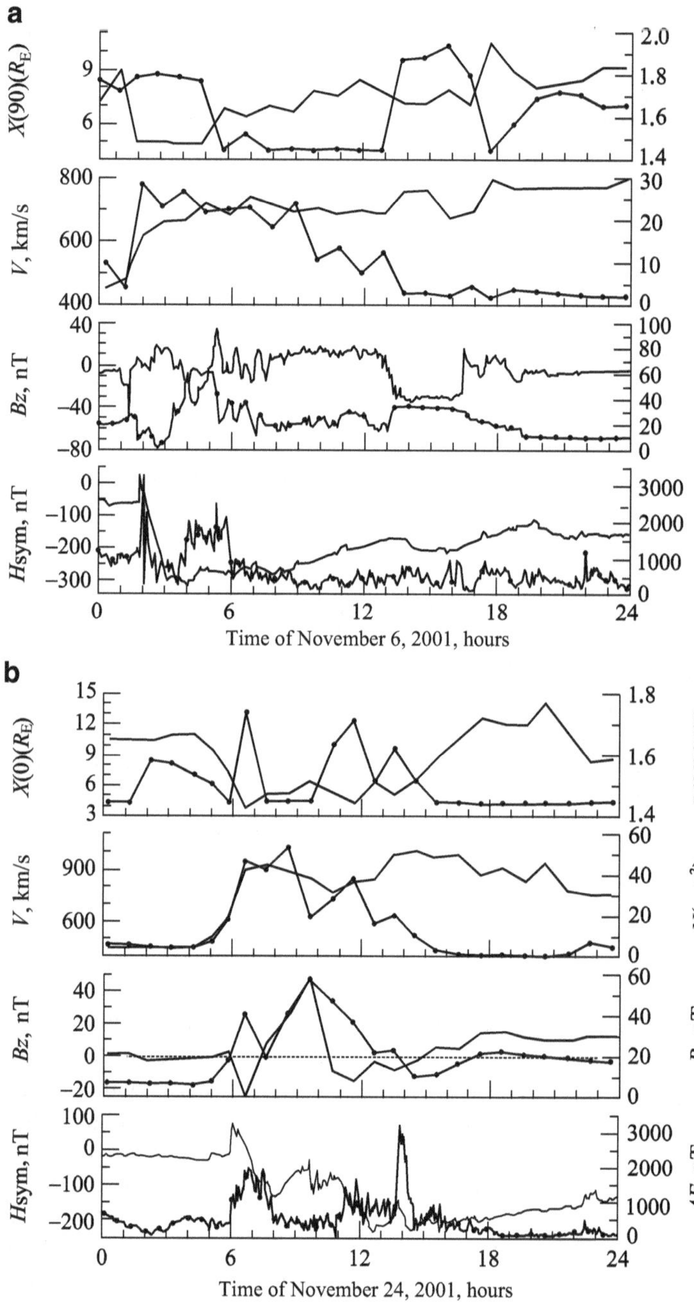


**Fig. 1** Disturbances in the magnetosphere and in the interplanetary space during the magnetic storms of November 6, 2001 (**a**) and November 24, 2001 (**b**), the legend is presented in text

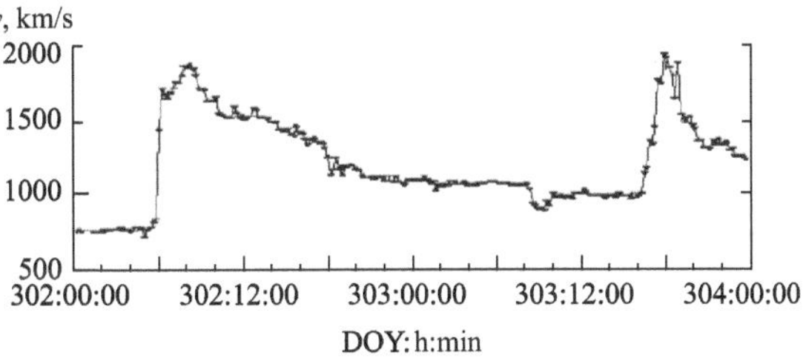


**Fig. 2** Speed of alfa-particles of the solar wind for October 29–30, 2003. From [7]

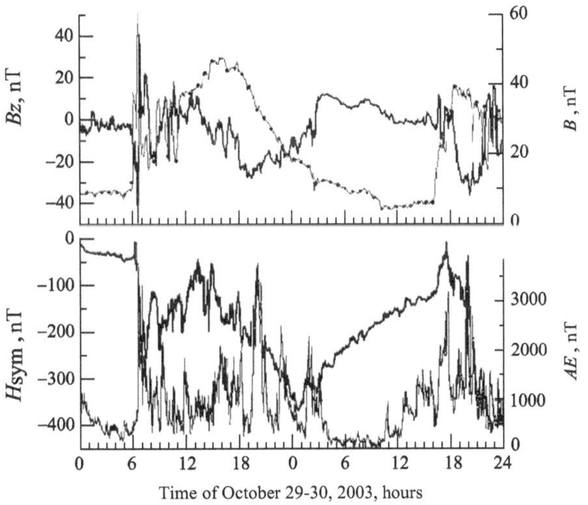


**Fig. 3** Disturbances of the magnetosphere and the solar wind during October 29–30, 2003 (*Bz* and *Hsym—thick lines*, *B—thin line with dots*, *AE—thin line*)

The last of the storms studied in the present chapter happened on May 15, 2005. It started at 02:38, when its sudden commence was recorded (see Table 1). In Fig. 4 parameters of the Earth's magnetosphere, solar wind, interplanetary magnetic field and geomagnetic indices *Hsym* and *AE* are presented similar to Fig. 1a,b. Unfortunately, there is no information about the speed and the density of solar wind for the moment of the storm's start. This storm differs from the rest studied in this chapter by long (approximately 3 h 40 min) first phase with slightly varied values of *Hsym*-index.

The first received information about the SW's parameters relates to the end of the first phase of the storm. It proves an assumption of coming of a high-speed SW flux with high density and positive value of *Bz*. Approximately at 06:00 *Bz* value changed to negative and in 20 min the main phase of the storm started. It finished with *Bz* sign change, approximately at 09:00. During this *Bz* sign change $X(0)$ increased from $5R_E$ up to $8R_E$. The moments of magnetic substorms' appearance relate to reduction of magnetosphere after the sudden commence of the magnetic storm, beginning of the main phase and recovery phase of the storm. In general this storm conforms the classic description.

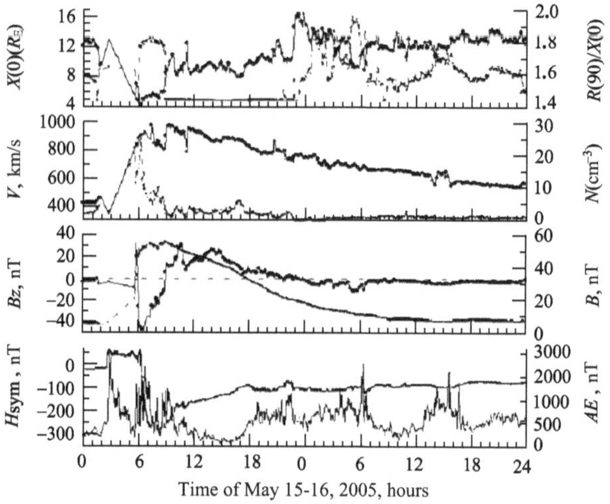


**Fig. 4** Disturbances of the magnetosphere and in the interplanetary space during the magnetic storm of May 15, 2005 (legend and the order of the parameters are the same as in Fig. 1)

## Dynamics of the Outer Radiation Belt

Now we'll review the dynamics of the ERB electron flux during the above-mentioned magnetic storms in detail. As far as during the magnetic storms of November 24, 2001 and May 15, 2005 the first phase of the storm was well defined we'll compare the variations of the outer ERB during these two events (see Fig. 5a, b). In these figures the values of the electrons of energy from 300 keV up to 6 MeV at different $L$ during the first phase of the storms of November 24, 2001 (Fig. 5a) and May 15, 2005 (Fig. 5b) are denoted by a solid line. Dashed line specifies the background values of electron flux for 23.11.2001 and 14.05.2005, respectively. It's clearly seen that for the both events at the first phase of the magnetic storm relative electron flux increases at $L > 3,5$–4.

The nature of the observed variation of ERB electron flux can be explained in a following manner. Electrons recorded at the altitude of the CORONAS-F satellite's orbit (400–500 km) at $B = 25\,000$–$40\,000$ nT, could not be essentially accelerated at the increasing of the field for 50–100 nT. The particles reflected near the equator plane are accelerated. Flux of the particles with energy of tens–hundreds keV exceeds the threshold of cyclotron instability and pitch-angle diffusion of the particles rises. Electrons of several MeV energy reflected at low altitude turn out to be at parasitic resonance with electromagnetic radiation of electrons of $\sim$ 10–$10^2$ keV energy. Analysis of the information on the SCR electrons penetration into the polar cap has shown that the penetration boundary for such electrons is situated at $L \sim 20$ and higher; therefore, they can't be a reason for this increase.

According to the presented in Fig. 6 comparison of the outer belt's structure before the storm, at the end of the main phase and day after the end of the storm of November 24, 2001 belt's recovery started from $L \sim 3$—the region of the ring

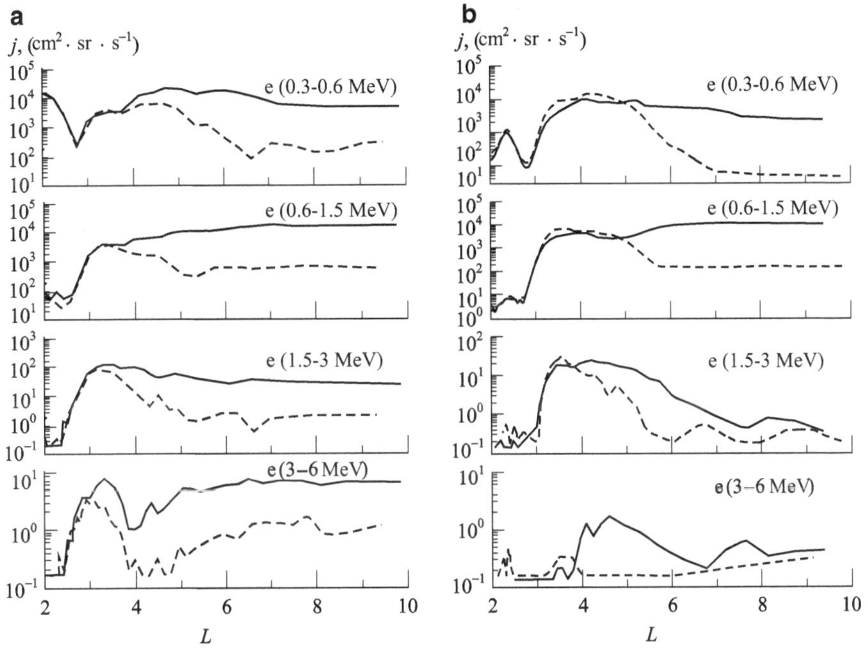


**Fig. 5** Comparison of the increases of electron flux of 300 keV–3 MeV energy during the first phase of the storms of November 24, 2001 (**a**) and May 15, 2005 (**b**). Background values of the electron flux are denoted with *dashed line*

current forming. As seen from Fig. 6 the day after the storm relativistic electron belt ($> 1,5$ MeV) was still absent. It should also be noted that first of all after the storm the electron belt of energy from several hundreds keV up to several MeV develops at $L \sim 3$ and recovers at $L \sim 4$.

The contour lines of common logarithm of the electron belt for the period of November 4–30, 2001 are presented in (time—$L$)-coordinates in Fig. 7. Previously it was published by Kuznetsov et al. in the paper [5]. $L$-axis for invariable latitude $\Lambda$ is chosen linear in order to reproduce the regions of the inner ($L \leqslant 2.5$) and outer ($L > 2.5$) belts sharply. It is evident that till November 6 electrons with energies 0.3–3 MeV formed the outer radiation belt with a maximum at $L \sim 4$, and the slot between the belts situated at $L \sim 2.5$.

Figure 7 also shows that during 17 days elapsed from the storm of November 6 by the storm of November 23 the following changes have happened. Electrons with energies of 0.3–0.6 MeV formed a classical outer belt with a maximum at $L \sim 4$, which to all seeming appears as a result of electrons diffusion from the capture boundary. Slow increase of electron flux of the outer belt can be related to the acceleration of electrons at the outer boundary of the belt caused by magnetic pumping [1, 8]. The upper panel of Fig. 7 also shows that during the quiet period after the storm of 06.11.2001 relativistic electrons with energies of 3–6 MeV formed

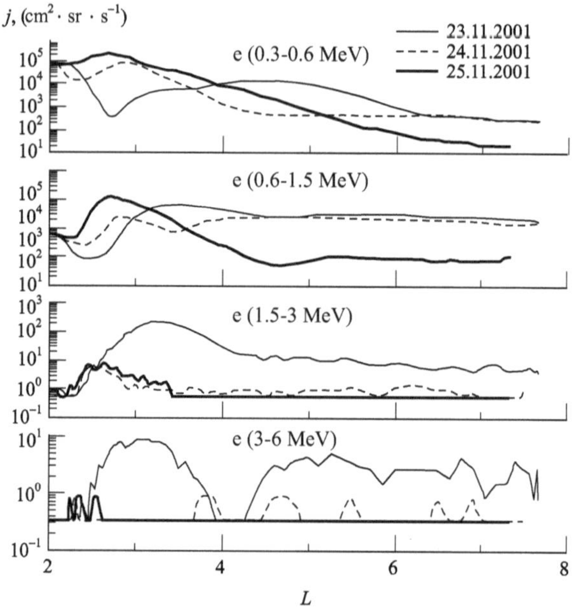


**Fig. 6** Comparison of the outer belt's structure before the storm, at the end of the main phase and a day after the storm of November 24, 2001

a belt with a maximum at $L \sim 3$ presumably resulted from acceleration of the electrons injected during the storm.

After the November 24 storm the maximum of the electron belt with energies of 0.3–3 MeV was observed significantly closer to the Earth, at $L \sim 3$. By November 30 the maximum of the outer ERB flux of electrons with energies of 0.6–1.5, 1.5–3, and 3–6 MeV was formed at $L \sim 2.7$. It is interesting that the flux of electrons with energies of 0.6–1.5 MeV has a structure with three maximums at $L = 1.5$; 1.9; and 2.7.

Dynamics of the belt during the October, 2003 storms can be evaluated according to the experimental data presented in Fig. 8. Information recorded during the daytime of MLT in the South-Atlantic Anomaly is presented in the left-hand panel, in the evening time of MLT—in the right-hand panel. Dashed line indicates the data related to a previous day, thin lines—electrons of 0.3–0.6 MeV energy, thick lines—electrons of 1.5–3 MeV energy. Information recorded after the main phase of the first storm is presented in the left-hand upper panel. Data related to the end of the main phase of the second storm is shown in the right-hand panel. Figure shows that the belt's size decreases and maximum of the belt moves to the lower $L$.

The second upper left panel where the information related to the recovery phase after the main phase of the second storm shows expansion of the capture area. The second upper right panel presents information obtained at the end of the main phase of the third storm. It is clearly seen that the size of the belt decreases and maximum shifts to the lower $L$-shells again.

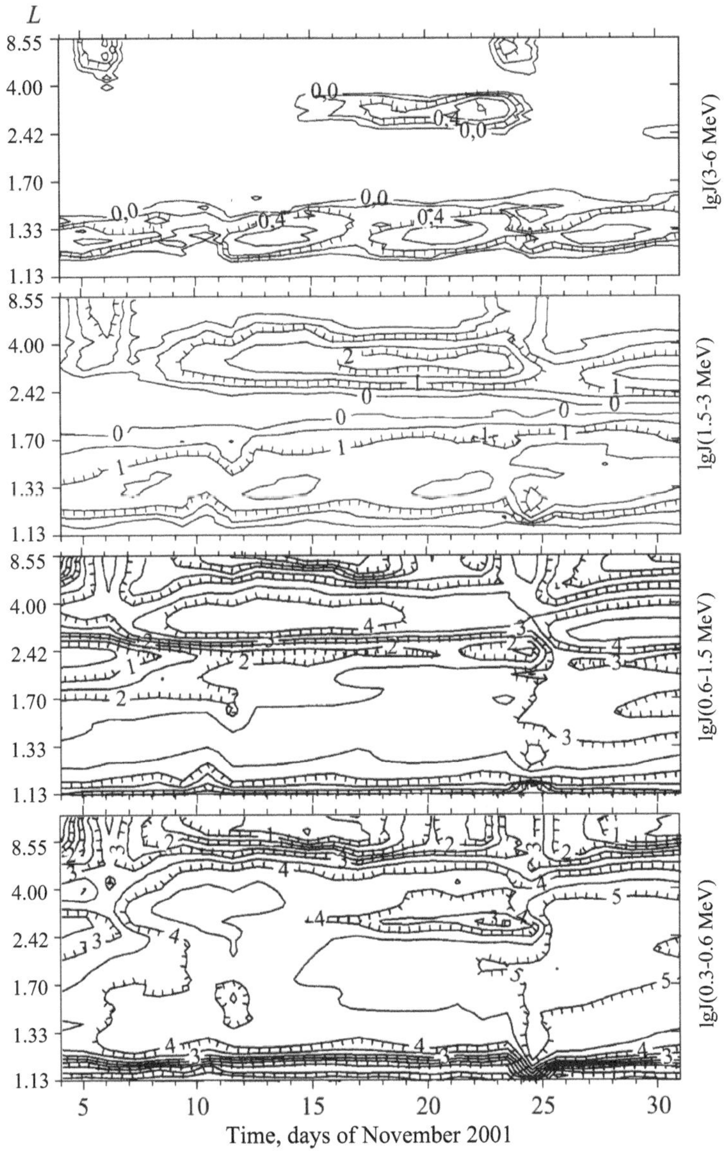


**Fig. 7** Variation of the structure of the electron belt during the storms of November 4–30, 2001. The *perpendicular dashes* at the contours show the direction of the decrease of the particle flux intensity. From [4]

In the lower panels the structure of the belt at the developed recovery phase is presented. Expansion of the belt of electrons with energies of hundreds keV at $L \sim$ 5.5–6 and recovery of relativistic belt recovery are evident.

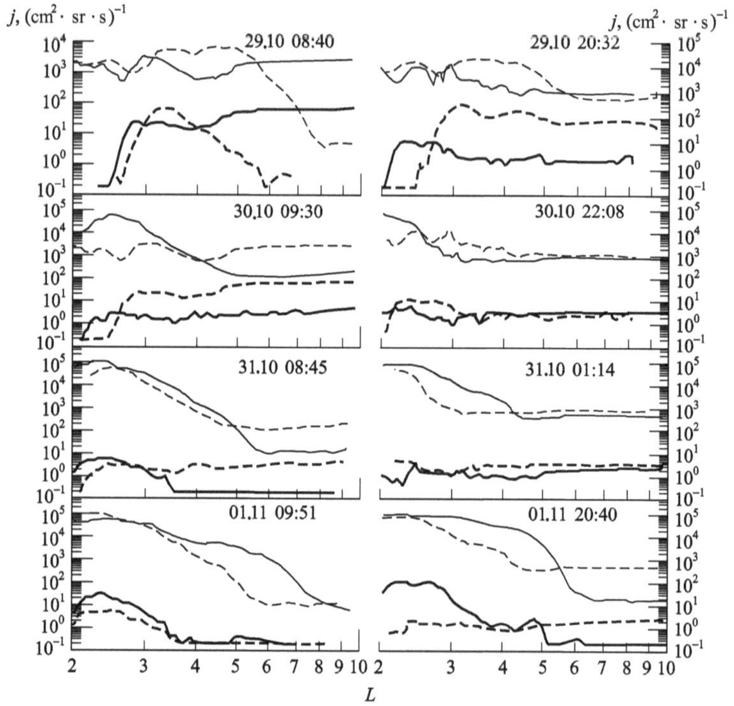


**Fig. 8** Variations of the outer belt structure during the magnetic storms of October 2003. At the left-hand panels—information obtained in the daytime, at the right-hand—data obtained in the evening

## Conclusions

The studies of the variation of electron flux for energies of 0.3–6 MeV happened in the outer ERB during a number of the most strong magnetic storms from August 2001 through May 2005 led to conclude the following:

1. During the first phase of the magnetic storm electron flux in the outer belt increases. In our opinion it is related to the acceleration of the particles in the equator plane and to the increase of their flux over the level of cyclotron instability.
2. At the main phase of the magnetic storm the size of the magnetosphere and of the outer belt decreases, polar cap increases, and relativistic electron belt actually disappears. This time the size of the closed L-shells decreases and ring current is formed.
3. During the recovery phase two regions can be separated in the area of the outer belt: electron flux at $L \sim 3$ increases during several hours, electron flux at $L \sim 4.5$ increases during several days or even weeks.

K. Kudela wishes to acknowledge the grant agency VEGA, project 2/0040/13, for support.

## References

1. Bakhareva, M.F.: Nonstationary statistical acceleration of relativistic particles and its role during geomagnetic storms. Geomagnetism Aeronomy **43**(6), 687 (2003)
2. Emelyanenko, S.P., Kuznetsov, S.N., Stolpovsky, V.G.: The outer radiation belt during the strong magnetic storm. Cosmic Res. **16**(4), 529 (1978) (in Russian)
3. Kuznetsov, S.N., Suvorova, A.V., Dmitriev, A.V.: Shape and size of the magnetopause associated with parameters of the interplanetary medium. Geomagnetism Aeronomy **38**(6), 697–702 (1998)
4. Kuznetsov, S.N., Myagkova, I.N., Yushkov, B.Yu., et al.: Dynamics of the Earth radiation belts during strong magnetic storms based on CORONAS-F data. Sol. Syst. Res. **41**(4), 338–347 (2007)
5. Kuznetsov, S.N., Yushkov, B.Yu., Kudela, K., Myagkova, I.N., et al.: Dynamics of the Earth's radiation belts during the magnetic storm of November 6th, 2001. Adv. Space Res. **36**(10), 1997–2002 (2005)
6. Li, X., Baker, D.N., Temerin, M., Cayton, T.E., et al.: Multisatellite observations of the outer zone electron variation during the November 3–4, 1993, magnetic storm. J. Geophys. Res. **102A**, 14123–14140 (1997)
7. Panasyuk, M.I., Kuznetsov, S.N., Lazutin, L.L., et al.: Cosmic Res. **42**, 489 (2004)
8. Summers, D., Ma, C.: A model for generating relativistic electrons in the Earth's inner magnetosphere based on gyroresonant wave-particle interactions. J. Geophys. Res. **105**(A2), 2625–2639 (2000)
9. Vakulov, P.V., Kovrygina, L.M., Mineev, Yu.V., Tverskaya, L.V.: Variation in intensity and spectrum of energetic electrons in Earth's radiation belts during strong magnetic disturbances. Space Res. **16**, 529–542 (1976) (in Russian)
10. West, A.I., Buck R.M., Davidson G.T.: The dynamics of energetic electrons in the Earth's outer radiation belt during 1968 as observed by the Lawrence Livermore National Laboratory's spectrometer on OGO-5. J. Geophys. Res. **86**, 2111–2138 (1981)
11. Williams, D.J., Arens, J.F., Lanzerotti, L.J.: Observations of trapped electrons at low and high altitudes. J. Geophys. Res. **73**, 5673–5696 (1968)

# Solar Protons in the Earth's Magnetosphere According to Riometric and Satellite Data During the Magnetic Storms of October 2003

L.L. Lazutin, S.N. Kuznetsov[1], Yu. Manninen, A. Ranta, S.N. Samsonov, A.V. Shirochkov, and B.Yu. Yushkov

Fluxes and boundaries of penetration of solar cosmic rays recorded by the CORONAS-F satellite during the super storms of October 2003 are compared with the measurements of riometric absorption by the world-wide riometer network. Dynamics of the polar cap boundaries is studied at different phases of the magnetic storms. Behavior of the absorption value depending on time of the day and solar protons spectrum is calculated for different phases of the solar cosmic rays flare.

## Introduction

Solar cosmic rays (SCR) with energies from few up to hundreds MeV generated during the chromospheric solar flares relatively free penetrate into the tail of the Earth's magnetosphere, which in ionosphere is projected to the polar cap. Solar protons essentially increase the ionization level in the $D$-layer of the ionosphere and it leads to the increase of radiowaves absorption within the frequency range

---

[1]Deceased

L.L. Lazutin (✉) · S.N. Kuznetsov · B.Yu. Yushkov
D.V. Skobeltsyn Institute of Nuclear Physics M.V. Lomonosov Moscow State University (SINP MSU), Vorobyevy Gory, Moscow 119991, Russia
e-mail: lll@srd.sinp.msu.ru

Yu. Manninen · A. Ranta
Sodankyla Geophysical Observatory, Finland

S.N. Samsonov
Yu. G. Shapher Institute of space research and aeronomy SB RAS, Yakutsk, Russia

A.V. Shirochkov
Arctic and Antarctic Research Institute, S.-Petersburg, Russia

V. Kuznetsov (ed.), *The Coronas-F Space Mission*, Astrophysics and Space Science Library 400, DOI 10.1007/978-3-642-39268-9_13,

of 10–50 MHz. This type of absorption is named a polar cap absorption (PCA), in distinction from analogous effects caused by auroral electrons [1–4].

Riometer continuously monitoring the level of space radionoise absorption is the most widespread instrument. Although riometric absorption does not carry any information about proton spectrum variations and often is "mixed" with absorption caused by auroral electron precipitation and radio-bursts both of natural (the Sun) and of anthropogenic origin, presence of a wide network of riometers allows to use its data in order to study time variations of SCR both separately and combined with the data of low-altitude satellites.

Previous comparison of the integral flux of solar protons and PCA has shown that proton flux is proportional to the absorption (dB) squared and that the basic input into the absorption is made by the protons with energies over 10–15 MeV [5, 6]. As SCR spectra vary both for different events and during each SCR flare and particles detectors characteristics are different for different satellites, the results of the comparison also differ essentially. Therefore for the quantitative specification it is necessary to conduct individual comparisons for a specific satellite.

The results of the comparison of PCA with solar protons measurements onboard the CORONAS-F satellite during a series of SCR flares and accompanying them extreme geomagnetic activity in October 2003 are reported in the present chapter. The main purpose of this work is to determine the coupling coefficients for the specific channels of the particles detectors onboard the CORONAS-F satellite and to find what proton energy ranges are the most effective concerning ionization of $D$-layer of the ionosphere. Besides, dynamics of proton flux and motion of the boundary of protons penetration into the Earth's magnetosphere can be monitored by satellite data and by longitudinal chains of riometers; comparison of these methods is of particular interest.

## Experimental Results

The period of the increased SCR intensity started on October 26 and lasted through November 6, 2003. Series of October storms and the processes on the Sun and in the interplanetary space, which caused them, were analyzed in detail in collective surveys by collaborations of Russian scientists [7, 8] and separate articles published in special issues of "Cosmic Research." The results of the analysis of time profile of SCR protons in three energy channels for the period of 26.10.2003–01.11.2003 are presented in Fig. 1.

The orbit of the CORONAS-F satellite is almost circular with inclination of 82.5° and altitude of 415–445 km. Detectors whose data was used in the present analysis are described in [9] and in the chapter "Scientific Set of Instruments ≪Solar Cosmic Rays≫" of the present book. The analysis is based on riometric data of longitudinal chain of Scandinavia: Abisko (International code ABK; corrected geomagnetic latitude 65.3°), Hornsund (HOR; 73.8°), Ivalo (IVA; 65.0°), Sodankyla (SOD;63.9°), Rovaniemi (ROV; 63.2°), Oulu (OUL; 61.5°) and Juviskyla (JYV; 58.8°) and Tiksi observatory (TIK; 65.65°) of the Yakutsk meridian. The frequency of riometers

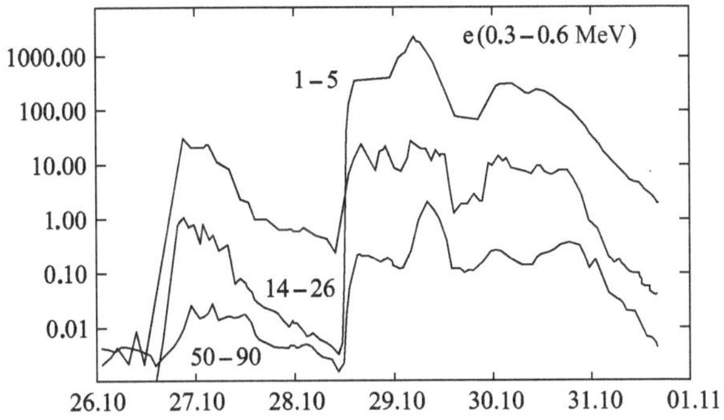


**Fig. 1** Time variation of solar protons for the period of October 26–November 1, 2003 (CORONAS-F). Numbers near the curves indicate $E_p$, MeV

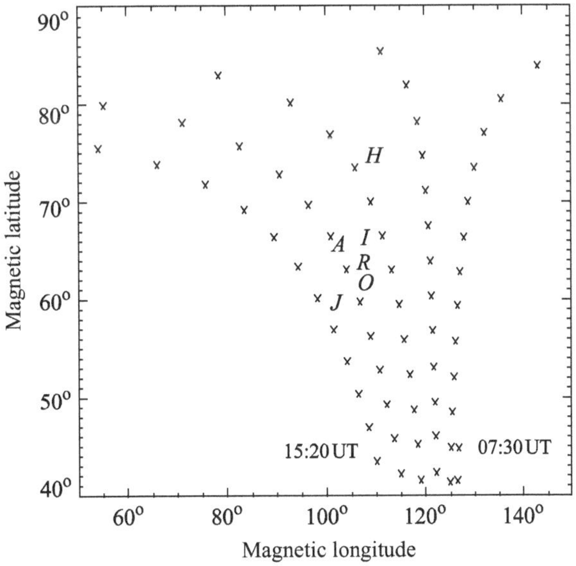


**Fig. 2** Relative position of Scandinavian chain of riometers (capital letters of the stations Abisko (*A*), Hornsund (*H*), Ivalo (*I*), Sodankyla (*S*), Rovaniemi (*R*), Oulu (*O*), and Jyviskula (*J*)). CORONAS-F satellite's trajectory is shown by crosses in magnetic system of coordinates on October 30, 2003. From [13]

receiver is 30–32 MHz (Sodankyla—25 and 40 MHz). Location of riometric stations of Scandinavian chain and several trajectories of the CORONAS-F satellite on October 30 in magnetic coordination system is presented in Fig. 2. It is seen that the satellite's trajectories are associated with location of the stations of Scandinavian sector and it is just eligible to compare data. Summary graph of riometric absorption at the Tiksi (Yakutsk network) and Sodankyla (Scandinavia) stations is presented in Fig. 3. Absorption variations indicate both real time dependence of proton flux and daily wave of absorption with maximum at day time (local noon in Sodankyla—at 10:30 UT, in Tiksi—at 03:40 UT).

As far as during the analyzed period the level of substorm activity was very high in order to analyze PCA variations it is necessary to separate input from auroral electrons precipitation (auroral absorption—AA) into total absorption. There is no standard procedure for it. The authors used morphological differences between PCA and AA: the latter are short and at night side coincide with bay-form disturbances of magnetic field. Difference of the morphological characteristics of these two

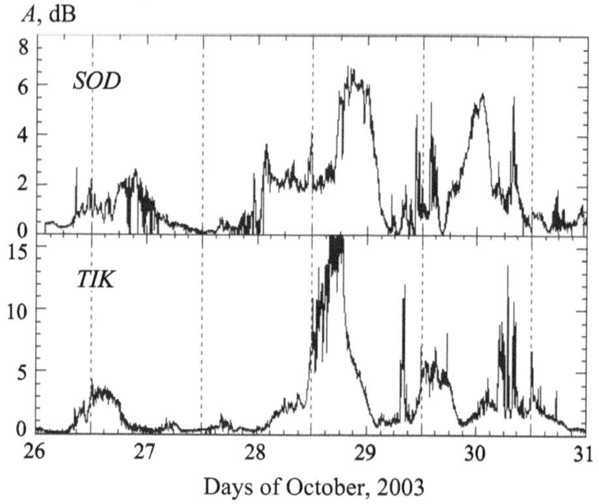


**Fig. 3** Riometric absorption in Sodankyla (*SOD*) and Tiksi (*TIK*). From [13]

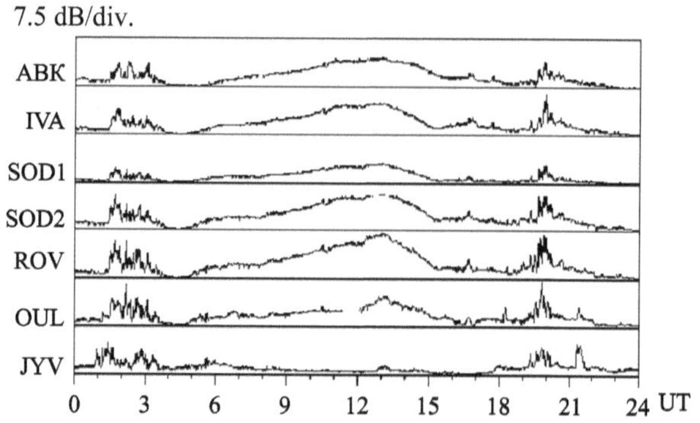


**Fig. 4** Riometric absorption measured by Scandinavian chain on 30.10.2003

types of absorption is seen in Fig. 4, where absorption registergrams of 30.10.2003 according to the Finnish stations chain are presented. At the night side auroral bays of absorption predominate, at the morning and the day sides—PCAs with marked variation of absorption depending on sun lighting. At daytime it is more difficult to separate AA because absorption bays are more smooth, they last up to several hours and are delayed in time relatively substorm activations at the night sector. Although additional selection criteria, i.e. time dependence of solar protons intensity recorded by the CORONAS-F satellite helps, nevertheless at separate moments identification of short-term variations source stays uncertain, for instance, during the main phase of the October 29 storm.

## Relation of PCA to the Solar Protons Flux

Analysis of physical processes which result in radiowaves' absorption in the ionosphere (see, for instance, [1]) predicts linear interconnections of the absorption value squared and proton flux $J_m$:

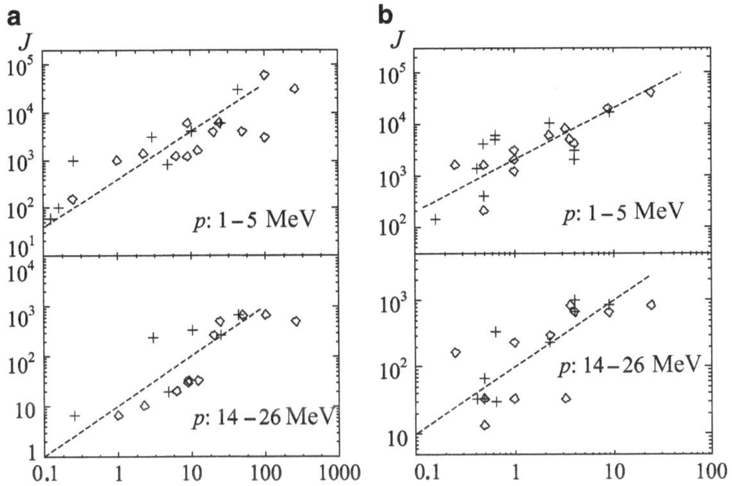


**Fig. 5** Relation of riometric absorption of polar cap type and proton flux measured by two channels of the CORONAS–F satellite's detector for the day (**a**) and night (**b**) sectors. *Crosses* indicate measurements in Scandinavian sector, *diamonds*—in the Yakutsk one. From [13]

$$J_m = K \cdot A^2, \tag{1}$$

where $A$—absorption (dB) at the frequency of 30 MHz, $K$—coefficient depending on the operation frequency of the riometer, antenna's parameters, proton spectrum, and state of the ionosphere.

Figure 5 presents the results of the comparison of riometric absorption and proton flux in the polar cap for the day and night absorption by our data. Accuracy of PCA's value calculation within the range of up to 1 dB is determined by the accuracy of the determination of diurnal variation of quiet level of radio-noise and can be less than 0.1 dB. But the level of noise increased during the last decades and complication of the account of electron precipitation into absorption result in a more essential dispersion of points. Taking into account the accuracy and possible mistakes in absorption calculations we'll note good coincidence of measurements made by two chains and proton flux measurements onboard the satellite. Dashed lines on the graphs are described with expression (1), coefficients $K$ for the day and night absorption and proton energy of 1–5 MeV are equal to 400 and 2000, and for the energy range of 14–26 MeV are equal to 10 and 100, respectively.

In most papers it is supposed that the main input into absorption is made by protons with energies over 10–15 MeV; therefore, good relation of PCA and proton flux of 1–5 MeV energy is rather unexpected. This result is not a simple consequence of synchronous variations of proton flux in general for the spectrum as a whole. Figure 1 shows that correspondence of count rate in different energy channels changes, and for several intervals time variations diverge. For instance, Fig. 6 shows

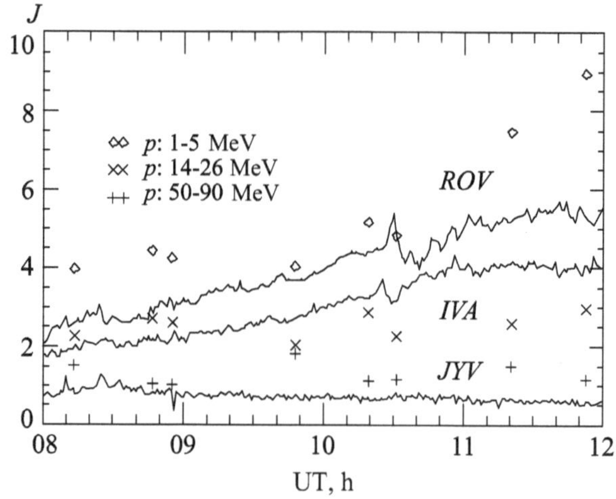


**Fig. 6** Time variation of absorption and proton flux on October 30, 2003. Indices near the curves indicate stations' international codes. From [13]

that for the time interval of 08–12 UT of October 30 absorption increase relates to the increase of proton flux within the energy range of 1–5 MeV, because for the rest channels flux does not increase. On October 29 during the period of 02–07 UT the same situation was observed. Character of proton variations diverges again: energy proton flux decreases, and low-energy proton flux increases. By riometer data absorption for these sections also increases, and it proves essential input of protons with energies of 1–5 MeV into absorption.

It should be noted that numeric dependence of the night PCA from solar proton flux was obtained for the first time due to extremely high intensity of the analyzed SCR flare. Among the earlier papers there is only one message about approximately five-fold decrease of absorption at the night sector comparing with the day one [1]. Above-mentioned dependencies show that on the average this ratio reaches 3.7 at the equal proton flux with energies of 14–26 MeV and reaches 2.3 at the equal proton flux with energies of 1–5 MeV. Difference of the ratio value is apparently related to the change of the shape of proton energy spectrum.

By means of riometric data it is possible to recover the time variation of proton flux only for separate sectors of day and nighttime, absorption at intermediate intervals is controlled by the angle of the Sun's elevation and other factors, which can be hardly taken into account. It is also necessary to dismiss the intervals of strong auroral activity, which for this time period were both numerous and intensive.

Comparison of the measured and calculated by riometric absorption time variations of proton flux during the beginning of the SCR flux increasing on October 28 is presented in Fig. 7. Calculated results for the high-latitude riometer (Abisko) is in good correspondence with direct measurements of proton flux. Diminished values of the flux calculated for two other riometers demonstrate their location at the boundary of proton penetration, where the flux is depressed significantly.

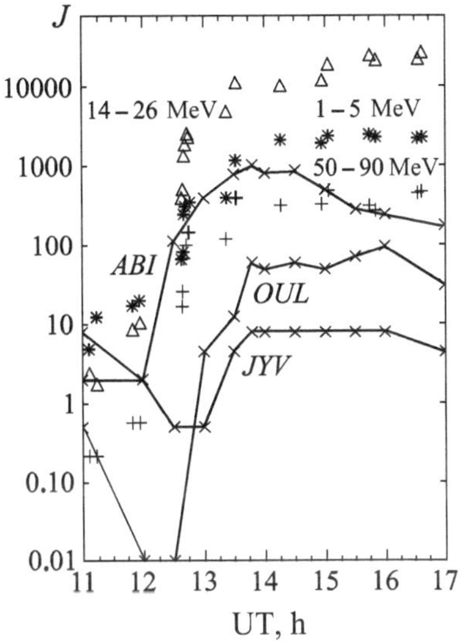


**Fig. 7** Comparison of proton flux with energies of 1–5 MeV calculated by riometric absorption for the stations of Abisko, Yuviskyla and Oulu (*solid lines*) with the values measured onboard the CORONAS-F satellite at the stage of SCR flux increase on October 28, 2003. From [13]

## Absorption at the Polar Cap Boundary

Penetrations of solar protons into the Earth's magnetosphere and, respectively, into the polar cap and auroral zone is regulated by the magnetic activity level. During the main phase of the magnetic storm the penetration boundaries significantly move to the equator (there were three such intervals during the studied period), and during the recovery phase—from the equator in alignment with *Dst*-variation [10–12].

Figure 8 presents the shift of the penetration boundary for the protons with energies of 2–4 MeV during the magnetic storms of October 2003 by records onboard the CORONAS-F satellite.

It is interesting to compare the character of proton flux decrease at the SCR penetration boundary by direct measurements and riometric data. Dependence of proton flux with energies of 14–26 MeV on magnetic latitude for two passes of the CORONAS-F satellite on October 28 during the time interval of 13:20–13:40 UT across the north polar cap and during the interval of 14:06–14:30 UT across the south polar cap is presented in Fig. 9. It can be seen that the day boundary is moved to the pole relatively the night one, which is analogous to the aurora oval shift. Besides, asymmetry of the location of the south and the north polar caps boundaries is observed: in the evening protons penetrate deeper at the north cap than at the south one, and in the morning—vice versa.

That time Scandinavian chain of riometers was at the day side of the Earth, where the value of the absorption is high, and it allowed to compare calculated values with direct measurements. In Fig. 9a x-coordinates of the letters defining the name of the stations correspond to the magnetic latitude of the stations, y-coordinates—to the calculated value of the proton flux with energies of 14–26 Mev according

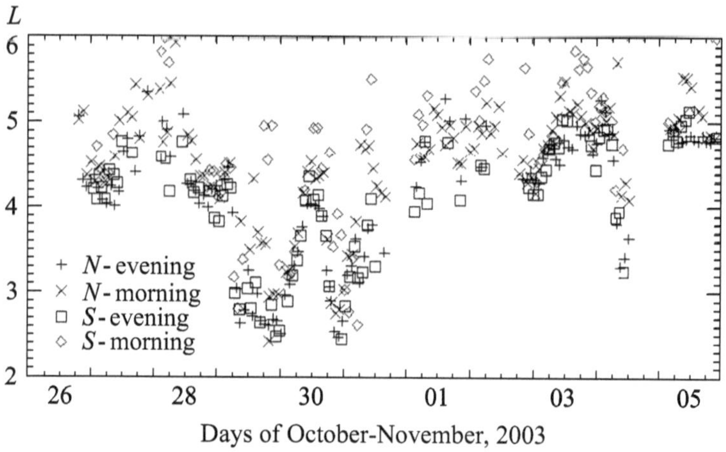


**Fig. 8** Dynamics of SCR penetration boundary during 11 days of extreme magnetic storms of October–November 2003 by records onboard the CORONAS–F satellite. Different signs indicate boundaries determined according to the evening and morning passes of the North (*N*) and South (*S*) polar caps. From [13]

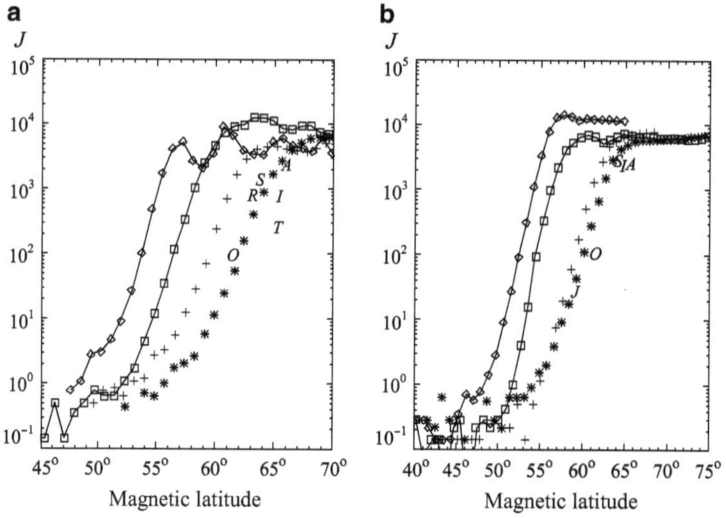


**Fig. 9** Latitudinal course of the solar protons penetration boundary and space radionoise absorption on October 29, 2003, at 14:00 (**a**) and on October 30, 2003 at 12:00 (**b**). The following signs are used: *solid lines*—evening-night passes (*diamonds*—north, *squares*—south), separate signs—morning-day passes (*stars*—north, *crosses*—south), *letters*—proton fluxes, calculated by riometric absorption, *T*—Tiksi, the rest legend see in Fig. 1. X-coordinate of the signs coincide with the magnetic latitude of the station. From [13]

to expression (1). Good coincidence of the latitudinal course of SCR penetration boundary with a profile calculated on the basis of riometer chain data at the day sector.

Yakutsk chain was that time at the night side. As seen from the figure the Tiksi station is situated in the zone of free penetration of the protons. But maximum value of absorption in Tiksi did not exceed 0,5 dB which gives the calculated value of the flux an order of magnitude lower than measured by the satellite's instruments both in the south and in the north hemispheres (letter *T*). Apparently, at that moment an asymmetry of the polar cap existed, but it was not observed at the pass of one of the satellites. Forty minutes later absorption in Tiksi increased up to the standard value, which is comparable with particles' flux measured in the cap.

Analogous graph for the passes of October 30 at about 12:00 UT is presented in Fig. 9b. That time it was a break in storms, so magnetic activity was limited. There is no difference between the location of the day boundaries of the south and the north hemispheres, and shift of the night boundaries kept with the same sign, although it diminished approximately to 2°. Riometric measurements are in a good agreement with direct measurements of protons. Relative location of Scandinavian chain of riometers and satellite's trajectory were advantageous for comparison which is seen from Fig. 2.

It should be noted that in this chapter we did not raise a problem of studying of penetration boundaries location at different longitudinal sectors. In order to compare them correctly it is necessary to use $L$–$B$ coordinates or invariant latitude. Using of geomagnetic coordinates is imposed by the task of comparison of direct measurements with riometric absorption.

## Conclusions

The period of intensive SCR flares during the strong magnetic storms of the end of October—beginning of November, 2003 provides a good opportunity for the studies of the relation between riometric absorption of polar cap type and direct measurements of solar protons. The conducted analysis allowed to confirm formerly found regularities and to show up new ones.

1. PCA in daytime (Sun-lighted ionosphere) is associated with proton flux by quadratic law. Coupling coefficients for absorption dependence of proton flux measured by the instruments onboard the CORONAS-F satellite were found.
2. It was shown that not only protons with energies over 10–15 MeV but also protons with energies of 1–5 MeV contribute essentially to absorption. In two cases when time course in the low channel diverged from the higher channels riometric absorption followed the variations in the low-energy channel.
3. Numeric relation between the proton flux and PCA absorption at the night (unlighted) side of the Earth was found. It was shown that ratio of the values of the absorption at the day and the night parts of the polar cap at equal solar proton flux varies within the range of 2–4.

This work was supported in part by the grants of INTAS (Grant 03-51-5359) and RFBR (Grants 03-05-65670 and 06-05-64225).

## References


1. Driatsky, V.N.: The nature of abnormal absorption of space radiation in the lower ionosphere of high latitudes. — L.:Hydrometeoizdat. 1974 (in Russian)
2. Hargreaves, J.K., Shirochkov, A.V, Farmer, A.D.: The polar cap absorption event of 19–21 March 1990: recombination coefficients, the twilight transition and the midday recovery. J. Atmos. Terr. Phys. **55**, 857–862 (1993)
3. Ranta, H., Yamagishi, H., Stauning, P.: Twilight anomaly, midday recovery and cutoff latitudes during the intense polar cap absorption event of March 1991. Ann. Geophys. **13**, 262–276 (1995)
4. Lazutin, L.L., Driatsky, V.N., Borovkov, L.P.: The absorption of polar cap as an indicator of solar cosmic rays. Proceedings of the Int. seminar on space res. 1969. — L.: USSR AC. p. 68 (in Russian)
5. Patterson, J.D., Armstrong, T.P., Laird, C.M., et al.: Correlation of solar energetic protons and polar cap absorption. J. Geophys. Res. **106A**, 149–163 (2001)
6. Croom, D.L.: Solar microwave bursts and polar cap absorption. Planet. Space Sci. **21**(4), 707–709 (1973)
7. Veselovsky, I.S., Panasyuk, M.I., Avdyusin, S.I., et al.: Solar and heliospheric phenomena in October–November 2003: causes and effects. Cosmic Res. **42**(5), 435–488 (2004)
8. Panasyuk, M.I., Kuznetsov, S.N., Lazutin, L.L. et al.: Magnetic storms in October 2003. Cosmic Res. **42**(5), 489–535 (2004)
9. Kuznetsov, S.N., Bogomolov, A.N., Gordeev, Yu.P., et al.: Preliminary results of the experiment on the SKL instrument onboard the CORONAS-I satellite. Bull. Russ. Acad. Sci. Phys. **59**(4), 2–6 (1995) (in Russian)
10. Darchieva, L.A., Ivanova, T.A., Sosnovets, E.N., Tverskaya, L.V. Dynamics of equatorial and polar boundaries of penetration of solar protons with energies of > 1 MeV into the magnetosphere during the strong magnetic storm. Geomagnetism Aeronomy **30**(5), 856 (1990) (in Russian)
11. Ivanova, T.A., Kuznetsov, S.N., Sosnovets, E.N., Tverskaya, L.V.: Dynamics of the low-latitude boundary of penetration of low-energy solar protons into the magnetosphere. Geomagnetism Aeronomy **25**(1), 7 (1985) (in Russian)
12. Kuznetsov, S.N., Yushkov, B.Yu., Denisov, Yu.I. et al.: Dynamics of the Boundary of the Penetration of Solar Energetic Particles to Earth's Magnetosphere According to CORONAS-F Data. Sol. Syst. Res. **41**(4), 348 (2007)
13. Kuznetsov, S.N., Lazutin, L.L., Manninen, U., et al.: Sol. Syst. Res. **41**(5), 433–438 (2007)

# Spectrometer IRIS: Investigation of the Time Structure and Energy Spectra of X-Ray Emission from Solar Flares

G.A. Matveev[1], P.B. Dmitriev, I.V. Kudryavtsev, V.P. Lazutkov, M.I. Savchenko, D.V. Skorodumov, and Yu.E. Charikov

## The Goal of the Research Program of the IRIS Experiment

The research program underlying the IRIS experiment (Russian acronym for Investigation of Solar X-Ray Emission) encompasses investigation of the characteristics of X-ray emission from solar flares and of their precursors in the 2–200-keV range, which would form a basis for development of physical models describing the process of buildup and explosive release of energy in a solar flare, as well as accumulation of experimental data to serve in devising techniques for forecasting the solar flare activity [23, 30].

Experimental data on the spectra and time structure of X-ray emission from solar flares contain direct information bearing on the processes involved in explosive energy release during a flare and an indirect one, on the parameters of the emitting flare plasma. Combined with observations of the Sun in the ultraviolet, optical, and radio-frequency ranges, they provide a basis underlying construction of physical models of solar flares. Particularly useful insight into the dynamics of solar flares can be gained from measurements of the characteristics of X-ray emission conducted with a high sensitivity and high time resolution. The IRIS spectrometer equipped with large aperture-area detectors and, therefore, detecting solar X-ray emission with a high sensitivity and high time resolution is capable of contributing valuable information in studies of low-intensity soft X-ray fluxes not only during solar flares but in the pre- and post-flare stages too, as well as of the fine temporal structure of hard X-ray fluxes emitted during the explosive phase of a flare.

---

[1]Deceased

G.A. Matveev · P.B. Dmitriev · I.V. Kudryavtsev · V.P. Lazutkov · M.I. Savchenko · D.V. Skorodumov · Yu.E. Charikov (✉)
A.F. Ioffe Physicotechnical Institute, Russian Academy of Sciences, St. Petersburg, Russia
e-mail: Yuri.Charikov@mail.ioffe.ru

V. Kuznetsov (ed.), *The Coronas-F Space Mission*, Astrophysics and Space Science Library 400, DOI 10.1007/978-3-642-39268-9_14,

**Table 1** Main physical characteristics of the IRIS instrument

| Characteristic | Soft X-rays | | Hard X-rays | | |
|---|---|---|---|---|---|
| Energy range (keV) | 2.0–15 | | 15–250 | | |
| Detector | Two four-section proportional counters (xenon 1 atm, beryllium 130 μm) | | Two scintillation counters (diam. CsI(Na)—80 mm) | | |
| Effective area ($cm^2$) | ∼10 and ∼0.1 | | ∼100 | | |
| Flux sensitivity ($erg\,cm^{-2}\,s^{-1}$) | $\sim 10^{-6}$ | | $\sim 10^{-7}$ | | |
| Time resolution (s) | 2.56 | 1.0 | 2.56 | 1.0 | 0.01 |
| Number of energy channels | 6 | 32 | 6 | 32 | 4 |

## The Purpose and Main Physical Characteristics of the IRIS Equipment

IRIS, a solar X-ray spectrometer (see Table 1 for the physical characteristics of the instrument), is intended for recording energy spectra of the soft and hard X-rays emitted in solar flares in the energy range of 2–250 keV [11, 22, 23]. In the soft X-ray range, the background radiation of the Sun is detected too.

The time resolution of the IRIS instrument depends on the actual operating mode of the equipment.

At background-level intensities of the X-ray emission, the instrument operates in the "patrol" mode with a time resolution of 2.5 s. The information accumulated in this mode is loaded into the on-board telemetry system (RTS).

On detection of a fast rise of radiation intensity above the background level, the instrument switches over to the "burst" mode, to measure, in addition to the patrol information, 64-channel, 2–150-keV spectra with a resolution of 1 s, and four-channel, 24–180-keV spectra with a resolution of 10 ms. This information is stored in the science data collection system (SSNI).

The sensitivity of the IRIS equipment varies from $10^{-5}$ to $10^{-8}\,erg\,cm^{-2}\,s^{-1}$, depending on the energy range covered. The high sensitivity is needed to record soft radiation emitted from weak flares of GOES X-ray classes A and B, and flare precursors, as well as to study the fine time structure of hard X- ray fluxes with a 10-ms resolution. The sensitivity of the IRIS equipment to fluxes of the solar X-rays is determined by the effective detector area, measurement duration, and background radiation level. In the soft X-ray range, the background is a sum of the detector instrumental background, background of magnetospheric origin, and the background radiation of the Sun.

To detect soft X-rays, the IRIS instrument is provided by two identical proportional counters which can be connected to the measurement channel in turn. Each proportional counter contains four sections of the same geometry mounted a common gas volume. Two sections of the four are intended for detection of solar X-rays. In one of them, the beryllium entrance window is fully open to form the "large aperture," ~10 $cm^2$ in area, while in the second section, the window is stopped down by an aperture ~$10^{-1}$ $cm^2$ in area to produce a "small aperture." The third (background) section does not view the Sun and detects background radiation arriving from other directions. The fourth section is used for in-flight calibration of the instrument. It is protected from external radiation. This section is provided by an Fe-55 source of X-rays. With due allowance for the background, the sensitivity of the equipment at the $3\sigma$-level to soft X-rays is, with a 1 s accumulation time (the "burst" mode), ~$2{\cdot}10^{-5}$ erg/($cm^2$ s) for the small aperture, and ~$2{\cdot}10^{-6}$ erg/($cm^2$ s) for the large one. The sensitivity in the hard X-ray range is ~$2{\cdot}10^{-8}$ erg/($cm^2$ s).

For measurement of hard X-rays, the IRIS instrument is provided by two scintillation detectors representing CsI (Na) crystals arranged in the phoswich assembly, with a total area of ~$10^2$ $cm^2$. For their in-flight calibration and gain auto-stabilization, pulsed light sources based on high-stability LEDs were provided, whose performance was constant within a broad temperature range. The scintillation detectors could be turned on by command from the Earth, both simultaneously and separately.

To ensure proper stability of the physical characteristics of the proportional counters (gas-amplification factor and energy resolution) during prolonged operation at high levels of radiation, pure xenon, an inert gas (without quenching admixtures), was chosen for filling the counters. The counters were assembled by special technology, without the use of cemented joints. This ensures adequate purity of the operating gas for a long time. Prior to filling the counters with xenon, they are evacuated for a long time (a month and more) without interruption at a pressure of ~$10^{-6}$ mmHg, combined with heating at up to +70 °C.

The in-flight stability of the physical characteristics of the counters is maintained by a system of gain auto-stabilization, with spectra of the Fe-55 isotope supplied from the calibration sections used as reference. The stability of the performance of the proportional counters in the course of the experiment is monitored from the 32-channel calibration spectra stored in the SSNI.

In-flight monitoring of the performance of the scintillation counters is effected from the 32-channel spectra referenced against the calibration spectra. These spectra are also loaded periodically into the SSNI and used in the gain stabilization system of the scintillation counters in the IRIS instrument.

## Information Telemetered from the IRIS Instrument

In the course of operation of the CORONAS-F satellite, the most regular data flow from the IRIS instrument to the SSNI was observed to occur in the "burst" and "calibration" modes. Altogether, during the operation of the IRIS instrument in the orbit, 2,961 downlink sessions with the SSNI were effected (with 256 sessions performed in 2001, 865 in 2002, 758 in 2003, 609 in 2004, and 470 in 2005).

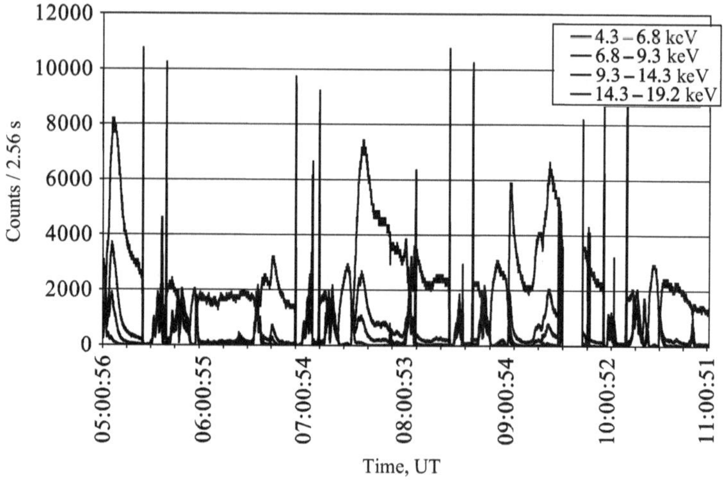


**Fig. 1** Time profile of the solar soft X-ray intensity measured in the patrol mode on August 22, 2001

On decoding, the information was translated into graphs relating the X-ray intensity to time. In this stage of information processing, the analysis of the session data amassed and preparation of the basic data for construction of the time profiles of X-ray emission intensity were performed with the use of the `vif.exe` program which transforms the SSNI codes into count rates in the various IRIS energy channels. The intensity profiles of the soft and hard X-rays were constructed with the `sx32clb.exe` code.

The time structures of the flares recorded by the IRIS instrument were derived from the 5.0–7.9-keV total-energy range. This was followed by a careful inspection of the data and selection of events of tentatively solar origin. The criterion underlying the selection was the pattern of the X-ray time profiles and the ratio of the radiation intensity in different energy ranges. Subsequent processing assessed the physical parameters of the flare region from the characteristics of the recorded X-rays; more specifically, one calculated the temperature and emission measure of the flare region, studied the periodic variations of the soft X-rays, and analyzed the spectral evolution of the hard X-rays in the course of the flare.

Patrol information telemetered to the Earth is contained in 398 communication sessions. The sessions cover the period from August 15, 2001 to August 31, 2004. Figure 1 exemplifies the intensity of soft solar X-ray emission recorded in the patrol mode.

This recording presents time structures of the solar flares of X-ray C6.2 class (maximum count rate was reached at 05:07 UT), class C4.2 (peak rate at 07:42 UT), and class C3.9 (peak count rate at 09:37 UT). The other periodic intensity peaks are connected with the satellite entering the Earth's shadow. As the satellite is entering the Earth's shadow, the radiation intensity is observed to fall off gradually, until it reaches the threshold level, at which the detector switches over automatically from the small to the large aperture. The processes reverse sense as the satellite exits the Earth's shadow. The short periodic bursts in the count rate detected in the hardest-energy channels of the instrument should be assigned to the satellite's entering the Earth's radiation belts. As evident from Fig. 1, the duration of recording of the soft solar X-rays for the region considered amounts to $\sim$60 % of the total measurement time.

The number of the flares recorded by the IRIS instrument in the "patrol" mode exceeds by far that observed in the "burst" regime. Indeed, in 8 days (August 16–23, 2001) the instrument recorded in the patrol mode 42 flares which coincided in time with the flares in the catalogue of X-ray events recorded on GOES satellites (http://spidr.ngdc.noaa.gov). In the "burst" mode, the number of recorded events is only about one tenth of the above, which should be attributed to limitations imposed on the volume of information from the IRIS instrument that could be loaded into the SSNI memory.

## Method Used to Study Quasi-Periodic Oscillations in the X-Ray Emission

A modified method of spectral analysis [13] was applied to the data obtained from the IRIS instrument to derive information required for a comprehensive analysis of the time structure of solar flare X-rays. The modification of the traditional spectral analysis consisted essentially in the following. Selected assessment of the normalized spectral density [16] for the original time series was calculated as a function not of the frequency but rather of a trial period, which followed from the formulation itself of the problem, namely, revealing a hidden periodicity in the original data. In addition, the original time series were subjected to preliminary high-frequency filtration [1] with a preset filter cutoff frequency at one half the signal power, to which in the time domain corresponds the length of the separating period $T_f$. Filtration of the original data is motivated by a wish to free them of a possible trend and higher-power low-frequency components. After this, for each series filtered with its specific $T_f$ parameter of the high-frequency component, one again estimated the normalized spectral density as a function of the period. All these estimates calculated for different values of the $T_f$ parameter were superimposed one on the other in the same field of the graph to form a combined spectral periodogram (CSP).

This modification of spectral analysis opens a way to studying the stability of the position of a detected period in a periodogram, i.e., verifying the independence of the revealed magnitude of a hidden periodicity on the parameters of the original time series which may affect the mathematical method of treatment chosen, and makes it possible to detect in the original signal shorter periods with small amplitudes. This is a result of elimination from the original signal of a trend and of more powerful large-period components which dominate signal dispersion. Therefore, the only contribution to dispersion of the filtered high-frequency signal component comes from weak, short-period components; normalization of spectral power makes this contribution to the combined periodogram comparable with that due to longer and stronger components of the signal.

An analysis of such periodograms raises immediately the question of whether the quasi-periodic components revealed in the latter are truly significant; said otherwise, what is the reliability of the positions of the peaks they show and of their amplitudes. The first of the questions involves the selectivity of the method and the magnitude of the step $\Delta T$ set in the periodogram on the axis of the trial periods $T$. In our case, for the CSP it is $\Delta T \approx 2 \times T^2/N$ [39]. For $N$, the length of the original time series equal to 644 points (i.e., the total length of the series expressed in units of the argument, i.e., the time, is $644\varDelta_t$, where $\varDelta_t$ is the step in argument), and a step $\Delta T$ in trial period $T$, which is equal to the same magnitude of $\varDelta_t$, the resolution in CSP below the trial periods $T \approx 18\varDelta_t$ will be insufficient, and above it, excessive. Therefore, the CSP peaks for small trial periods are more narrow than those for larger ones.

As for their magnitude, the significance of the maximum peak amplitudes in the corresponding periodograms of filtered components exceeds the significance level from $4\sigma$ to $7\sigma$, because the normalized spectral periodograms were constructed not only for the original data but for their high-frequency components as well (which were filtered with different separating periods $T_f$).

Figure 2 illustrates application of the above method to test data sequences, which contain both pure harmonic and noise components as well. Figure 2a illustrates a CSP constructed for a test signal consisting of one harmonic only: $Y(t_i) = \cos(2\pi/T_1 \cdot t_i)$, where $Y(t_i)$ are the values of the test sequence, $T_1 = 13\varDelta_t$ is the harmonic period, and $t_i = \varDelta_t \cdot i$ ($i = 1, 2, \ldots, 200$) are the values of the argument of the test function time set with a step $\varDelta_t = 1$ at 200 points of the region of identification. Shown in Fig. 2b are CSPs for the test signal drawn under the assumption that within the signal identification region the period of the harmonic component "floats," i.e., it is not constant and varies with time as $Y(t_i) = \cos(2\pi/T_1 \cdot t_i)$, where $T_1 = 13\varDelta_t$ for $t_i = \varDelta_t \cdot i$ ($i = 1, 2, \ldots, 49$); $T_1 = 16\varDelta_t$ for $t_i = \varDelta_t \cdot i$ ($i = 50, 51, \ldots, 99$); $T_1 = 19\varDelta_t$ for $t_i = \varDelta_t \cdot i$ ($i = 100, 101, \ldots, 149$); and $T_1 = 22\varDelta_t$ for $t_i = \varDelta_t \cdot i$ ($i = 150, 151, \ldots, 200$). Figure 2c visualizes a CSP for a test signal consisting of four harmonic components with equal amplitudes and different periods: $Y(t_i) = \cos(2\pi/T_1 \cdot t_i) + \cos(2\pi/T_2 \cdot t_i) + \cos(2\pi/T_3 \cdot t_i) + \cos(2\pi/T_4 \cdot t_i)$, where $T_1 = 13\varDelta_t$, $T_2 = 16\varDelta_t$, $T_3 = 19\varDelta_t$, and $T_4 = 22\varDelta_t$ for $t_i = \varDelta_t \cdot i$ ($i = 1, 2, \ldots, 200$). After this, a random series of numbers (numerical analog of "white noise") was superposed on these test series with a dispersion equal to that of the corresponding harmonic signal. The CSPs plotted for the noise-superposed test sequences are plotted in Fig. 2d, e, f, respectively.

We readily see that the harmonic component of the signal experiences the strongest perturbation at the change of the period (phase perturbation) in the region of signal processing (see Fig. 2b). By contrast, perturbation of this harmonic

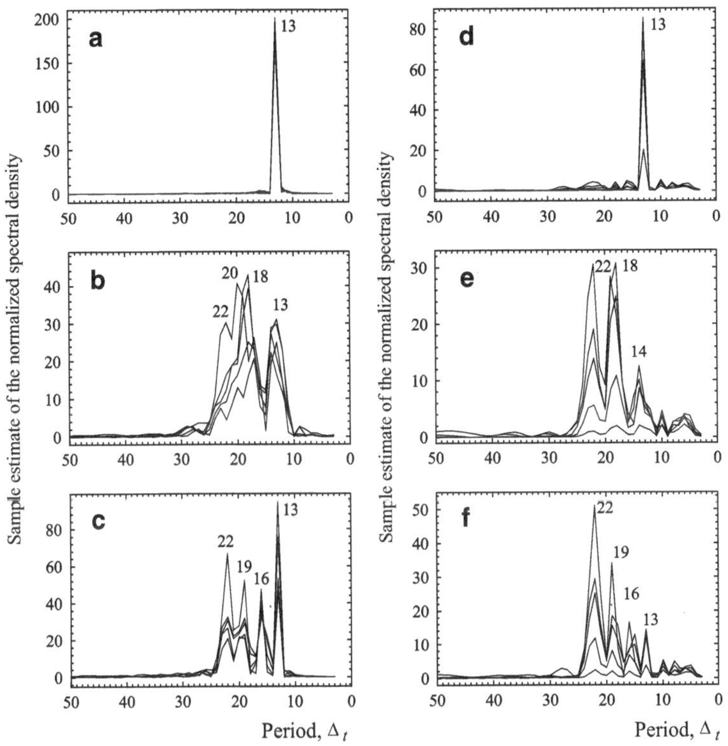


**Fig. 2** Application of the modified method of spectral analysis to test series of data containing pure harmonic and noise components: (**a**) CSP of harmonic signal with $T_1 = 13\Delta_t$ ($t_i = \Delta_t \cdot i$, $i = 1, 2, \ldots, 200$); (**b**) of harmonic signal with a "floating" period: $T_1 = 13\Delta_t$ ($t_i = \Delta_t \cdot i$, $i = 1, 2, \ldots, 49$), $T_1 = 16\Delta_t$ ($t_i = \Delta_t \cdot i, i = 50, 51, \ldots, 99$), $T_1 = 19\Delta_t$ ($t_i = \Delta_t \cdot i$, $i = 100, 101, \ldots, 149$), $T_1 = 22\Delta_t$ ($t_i = \Delta_t \cdot i, i = 150, 151, \ldots, 200$); (**c**) CSP of a test signal consisting of four harmonics with equal amplitudes and different periods: $T_1 = 13\Delta_t$, $T_2 = 16\Delta_t$, $T_3 = 19\Delta_t$, $T_4 = 22\Delta_t$ ($t_i = \Delta_t \cdot i, i = 1, 2, \ldots, 200$); (**d–f**) CSP of test harmonic signals **a**, **b**, **c** (respectively), with superposed "white noise" with dispersion equal to that of the corresponding test harmonic signal. Filtration of the high-frequency component for the test series was performed with the parameter $T_f = 7, 13, 19, 23\ \Delta_t$

component by "white noise" with the dispersion of the same magnitude as that of the harmonic proper is insignificant (Fig. 2d).

## Millisecond Time Structure of Solar Flares

In the course of in-flight operation, the IRIS spectrometer recorded intermittently hard solar-flare X-rays with a fine time structure [8, 9, 13, 29]. We are going to consider here a particular example of the evolution in time of the flux of hard X-

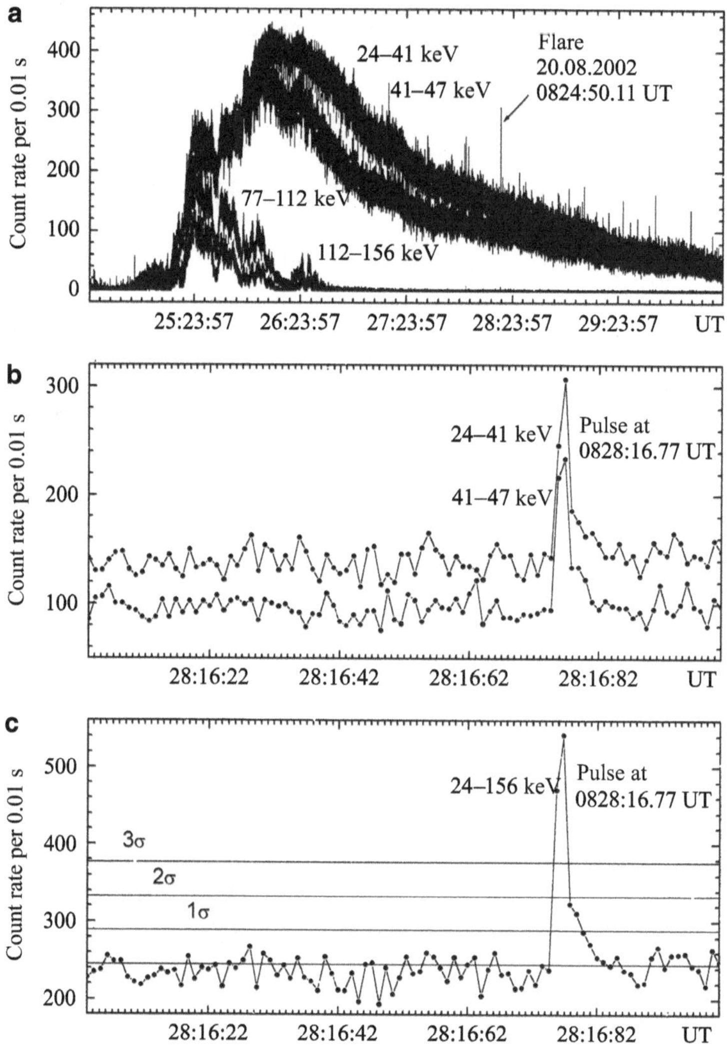


**Fig. 3** (**a**) Time profiles of the X-ray flare emission on August 20, 2002 and their fragments with the most powerful burst at 08:28:16 UT, (**b**) energy ranges 24–41 and 41–77 keV, and (**c**) energy range 24–156 keV. *Horizontal lines* refer to the significance of the burst at levels from $1\sigma$ to $3\sigma$

rays which were emitted from a flare recorded on August 20, 2002 at 08:24:50 UT. Figure 3a visualizes time profiles of the flare in four energy channels (keV)—24–41, 41–77, 77–112, and 112–156.

As evident from Fig. 3a, the total duration of the X-ray emission from the flare amounted to ~5 min in the 24–41 and 41–77 keV channels, and 1.5 min in the more energetic channels. The optical importance of the flare was 1 B, its coordinates

S 10, W 38, and the X-ray class M 3.4. Thus, this flare is powerful enough over all the wavelength range covered. The time profiles are seen to consist of a smoothly varying "pedestal" superposed by numerous fine structure bursts. The burst structure was particularly pronounced in the decay phase of the radiation, although this feature is not characteristic of all flares. Examine in more detail the most powerful X-ray burst recorded at 08:28:16.77 UT (Fig. 3b, c). This burst was detected only in two energy channels, 24–41 and 41–77 keV. By that time, emission of harder radiation had stopped completely (Fig. 3a). As seen from Fig. 3b, the count rate in the burst is extremely high, reaching 300 counts in 0.01 s, while that of the "pedestal" does not exceed 150 in the same time interval. The count rate at the peak exceeds the $6\sigma$ significance level (Fig. 3c), and the time profile is plotted by eight points separated by 10.49 ms intervals. All these features argue for the physical nature of this radiation burst. The time profile is slightly asymmetric, with the growth phase lasting for 20 ms and being shorter than the decay, which is 60 ms long. Note, however, that the last four time points may belong to the pedestal rays. One should therefore analyze not the total pulse length (80 ms) but rather the duration at half maximum count rate, which is 30 ms. A similar burst was recorded earlier on the SMM satellite [19]. There is no sense to speak here about the energy spectrum of this burst, its having been detected in two energy channels only. The same flare produced a number of weaker bursts with count rates above the $3\sigma$ level, and duration at half maximum of hundreds of milliseconds. Similar bursts were recorded in the rising phase of the flare emission, at the peak, and in the decay stage.

Observation of a millisecond-range spike structure provides information on the duration of the process in which an X-ray burst forms, i.e., on the electron acceleration time in the course of evolution of the flare process. As another example, Fig. 4 demonstrates time profiles and power spectra of the X-rays emitted in the August 26, 2002 flare.

The different fragments presented (Fig. 4b–d) relate to different time intervals of the flare emission.

We note that while a fine time structure with individual $\sim$20-ms long bursts was recorded in all stages of the flare evolution, the most reliable are the bursts measured in the region of the maximum, which occurred at 41:22:34 UT and 41:23:09 UT (Fig. 4c). The count rate in these bursts exceeds the $3\sigma$ level and the profile is traced with 6–7 points. The spectral analysis of individual time fragments performed in different stages of the flare emission suggests a conclusion that acceleration of charged particles during the flare is quasi-periodic. Figure 4e–g illustrate the results obtained in an attempt at revealing quasi-periodic components in the 24–156-keV X-rays by the above technique. As shown earlier (see Fig. 2), an analysis of a spectrogram should include both single peaks corresponding to specific frequencies (the periods of harmonic components), and the groups of peaks closely lying in frequency. The separation of frequencies within such groups is most probably accounted for by the "main" period in this group varying during the event rather than by the existence in the group of a variety of harmonics. An analysis of the spectrograms displayed in Fig. 4e–g reveals quasi-periodic components with

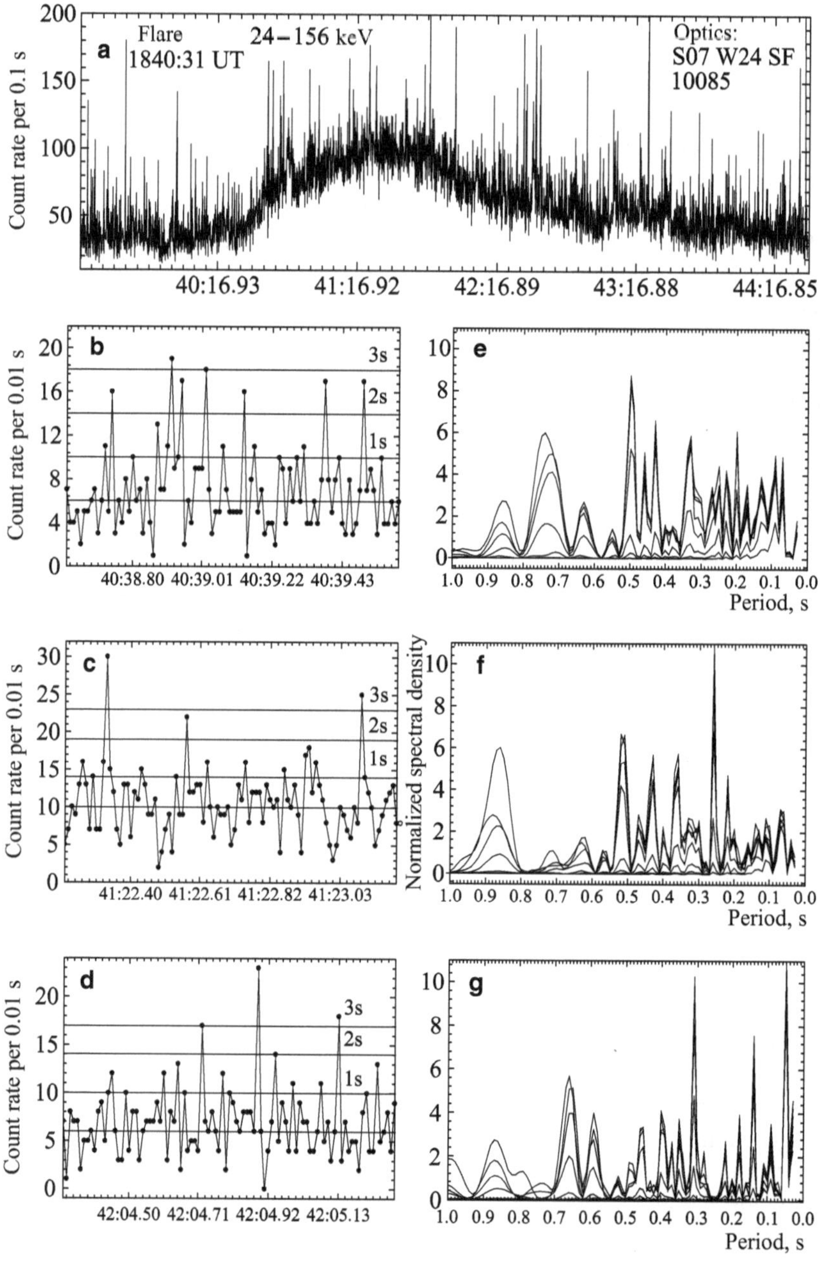


**Fig. 4** (**a**) Time profile of hard X-ray emission from the flare on August 26, 2002, and its fragments (**b–d**). *Horizontal lines* refer to the significance of the bursts at levels from 1$\sigma$ to 3$\sigma$. CSPs of some time fragments in the evolution of the flare emission in the 24–156-keV energy range: (**e**) 1840:36 to 1840:46 UT; (**f**) 1841:17 to 1841:27 UT; (**g**) 1841:59 to 1842:09 UT. Filtration of the high-frequency component for all time fragments was performed with the parameter $T_f = 0.07, 0.11, 0.19, 0.31, 0.57, 0.79, 0.93$ s

characteristic periods ranging from a few tens to hundreds of milliseconds in different stages of the flare development. Thus, the time sequence of hard X-ray bursts in this flare should be considered quasi-periodic with a characteristic period varying from tens of milliseconds to a second.

Let us discuss, necessarily briefly, the possibilities available for interpretation of the fine structure of hard X-ray emission. We note that millisecond-range structure is observed in far from all the flares. Indeed, as follows from numerous observations performed on the BATSE space station the peak in the time distribution of more than 5,000 bursts falls on the 300–1,000-ms interval (a summary of the latest results on recording of a fine structure in hard X-rays can be found in the review [2]). The most common among the time structures of hard flare X-rays appears presently to be that of flares with bursts about a few hundreds rather than tens of milliseconds long [4,5]. Single X-ray bursts 20–30 ms long are rather exclusions to the rule.

Processes involving the tearing-mode magnetic field reconnection are considered presently to be the most appropriate of the widely discussed mechanisms that could account for the fast energy release. The tearing mode as applied to solar flares was examined by many authors [20, 43, 44]. The build-up time of a long-wavelength tearing mode is given by the expression $\tau_{\mathrm{min}} = (\tau_{\mathrm{d}}\,\tau_{\mathrm{a}})^{1/2}$, where $\tau_{\mathrm{d}}$ is the time required for magnetic field to diffuse to the reconnection region, and $\tau_{\mathrm{a}}$ is the Alfven time expressed through the Alfven velocity. The time taken up by energy release initiated by tearing-mode instability can be estimated as ~0.5 s, if the plasma parameters in the flare region correlate with the mean values of the coronal plasma parameters, more specifically, if the particle concentration is $10^{10}\,\mathrm{cm}^{-3}$, temperature $2{\cdot}10^6$ K, magnetic field induction 200 G, and the characteristic current layer width $7{\cdot}10^3$ cm [20]. Tearing mode is known to produce filaments in the current layer, which, if located close to one another, are capable of coalescing [38] and releasing the larger part of the current layer energy in the process [31]. The characteristic coalescence time can be assessed through the scale of coalescence, Alfven velocity, and the velocity with which the current layers approach one another [38]. For the above coronal conditions, estimation of the coalescence time yields 0.2–2 s, i.e., values of the same order of magnitude as the tearing instability. Most probably, both processes, namely, the tearing-mode instability and fiber coalescence, are interrelated and culminate in explosive reconnection [21,31,37,38]. Tearing instability brings about formation of many fibers, which, in their turn, are unstable against subsequent tearing instability, and so on. This chain of events initiates modulation of electron acceleration on the subsecond time scale, which are detectable in hard X-ray emission. The duration of single X-ray bursts on the order of 10 ms long can be interpreted by properly varying the parameters in the electron acceleration region or by revising the presently accepted physical concepts underlying the mechanism of energy release and electron acceleration in solar flares.

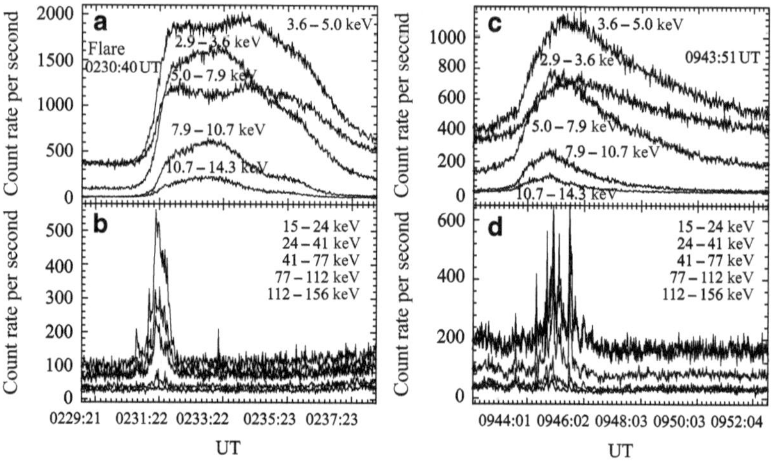


**Fig. 5** Time profile of soft and hard X-ray emission of solar flares on December 19, 2001 (**a**, **b**) and August 10, 2002 (**c**, **d**)

## Dynamics of the Energy Release Process in a Flare with Second-Scale Resolution

Evolution of the process of energy release on time scales from 1 s and up during a solar flare is exemplified by the solar events recorded by the IRIS X-ray spectrometer on December 19, 2001 (Fig. 5a, b), August 10 (Fig. 5c, d), and October 29, 2002 (Fig. 7), with their time structure and hard X-ray spectrum being of prime interest.

The December 19, 2001 solar flare (Fig. 5a, b) started at 02:30:40 UT with a count rate peak in the energy channels with energies above 15 keV. The flare radiation was recorded in all 64 instrument channels, but in order to increase the statistical significance of the results of the data treatment, the count rates of 32 channels in the hard X-ray range were combined to make five broader energy ranges of 15–24, 24–41, 41–77, 77–112, and 112–156 keV, and those of 32 soft X-ray channels, combined to record count rates in the 2.9–3.6, 3.6–5.0, 5.0–7.9, 7.9–10.7, and 10.7–14.3 keV windows. The count rate for the December 19, 2001 flare in the 3.6–5.0 keV window (Fig. 5a) can be identified with the flare GOES X-ray class C 4.9. The time profiles of both the soft and the hard X-rays exhibit bursts separated by about 10 s. These bursts become most revealing in the 24–41 keV channel (Fig. 5b), with the flare having started exactly with an explosive liberation of energy at 02:30:40 UT. Subsequent growth of the count rate followed by its decay were accompanied by the appearance of single peaks. The dynamics of evolution of the hard X-rays was dominated by generation of single overlapping bursts with relatively shallow drops, and that of the soft X-rays, by superposition of continuous radiation and single bursts, the burst component revealing in this case much smaller drops than those observed in the case of hard radiation.

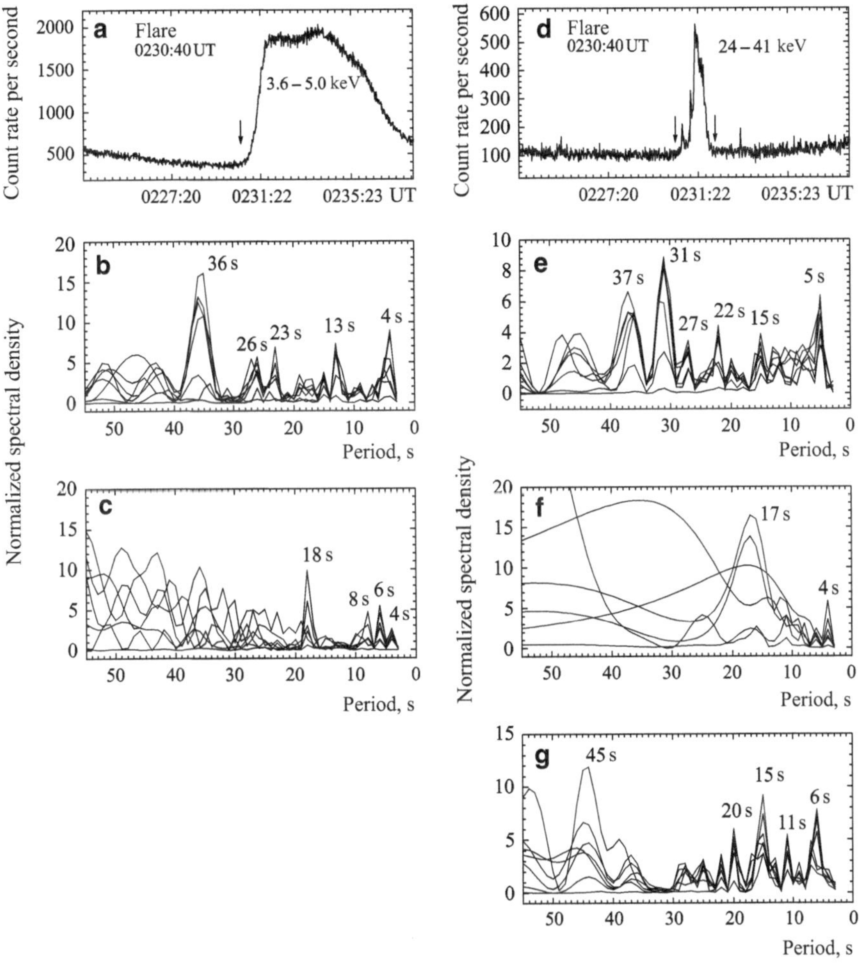


**Fig. 6** December 19, 2001 solar flare. Count rates in the energy ranges: (**a**) 3.6–5.0 keV, (**d**) 24–41 keV. CSP calculated for: (**b**) pre-flare emission, (**c**) flare emission in the 3.6–5.0 keV range; and (**e**) pre-flare, (**f**) flare, (**g**) post-flare emission in the 24–41-keV range. Filtration of the high-frequency component for all flare stages was performed with the parameter $T_f = 13, 29, 53, 73, 97$, and 113 s

A similar time structure is observed in the August 8, 2002 flare as well (Fig. 5c, d), but in this case the drops of single bursts of hard X-rays are considerably deeper. Here single bursts with larger drops in between follow one another every 15–20 s.

The results obtained by application of the above technique of revealing hidden periodicities to the December 19, 2001 event are shown in considerable detail in Fig. 6. CSP graphs for the count rate in the 3.6–5.0 keV channel and of the filtered high-frequency components with the parameter $T_f = 13, 29, 53, 73, 97$, and 113 s are displayed in Fig. 6b, c. The first of the panels visualizes power spectra of the pre-flare stage 6 min long (the end of the interval is identified by an arrow). In this stage, one immediately sees a powerful component with a period of 36 s and several weaker ones with periods of 4, 13, 23, and 26 s. During the flare, the structure of the radiation profile evolves, starting with 02:30:40 UT, to reveal components with periods of 6, 8, and 18 s, with the latter being the most powerful; the oscillations with the 4-s period persist, whereas the other components die out.

Standing out in the hard X-rays of 24–41 keV (Fig. 6d) in the pre-flare stage are the periodic components with periods of 5, 15, 22, 27, 31 (the most powerful), and 37 s (Fig. 6e), which coincide, to within the shortest time unit in the X-rays recording system of 1 s, with similar components observed in soft X-rays. During the flare, just as in the case of soft radiation, a dominant powerful component with a period of 17 s, and a weaker, 4-s one stand out (Fig. 6f). An analysis of the post-flare activities whose count rate is shown to the right of the arrow in Fig. 6d reveals the presence of fairly strong periodic components with 6, 15, and 45-s periods, and two weaker ones of 11 and 20 s (Fig. 6g).

Summing up the results of the above spectral analysis, it appears pertinent to note that the liberation of energy during the flare followed a quasi-periodic pattern with a period $\approx$ 18 s; interestingly, in the pre-flare stage this component was not detected. The pre-flare phase exhibits, however, the most powerful component with a period of 36 s, a dominant harmonic with respect to the 16-s oscillation appearing during the flare. Significantly, the 16-s periodicity appears simultaneously in both the soft and the hard energy ranges, which suggests possible explosive regime of electron acceleration in this flare over a broad energy range, starting with a few keV.

The October 29, 2002 flare started at 21:47:02 UT in energy channels of the 15–156-keV hard X-ray range, and, somewhat later, at 21:48:49 UT, in the soft radiation channels covering the 2.9–14.3-keV interval [12]. The time profiles of hard (24–41 keV) X-rays (Fig. 7a) and soft (3.6–5.0 keV) X-rays (Fig. 7e) exhibit all through the event bursts of about 3–5 s long standing out against the background of longer, about 10 s, ones. In the 112–156-keV interval one can observe also variations in the X-rays emission with periods of about 20 s.

CSP graphs for the count rate in the 112–156-keV channel and for its filtered high-frequency components with $T_{\mathrm{f}} = 7, 13, 17, 23, 37$, and 43 s are plotted in Fig. 7b, d, and c for the pre-, post-, and flare time intervals, accordingly. The first of them shows normalized spectral density as a function of the 2.2-min long "trial" period of the pre-flare stage. As in all the subsequent cases, the resolution of the periodogram is 0.25 s; therefore, periods are derived from the periodograms presented here to within one quarter of a second. Note also in this stage the most powerful components with periods of 8 and 22 s and several weaker components with periods of 3.5, 5, 12, and 16 s. During a 1-min long flare the time structure changes, revealing a component with a period of 6 s and a strong growth in amplitude of the component with a period of 12 s, whose power exceeds nearly twofold that of the persisting oscillations with periods of 4 and 5 s, with the most powerful 8-s component of the pre-flare stage disappearing altogether. In the post-flare stage lasting 5 min, almost all of the above components survive, but the shape

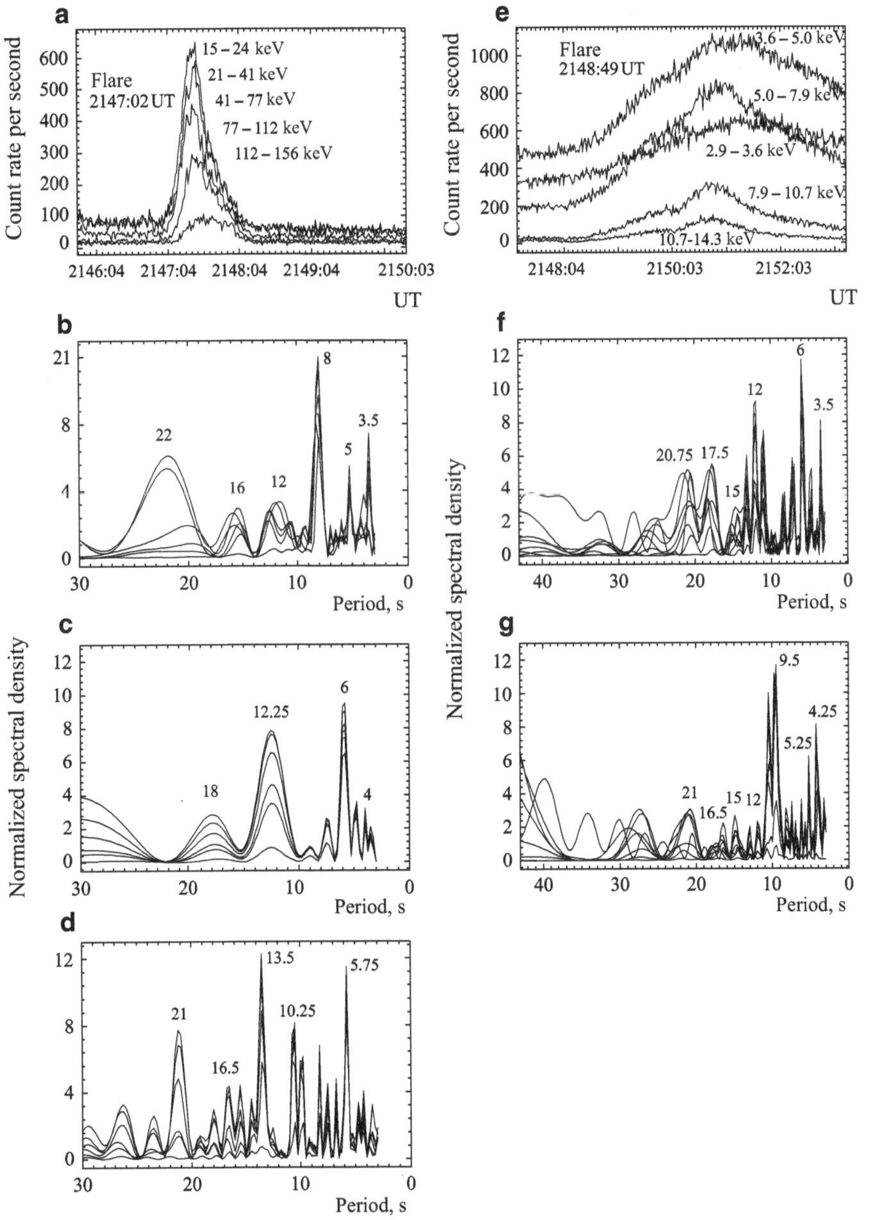


**Fig. 7** (**a**) Time profiles of the October 29, 2002 solar flare emission as recorded in five hard and (**e**) five soft X-ray ranges. CSP drawn for pre-flare radiation: (**b**) 112–156 keV range, (**f**) 3.6–5.0 keV; for flare emission: (**c**) 112–156 keV, (**g**) 3.6–5.0 keV range; for post-flare radiation: (**d**) 112–156 keV range. Filtration of the high-frequency component for all flare stages was performed with the parameter $T_f = 7, 13, 17, 23, 37$, and 43 s

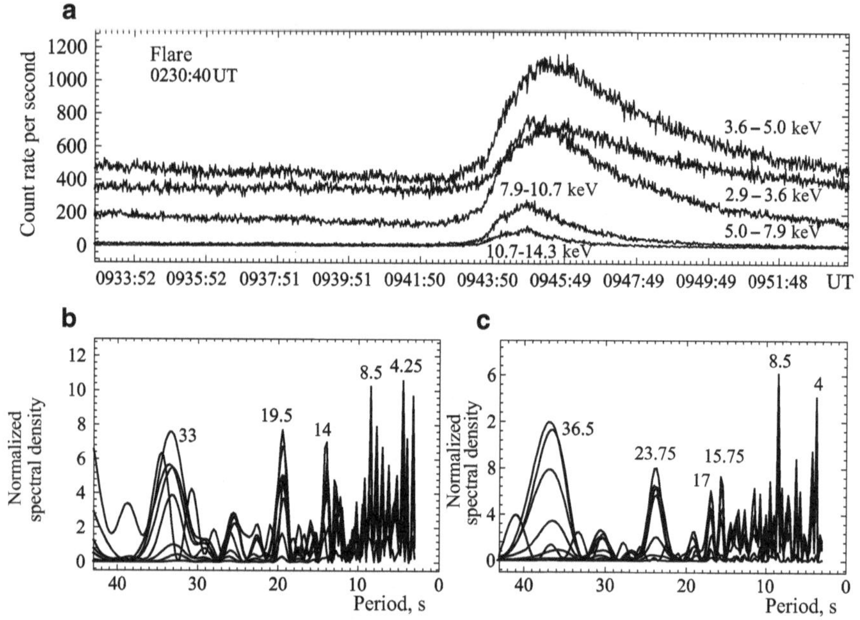


**Fig. 8** (**a**) Time profiles of the August 10, 2002 solar flare emission recorded in five soft X-ray ranges. CSP drawn for the count rate in the 3.6–5.0 keV energy range: (**b**) for pre-flare emission; (**c**) for the flare. Filtration of the high-frequency component for all flare stages was performed with the parameter $T_f = 7, 13, 19, 29, 43, 59$, and 73 s

and structure of the periodogram change sharply; indeed, almost all components split into two and more constituents with peak halfwidths noticeably smaller than those observed in the preceding stages of the event, and with most of the radiation power becoming concentrated in the region of short periods (the higher-frequency part of the spectrum), so that the spectrum acquires progressively the "white noise" pattern.

In the soft X-ray range, 3.6–5.0 keV, the pre-flare stage is characterized by periodic components with periods of 6 s (the most powerful), 3.5, 12, 17.5, and 20.75 s (Fig. 7f). During the flare, the most powerful component with 9.5-s period appears, the others differing from their pre-flare periods by one to two fourths of a second (Fig. 7g).

Burst structure of the above type was observed repeatedly in other flares as well, for instance, in the flares of August 10 and 26, 2002 [14]. In this case, in each of the five soft X-ray ranges (Figs. 8a and 9a, respectively) one could observe, besides second-range bursts, count rate variations of the order of 10 s.

To study the possibility of existence of quasi-periodic oscillations in the soft X-ray emission of the solar flares detected by the IRIS instrument during the period from December 2001 to December 2003, 13 events were selected out of 145 X-ray events, which had a fairly long ($\sim$5 min) pre-flare section of background

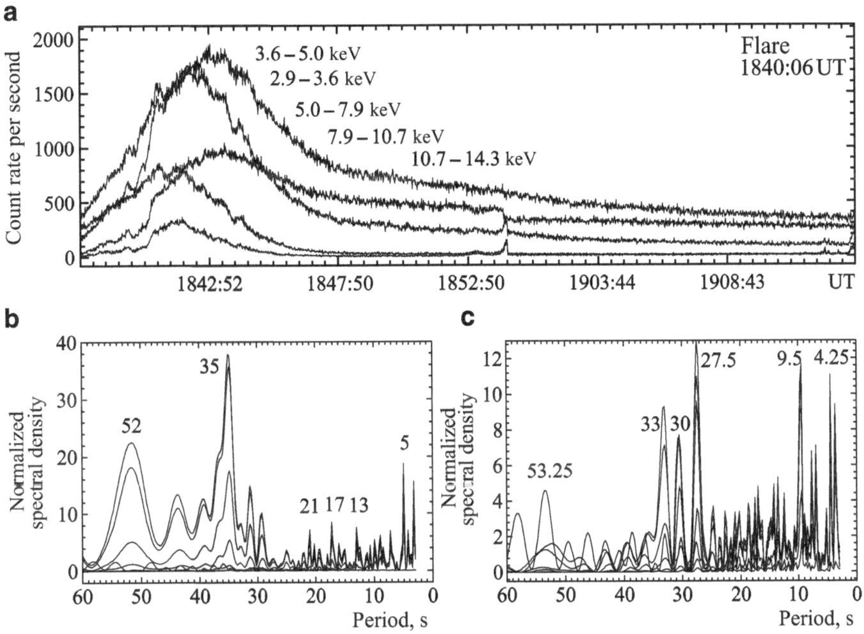


**Fig. 9** (**a**) Time profiles of the August 26, 2002 solar flare emission for five soft X-ray ranges. CSP drawn for the count rate in the 3.6–5.0-keV energy range: (**b**) for the flare; (**c**) for post-flare radiation. Filtration of the high-frequency component for all flare stages was performed with the parameter $T_f = 7, 11, 19, 31, 47$, and 59 s

radiation adjoining directly the flare, and the same duration as the background. For the pre-flare and flare radiation, CSP profiles were constructed for each flare in the 3.6–5.0 keV interval, and quasi-periodic components were selected from their amplitudes at a significance level of $\geq$95 %.

A comparison of the behavior of the revealed quasi-periodic components before and during the flare suggests that in 11 out of the 13 flares studied a "new" component with a period of 9–33 s appears in the flare emission, and in the remaining two, two "new" components in the same time interval come to life.

In Table 2 the quasi-periods of these components are singled out in bold. The "+" sign indicates that the average values of these components are equal to that of the corresponding component from the neighboring column, which may mean either that the pre-flare component split into two during the flare or that they coalesced to form one.

An analysis of the CSP structures of the events under consideration showed that the structure of the spectral periodogram (see Fig. 10) either consists of single and widely spaced peaks (b, c), as was the case with the August 26, 2002 event (see Fig. 10a), or of more narrow, closer lying peaks which form in the periodogram groups characterized by a wider separation than that of their

**Table 2** Quasi-periodic oscillations in the 3.6–5.0-keV soft X-ray range measured in the pre-flare and flare emission of 13 solar flares

| Flare start time | | Measured quasi-periods (s) | | |
|---|---|---|---|---|
| Date | UT | In emission before flare | In flare emission | |
| 19.12.01 | 0230:40 | 5; 13.25; 15.25; 26.25; 35.5 | 6.5; **9.75**; 13.75; 14.75;**18**; 28; 36 | C4.9 |
| 10.08.02 | 0943:48 | 4.25; 8.5; 14; 19.5; 33 | 4; 8.5; 15.75; 17; **23.75**; 36.5 | C2.0 |
| 20.08.02 | 1735:18 | 8; 11.5; 13; 16.25 + 18.5; 23.5 | 6.75; 10; 13; 17; **21.5**; 25; **33** | C9.9 |
| 26.08.02 | 1622:09 | 6; 10.25; 14; 27; 35 | 6.5; 10; 13.25 + 15.5; **19**; 26; 39.5 | C7.6 |
| 03.09.02 | 0744:06 | 4; 6.75; 12.25; 18.75; 30; 37.5 | 5; 6.25; **9**; 13.75; 20.5; 27; 37 | C2.6 |
| 04.09.02 | 0336:44 | 3.5; 6.75; 9.5; 11.75; 18.75; 24 | 3.75; 6; 8; 11.25; **16.25**; 19; 24.5 | C1.7 |
| 26.09.02 | 0206:21 | 4.5; 9.5; 14.5; 18.5; 23.75; 26.5 | 4; 8.5; 15.5; 18.5; **21.25**; 24.75; 27 | C2.9 |
| 04.10.02 | 1025:53 | 4.5; 10.5; 14.5; 17.5; 19.25; 32 | 5; 9.75; 14.75; 16.5; 19.5; **25**; 31.5 | C6.2 |
| 04.10.02 | 1200:47 | 6.5; 8; 12; 14.5; 27.5 + 36.5 | 4; 6.75; 12.75; 15; **24.5**; 32 | C3.2 |
| 15.10.02 | 0914:31 | 3.5; 7.5; 10.75; 16.5; 22; 32.25 | 5.75; 7.25; 11; 17.5; 24; **27**; 31.5 | C1.6 |
| 29.10.02 | 2148:49 | 3.5; 6; 12; 15; 17.5; 20.75 | 4.25; 7.5; **9.5**; 12; 15; 16.5; 21; **27** | C1.8 |
| 20.01.03 | 2144:11 | 6; 8; 11; 14; 21; 31.5 + 38.5 | 5.25; 8.5; 10.5; 15; 20; **28.5**; 35 | C1.5 |
| 19.12.03 | 0137:23 | 4.5; 7.75; 11.5; 14.5; 33.5 | 4.75; 7.5; 10.5; 14; **19**; 31.5 | C2.0 |

components (Fig. 10e, f), as in the particular example of the September 3, 2002 flare (Fig. 10d). Numerical simulation (see Fig. 2) suggests that the existence of these groups should be rather attributed to a floating period (and not of the amplitude) of a harmonic within the signal treatment region, rather than to the presence of a large number of closely lying harmonics, even if the noise component is strong. Therefore, if the CSP structure contained such spectral groups, one selected from them the largest peak, and its value was assigned to the group as a whole (see Fig. 10e, f). Figure 10 illustrates the principle underlying the selection of the revealed periods in compiling Table 2.

An analysis of the general pattern of the formation and evolution of various oscillations in different stages of the flare development suggests the existence of a system of "coupled oscillators" which are linked with one another and can vary the frequency and amplitude of their oscillations during the evolution; it appears reasonable to assume that in the course of the flare development energy most probably can transfer from one oscillating magnetic structure to another (arch-shaped and flaring loop structures of active regions in the solar atmosphere), which persist or recover in the post-flare stage as well.

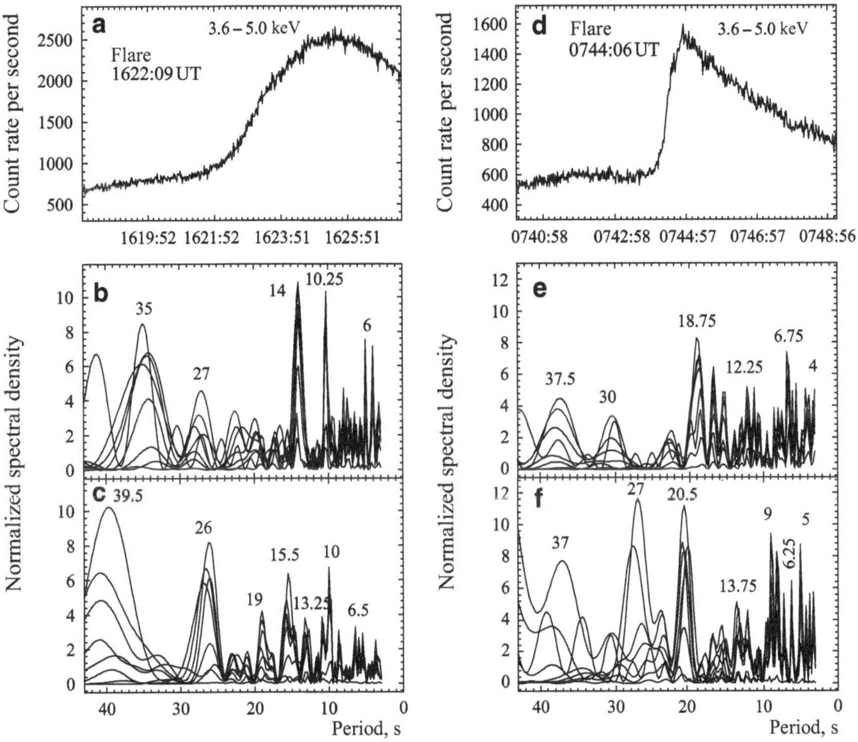


**Fig. 10** Time profiles (3.6–5 keV range) of the solar flares on August 26 (**a**) and September 3, 2002 (**d**), and CSP for pre-flare count rate in the August 26 (**b**) and September 3 (**e**) flares and flare emission (**c, f**). Filtration of the high-frequency component for all flare stages was performed with the parameter $T_f = 7, 13, 19, 29, 43, 59$, and 73 s

# Evaluation of Plasma Parameters in the Flaring Region of the Solar Atmosphere

The temperature and emission measure of the flaring region in the Sun can be calculated from the recorded X-rays in terms of the "quasi-thermal" model of the energy spectrum, i.e., under the assumption of X-rays being generated by the bremsstrahlung mechanism in the optically thin thermal plasma. In this case, the differential flux of X-rays at a distance of 1 AU from the emitting object, which is a compact hot region in the flaring arch filled by strongly ionized plasma with the characteristic chemical composition of the solar atmosphere can be written in the form [10]

$$I(E) \approx 1.07 \cdot 10^3 \frac{1.2\text{EM}}{E\sqrt{T}} g_{ff}(E, T) e^{-E/T},$$

where $I(E)$ is differential flux of X-rays in units of $\mathrm{cm^{-2}\,s^{-1}\,keV^{-1}}$; $E$ is energy and $T$ is the temperature in keV units, $\mathrm{EM} = n_\mathrm{e}^2 V$ is the emission measure of radiating electrons in units of $10^{45}\,\mathrm{cm^{-3}}$ ($n_\mathrm{e}$ is the electron density, and $V$ is the volume of the emitting region), and $g_{ff}(E,T)$ is the Gaunt factor.

The Gaunt factor can be approximately defined to within 2 % with the expression $g_{ff}(E,T) \approx 0.90(T/E)^\alpha$, where $\alpha = 0.37(30\,\mathrm{keV}/T)^{0.15}$ [33]. Hence, for the differential flux of X-rays we have a fairly accurate (to within 2 %) expression

$$I(E) \approx 1.16 \cdot 10^3 \frac{\mathrm{EM}}{E\sqrt{T}} \left(\frac{T}{E}\right)^\alpha e^{-E/T}.$$

The count rate $\mu_j$ measured in the time interval $\Delta_t$ for each of the five energy channels $[E'_j, E'_{j+1}]$ ($j = 1, 2, \ldots, 5$) of the IRIS instrument can be expressed through the integral of the differential radiation flux of the source emitting in an energy range $[E_1, E_2]$ and the instrument response function $F(E, E')$, which yields an excessive system of nonlinear equations for the $T$ and EM parameters [15]:

$$\mu_j = \Delta_t \int_{E_1}^{E_2} dE \int_{E'_j}^{E'_{j+1}} I(E) F(E, E')\, dE', \quad j = 1, 2, \ldots, 5.$$

On linearization, this system is solved by successive approximations for the unknowns $T$ and EM. As the initial approximation, one can assume the estimation of these parameters with the use of a simpler but less accurate expression for the differential radiation flux [10]

$$I(E) = 1.3 \cdot 10^3 \frac{\mathrm{EM}\, e^{-E/T}}{E^{1/4} T^{0.1}},$$

where the Gaunt factor is approximated by the expression $g_{ff}(E,T) = (T/E)^{0.4}$ to within 20 %.

The variation with time of the temperature parameters and of the emission measure calculated by the above technique with the "quasi-thermal" model of the energy spectrum of soft X-rays for the four flares considered above is presented in Fig. 11 for the December 19, 2001 flare, in Fig. 12 for the August 10, 2002 flare, in Fig. 13, for the August 26, 2002 flare, and in Fig. 14, for the event that occurred on October 29, 2002.

The time structure of the August 10, 2002 flare is the simplest of all the above events. As seen from Fig. 12, it was generated most probably by a single large energy release. This makes the flare most convenient for interpretation of the behavior of temperature and of the emission measure in different phases of the flare process. As seen from Fig. 12c, the first thing that happens is a fast heating of a compact region to a high temperature, with the latter rising rapidly. This flare stage corresponds to the very beginning of the growth of soft X-ray emission (Fig. 12a). In the next phase, both the temperature (Fig. 12c) and the emission measure (Fig. 12b) increase, which

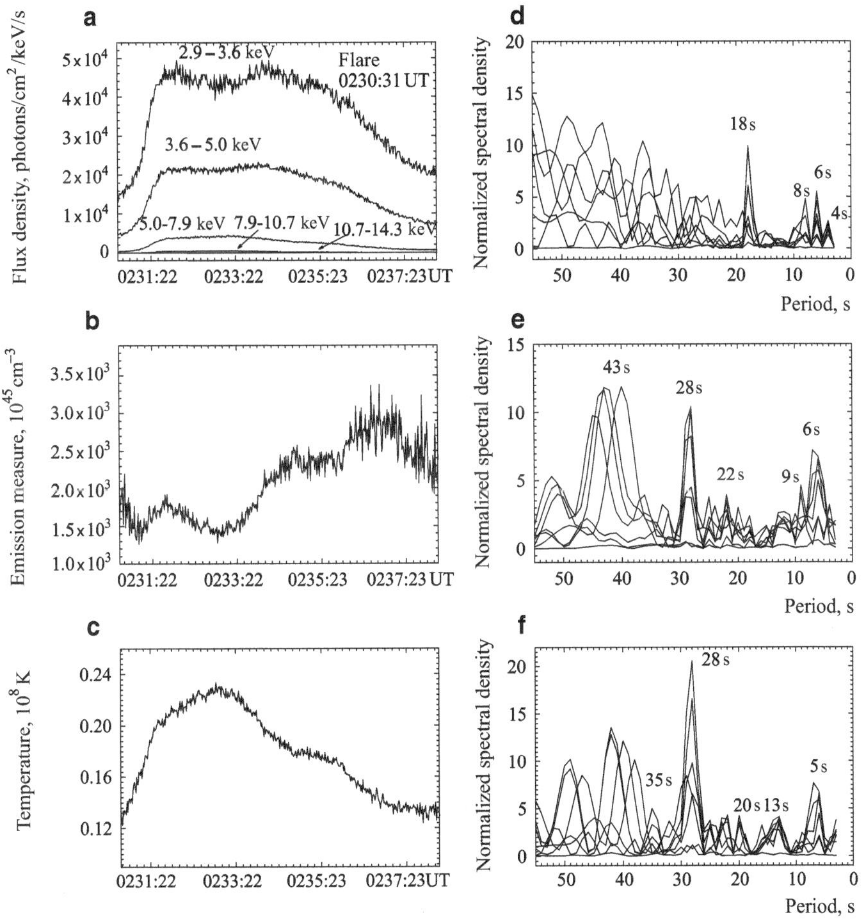


**Fig. 11** (**a**) Time structures of soft X-ray emission from the December 19, 2001 solar flare and time profile of the parameters of the "quasi-thermal" model developed for the energy spectrum of soft X-ray emission of this flare: (**b**) emission measure, (**c**) temperature. CSP for: (**d**) 3.5–5.0-keV X-rays, (**e**) emission measure, (**f**) temperature. Filtration of the high-frequency component for the radiation, emission measure, and temperature was performed with the parameter $T_f$ = 13, 29, 53, 73, 97, and 113 s

evidences simultaneous heating and expansion of the flare region. The growth of temperature in this phase is substantially slower than that in the first one, which reflects further increase of the flux of soft X-rays from the flare (Fig. 12a). The next to set in is the phase in which the temperature of the emitting region starts to fall off (Fig. 12c), while the emission measure continues to rise (Fig. 12b). This phase can be identified with the flat maximum in the soft X-rays (Fig. 12a). And in the last phase of the flare, as the temperature continues to fall off (Fig. 12c), the emission

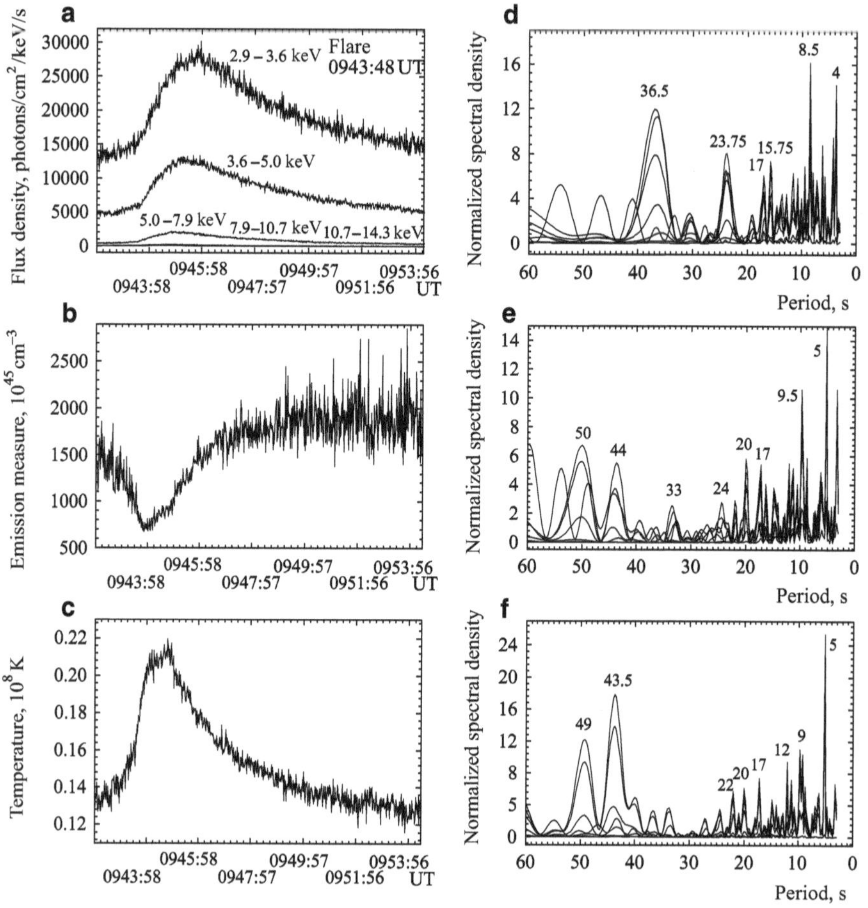


**Fig. 12** (**a**) Time structures of soft X-ray emission from the August 10, 2002 solar flare and time profile of the parameters of the "quasi-thermal" model developed for the energy spectrum of soft X-ray emission of this flare: (**b**) emission measure, (**c**) temperature. CSP for: (**d**) 3.5–5.0-keV X-rays, (**e**) emission measure, (**f**) temperature. Filtration of the high-frequency component for the radiation, emission measure, and temperature was performed with the parameter $T_f$ = 7, 13, 19, 29, 43, 59, and 73 s

measure practically ceases to grow (Fig. 12b). This phase corresponds to decay of the X-ray flux from the flare (Fig. 12a).

Compared with the time structure of the above flare, that of the X-rays from the December 19, 2001 event is more complicated (Fig. 11a); this flare reveals at least two large explosive energy releases, thus licensing us to consider this flare as a superposition of more than one flare processes. As a result, the time structure of the temperature (Fig. 11c) and emission measure (Fig. 11b) in this flare are more complicated and specific.

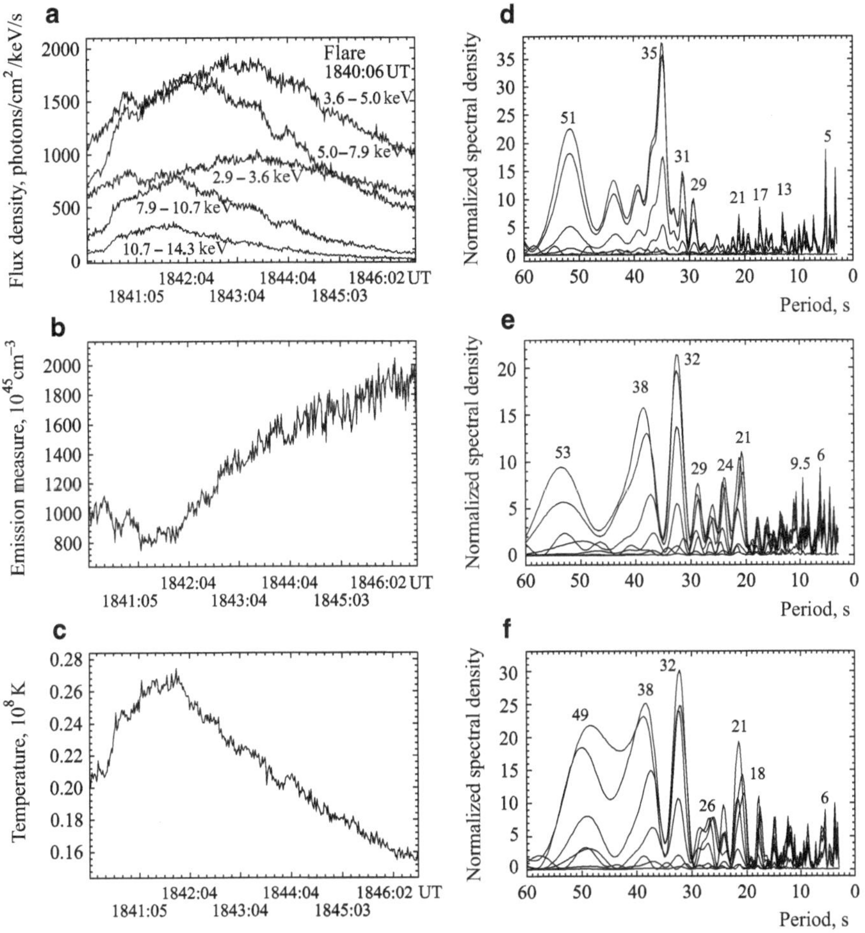


**Fig. 13** (**a**) Time structures of soft X-ray emission from the August 26, 2002 solar flare and time profile of the parameters of the "quasi-thermal" model developed for the energy spectrum of soft X-ray emission of this flare: (**b**) emission measure, (**c**) temperature. CSP for: (**d**) 3.5–5.0-keV X-rays, (**e**) emission measure, (**f**) temperature. Filtration of the high-frequency component for the radiation, emission measure, and temperature was performed with the parameter $T_f = 7, 13, 19, 29, 43, 59$, and 73 s

Similar results were obtained also for other flares with close to or rather simple time structures of soft X-rays. Consider, for instance, the flares on August 26, 2002 (Fig. 13, X-ray class C3.4), October 29, 2002 (Fig. 14, class C1.8), January 20, 2003 (21:46 UT, class C1.5), and February 13, 2003 (23:32 UT, class C2.2). Some slight differences may be observed in the second phase of the flare, which is characterized by low rates of rise of the temperature and emission measure. In some flares, the rates of variation of one of these parameters is so small that they may be considered not to change during the phase at all.

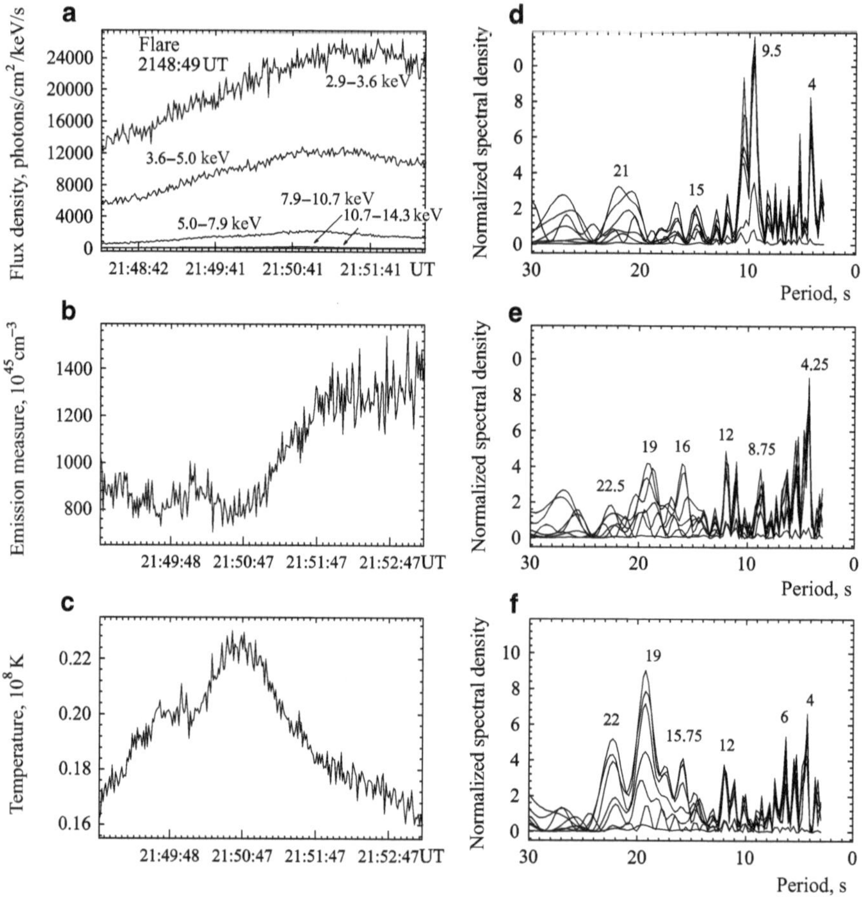


**Fig. 14** (**a**) Time structures of soft X-ray emission from the October 29, 2002 solar flare and time profile of the parameters of the "quasi-thermal" model developed for the energy spectrum of soft X-ray emission of this flare: (**b**) emission measure, (**c**) temperature. CSP for: (**d**) 3.5–5.0-keV X-rays, (**e**) emission measure, (**f**) temperature. Filtration of the high-frequency component for the radiation, emission measure, and temperature was performed with the parameter $T_f = 7, 13, 19, 29, 43, 59$, and 73 s

We applied the modified method of spectral analysis to the calculated values of the temperature and emission measure of these flares too, in order to reveal a periodic structure and compare it with similar structures in the count rate (Figs. 11–14d–f).

Interestingly, both the temperature variation (Fig. 11f) and the emission measure (Fig. 11e) in the December 19, 2001 event exhibited a component with a period of 28 s, which differs from the 18-s period characteristic of the flare (Fig. 11d). The absence of the 18-s component in the temperature and emission measure data, as contrasted with the behavior of the count rate, offers one more indirect evidence for the explosive nature of generation of X-ray bursts, a point already raised above.

In the August 10, 2002 flare, the most powerful components, both in the temperature variation (Fig. 12f) and in the emission measure (Fig. 12e), are those with periods of 5, 9–9.5, 43.5–44, and 49–50 s, and the weakest ones, with periods of 17–35 s. The structure of the power spectrum for the count rate in this flare is different (Fig. 12d); indeed, here the most powerful components have periods of 8.5 s (as in the case of emission measure and temperature) and 23.75 and 36.5 s, which in the preceding case were very weak or absent altogether. Similar relations among the elements of the structure in the spectra of emission measure and temperature are observed in the August 26, 2002 flare as well (Fig. 13). These spectra reveal very powerful components with a period of 32–35 s, a slightly weaker one with 49–53 s, and numerous still weaker components with periods of 5–30 s.

The pattern of the October 29, 2002 flare (Fig. 14) is totally different from those observed in the preceding cases. The temperature series is dominated by the most powerful 19-s component (Fig. 14f), which is not detected in the count rate spectrum (Fig. 14d) and is noticeably weaker in the emission measure spectrum (Fig. 14e). By contrast, the count rate spectrum is dominated by a 9.5-s harmonic, which is much weaker in the emission measure spectrum and close to undetectable in the temperature spectrum. This may be interpreted as resulting from pulsed "injection" of a beam of particles into the radiation region with a frequency of 9 s, which becomes manifest in a "heating" or periodic increase/decrease of the volume of the emission region as a 19-s oscillation. In the two preceding flares, both "injection" and "heating" occur in parallel but at a frequency characteristic of the second process. This suggests a conclusion that the time scale of about 20 s is a sort of "transitional" for "thermal" excitation of the emission region in the flare plasma by particles "injected" into it more frequently, whereas the 30–50-s interval is rather a characteristic time for the oscillating "thermal mode" in the heating and radiation region [27, 28].

## Dynamics of Variation of the Energy Spectrum of Hard Solar Flare X-Rays Measured with Second-Scale Time Resolution

The key problem in the theory of solar flares can be formulated as "How the energy stored in a magnetic field is transferred to the charged plasma particles?" The mechanisms accounting for heating of the plasma to high temperatures ($>10^8$ K) and acceleration of charged particles by electric fields are capable of offering the most probable answers to this question. The hard X-rays emitted from solar flares bears direct information on the high-energy electrons of the flare plasma and, thus, can shed light on this situation, namely, which of the mechanisms is realized with the higher probability in a solar flare; or maybe they both operate in nearly each case? For some flares the spectra of hard X-rays can be fitted by

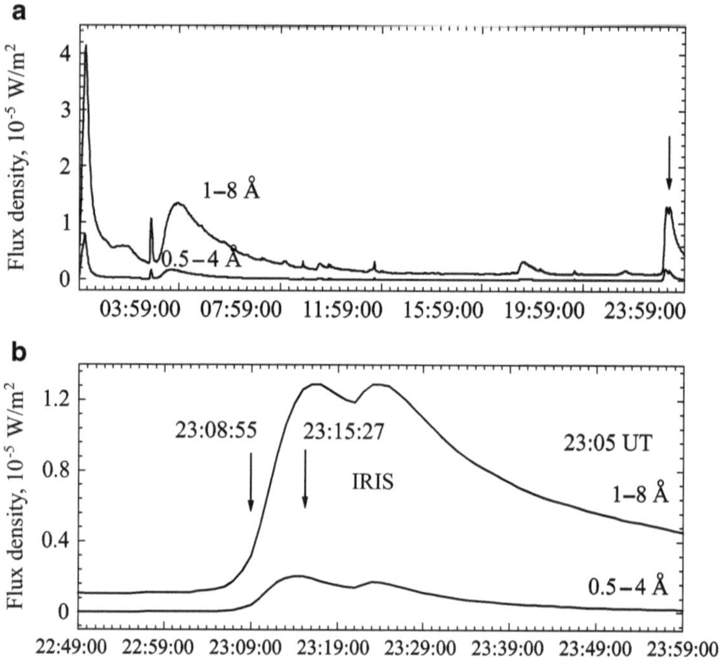


**Fig. 15** (**a**) Solar activity in the soft X-rays (1–8, 0.5–4 Å) on April 15, 2002. Data of the GOES satellite: http://spidr.ngdc.noaa.gov (free access); (**b**) the arrows show time of switch on/off of the "burst"—mode of the IRIS operation

a thermal spectrum at plasma temperatures above $10^8$ K [27, 28]. In other cases, X-ray spectra are described by a power law, which suggests that this radiation is the bremsstrahlung radiation of electrons accelerated in flares and reflects the dynamics of the acceleration and propagation of these electrons in the flare plasma. The high time and energy resolution of the IRIS spectrometer permits one to study the dynamics of the energy spectra of hard flare X-ray emission with a 1-s time resolution. Consider the evolution of the energy spectra of some recorded flares differing from one another in the spectrum behavior pattern.

The solar flare that occurred on April 15, 2002 [12], the strongest of the above, started at 23:05 UT and was assigned to the X-ray class M1.2. Figure 15a shows the pattern of solar activity (X-ray flux in W/m$^2$ plotted with 1-min time resolution) on April 15, 2002 in the soft X-ray wavelength range of 1–8 and 0.5–4 Å, as measured with the GOES-10 geostationary satellite (http://spidr.ngdc.noaa.gov). The arrow identifies the April 15, 2002 flare which was recorded by the IRIS spectrometer simultaneously with GOES-10. Figure 15b presents this flare on an enlarged time scale, with the arrows defining the time interval in which solar X-rays were studied in the soft X-radiation range simultaneously by the IRIS instrument and the GOES-10 satellite. Shown in Fig. 16 is the time profile of hard X-rays from the solar flare recorded in five energy ranges with second-scale resolution.

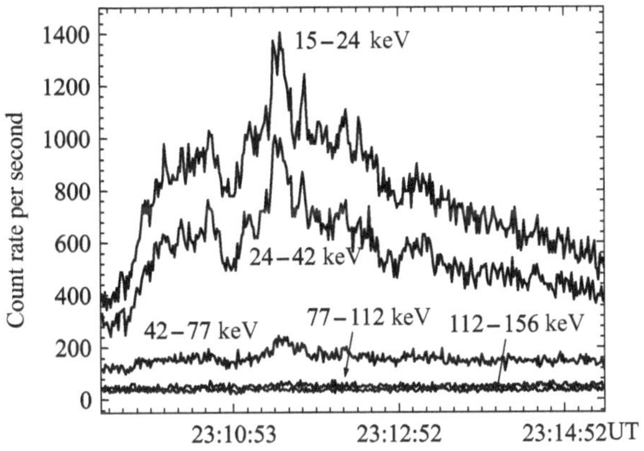


**Fig. 16** Time profile of X-ray emission from the April 15, 2002 flare

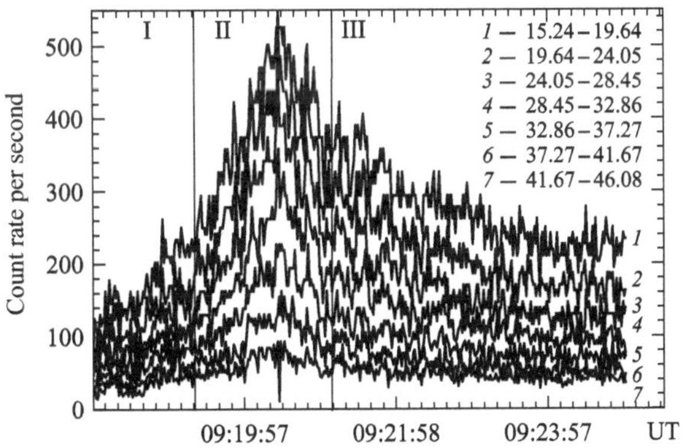


**Fig. 17** Time profile of hard X-ray intensity from the July 26, 2002 solar flare obtained in seven energy ranges of the instrument (keV): 1, 15.24–19.64; 2, 19.64–24.05; 3, 24.05–28.45; 4, 28.45–32.86; 5, 32.86–37.27; 6, 37.27–41.67; 7, 41.67–46.08

The emission spectrum of this flare can be approximated by a power-law relation in all stages of the flare development, with the spectral index being time dependent. The spectrum of the background before the flare can be described by a power law function with an exponent $\gamma \approx -2.7$, which decreases in the phase of growth to reach the value $\gamma \approx -3.6$ at the maximum of intensity, to grow again during the decay. The power-law pattern of the energy spectrum suggests that the hard X-rays of the April 15, 2002 solar flare is actually bremsstrahlung radiation of the flare-accelerated electrons.

Figure 17 demonstrates time profiles of hard X-ray emission of the July 26, 2002 solar flare recorded in seven energy channels [8, 13]. Three intervals can be isolated in the time profile of the flare, I, II, and III. In the very beginning of the growth in intensity (I), the energy spectrum of X-rays changes its power-law pattern and recovering it again. Starting with 09:19:40 (II), the spectrum assumes a non-power-law shape (Fig. 18a), to restore again a power law behavior at 09:21:05 (Fig. 18b). A specific feature of this non-power-law behavior consists in a faster rise in the number of quanta with lower energy in the growth phase compared with that of high-energy quanta, which reflects most probably some specific features of the electron acceleration process during a flare.

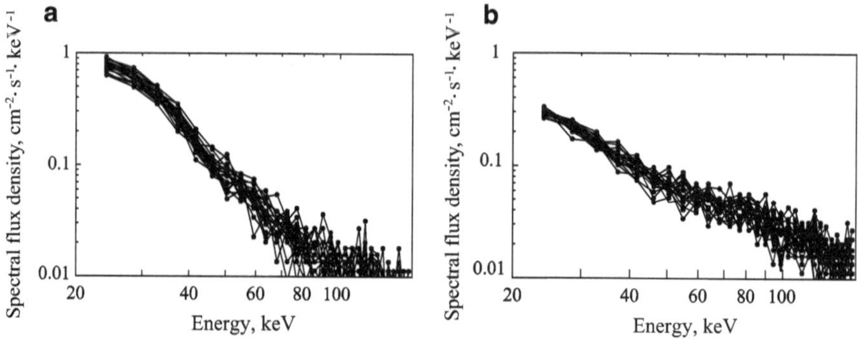


**Fig. 18** Energy spectra of hard X-ray emission obtained in the maximum (**a**) and decay (**b**) stages of the July 26, 2002 flare with a 1-s time resolution

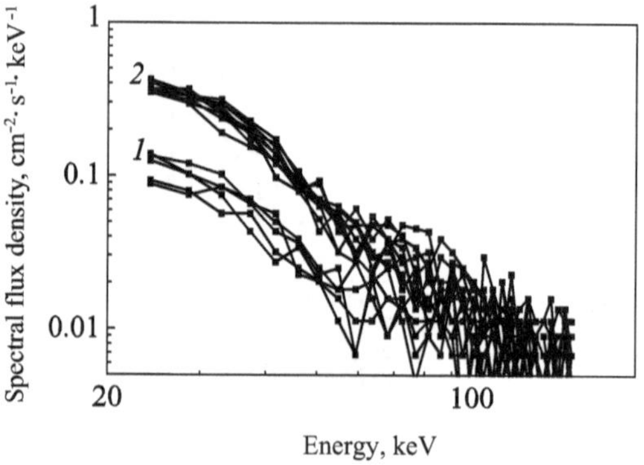


**Fig. 19** Spectra of hard X-ray emission obtained in the stages of growth (1) and maximum (2) intensity in the December 19, 2001 flare with a 1-s time resolution

Unlike the July 26, 2002 flare, the energy spectrum of hard X-rays in the December 19, 2001 event [13, 26] exhibits other than the power-law shape in all stages of the flare (Fig. 19).

Figure 20 presents the measured energy spectrum of the flare after the background subtraction at the maximum of emission at 02:31:23, and the spectra calculated from the thermal and nonthermal models with inclusion of the instrumental response of the IRIS spectrometer. We readily see that the spectrum of hard X-rays of the flare can be fitted with both the thermal and nonthermal models, thus making discrimination between them a difficult problem.

To describe hard X-ray emission in terms of the thermal model, however, one would allow plasma heating to extremely high temperatures, in excess of $10^8$ K. The problem of plasma heating to such temperatures in solar flares still awaits solution. As pointed out above, hard X-ray emission from a flare can be fitted with a nonthermal model as well. We are going to demonstrate this in the particular example of the bremsstrahlung radiation of fast electrons with a flat-top distribution

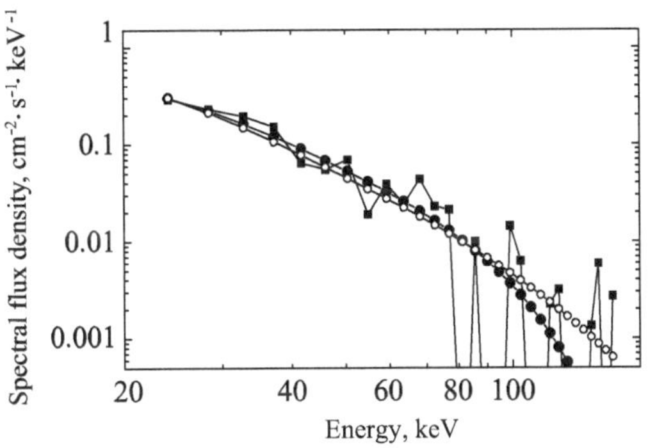


**Fig. 20** Measured spectrum (*filled squares*) of hard X-ray emission from the December 19, 2001 flare obtained at the maximum of the event at 02:31:23; calculated spectrum (*filled circles*) for the same moment of time obtained with a "flat-top" distribution function of injected electrons with $E_{max} = 140$ keV, and (*open circles*) spectrum of X-ray emission calculated with the thermal model with the plasma temperature 35 keV

function in velocity injected into the plasma. The intensity of the hard X-rays was calculated in terms of the universally accepted thick target model, with the bremsstrahlung radiation cross section taken in the Born approximation, as this was done in [3, 7, 18, 45]. The distribution function of fast electrons in the flare plasma was derived with the use of a kinetic equation including Coulomb collisions of fast electrons with plasma particles, which result in variation of their energy and angular distributions [42]. The initial distribution function of fast electrons injected into the plasma was assumed to be plateau-shaped, with a drop at the maximum electron energy $V_{max}$. The calculations of the energy spectra of X-rays were performed for different values of the maximum energy of accelerated electrons $E_{max}$. As evident from Fig. 20, the calculated spectrum of the bremsstrahlung radiation of electrons with the starting function of this kind is capable of accounting for the recorded radiation spectrum of the December 19, 2001 flare with $E_{max} \approx 140$ keV.

The flat-top distribution function of fast electrons with a sharp drop can form in their resonant interaction with plasma oscillations. It is well known that quasi-linear relaxation of an electron beam culminates in formation of a flat-top distribution function. Taking into account the nonlinear scattering of plasma waves by ions of the background plasma translates into transfer of the energy of oscillations into the domain of high phase velocities [17]. Subsequent resonant interaction of fast electrons with oscillations of this spectrum initiates additional acceleration of a part of electrons beyond the current layer, bringing about a sharp drop in the distribution function at a certain maximum velocity $V_{max}$ [24, 25]. This additional acceleration of a part of electrons was observed in laboratory experiments as well [32].

This means that the distribution function of fast electrons which forms in interaction with plasma waves is capable of accounting for generation of bremsstrahlung radiation with the spectrum measured with the IRIS spectrometer on December 19, 2001. Significantly, the acceleration of electrons in this flare reveals a clearly

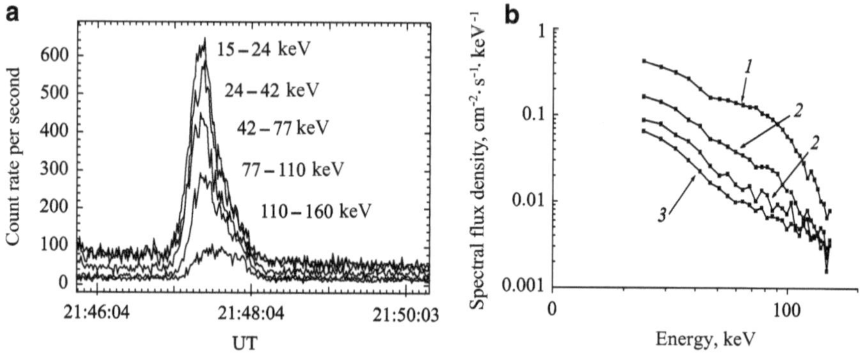


**Fig. 21** Time profile (**a**) and spectra (**b**) of hard X-ray emission from the October 29, 2002 flare. (**b**) 1, spectrum of the flare emission in the flare maximum stage averaged over 10 s; 2, spectra of the flare emission in the growth stage averaged over 10 s; 3, spectrum of the background emission preceding the flare

pronounced pulsed character, a factor that apparently is conducive to excitation of plasma waves.

Thus, the spectrum of radiation emitted from a flare cannot be described by any one pattern of fast electron distribution; indeed, electrons with both the "Maxwellian" and the "flat-top" distribution functions can generate radiation with the measured spectrum.

It thus follows that selection of a physical model which would most closely fit experimental data should be preceded by solution of the following problems: heating of the flare plasma to temperatures above $10^8$ K and formation of the distribution function of flare-accelerated electrons that would be capable of describing the measured radiation spectra.

In conclusion, consider the evolution of the energy spectrum the October 29, 2002 flare (which started at 21:47:02) whose time profile is visualized in Fig. 21a in four energy ranges [12]. The total duration of the flare X-ray emission was ~1 min, the flare X-ray class was C1.8.

Figure 21b displays spectra measured in the phases of growth and maximum intensity with an interval of 1 s and averaged over 10 s. We readily see that the spectra forming in the phases of growth and maximum intensity have a non-power-law shape. In the maximum phase, the energy spectrum exhibits a break in the region of 45 keV. The low-energy spectrum to the left of the break is probably generated by thermal emission of the hot plasma, and the spectrum at high energies by bremsstrahlung radiation of fast electrons. This shape of the spectrum persists in the decay phase as well. Another characteristic feature of the spectrum of the October 29, 2002 flare consists in its becoming steeper with increasing energy. This behavior of the spectrum finds a ready explanation if the distribution function of fast electrons generating bremsstrahlung X-rays drops sharply at an energy of ≈160 keV.

An analysis of the spectra of hard X-rays emitted in solar flares of different classes suggests that they can have different shapes and that their shape can vary during a flare. Some flares maintain a power-law or non-power-law spectrum of radiation throughout the event, which suggests the corresponding mechanism of radiation generation. The spectra of others may pass through different shapes in different stages of a flare [34–36]. This behavior of a spectrum suggests nonstationary propagation of fast particles in the flare plasma.

## References


1. Alavi, A.S., Jenkins, G.M.: An example of digital filtering. Appl. Stat. **14**, 70–74 (1965)
2. Aschwanden, M.: Particle acceleration and kinematics in solar flares - a synthesis of recent observations and theoretical concepts (Invited review). Space Sci. Rev. **101**(1–2), 227 (2002)
3. Brown, J.C.: The deduction of energy spectra of non-thermal electrons in flares from the observed dynamic spectra of hard X-ray bursts. Sol. Phys. **18**, 489–502 (1971)
4. Charikov, Y.E.: Time structure, directivity and polarization degree of hard X-rays in solar flares. In: Stepanov, A.V., Nagovitsyn, Yu.A. (eds.) Proceedings of Annual Conference on Physics of Sun, Solar and Solar–Terrestrial Physics, pp. 441–445. Pulkovo Observatory, St. Petersburg (2009) (in Russian)
5. Charikov, Y.E.: Subsecond time structure of hqard X-rays in solar flares. In: Stepanov, A.V., Nagovitsyn, Yu.A. (eds.) Proceedings of Annual Conference on Physics of Sun, Solar and Solar–Terrestrial Physics, pp. 449–451. Pulkovo Observatory, St. Petersburg (2009) (in Russian)
6. Charikov, Y.E., Sklyarova, E.M.: Conception of the nature of solar thermal flares Nauch.-Tech. Vedomosti SPbGPU Fiz.-Mat. Nauki. 1(116), 88–93 (2011) (in Russian)
7. Charikov, Yu.E., Guzman, A.B., Koudriavtsev, I.V.: Hard X-ray emission of solar flares and non-stationary kinetics of electron beams. Astron. Astrophys. **308**, 924–928 (1996)
8. Charikov, Yu.E., Dmitrijev, P.B., Koudriavtsev, I.V., et al.: Solar flare hard X-rays measured by spectrometer “IRIS”: spectral and temporal characteristics. In: Stepanov, A.V., Benevolenskaya, E.E., Kosovichev, A.G. (eds.) Proceedings of International Union Multi-wavelength Investigations of Solar Activity. IAU Symposium 223, 14–19 June 2004, St. Petersburg, Russia, pp. 429–432
9. Charikov, Y.E., Ogurtsov, M.G, Kostuchenko, I.G.: Spectral analysis of solar flare hard X-rays, registered by CORONAS-F and CGRO satellites. In: Stepanov, A.V., Nagovitsyn, Yu.A. (eds.) Proceedings of Annual Conference on Physics of Sun, Solar and Solar–Terrestrial Physics, pp. 377–380. Pulkovo Observatory, St. Petersburg (2012) (in Russian)
10. Crannell, C.J., Frost, K.J., Matzler, C., et al.: Impulsive solar X-ray bursts. Astrophys. J. **223**, 620–637 (1978)
11. Dmitrijev, P.B., Matveev, G.A., Charikov, Yu.E. et al.: Impulsive structure of energy release in solar flare 19 December 2001. In: Active Processes in the Sun and Stars. Proceedings of Scientific Conference of CNS and Baltic States, St. Petersburg, 1–6 July 2002, pp. 195–198
12. Dmitrijev, P.B., Koudriavtsev, I.V., Lazutkov, V.P., et al.: X-ray emission peculiarities of the solar flares on April 15 and October 29, 2002. In: Proceedings of International Conference Pulkovo Observatory, Pulkovo, 4–9 July, 2005, pp. 509–516 (in Russian)
13. Dmitrijev, P.B., Koudriavtsev, I.V., Lazutkov, V.P., et al.: Solar flares registered by the IRIS spectrometer onboard the CORONAS-F satellite: Peculiarities of the X-ray emission. Sol. Syst. Res. **40**(2), 142–152 (2006)
14. Dmitrijev, P.B., Koudriavtsev, I.V., Lazutkov, V.P., et al.: Characteristic properties of time and energy spectra of solar flares measured by IRIS X-ray spectrometer during 2001–2003. Izvestia RAS Phys. **70**(10), 1453–1456 (2006) (in Russian)

15. Hoyng, P., Stevens, G.A.: On the formation and unfolding of pulse height distributions. Astrophys. Space Sci. **27**, 307–321 (1974)
16. Jenkins, G., Watts, D.: Spectral Analyses and Its Applications, 525 pp. Holden-Day, San Francisco (1968)
17. Kaplan, S.A., Tsytovich, V.N.: Plasma Astrophysics. Pergamon Press, Oxford (1973)
18. Kel'ner, C.R., Skrynnikov, Yu.I.: Polarization and directivity of hard X-ray bremsstrahlung in solar flares. Sov. Astron. **29**(2), 445–449 (1985)
19. Kiplinger, A.L., Dennis, B.R., Emslie, A.G. et al.: Millisecond time variations in hard X-ray solar flares. Astrophys. J. **265**, L99–L104 (1983)
20. Kliem, B.: Particle orbits, trapping, and acceleration in a filamentary current sheet model. Astrophys. J. Suppl. Ser. **90**, 719–728 (1994)
21. Kliem, B., Karlicky, M., Benz, A.O.: Solar flare radio pulsations as a signature of dynamic magnetic reconnection. Astron. Astrophys. **360**, 715–728 (2000)
22. Kocharov, G.E., Matveev, G.A., Charikov, Yu.E., et al.: The study of the physical parameters and the calibration of the IRIS instrument. The report of A.F. Ioffe Physico-technical institute of RAS, N 13037, 44 pp. (1996) (in Russian)
23. Kocharov, G.E., Lazutkov, V.P., Matveev, G.A., et al.: Experement "IRIS" on solar X-ray exploration on aboard of CORONAS satellite. In: Active processes in the Sun and stars. Proceedings of Scientific Conference of CNS and Baltic States, St. -Petersburg, 1–6 July 2002, pp. 314–317
24. Koudriavtsev, I.V., Charikov, Yu.E.: Nonlinear relaxation of an electron beam and reacceleration of fast electrons in solar flares. Astron. Rep. **76**(6), 400–405 (1999)
25. Koudriavtsev, I.V., Charikov, Yu.E., Kal'tman, T.I.: Dynamics of electron beams, radio- and x-ray emission in solar flares. In: Active Processes in the Sun and Stars. Proceedings of Scientific Conference of CNS and Baltic States, St. Petersburg, 1–6 July 2002, 265–268
26. Koudriavtsev, I.V., Charikov, Yu.E., Matveev, G.A., et al.: Hard X-ray modeling of solar flare 19 December 2001 year. In: Modern Problems in the Solar and Stars Activity, Proceedings of Scientific Conference of CNS and Baltic states, Nizhniy Novgorod, 2–7 June 2003, pp. 98–100
27. Kovalev, V.A., Valchuk, T.E., Ishkov, V.N., Kostuchenko, I.G., Savchenko, M.I., Charikov, Y.E.: Regimes of heating and "evaporation" of plasma during solar flare 05.07.2010. In: Stepanov, A.V., Nagovitsyn, Yu.A. (eds.) Proceedings of Annual Conference on Physics of Sun, Solar and Solar–Terrestrial Physics, pp. 209–213. Pulkovo Observatory, St. Petersburg (2010) (in Russian)
28. Kovalev, V.A., Valchuk, T.E., Ishkov, V.N., Kostuchenko, I.G., Charikov, Y.E.: Research of X-ray of solar flare by differential technique. Izv. Crimea Astr. Obs. **107**(1), 209 (2011) (in Russian)
29. Kuznetsov, V.D., Charikov, Yu.E., Kotov, Yu.D., et al.: A review of the solar results from CORONAS-F satellite. In: Stepanov A.V., Benevolenskaya E.E., Kosovichev A.G. (eds.) Proceedings of International Union Multi-wavelength Investigations of Solar Activity. IAU Symposium 223, 14–19 June 2004, St. Petersburg, Russia, pp. 357–366
30. Kuznetsov, V.D., Sobelman, I.I., Zhitnik, I.A., Kuzin, S.V., Kotov, Yu.D., Charikov, Yu.E., Kuznetsov, S.N., Mazets, E.P., Nusinov, A.A., Pankov, A.M., Sylwester, J.: Results of solar observation by the CORONAS-F payload. Adv. Space Res. **47**(9), 1538–1543 (2011)
31. Leboeuf, J.N., Tajima, T., Dawson, J.M.: Dynamic magnetic x points. Phys. Fluids **25**, 784 (1982)
32. Levitskii, S.M., Shashurin, I.P.: Spatial development of plasma-beam instability. J. Exp. Theor. Phys. **25**, 227–232 (1967)
33. Matzler, C., Bai, T., Crannell, C.J., Frost, K.J.: Adiabatic heating in impulsive solar flares. Astrophys. J. **223**, 1058–1071 (1978)
34. Motorina, G.G., Koudriavtsev, I.V., Lazutkov, V.P., Matveev, G.A., Savchenko, M.I., Skorodumov, D.V., Charikov, Y.E.: The reconstruction of energy distribution of electrons accelerated in solar flare 26.07.2002 on the basis of the hard X-ray data. In: Stepanov, A.V., Nagovitsyn, Yu.A. (eds.) Proceedings of Annual Conference on Physics of Sun, Solar

and Solar–Terrestrial Physics, pp. 171–175. Pulkovo Observatory, St. Petersburg (2011) (in Russian)
35. Motorina, G.G., Koudriavtsev, I.V., Lazutkov, V.P., Matveev, G.A., Savchenko, M.I., Skorodumov, D.V., Charikov, Y.E.: The reconstruction of energy spectrum of electrons accelerated in solar flare 15 April 2002. In: Stepanov, A.V., Nagovitsyn, Yu.A. (eds.) Proceedings of Annual Conference on Physics of Sun, Solar and Solar–Terrestrial Physics, pp. 301–304. Pulkovo Observatory, St. Petersburg (2012) (in Russian)
36. Motorina, G.G., Koudriavtsev, I.V., Lazutkov, V.P., Matveev, G.A., Savchenko, M.I., Skorodumov, D.V., Charikov, Y.E.: On the reconstruction of the energy distribution of electrons accelerated in solar flares. Tech. Phys. **57**(12), 1618–1622 (2012)
37. Priest, E.R.: The magnetohydrodynamics of current sheets. Rep. Prog. Phys. **48**, 955 (1985)
38. Pritchett, P.L., Wu, C.C.: Coalescence of magnetic islands. Phys. Fluids **22**, 2140–2146 (1979)
39. Serebrennikov, M.G., Pervozvanskiy, A.A.: The Findings of Hidden Periodicities, 244 pp. M. Nauka, Moscow (1965) (in Russian)
40. Sklyarova, E.M., Charikov, Y.E.: Thermal flares: charge particle acceleration and plasma heating. In: Stepanov, A.V., Nagovitsyn, Yu.A. (eds.) Proceedings of Annual Conference on Physics of Sun, Solar and Solar–Terrestrial Physics, pp. 79–383. Pulkovo Observatory, St. Petersburg (2010) (in Russian)
41. Sklyarova, E.M., Charikov, Y.E.: X-ray radiation of thermal flares on the sun. In: Stepanov, A.V., Nagovitsyn, Yu.A. (eds.) Proceedings of Annual Conference on Physics of Sun, Solar and Solar–Terrestrial Physics. Pulkovo Observatory, St. Petersburg, pp. 187–191 (2011) (in Russian)
42. Somov, B.V.: The storage and liberation of energy in solar flares. In: Proceedings of XII Seminar on Cosmophysics, LIIYaF, Leningrad, 6–8 February, 1982, pp. 6–49 (in Russian)
43. Somov, B.V., Vernetta, A.J.: Magnetic reconnection in a high-temperature plasma of solar flares. IV - Resistive tearing mode in non-neutral current sheets. Sol. Phys. **120**, 93–115 (1989)
44. Sturrock, P.A.: Model of the high-energy phase of solar flares. Nature. **211**(5050), 695–704 (1966)
45. Syrovatskii, S.I., Shmeleva, O.P.: Heating of plasma by high-energy electrons, and nonthermal X-ray emission in solar flares. Sov. Astron. **16**(2), 273–283 (1972)

# Study of Solar Flares and Gamma-Ray Bursts in the Helicon Experiment

**E.P. Mazets[1], R.L. Aptekar, S.V. Golenetskii, V.N. Il'inskii, V.D. Pal'shin, Z.Ya. Sokolova, D.D. Frederiks, and M.V. Ulanov**

Detailed data on temporal profiles, energy spectra, and spectral variability of hard X-ray and gamma-ray flares of solar origin have been obtained in the energy range from 10 keV to 10 MeV in the Helicon experiment for the entire period of observations, beginning with the solar activity maximum in 2001.

Synchronously with the KONUS–WIND experiment onboard the U.S. Wind spacecraft, the Helicon experiment has also investigated cosmic gamma-ray bursts and soft gamma repeaters. The Helicon instrument performed unique observations of the giant flare from SGR1806-20 soft gamma repeater and of its reflection from the Moon. The intensity, history, and energy of the initial pulse of the giant flare from the gamma repeater are reconstructed for the first time with high reliability. A natural location of the Moon by a high-intensity flux of X-ray and gamma radiation has been recorded for the first time in the history of the extraterrestrial astronomy.

## Scientific Purposes of the Helicon Experiment

The main scientific aim of the Helicon experiment was to investigate the temporal profiles and energy spectra of hard X-ray and gamma-ray flares of solar origin in the energy range from 10 keV to 10 MeV. According to the existing concepts, the temporal and spectral characteristics of the hard electromagnetic radiation of solar flares are directly related to the pulsed energy release and particle acceleration

---

[1]Deceased

E.P. Mazets · R.L. Aptekar (✉) · S.V. Golenetskii · V.N. Il'inskii · V.D. Pal'shin · Z.Ya. Sokolova · D.D. Frederiks · M.V. Ulanov
Ioffe Institute, St. Petersburg, Russia
e-mail: aptekar@mail.ioffe.ru

V. Kuznetsov (ed.), *The Coronas-F Space Mission*, Astrophysics and Space Science Library 400, DOI 10.1007/978-3-642-39268-9_15,

processes in the course of a solar flare. Combined with the results of solar observations in the ultraviolet, optical, and radio-frequency spectral ranges, these data can serve as a basis for construction of solar flare models.

An additional scientific aim of the experiment was to study cosmic gamma-ray bursts synchronously with the KONUS–WIND experiment onboard the U.S. Wind spacecraft. Recording the gamma-ray burst radiation by identical detectors situated at different points of the space markedly improves the reliability and validity of the information about fine details of the temporal profiles and energy spectra of the bursts.

## Helicon Scientific Instrument

The Helicon scientific instrument was a highly sensitive and highly informative gamma-ray spectrometer developed and fabricated at the Ioffe Institute, Russian Academy of Sciences, to analyze the temporal and spectral characteristics of the solar X-ray and gamma-ray flares, cosmic gamma-ray bursts, and activity of soft gamma repeaters in a wide energy range from 10 keV to 10 MeV.

The instrument comprised two scintillation spectrometric detectors and an electronic unit for data acquisition and processing, which also included interfaces with the onboard systems of the spacecraft. Each detector contained an NaJ(Tl) crystal with a diameter of 130 mm and height of 75 mm, placed in a thin-walled aluminum container with a beryllium entrance window. The scintillation crystal is viewed with an FEU-167 photomultiplier through the exit window of the container, made of highly transparent lead glass, which provides effective protection of the detector from the spacecraft's background in the soft spectral range. This detector provides a low energy threshold of recording of the 10–12 keV radiation with the upper measurement limit of 10 MeV and a burst detection sensitivity of $\sim 10^{-7}$ erg cm$^{-2}$. Each detector's field of view was $2\pi$ sr. The axis of the field of view of one of the detectors was directed along the spacecraft axis oriented toward the Sun. The other detector viewed the anti-solar hemisphere.

The instrument operated in two main modes: background and burst. In the background mode, the hard solar X-ray radiation was monitored in eight energy windows in the range from 10 keV to 200 keV with a temporal resolution of 1 s. The detector oriented toward the Sun was equipped with a multiple-channel analyzer permanently measuring energy spectra in the range from 200 keV to 10 MeV, with an accumulation time of 32 s. The data obtained were uploaded into the onboard memory without interruption of measurements.

The instrument automatically switched over to the burst mode just after a statistically meaningful rise in the counting rate was recorded by the burst detection cell. The temporal profiles of solar flares and gamma-ray bursts were recorded in the following energy ranges: 10–50 keV, 50–200 keV, and 200 keV to 1 MeV with a time resolution of 2 ms to 0.25 s and total recording time of 240 s. The profile recording software also provided a 2-ms resolution during a 0.5-s prehistory before the onset of the burst mode.

The energy spectra of the events were measured in two energy ranges, from 13 to 700 keV and from 200 keV to 10 MeV, with 63 channels and accumulation time adjusted by a special adaptive system from 0.25 to 4 s, depending on the current intensity of a burst. This made it possible to obtain data on the fast spectral variability even for weak events under limited-statistics conditions.

In development and fabrication of the Helicon instrument, measures were taken for providing a stable operation of linear measuring circuits. The detectors were equipped with a specially developed magnetic-protection system, which ensured stable characteristics of the instrument in its motion in the Earth's magnetic field. The photomultipliers were trained for many months and were carefully selected in relation to the amplification coefficient and noise level. These measures, combined with the control over the amplification of the detectors by the position of reference lines in background energy spectra and its possible correction with digital radio commands provided stable operation of the measuring circuits of the spectrometric detectors in the Helicon instrument during many years of observations onboard the CORONAS-F spacecraft.

In measurements of the X-ray and gamma radiation fluxes in near-Earth orbits, there are two types of false events whose recording wastes the useful capacity of the onboard memory unit and thereby restricts the possibility of making observations without loss of solar flares and gamma-ray bursts. One of these events is associated with transit of multiply charged nuclei of cosmic-ray origin, which results in that, because of the high energy release in the scintillation crystal, events are recorded, named imitations of flares and bursts. A system has been developed for suppression of events of this kind by using signals from one of the dynodes closest to the photocathode of the photomultiplier for their inhibition. Another type of false flares and bursts is associated with transient processes in captured-radiation zones and precipitation of charged particle fluxes at high latitudes. These events were discriminated by inhibiting the burst mode on the basis of the counting rate of charged particles at high geomagnetic latitudes. The whole set of these measures provided the optimal use of the capacity of the onboard memory unit, allocated for the Helicon experiment.

Thus, the Helicon instrument provided temporal ($\sim$1 ms) and spectral ($\sim$6–8 %) resolutions adequate to the scientific goal of the experiment in measurements of both stationary and variable fluxes of X-ray and gamma radiation in solar flares, transient sources, and cosmic gamma-ray bursts. The optimal, as regards the technical capacity of the CORONAS-F spacecraft, software for recording the temporal and spectral characteristics of the solar flare related X-ray and gamma radiation and cosmic gamma-ray bursts was used in the Helicon instrument. This software made it possible to obtain, in the course of many years of permanent observations, new data on the pulse phase of solar flares, cosmic gamma-ray bursts, and activity of soft gamma-repeaters.

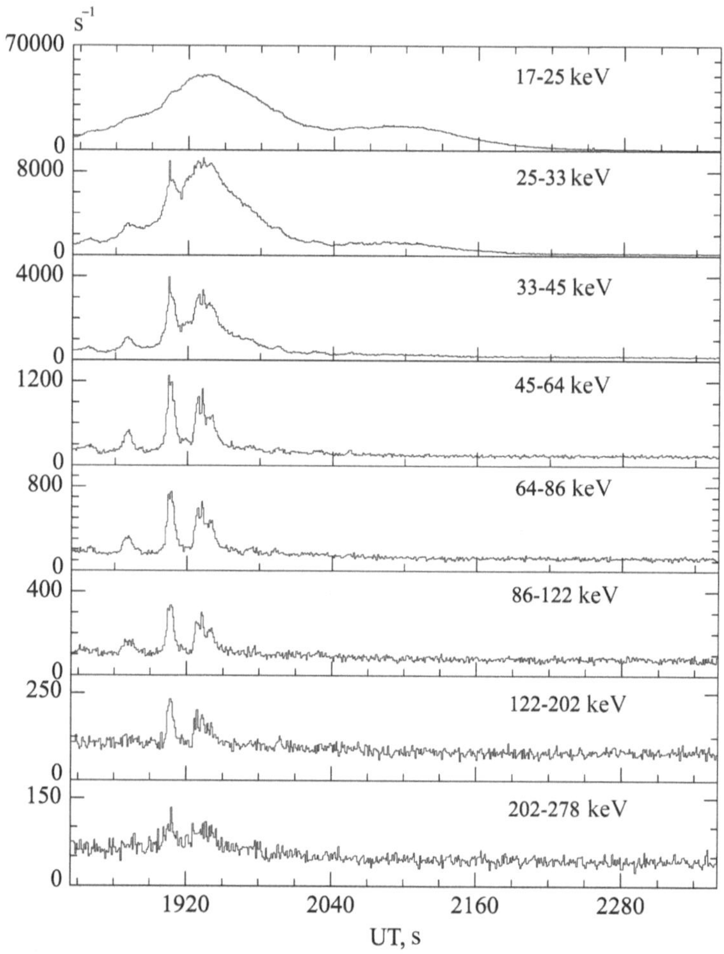


**Fig. 1** Temporal profile of solar flare on December 24, 2001

## Studies of the Pulse Phase of Solar Flares with Helicon Gamma-Spectrometer

The Helicon experiment furnished, in permanent observations during more than 4 years, a vast body of data on manifestations of the solar flare activity in the range of hard X-ray and gamma radiation. A total of 73 solar flares in the burst mode and more than 3,100 events in the background mode have been recorded during the whole period of observations. In the orbit of CORONAS-F spacecraft, observations with a steady radiation background are made in the near-equatorial

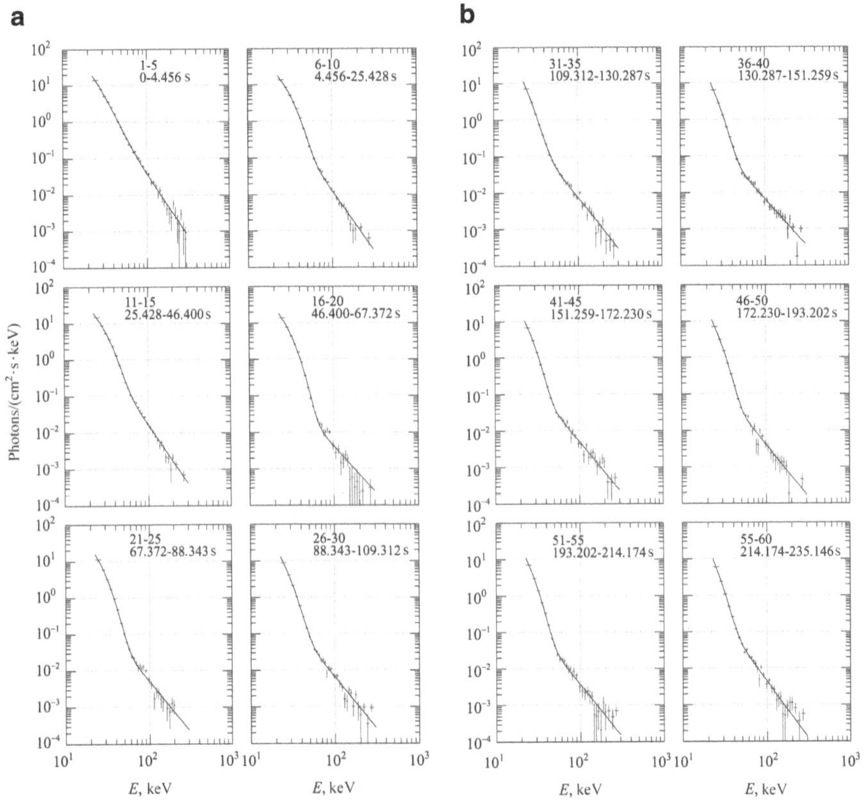


**Fig. 2** Energy spectra of flare on December 24, 2001, registered in different phases of the event ($T_0$ = 00:31:41.895 UT): (**a**) 0–109.312 s; (**b**) 109.312–235.146 s

parts of the orbit during 15–20 min and in the regions of the so-called *polar caps*, where a comparatively stable radiation background is observed during 7–10 min. The flare activity was measured by the Helicon instrument just in these parts of the orbit. The burst signal switching the instrument to the mode of detailed temporal and spectral measurements was formed in the energy range from 50 to 150 keV. Many solar flares have a considerably softer spectrum and were recorded by the Helicon instrument only in the background mode simultaneously in eight channels in the energy range from 10 to 200 keV with a temporal resolution of 1 s and were supplemented with spectral measurements in the energy range from 200 keV to 10 MeV. Such an observation approach proved to be the optimal for analysis of solar flares in the hard X-ray range at event durations of several minutes to tens of minutes and made it possible to obtain new important information about the pulse phase of solar flares.

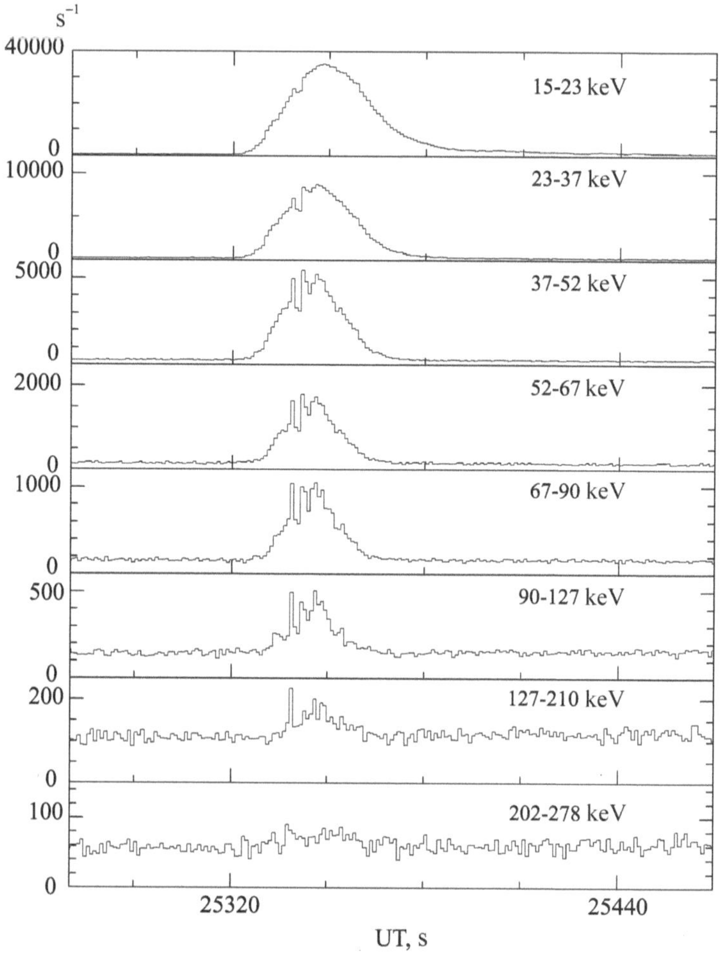


**Fig. 3** Temporal profile of solar flare on October 25, 2002 ($T_0$ = 25331.918 s UT)

Figures 1–6 demonstrate the capacity of the Helicon instrument in studies of the temporal profiles and energy spectra of the X-ray and gamma radiation in the pulse phase of solar flares. Figure 1 shows the temporal profile of the flare on December 24, 2001, measured in the background mode in eight energy channels. Figure 2a and b present successively recorded photon spectra of the flare for time intervals 0–109 and 109–235 s from the beginning of the event. Figure 3 shows the temporal profile of the solar flare on October 25, 2002. The energy spectrum at the peak of this event is presented in Fig. 4. Figures 5 and 6 show the temporal profiles of the solar flares on April 9 and June 12, 2003.

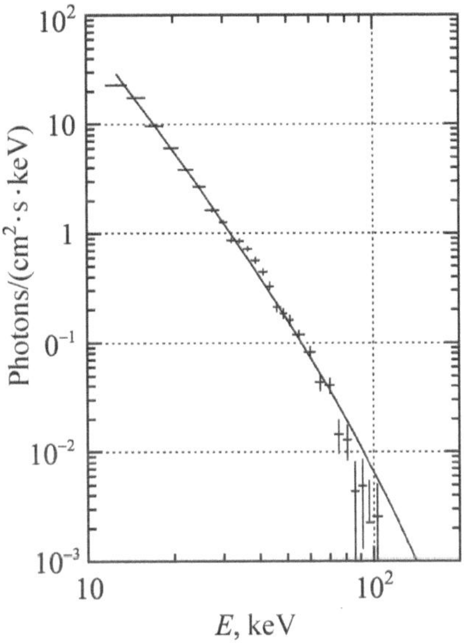


**Fig. 4** Energy spectrum of solar flare on October 25, 2002, at the peak of the event ($T_0$ = 25332.226 s UT)

## Studies of Cosmic Gamma-Ray Bursts and of the Activity of Soft Gamma Repeaters

In studies of cosmic gamma-ray bursts and soft gamma repeaters, the main task of the Helicon instrument was to perform synchronous observations with the KONUS–WIND instrument onboard the U.S. Wind spacecraft. During the operation time of the Helicon instrument, 89 cosmic gamma-ray bursts and 39 repeated bursts from the SGR1900-14 and SGR1806-20 gamma repeaters have been detected and analyzed in detail. The information about the temporal and spectral characteristics of the most interesting events has been published in 32 international electronic circulars by the GCN (Gamma-Ray Burst Coordinate Network, gcn.gsfc.nasa.gov/gcn/gcn3_archive.html).

On December 27, 2004, the Helicon instrument performed, synchronously with KONUS–WIND, unique observations of a giant flare from the SGR1806-20 gamma repeater. The giant flare from the gamma repeater is a short, exclusively high-intensity initial pulse of gamma radiation, which gives way to a weaker, decaying in ~500 s, pulse with a tail pulsating with a period characteristic of a neutron star period. The detectors of the instrument are fully overloaded ("saturated") by the initial pulse. Instead of a precise measurement of the energy characteristics of the initial pulse, only rough estimates become possible. The unique nature of the observations of the event on December 27, 2004 consists in that the CORONAS-F spacecraft was screened from the radiation source by the Earth, but its instruments

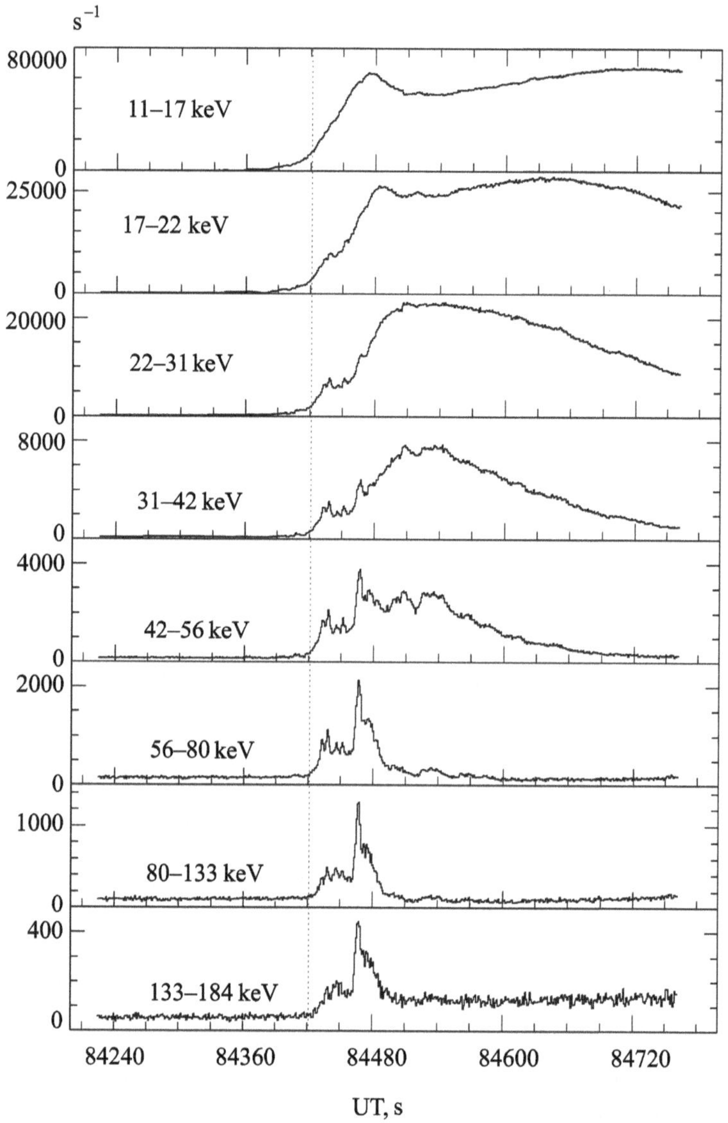


**Fig. 5** Temporal profile of solar flare on April 9, 2003

were exposed to the full Moon (Fig. 7). The detector of the Helicon instrument, identical in its characteristics to the detectors on Wind, clearly detected the inverse Compton reflection of the giant pulse from the Moon and measured the temporal profile of the signal and its energy spectrum. The detection of the signal reflected from the Moon made it possible to reconstruct for the first time with high reliability

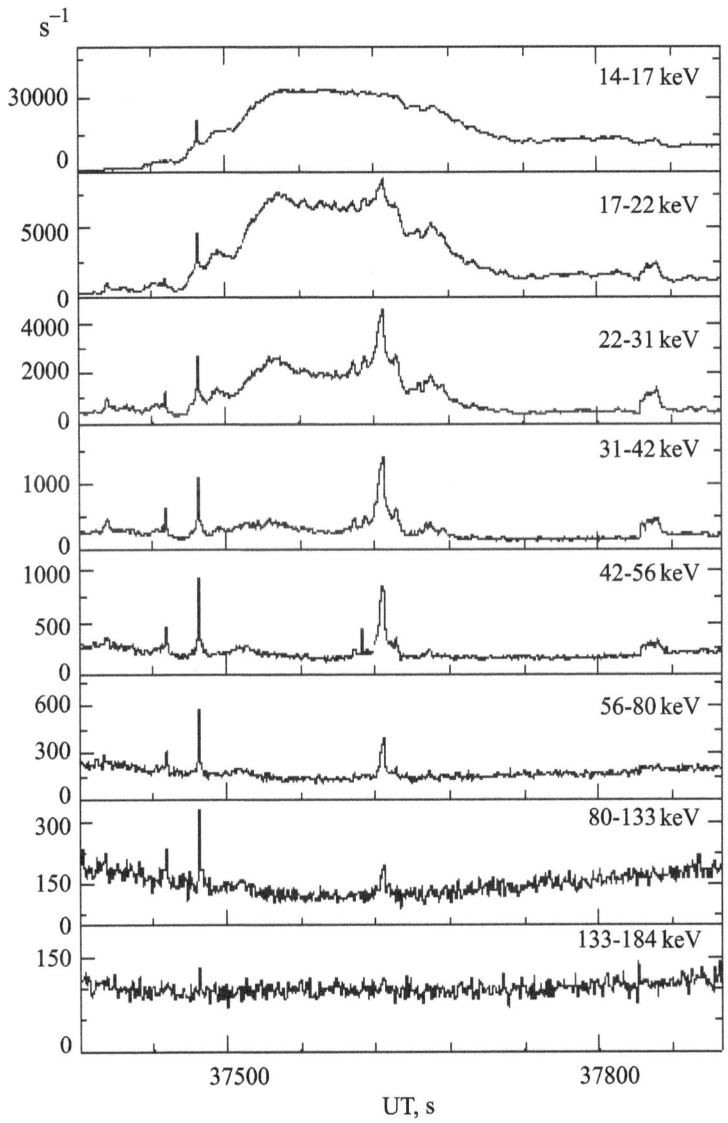


**Fig. 6** Temporal profile of solar flare on June 12, 2003

the intensity, temporal history (Fig. 8), spectrum, and energy parameters of the giant flare from the gamma repeater. It should be noted that natural location of the Moon by a high-power flux of X-ray and gamma radiation from a cosmic source has been observed for the first time in the history of the extraterrestrial astronomy [1].

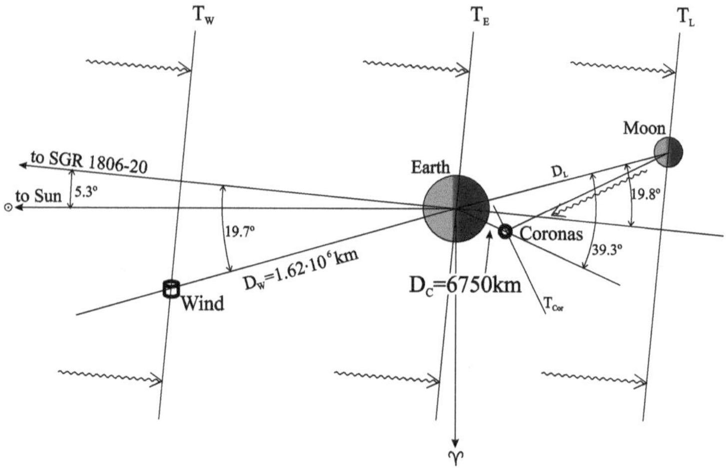


**Fig. 7** Scheme illustrating the Helicon and KONUS–WIND observations of the giant flare. The leading edge of the flare from SGR 1806-20 arrives at Wind at time $T_{\mathrm{W}}$, passes by the Earth at $T_{\mathrm{E}} = T_{\mathrm{W}} + 5.086\,\mathrm{s}$, reaches the Moon and is reflected from it, and, finally, the reflected emission reaches the Helicon detector at $T_{\mathrm{Cor}} = T_{\mathrm{W}} + 7.69\,\mathrm{s}$. From [1]

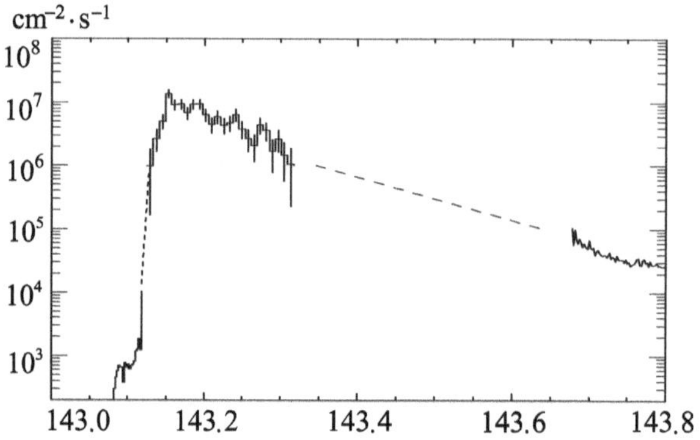


**Fig. 8** Reconstructed temporal history of initial pulse of the giant flare. From [1]

# Conclusions

The many years' Helicon experiment has been successfully performed onboard the CORONAS-F space observatory. The highly sensitive and informative gamma spectrometer was used to investigate the temporal and spectral characteristics of solar X-ray and gamma radiation flares and cosmic gamma-ray bursts in a wide energy range from 10 keV to 10 MeV. Unique observations of the giant flare from the SGR1806-20 source of soft recurrent bursts, reflected from the Moon, made it possible to reliably reconstruct for the first time the intensity, light curve, spectrum, and energy parameters of the giant flare from the gamma repeater. A natural location of the Moon by an exceedingly intense flux of X-ray and gamma radiation from a source in the space was observed for the first time in the history of the extraterrestrial astronomy.

## Reference

1. Frederiks, D.D., Golenetskii, S.V., Palshin, V.D., Aptekar, R.L., Ilyinskii, V.N., Oleinik, F.P., Mazets, E.P., Cline, T.L.: Giant flare in SGR 1806-20 and its Compton reflection from the Moon. Astron. Lett. J. Astron. Space Astrophys. **33**, 1–18 (2007)

# RPS-1 Experiment

**V.M. Pankov[1], V.L. Prokhin[2], N.G. Khavenson, A.A. Gusev, Yu.D. Kotov, A.S. Glyanenko, A.N. Afanas'yev, and A.A. Karapet'yants[3]**

This chapter deals with the results of observation of the Sun and the Earth upper nighttime atmosphere obtained in the X-ray energy range of 3–31.5 keV with the aid of a semiconductor CdTe spectrometer. The energy spectra of some weak solar X-ray flares at different stages of their evolution are shown along with the spectra of the pre-flare X-ray background. The X-ray glow of the nighttime atmosphere is displayed on the Earth blank maps by plotting the mean counting rates of the device in four energy ranges (3–5, 3–8, 8–16, and 16–31.5 keV) over a chosen observation interval recorded along the satellite trajectory.

[1]Deceased

[2]Deceased

[3]Deceased

V.M. Pankov · V.L. Prokhin · N.G. Khavenson (✉) · A.A. Gusev
Space Research Institute RAS, Moscow, Russia
e-mail: xaba@rssi.ru

Yu.D. Kotov · A.S. Glyanenko
Institute of Astrophysics, National Research Nuclear University "MEPhI", Moscow, Russia

A.N. Afanas'yev
Pushkov Institute of Terrestrial Magnetism, Ionosphere, and Radio Wave Propagation (IZMIRAN), Troitsk, Moscow, Russia

A.A. Karapet'yants
State Design Office, Dnepropetrovsk, Ukraine

V. Kuznetsov (ed.), *The Coronas-F Space Mission*, Astrophysics and Space Science Library 400, DOI 10.1007/978-3-642-39268-9_16, 

## Introduction

An objective of the completed CORONAS-F space mission was the observation of the Sun in the X-ray wavelength range with the aim to improve the diagnostics and forecast of solar activity and to study the solar–terrestrial coupling. At present, patrol observations of the Sun in the soft X-ray range (1.5–12.5 keV) are carried out for this purpose onboard the Geostationary Operational Environmental Satellite (GOES). The GOES data are used to classify the solar X-ray flares. The class of a flare is determined by the flux emitted at the maximum of its evolution. A drawback of this classification is the lack of data on the energy spectra and incomplete information on the temporal characteristics of solar flares, which are important for understanding the physics of the solar activity. Experimental data on the X-ray spectra of solar flares in the energy intervals covering the soft and hard components are rather scarce [3,9,10]. Onboard the CORONAS-F mission, the solar X-ray radiation in the energy range of 3–31.5 keV was measured with the RPS-1 semiconductor spectrometer, in which a cadmium telluride detector cooled by a semiconductor microcooler was used for the first time for fine spectrometry under cosmic conditions [5, 6]. The detector has high recording efficiency and radiation resistance, which is of great importance for long-term space experiments.

In the course of the experiment, the spacecraft entered the Earth shadow and recorded variations of the soft X-ray radiation of the upper nighttime atmosphere and the response of the spectrometer to energetic charged particles of the Earth radiation belts (ERBs) in the high-latitude regions and South-Atlantic magnetic Anomaly (SAA).

The X-ray radiation of the Earth upper atmosphere is secondary by its nature. In the energy range under consideration, it is mainly due to the albedo bremsstrahlung of electrons precipitating from the magnetosphere. The most intensive particle fluxes precipitate in the auroral zones (auroral oval). The X-ray radiation from these zones was detected in near-Earth space [4, 8].

Long-term high-sensitivity observations of the nighttime atmosphere carried out with RPS-1 made it possible to gain a global pattern of the atmospheric radiations and represent them in the form of X-ray maps of the Earth.

## RPS-1 Device and Its Basic Parameters

The base of the RPS-1 spectrometer was a cadmium-telluride detector not deeply cooled by a semiconductor thermoelectric microcooler to ensure the best energy resolution, which allowed fine spectrometry of the recorded X-ray radiation. In this experiment, a cadmium-telluride detector was used for the first time for spectrometric measurements in space.

RPS-1 was composed constructively of two separate instruments (units): the detecting unit RPS-1BD and the unit of processing electronic equipment of the time–amplitude spectrometer (AVS-F), which formed part of the CORONAS-F science payload [2]. RPS-1BD recorded, amplified, and formed signals triggered by the radiation incident onto the detector. These signals arrived at the input of AVS-F, whose tasks included the amplitude analysis and signal integration in the 32-channel analog-to-digital converter (ADC). The information accumulated during the chosen exposure was output to the telemetry of the research data acquisition system (SSNI) and was transmitted to the ground during communication sessions with the satellite.

The RPS-1BD instrument consisted of separate functional units. The main one was the cadmium telluride detecting unit (CTDU); the other units formed the infrastructure, which ensured its normal operation. The incident X-ray radiation passed through a thin beryllium window of CTDU and interacted with the material of the detector, located at the cold junction of the thermoelectric cooler. The signals from the detector were amplified by a charge-sensitive preamplifier, which was also incorporated in the CTDU.

The main parameters of the detecting unit employed in the experiment are listed in the table.

Basic parameters of the cadmium telluride detecting unit CTDU

| Parameter | Value |
|---|---|
| Interval of detected energies (keV) | 3–31.5 |
| Thickness of the beryllium window (μm) | 50 |
| Detector area ($mm^2$) | 46 |
| Thickness of the detector (mm) | 1.39 |
| Operating voltage (V) | 100 |
| Operating detector temperature (°C) | −35 |
| Energy resolution at $10^3$ counts $s^{-1}$ (keV): | |
| in the 5.9 keV $Fe^{55}$ line | 0.74 |
| in the 13.87 keV $Am^{241}$ line | 0.88 |
| Degradation of energy resolution at $10^4$ counts $s^{-1}$ in the 5.9 keV line (%) | 20 |
| Effective detector area in the 5.9 keV line ($mm^2$) | 33 |
| Conversion coefficient of the CTDU (μV/keV) | 1,670 |
| Instability of the CTDU conversion coefficient during 8 h of continuous operation (%) | 0.36 |
| Power consumption of the TEMC at the ambient temperature of +30 °C (W) | 5.2 |
| Dimensions without cable (mm) | 112 × 52 |
| Mass (kg) | 0.38 |

The CTDU infrastructure included the following functional units: the driving amplifier, which received pulses from the charge-sensitive preamplifier and sent its output to the input of AVS-F, the secondary power supply of CTDU electronics, and the driving amplifier, which supplied operating voltage to the detector. The infrastructure also comprised a heat-control unit fed directly from the onboard power system to maintain (via a thermal resistor located at the cold junction of the microcooler) the detector's temperature ensuring the best energy resolution, and a radiator for heat abstraction.

The RPS-BD had a mass of 1.8 kg, a power consumption of 8.5 W, and dimensions $190 \times 215 \times 330\,mm^3$.

The tests of RPS-BD showed that an anti-interference filter had to be introduced into the circuit of the thermoelectric cooler; such a filter was then incorporated into the instrument.

To ensure normal long-term operation of the spectrometer, the detector's power supply was switched off for several minutes every 8 h in order to eliminate the negative effect of the accumulated bulk charge (depolarization).

Radioactive sources were used to calibrate the energy range of the spectrometer [1], i.e., the corresponding energies of measured radiation were attributed to the channels of AVS-F ADC. The energy interval of 3–9 keV fell within the first 12 channels 0.5 keV wide each. The other 12 channels had the width of 1 keV each, and the remaining 8 channels 1.3125 keV wide each shared the energies from 21 to 31.5 keV.

## Some Results of the Spectrometry of Solar X-Ray Emission

Information on the RPS-1 experiment was transmitted during the communication sessions with the satellite in the form of the distribution of counts recorded and accumulated by the ADS channels of AVS-F in every 16-s operation interval (measuring frame). Each measuring frame was tied to UT and the geographic coordinates of the satellite.

To analyze the information received during the communication sessions and to identify the recorded flare events, we used the time dependences of the count rates in the energy intervals of 3–5, 5–8, 8–16, and 16–31.5 keV (survey plots). The signal count rate was determined by dividing the total number of counts of the device recorded in the given energy range by the measuring frame time, i.e., by 16 s. The survey plots displayed on the computer allowed online search and identification of flare-related signals (bursts of count rate), which were compared with GOES data. The identified flares were assigned the corresponding classes.

Figure 1 represents a part of the survey plot based on the data of communication session no. 21165 (the number of SC rotation when the communication session was held) on 02.05.2005. The figure illustrates a series of weak flares of class B and the response of the detector when it crossed the ERB. The figures at the peaks of the curve of count rates in the energy range of 3–5 keV are the flux vales at the maximum of flare evolution in the units $10^{-7}\,W/m^2$ (or $10^{-4}\,erg/cm^2\,s$). The flare time profiles are also clearly seen on the count-rate curve in the energy range of 5–8 keV. The emission recorded in the 8–16 keV channel was insignificant. It is noticeable in the figure, yet the count rate is not specified. It should be noted that all flares represented in Fig. 1 occurred in or near one solar region with the coordinates S08E83.

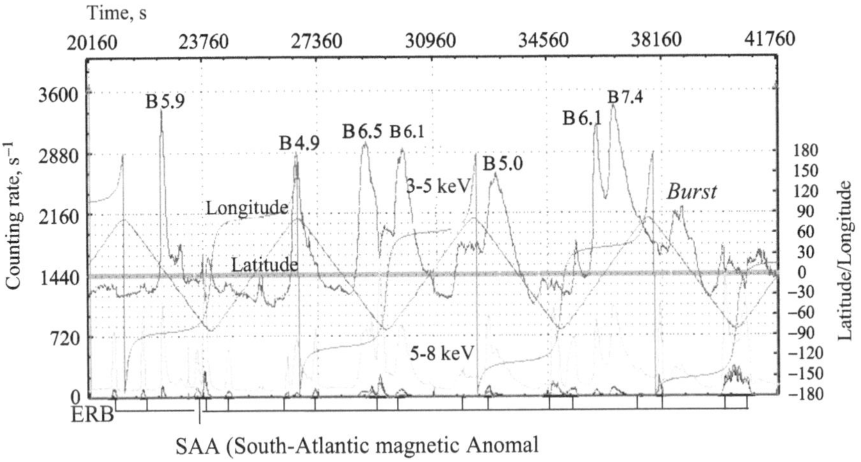


**Fig. 1** A series of solar flares of importance B, a burst event not recorded by GOES, and the response of the instrument to the passage through ERB (02.05.05, from 04:24)

Figure 1 shows also a weak flare event (burst), which was recorded at the decay of flare B7.4, but was not confirmed by GOES data. The burst started after the satellite had left the ERB region over the North Pacific (the instrument response to ERB was recorded) and ended when it was over the South Pacific. This event appears to have been caused by the bremsstrahlung of electrons precipitating from the magnetosphere as a result of changing geomagnetic conditions. Such magnetospheric events were recorded occasionally during the experiment, mainly when the satellite was passing the lighted portion of the orbit over the Pacific. These events are interesting to the study of precipitating particle fluxes and their spatial distribution, i.e., to the study of solar–terrestrial coupling.

The weak solar flares like those shown in Fig. 1 were recorded under low solar activity conditions determined by the number of sunspots, because the intensity and hardness of the solar X-ray background (without flares) increased with the increasing activity. For example, according to the counting rate in the 3–5 keV channel, the background radiation flux was $\sim 300\,s^{-1}$ under low activity conditions and could exceed this value by an order of magnitude or more at the maximum activity. The maximum energy of the background X-ray emission could change from ~6 to ~20 keV, as it did, e.g., in October 2003 (see Fig. 2).

Figure 2 represents the spectra of the solar background radiation recorded in the equatorial region in October 2003 when the level of solar activity changed dramatically. Each spectrum is provided with the number of the communication session, the date, the sunspot number $W$, and the background flux value according to GOES measurements expressed in the same units as the flare radiation. As is evident from both GOES and RPS-1 data, the solar background radiation flux and its hardness increased with the increasing solar activity.

During the experiment, a lot of weak flares (and magnetospheric events) were recorded as the satellite was passing outside the ERB zones. Besides, a number of

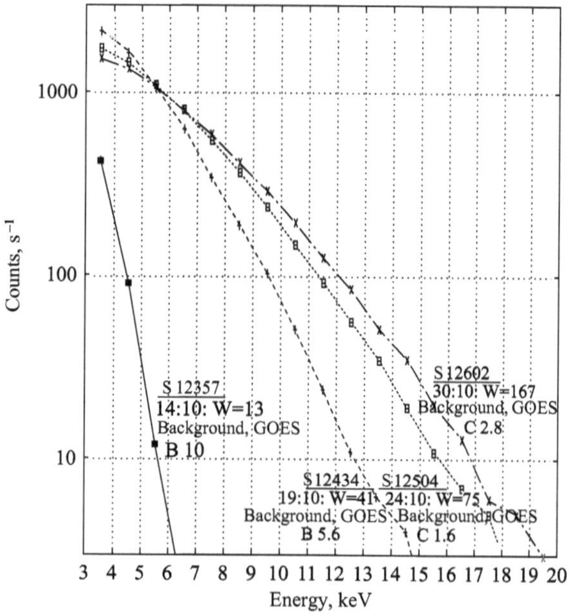


**Fig. 2** The spectra of solar background radiation (without flares) in October 2003 recorded in the equatorial region. Written at each curve are the number of the communication session, the date, the sunspot number $W$, and the background flux according to GOES data. From GOES Data from: http://www.swpc.noaa.gov/Data/goes.html (free access)

flares, whose time profiles were not distorted severely, were recorded inside these zones. Some of the flare and background radiation spectra are given in [7].

Unfortunately, the capabilities of ADC did not allow a detailed evolution study of major solar flares of class $>$ C 3.0, because the detector began to miscount at large loads as seen from the cut of spectra in the soft range in Fig. 2. The major flares were recorded by RPS-1 and were confirmed by GOES data, but their spectral variations and maximum energy could only be estimated in the initial and final phases of the flare evolution.

The weak flares recorded had their particular time and spectral characteristics. Represented here are some selected flares that occurred at different times during the experiment.

Figure 3 shows the time profiles of three bursts and their spectra at points 1–5 recorded along the lighted part of the satellite trajectory at a low level of background radiation (1) during the communication session 18019 on 13.10.2004. The first burst was identified with the weakest flare B1.0; the following two bursts were not recorded by GOES, most probably, because of insufficiently high time resolution. On the other hand, these bursts could hardly be attributed to precipitating magnetospheric electrons, since the latter used to cause much longer burst events.

We believe that the three X-ray flares under discussion manifest fast and abrupt temperature changes at the source. Since the signal counting rates at the maximum of the first and the third flares (channels 3–5 and 5–8 keV) were virtually equal (points 2, 5) and their spectra coincided, the last flare had to be a flare of class B1.0. The intermediate (second) one was a fortiori weaker and can be classified as a flare of class A with the maximum flux $< 10^{-7}$ W/m$^2$.

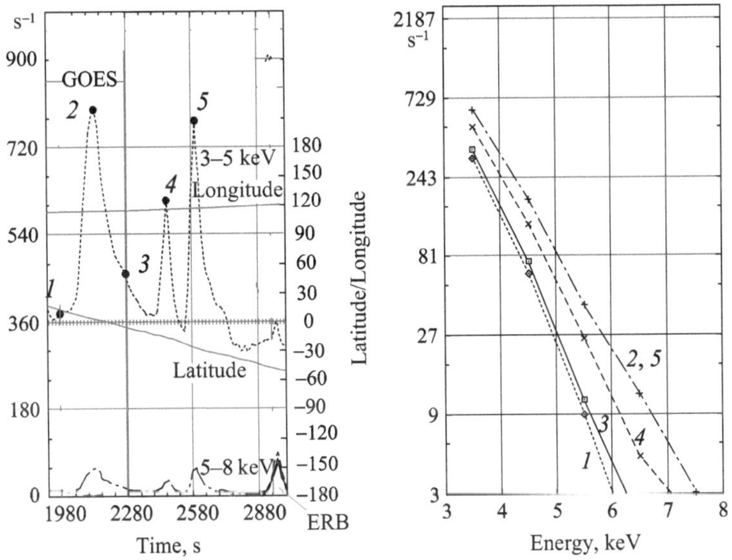


**Fig. 3** The time profiles in the energy ranges of 3–5 and 5–8 keV and the spectra at points 1–5 of three flares recorded by RPS-1 on 13.10.2004, of which only the first one was confirmed by GOES data (B 1.0; the beginning at 00:31)

Figure 4 illustrates the time profiles in three energy intervals and the spectra at points 1–3 for the flare of importance B 7.6 (stronger than that shown in Fig. 3) recorded by RPS-1 on 21.08.2003. As seen from the comparison of Figs. 3 and 4, the B 7.6 flare occurred against a more intensive and harder background (pre-flare) radiation (point 1), which, nevertheless, did not disguise its spectral and temporal characteristics. It also follows from Fig. 4 that the flare was characterized by a long decay (≪step≫), and its spectrum at the peak of evolution (point 2) was much harder than that of the B 1.0 flare (see 2 in Fig. 3).

A flare of B 9.5 class was recorded on 02.05.2005 against a relatively low radiation background. Its time profiles and spectra (points 1–4) are shown in Fig. 5. The beginning, maximum, and end of the flare (points 1, 3, 4) are given according to GOES data. RPS-1 did not fix the end of the flare, because the satellite entered the Earth's shadow. The maximum energy of this flare (spectrum at point 3) was above the upper boundary of the GOES measuring range—24.8 keV.

In the same way as above (Figs. 2–5), Fig. 6 illustrates a still more intensive flare C 1.2 (the flux at the maximum equal to $1.2 \cdot 10^{-6}\,\mathrm{W/m^2}$), which occurred on 08.11.2005 against a relatively low radiation background (point 1). Again, the beginning, maximum, and end of the flare are given according to GOES. Like in the previous case, RPS-1 did not fix the end of the flare because of the shadow-enter. As seen from Fig. 6, the evolution of this flare passed through two maxima, the second maximum having a harder spectrum than the first one. One can see that the signals in the highest energy interval (8–16 keV) appeared about a minute after the flare had begun. In the second maximum, the count rates began to increase virtually simultaneously with the appearance of the signal.

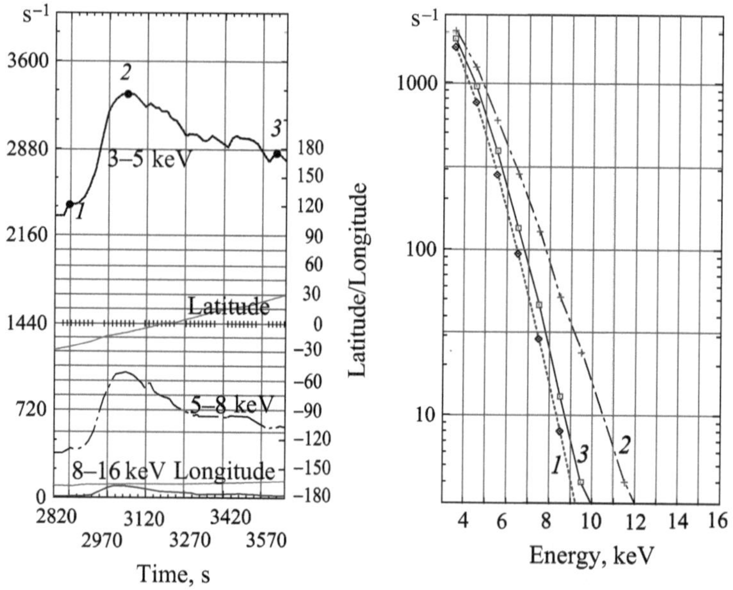


**Fig. 4** The time profiles in the energy ranges of 3–5, 5–8, and 8–16 keV and the spectra at points 1–3 of the class B 7.6 flare recorded by RPS-1 on 21.08.2003 (the beginning at 15:37)

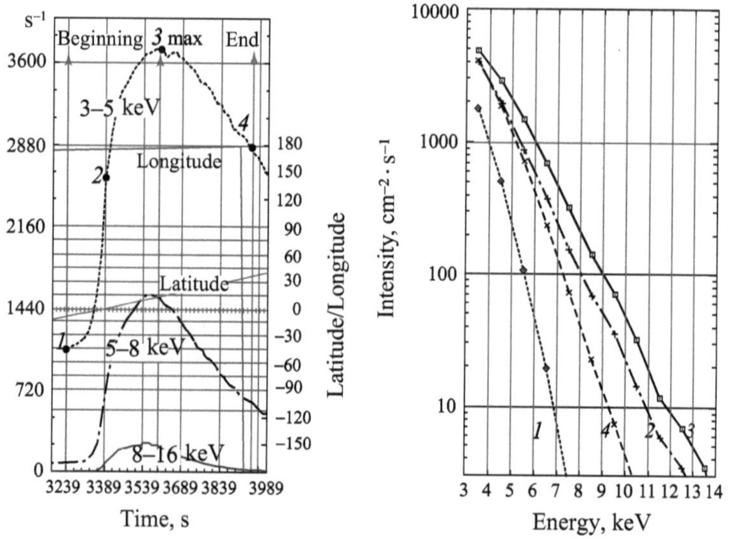


**Fig. 5** The time profiles in the energy ranges of 3–5, 5–8, and 8–16 keV and the spectra at points 1–4 of the class B 9.5 flare recorded by RPS-1 on 02.05.2005 (the beginning at 04:24)

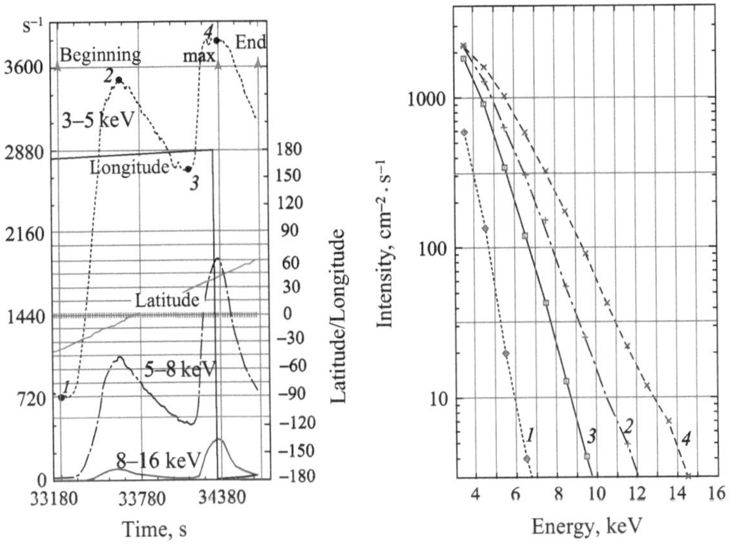


**Fig. 6** The time profiles in the energy ranges of 3–5, 5–8, and 8–16 keV and the spectra at points 1–4 of the class C 1.2 flare recorded by RPS-1 on 08.11.2005 (the beginning at 17:28)

Thus, Figs. 3–6 illustrate weak solar flares with the variety of their spectral and temporal characteristics. The hardness of the flare radiation spectrum at the maximum of evolution increases with the increasing flare class, and the hardness of the background spectrum, with the increasing solar activity.

## The Nightglow of the Earth Upper Atmosphere

The source of the Earth X-ray emission in near-Earth space is its upper atmosphere. This emission is produced as the atmosphere is affected by the solar radiation fluxes, charged particles precipitating from magnetospheric sources, and galactic cosmic rays GCR. The degree of this effect on the Earth atmosphere, i.e., the radiation emitted to near-Earth space, strongly depends on the level of solar and geomagnetic activity. The contribution of GCR to this radiation is insignificant, because the high-energy particles of GCR penetrate deep into the atmosphere (contrary to the less energetic magnetospheric particles), and the secondary radiation is absorbed intensively in the upper atmospheric layers.

The main part of the X-ray bremsstrahlung radiation from precipitating magnetospheric electrons propagates inside the atmosphere (particularly, high-energy electrons) and sideways, i.e., to the lower hemisphere. As the electron flux is moving downward, about 35 % of the bremsstrahlung radiation goes to the upper hemisphere at the electron energy $E_e = 10$ keV and about 20 %, at $E_e = 50$ keV. This is due to the fact that, with the increase of $E_e$, the preferential distribution of bremsstrahlung quanta is in the direction of the velocity vector of the parent electron.

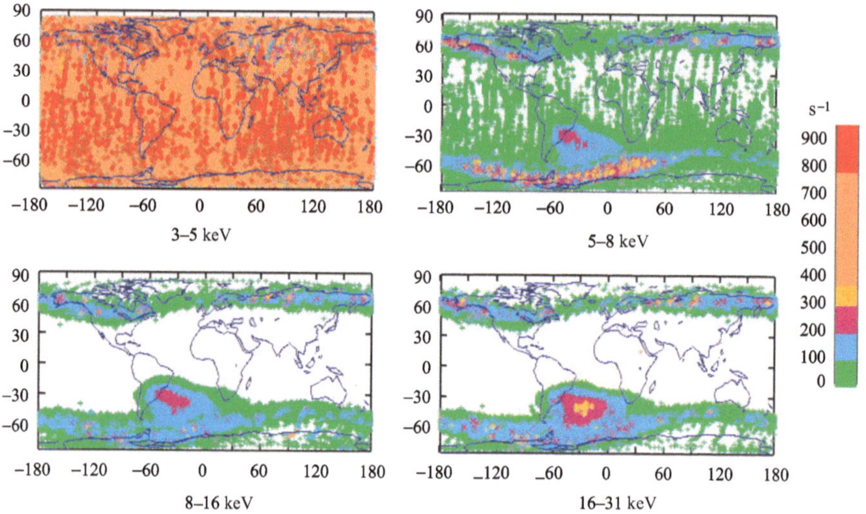


**Fig. 7** Atmospheric nightglow maps plotted in four energy ranges (keV) 3–5, 5–8, 8–16, and 16–31 based on observation data for an annual period from 23.03.2002 to 23.03.2003

It is to be noted that there might be a probability of the diffuse cosmic X-ray background falling within the view of RPS-1.

The most intensive particle fluxes precipitate at the latitudes of the auroral oval, i.e., in the auroral zone. Hence, this zone is the source of the most intensive X-ray emission coming from the atmosphere. The spectra of X-rays in the energy range of 1.4–8 keV emitted from auroral arcs were obtained from observations at nadir onboard the DMSP satellite at the altitude of 230 km [4].

X-ray images of the Earth in the range of 2–20 keV were taken onboard the Polar satellite. The brightest glow is observed in the auroral zone [8].

Outside the auroral zone, there are other regions of precipitation of magnetospheric particles. They are characterized by their own particularities that allow us to isolate the mid-latitude, low-latitude, and equatorial zones [11]. Most of the particles in the magnetispheric fluxes are usually electrons, which move mainly downwards. Hence, these zones are also sources (though weaker ones) of the atmospheric X-ray fluxes emitted upwards (to near-Earth space).

Long-term, high-sensitivity observations of the nighttime atmosphere with RPS-1 gave us an idea of the flux values and energy of X-ray quanta emitted from these sources to the upper hemisphere.

Figure 7 represents the atmospheric nightglow maps based on the measurements in four energy intervals of 3–5, 5–8, 8–16, and 16–31.5 keV taken in the period from 23.03.2002 to 23.03.2003. Examining the maps, one can see that signals with the energies > 8 keV were only recorded in the ERB zones.

Figure 8 shows the maps of global radiation of the Earth upper atmosphere based on the data for the period from 23.03.2002 to 23.09.2002 (summer 2002), and Fig. 9

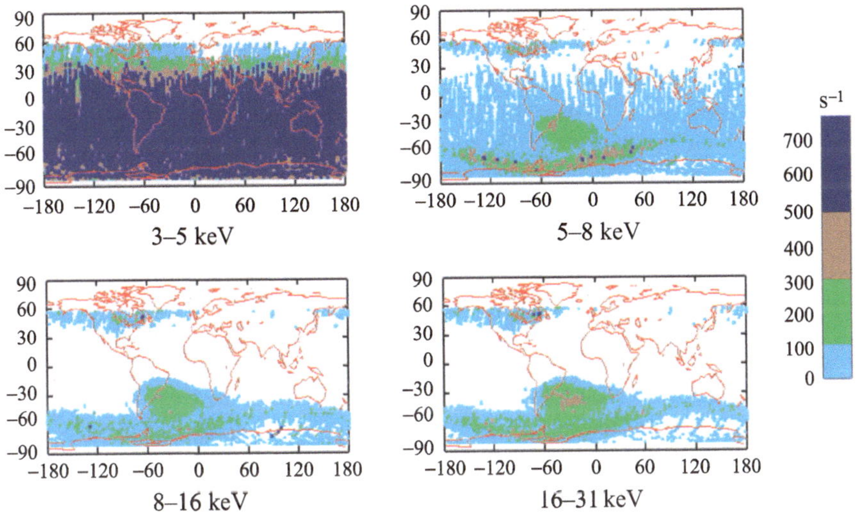


**Fig. 8** Atmospheric nightglow maps plotted in four energy ranges (keV) 3–5, 5–8, 8–16, and 16–31 based on observations for the period from 23.03.2002 to 23.09.2002 (summer)

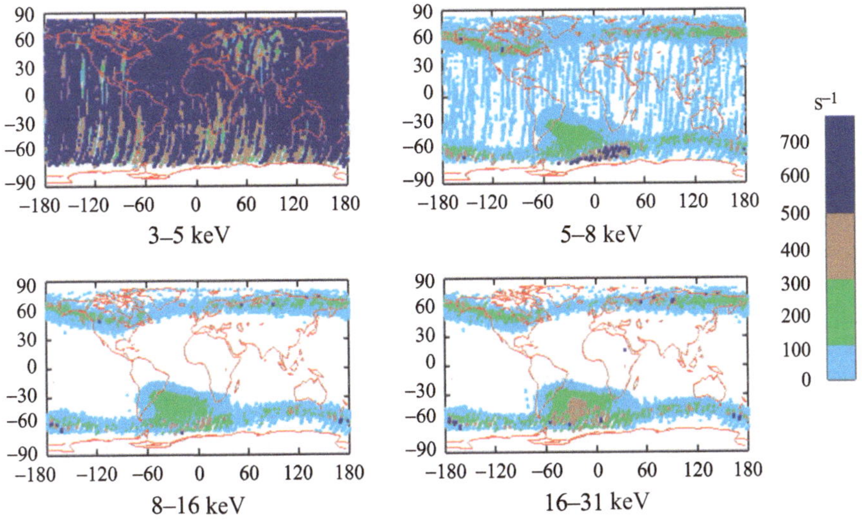


**Fig. 9** Atmospheric nightglow maps plotted in four energy ranges (keV) 3–5, 5–8, 8–16, and 16–31 based on observations for the period from 23.09.2002 to 23.03.2003 (winter)

represents similar maps for the period 23.09.2002 to 23.03.2003, i.e., for the time interval between the equinoxes. The comparison of the series of maps reveals, firstly, the boundary of the polar day on the maps in the energy range of 3–5 keV.

Secondly, the summer map in the range of 5–8 keV shows that the glow from the northern ERB (unlike that from the southern one) was not detected in the zone from

Europe eastward to the Chukchi Peninsula. This might imply that the field lines of the Earth magnetic field on the night side were either blown up to the height > 500 km or got disrupted as a result of high solar activity after the epoch of maximum of 2001. Indeed, the comparison with solar data shows that frequent and intensive proton flares and an increase of the solar-wind velocity and density were recorded in the period under discussion. These events might have caused ≪disruptions≫ in the northern ERB. If that was the case, the weak nightglow of the Earth atmosphere in the disruption region visible on the 3–5 keV maps (see Fig. 8) was due to the secondary processes of GCR effect (a contribution of diffuse cosmic radiation was also possible).

The atmospheric glow observed on the maps at the emission energies > 8 keV was, in our opinion, the response of the detector to penetration of energetic charged particles trapped by the Earth magnetic field (electrons with the energies > 100 keV and protons with the energies ⩾ 3 MeV), which might enter the device input window, and to the products of their reaction with the material of the detector and satellite. These particularities were not seen on the annual maps (23.03.2002 to 23.03.2003) because of the summation of data obtained in the time intervals under discussion.

## Conclusion

In the course of the RPS-1 experiment, detailed spectral and time characteristics of weak solar flares were obtained for the first time under relatively low solar activity conditions and at the satellite passage outside the ERBs. It was established that the increase of solar activity was accompanied by the hardening of the background radiation spectra (without flares). Observations of the nighttime Earth atmosphere visualized the ERB dynamics (disruption of the northern radiation belt in summer 2002) in the period following the cycle maximum. The energy of the nighttime X-ray emission of the upper atmosphere caused by precipitating magnetospheric electrons did not exceed 8 keV. The energy of GCR contribution to this emission did not exceed 5 keV.

A stable work of the CdTe device throughout the CORONAS-F lifetime showed that such detectors could ensure long-term observations of the Sun and space radiation sources.

## References

1. Glyanenko, A.S., Pavlov, A.A., Pankov, V.M., et al.: Calibration of the RPS-1 semi-conductor detector in the ≪CORONAS-F≫ project. MEPhI Sci. Sess. **7**, 37–38 (2001) (in Russian)
2. Kotov, Yu.D., Glyanenko, A.S., Arkhangelskii, A.I., Pavlov, A.A.: The results of preparation of the AVS-F experiment. MEPhI Sci. Sess. **7**, 33–34 (2001) (in Russian)

3. Lin, R.P., Schwarz, R.A., Pelling, R.M., Hurley, K.C.: A new component of hard X-rays in solar flares. Astrophys. J. **251**, L109–L114 (1981)
4. Mizera, P.F., et al.: Observation of auroral arcs, electrons, and X-rays from a DMSP satellite. J. Geophys. Res. **83**(A12), 5573–5578 (1978)
5. Pankov, V.M., Prokhin, V.L., Shkurkin, Yu.G., et al.: The X-ray semi-conductor spectrometer (RPS-1) in the ≪CORONAS-F≫ project. High-energy cosmic radiation. L: PhTI, pp. 196–199 (1989) (in Russian)
6. Pankov, V.M., Prokhin, V.L., Shkurkin, Yu.G., et al.: X-ray semiconductor (CdTe) spectrometer for solar flare and preflare studies. Radiophys. Quantum Electron. **34**(11–12), 1002–1005 (1996)
7. Pankov, V.M., Prokhin, V.L., Khavenson, N.G.: The RPS-1 X-ray CdTe spectrometer onboard the CORONAS-F satellite. Sol. Syst. Res. **40**(4), 314–318 (2006)
8. PIXIE_Image: http://muir.spasci.com/pixie/homepage/img_dir/internal/Image_Def_Slide.gif
9. Pounds, K.A.: The solar X-radiation below 25 Å. Ann. Geophys. **26**(2), 555–565 (1970)
10. Turiel, J., MacCregor, A.: Observation of solar X-rays with a solid-state detector. Astrophys. J. **162**, 287–291 (1970)
11. Voss, H.D., Smith, L.G.: Global zones of energetic particle precipitation. J. Atmos. Terr. Phys. **42**(2), 227–239 (1980)

# The Impact of Solar Activity on the Earth Upper Atmosphere as Inferred from the CORONAS-F Scientific Experiments

S.I. Boldyrev[1], I.A. Egorov, I.A. Zhitnik[2], G.S. Ivanov–Kholodny, S.P. Ignat'yev, V.N. Ishkov, O.P. Kolomiitsev[3], S.V. Kuzin, V.D. Kuznetsov, and A.I. Osin

The chapter is devoted to the first results of processing and analysis of data on the absorption of solar XUV radiation in the Earth upper atmosphere measured on board the CORONAS-F space mission.

The variability of the Earth upper atmosphere associated with solar activity has been studied by analyzing the orbital evolution of the CORONAS-F satellite.

Experimental data have been compared with model calculations of the parameters of the upper atmosphere; and the preliminary results have been discussed. At present, the whole bulk of the CORONAS-F data on XUV absorption and density variations in the Earth upper atmosphere is being processed.

The mathematical model of the Earth upper atmosphere (WMA01) developed at IZMIRAN is described in general terms. This model allows us to determine the state of the upper atmosphere in on-line mode taking into account the changing heliogeophysical conditions. A list of active events on the Sun and associated processes in the Earth magnetosphere recorded during the CORONAS-F flight time (2001–2005) is presented.

[1]Deceased

[2]Deceased

[3]Deceased

S.I. Boldyrev · I.A. Egorov · G.S. Ivanov-Kholodny (✉) · V.N. Ishkov · O.P. Kolomiitsev · V.D. Kuznetsov · A. I. Osin
Pushkov Institute of Terrestrial Magnetism, Ionosphere, and Radio Wave Propagation RAS (IZMIRAN), Troitsk, Moscow, Russia
e-mail: gor@izmiran.ru

I.A. Zhitnik · S.P. Ignat'yev · S.V. Kuzin
Lebedev Physical Institute RAS (FIAN), Moskow, Russia

V. Kuznetsov (ed.), *The Coronas-F Space Mission*, Astrophysics and Space Science Library 400, DOI 10.1007/978-3-642-39268-9_17, 

The comparison of model calculations with the experimental satellite data shows that the Earth atmosphere models available need updating. The possible ways to attack this problem are discussed.

## Introduction

The effect of solar activity on the Earth upper atmosphere can be characterized by two components: the long-term integrated component determined by the cyclic nature of solar activity and, in particular, by the 11-year cycle; and the sporadic component determined by powerful nonstationary geoeffective events in the solar atmosphere. The Earth upper atmosphere is ionized and heated by both electromagnetic and corpuscular solar radiation. The corresponding spatial and temporal variations in the Earth atmosphere density and concentration of radio active gas, as well as variations of geomagnetic activity are controlled to a large extent by the solar activity and are among the main external factors that determine the orbital characteristics of satellites.

Any scientific space mission developed for years is a unique special-purpose space observatory. The instruments comprising its scientific payload are usually designed using the latest achievements in experimental technologies. Such projects are very costly and time-consuming, so that the active lifetime of each spacecraft is critical from the viewpoint of the efficiency of the conducted and planned experiments. The estimate of the satellite lifetime is based on a comprehensive knowledge of the structure of the Earth upper atmosphere and its response to external factors, such as the level of long-term solar activity and solar geoeffective events (solar flares, coronal mass ejections, coronal holes with open magnetic field, high- and low-speed solar wind streams).

The CORONAS mission is a scientific space project under the auspices of the Russian Academy of Sciences and Federal Space Agency aimed at the study of the Sun and solar-terrestrial coupling. The mission comprises three artificial Earth satellites launched successively. The first satellite of the series—CORONAS-I—was operating in the orbit during 1994–2001, the second mission—CORONAS-F—was realized in 2001–2005, and the third one—CORONAS–FOTON—was launched to a polar circular orbit on January 30, 2009 [1, 2].

In accordance with its research program, CORONAS-F was studying the structure and parameters of the Earth upper atmosphere and their dependence on the level of solar and geomagnetic activity in the period from August 2001 to December 2005. The study was aimed at updating the existing models of the upper atmosphere.

Investigations comprised systematic measurements of the atmospheric absorption of soft X-ray and ultraviolet solar radiation at the wavelengths of 8.42 Å, 175 Å, and 304 Åand the analysis of the satellite orbit evolutions in the low-orbit flight mode, which are very sensitive to the slightest density variations in the upper atmosphere. The orbital data of CORONAS-I, which was an analog of CORONAS-F both in design, mass, and initial orbital parameters, were also used [3–7].

Forecasting the "low-orbit" flight mode and deorbit of a satellite provides the parameters used to control the state (composition and density) of cosmic "debris" in the Earth environment—the so-called ecological monitoring of near-Earth space. A particular case of such control is the monitoring of the orbital evolution of the International Space Station (ISS), whose circular orbit has an altitude of ~ 350 km, and that of other spacecraft flying in near-Earth orbits or undergoing a soft landing. This is necessary to take preventive measures to correct their orbits or to refine the location of their landing (or impact) sites. We can mention by way of example the "Progress" supply spacecraft, whose low-orbit flight mode can be used to study the Earth upper atmosphere.

## The Study of the Earth Upper Atmosphere by Measuring the Absorption of the Solar UV Radiation

Observations were carried out with the SPIRIT instrument complex on board the CORONAS-F satellite launched on 31 July 2001 to a near circular polar orbit with the inclination of 82.5° and altitude of 490–550 km.

Altitude variations in the atmospheric absorption of the solar radiation at the wavelengths of 175 Å and 304 Å were studied by regular observations of the Sun along the illuminated and occulted parts of the spacecraft trajectory, as well as in the periods without occultation when the line of sight of the solar disk passed through the atmosphere at the heights of 200–500 km over the Earth surface.

The absorption measurements allow us to determine the parameters of the atmosphere (concentration of constituent gases), direction of the density gradient, and local absorption variations with altitude. The results of these measurements can be used both for updating the existing atmosphere models and for monitoring local disturbances in the upper atmosphere due to the natural (heliogeophysical processes) and man-made (launch of heavy carrier rockets, explosions in the atmosphere, etc.) factors.

In this section, we shall compare CORONAS-F measuring data with model calculations of the parameters of the Earth upper atmosphere.

The CORONAS-F/SPIRIT experiment and measuring techniques are described in [8].

Numerical simulations were made with the Working Model of the Atmosphere 2001 (WMA01) developed at IZMIRAN [9]. The WMA01 model was created within the frames of a numerical–analytical ionospheric model for the description of large-scale nonuniform features with low electron concentrations [10].

### *Method for Calculating the Short-Wave Absorption in the Atmosphere*

The WMA01 model involves the coefficients of the semi-empirical atmospheric model, MSIS [11, 12]. The coefficients were normalized using the available space-borne mass-spectrometric measurements and long-term satellite drag data. A disadvantage of this normalization method is the use of integral data obtained during the motion of satellite along a large part of its orbit instead of the local data. This results in some inaccuracy in the model description of conditions in the upper atmosphere. That is why it is important to develop a method for validating the results of model calculations against experimental data.

The absorption of the solar shortwave radiation in the Earth atmosphere is calculated using the equation from [13]:

$$I = I_0 \exp\{-((\sigma(O)H(O)[O] + \sigma(N_2)H(N_2)[N_2]))ChZ_{\odot}\}, \qquad (1)$$

where $I_0$, $\mathrm{erg \cdot cm^{-2} \cdot s^{-1}}$ is the radiation intensity outside the atmosphere; $I$, $\mathrm{erg \cdot cm^{-2} \cdot s^{-1}}$, is the same quantity at the given height $h$, km; $\sigma(\mathrm{O})$, $\mathrm{cm}^2$ is the absorption cross-section of atomic oxygen; $H(\mathrm{O})$, km, is the scale height for atomic oxygen; [O], $\mathrm{cm}^{-3}$ is the concentration of atomic oxygen; $\sigma(\mathrm{N_2})$ is the absorption cross-section of molecular nitrogen; $H(\mathrm{N_2})$, km is the scale height for molecular nitrogen; $[\mathrm{N_2}]$, $\mathrm{cm}^{-3}$ is the concentration of molecular nitrogen; and $Ch\,Z_{\odot}$ is the Chapman function for the solar zenith angle $Z_{\odot}$.

Relation (1) characterizes the absorption of solar radiation during its passage through the atmosphere at a certain height. Using (1) for two different wavelengths $\lambda_1$ and $\lambda_2$ and solving the system of two equations for oxygen O and nitrogen $\mathrm{N_2}$ with the use of simultaneous measurements of $I/I_0$ for these wavelengths, we can estimate the absorption separately for the main constituents of the upper atmosphere—oxygen and nitrogen.

Below, for preliminary estimates at the heights dominated by atomic oxygen (above 300–350 km), we use Eq. (1) with allowance for the oxygen absorption at $\lambda = 304$ Å alone. In this case, after simple transformations, Eq. (1) for [O] takes the form

$$[\mathrm{O}] = \frac{\lg I/I_0}{\sigma(\mathrm{O}) \cdot H(\mathrm{O}) \cdot \lg e \cdot ChZ_{\odot}}; \qquad (2)$$

here $H(\mathrm{O}) = \dfrac{k \cdot T}{m(\mathrm{O}) \cdot m_{\mathrm{H}} \cdot g}$ and $g = g_0 \cdot \left(\dfrac{R}{R+h}\right)^2$, where $k = 1{,}38 \cdot 10^{-16}$ erg/deg is the Boltzmann constant; $\sigma(\mathrm{O}) = 9{,}8 \cdot 10^{-18}$ is the absorption cross-section at the wavelength of 304 Å; $m(\mathrm{O})$, a.m.u. is the molecular weight of atomic oxygen; $m_H = 1.67 \cdot 10^{-24}$ g is the molecular weight of atomic hydrogen; and $g$, $\mathrm{cm \cdot s^{-2}}$ is the acceleration due to gravity at the height $h$.

The Chapman function $ChZ_{\odot}$ for $Z_{\odot} = 70°$ is taken from [13].

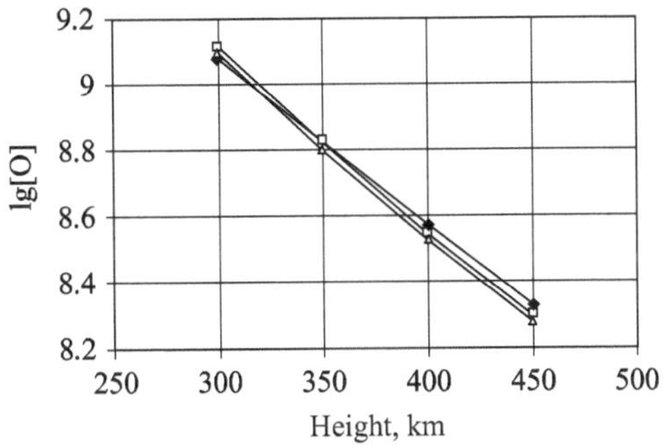


**Fig. 1** The concentration of atomic oxygen (Lg[O]) in the Earth upper atmosphere according to CORONAS-F measurements and numerical simulations for a quiet day on 26 November 2001: *diamond*—WMA01 model; *filled square*—MSIS model; *empty triangle*—measurements

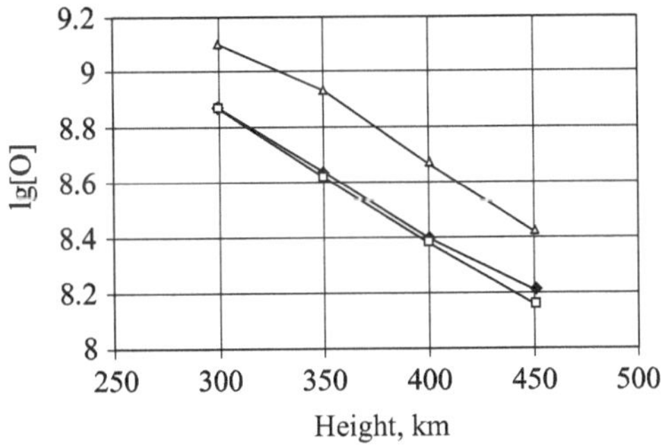


**Fig. 2** The concentration of atomic oxygen (Lg[O]) in the Earth upper atmosphere according to CORONAS-F measurements and numerical simulations for a disturbed day on 21 October 2001: *diamond*—WMA01 model; *filled square*—MSIS model; *empty triangle*—measurements

## *Results of Model Calculations and Comparison with Experimental Data*

The model calculations of concentration of the atomic oxygen and their comparison with the absorption of solar radiation at 304 Å measured on board the CORONAS-F spacecraft are represented in Figs. 1 and 2 [14].

Figure 1 presents an example for quiet conditions on November 26, 2001: $F_{10.7} = 170$, $(F_{10.7})_{90} = 217$, $A_p = 4.5$. Satellite measurements were taken at low latitudes in the northern hemisphere at 15° 36′ and 32°48′ N, 318°42′ and 309°12′ E at 20:07:38 and 20:12:45 UT, respectively.

Figure 2 shows the same characteristics for disturbed conditions on October 21, 2001: $F_{10.7} = 244.7$, $(F_{10.7})_{90} = 190.4$, $A_p = 67.0$. Measurements were taken at the southern polar latitudes at −75°06′ and −71°48′ S, 95°18′ and 83°36′ E at 14:27:38 and 14:33:34 UT, respectively.

One can see that the model concentrations of atomic oxygen are in good agreement with experimental data if the geomagnetic field is quiet, $A_p \approx 5$. Individual deviations of the calculated and measured values do not exceed 5.

The situation is quite different under disturbed conditions. In this case, at $A_p \approx 70$, the concentration of atomic oxygen is overestimated because of the neglect of molecular nitrogen in calculations. The concentration of atomic oxygen inferred from the measured absorption intensity is $\sim 50\,\%$ higher than the model value.

**Table 1** Height variation in the solar radiation intensity at 304 Å

| | $\lg I/I_o$ | |
|---|---|---|
| $h$, km | Quiet conditions, 26.11.01 | Disturbed conditions, 21.10.01 |
| 300 | −0.570 | −1.000 |
| 350 | −0.250 | −0.343 |
| 400 | −0.090 | −0.216 |
| 450 | −0.018 | −0.046 |

Table 1 shows the height variation in the solar radiation intensity at 304 Å due to absorption in the upper atmosphere for the quiet and disturbed heliogeophysical conditions

Our calculations are based on the assumption that the absorption of solar radiation at the wavelength of 304 Å at 300–400 km is mainly determined by atomic oxygen. Atomic oxygen is actually the dominating atmospheric constituent at these heights under normal conditions. Meanwhile, the concentration of molecular nitrogen abruptly increases during disturbed conditions. The estimates show that, under disturbed conditions, the ratio is $[O]/[N_2] \sim 1.0$.

In Fig. 3, the model sum concentrations $[N] = [O] + [N_2]$ are compared with the measurements obtained in a disturbed period at the heights of 300–450 km. The calculated and measured particle concentrations are seen to agree fairly well. The numerical concentration values were obtained using the WMA01 model [14].

It should be noted that a large amount of observation data was acquired during the five–year period of orbital operation of the CORONAS-F mission. At present, the data are processed in order to obtain information on the behavior of the Earth upper atmosphere under different heliogeophysical conditions and to find out how the atmospheric parameters depend on the height, latitude, longitude, season, solar and geomagnetic activity, to study cyclic variations of the parameters, etc. Special software was developed to allow the parameters of the upper atmosphere to be calculated promptly and compared with satellite observations. This will help us correct the atmosphere models available.

As mentioned above, the model described in this paper, as well as the other current atmospheric models is bounded from below at a height of 120 km, where all atmospheric constituents are assumed to have planetary averaged concentrations. However, as shown by the recent studies, the neutral atmosphere parameters at this height also depend on the solar and geomagnetic activity and on geographical coordinates (see, e.g., [11]). The solution of this problem is needed for updating the existing atmospheric models

And finally, note that the space-borne physical experiments, e.g., UV absorption measurements on board the CORONAS-F satellite, have a specific feature—the observer and the observed object move permanently about one another. This may be the source of calculation errors when processing the observation data. Some of the processing results are demonstrated below.

Four fragments of four measuring sessions realized under relatively quiet geomagnetic conditions were used to demonstrate the aforesaid. Some elements of these measurements are presented in Tables 2 and 3.

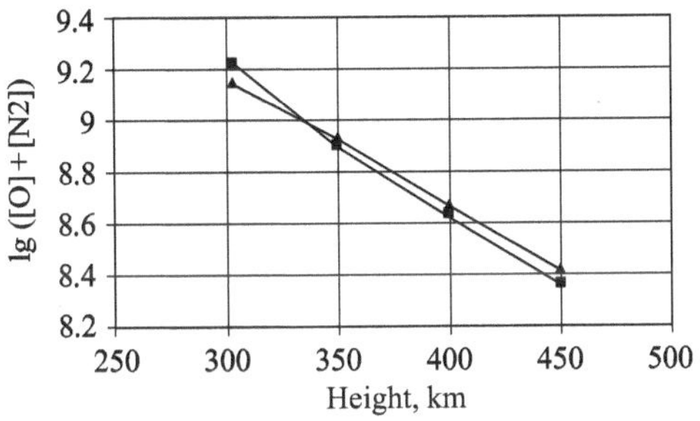


**Fig. 3** The sum concentrations of atomic oxygen and molecular nitrogen LgN =Lg([O] + [$N_2$]) in the Earth upper atmosphere according to CORONAS-F measurements and numerical simulation on a disturbed day, 21 October 2001: *filled square*—WMA01 model; *empty triangle*—measurements

**Table 2** UV absorption on board the CORONAS-F satellite

| | Latitude | | Longitude | | Time UT | | | |
|---|---|---|---|---|---|---|---|---|
| Session | Beginning | End | Beginning | End | Beginning | End | $F_{10.7}$ | $A_p$ |
| A. 07.05.2004 | −65°19′ | −72°40′ | 18°97′ | 82°12′ | 7:24:42 | 7:31:38 | 85 | 17 |
| B. 15.05.2004 | 15°66′ | −6°04′ | 9°77′ | 15°89′ | 4:55:16 | 5:01:13 | 115 | 9 |
| C. 26.04.2005 | −42°26′ | −57°56′ | 131°24′ | 140°01′ | 22:04:21 | 22:08:33 | 91 | 5 |
| D. 01.05.2005 | 37°03′ | 9°79′ | 171°16′ | 178°62′ | 17:44:56 | 17:52:16 | 112 | 26 |

**Table 3** Concentration of atomic oxygen and the temperature of the neutral atmosphere

| Fragment | Height $h$, km | Concentration of [O] at the height $h$, $cm^{-3} \cdot 10^8$ | $Z_\odot$, ° | $T$, K Beginning | End |
|---|---|---|---|---|---|
| A | 320 | 1.49–1.13 | 90.2–91.2 | 909 | 813 |
| B | 267 | 8.55–1.84 | 90.2–94.9 | 750 | 725 |
| C | 339 | 4.53–3.04 | 89.5–91.0 | 782 | 726 |
| D | 342 | 3.25–1.54 | 93.5–95.0 | 782 | 702 |

Table 2 provides details of the UV absorption measurements conducted on board the CORONAS-F satellite, including the date of the measuring session, coordinates, universal time, solar and geomagnetic activity(fragments of measuring sessions).

Table 3 provides the parameters that were determined from fragments A, B, C, and D of the observation sessions. It contains the concentration of atomic oxygen at the lower heights of [O] profiles for the maximal and minimal $Z_\odot$ and the temperature of the neutral atmosphere at the beginning and the end of each of the four measuring sessions. One can clearly see variations of $Z_\odot$ and temperature $T$ that occasionally may reach $\sim$ 5° of the zenith angle and $\sim$ 100 K, respectively.

In this section, the concentration of [O] was calculated at the mean zenith angles of the Sun and mean temperatures for all four observation sessions. The calculation results for session B when the maximum deviation of $Z_\odot$ was recorded for

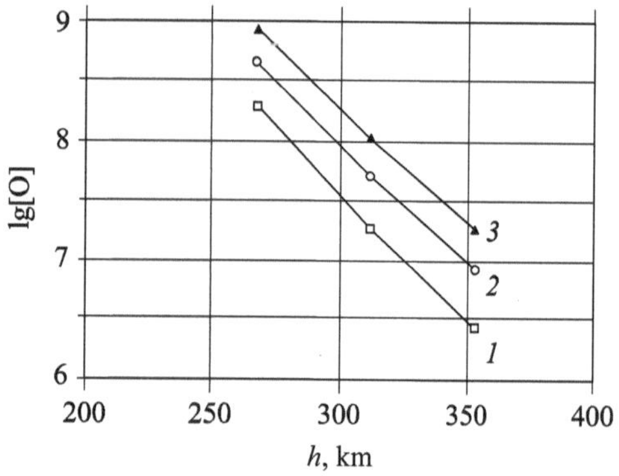


**Fig. 4** Concentration of atomic oxygen at different zenith angles of the Sun: (1)—maximal; (2)—medium; and (3)—minimal zenith angle

six minutes of observation are illustrated in Fig. 4. It is assumed that the mean curve represents the concentration of atomic oxygen measured in session B most adequately.

The result obtained is important to data processing in future, since it allows us to take into account the change of various parameters during the measuring session and to use these variations when calculating the concentration of the Earth atmospheric constituents.

## The Study of the Earth Upper Atmosphere Based on the Analysis of Spacecraft Orbital Evolution: The Probability of Long-Term Prediction of the Spacecraft Orbital Lifetime

An important property of the Earth upper atmosphere is its dependence on many variable factors. To forecast the state of the atmosphere, one must take into account not only the height, latitude, local time, and season but also the level of solar and geomagnetic activity, whose variations manifest themselves in the upper atmosphere through its ionization and heating by electromagnetic and corpuscular radiations. The response of the upper atmosphere to solar activity is rather complex. It was described by various authors [9, 15–21].

Spatial and temporal variations of density of the Earth atmosphere are one of the external factors that determine the orbital parameters of an artificial satellite. These density variations, like those of geomagnetic activity, are largely controlled by the level of geoeffective solar activity. In the period of active flight of CORONAS-F, significant variations of the planetary geomagnetic activity index, $A_p$, were recorded with the maximum monthly mean values up to $\sim$ 20 nT in mid-2003. During the well-known outstanding events on the Sun in October–November 2003, the peak *Dst* variations reached $\sim$ 450 nT [22]. The atmospheric density at the heights of the satellite orbit for different solar and geophysical conditions was estimated using the WMA01 mathematical model of the Earth upper atmosphere developed at IZMIRAN. For detailed description of the model, see chapter "Experiment with the

SPR-N Instrument on board the CORONAS-F Satellite: Polarization, Temporal and Spectral Characteristics of the Hard X-ray of the Solar Flares" and paper [9].

For our model calculations and analysis of heliogeophysical conditions, we used data published on the web sites of the World Data Centers:

http://www.sec.noaa.gov/today.html—Space Environment Center (SEC), Boulder, USA;

http://swdcdb.kugi.kyoto-u.ac.jp—World Data Center (WDC), Kyoto, Japan;

http://sohowww.estec.esa.nl—Solar Orbital Observatory (SOHO);

http:///www.izmiran.ru—Geophysical Situation Forecasting Center, Pushkov Institute of Terrestrial Magnetism, Ionosphere, and Radio Wave Propagation.

These data were used to predict solar and geophysical conditions for the period from 2005 to 2006, for which density variations of the upper atmosphere and orbital parameters of the Coronas-F satellite were computed.

## *Description of the CORONAS-F and CORONAS-I Orbital Data: Analysis of Density Variations in the Earth Upper Atmosphere*

When estimating the orbital lifetime of CORONAS-F, we had a unique opportunity to make use of the flight scenario of the CORONAS-I satellite. Both satellites are virtually identical in their construction and have the same initial orbital parameters. Table 4 gives information concerning these satellites and their orbits.

The CORONAS-I satellite was launched in March 1994 (at the minimum of the solar activity cycle) and was in flight until March 2001 (the year of maximum solar activity). Figure 5 shows year-by-year variations of the perigee and apogee of the satellite orbit for the whole flight period.

One can see that, before 1999, the perigee and apogee of the satellite decreased rather slowly, because it was orbiting high in the period of minimum solar activity, when the quiet Earth atmosphere had no substantial effect on the satellite orbit. Only after the level of solar activity had increased substantially did the curves of the satellite apogee and perigee go down steeply. The results of the analysis of the orbital parameters of the CORONAS-I satellite with allowance for the specific conditions in the Earth atmosphere were used to analyze and forecast the orbital parameters of the CORONAS-F mission.

Figure 6 shows the variations of the perigee and apogee heights of the CORONAS-F orbit in the period 2001–2005 (data of the NASA Goddard Space Flight Center downloaded from the web site of the Orbital Information Group (OIG)). These measurements were made by passive methods of observation of satellite motion. It is evident that, from the beginning of the flight of CORONAS-F, its apogee and perigee heights were decreasing faster than those of the CORONAS-I satellite. This difference is likely to be accounted for by different levels of solar activity during the two time intervals under consideration; i.e., the temperature of the upper atmosphere during the flight of the CORONAS-F satellite was higher, and so higher were the dense atmospheric layers. As the level of solar activity decreases, the atmospheric temperature decreases accordingly.

**Table 4** Orbital parameters of the CORONAS-F and CORONAS-I satellites

| Orbit | Geocentric quasi–solar–synchronous |
|---|---|
| Altitude at apogee | 545 km |
| Eccentricity | 0.08257 |
| Orbital period | 95 min |
| Inclination of the orbital plane | 82.5° |
| Precession of the orbital plane | 1.8 °/day |
| Spacecraft | AUOS–SM space platform |
| Total mass | 2260 kg |
| Mass of scientific payload | 395 kg |
| Constructive shape | barrel: $L = 2,400$ mm; $D = 1,360$ mm |
| Photoelectric panels | 8 panels (2400 mm × 640 mm) |
| Orientation of the spacecraft | single–axis sunward orientation |
| Accuracy of orientation | within 10′ |
| Drift in the axes perpendicular to the axis directed toward the Sun | within 0.005 °/s |

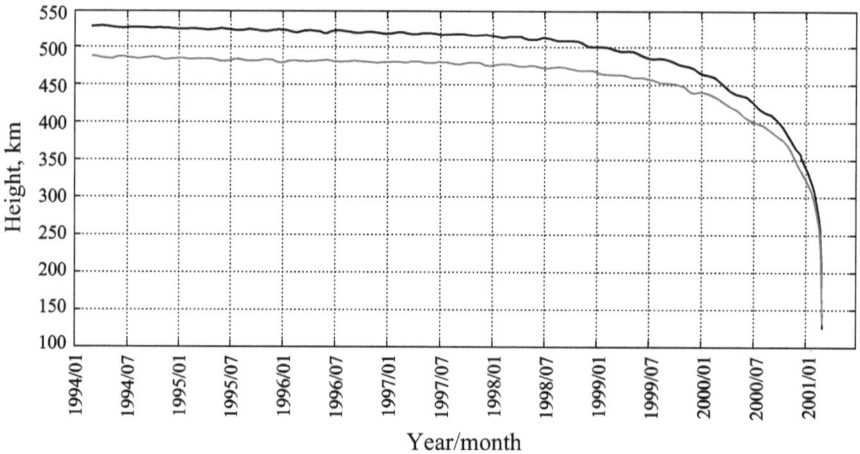


**Fig. 5** Dynamics of the CORONAS-I orbit according to the data of the NASA Goddard Space Flight Center for the period 1994–2001. The upper and lower curves show the decay of the apogee and perigee of the satellite orbit, respectively

It is appropriate to give here the ratio of the aerodynamic forces acting on the CORONAS-F and CORONAS-I satellites during the periods under consideration. In the aerodynamic approximation, the drag force $D$ acting on the satellite in the direction opposite to its velocity of motion $V$ with respect to the surrounding atmosphere of density $\rho$ can be written in the form:

$$D = 0,5\rho V^2 SC_D, \tag{3}$$

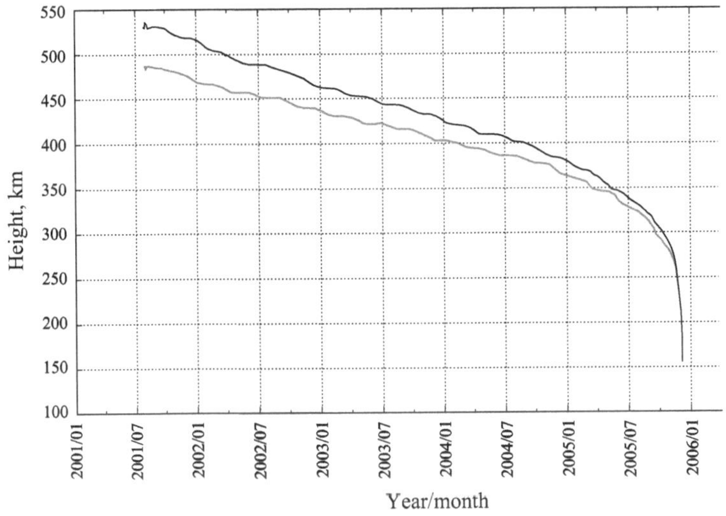


**Fig. 6** Dynamics of the CORONAS-F orbit according to the data of the NASA Goddard Space Flight Center for the period 2001–2005. The upper and lower curves show the decay of the apogee and perigee of the satellite orbit, respectively

where $S$ is the characteristic cross-section (area) of the object, and $C_D$ is the dimensionless drag coefficient.

The ratio of the drag forces acting on the CORONAS-F and CORONAS-I satellites reduces to the ratio of the atmospheric densities for the given orbital conditions: $D_I/D_F = \rho(I)/\rho(F)$.

For the flight periods in 2000 and 2004 when both satellites had the perigee heights of $\sim$ 370 km we find for the days of similar geomagnetic activity: $D_I/D_F \approx 10$.

Thus, the atmospheric drag of CORONAS-F was much stronger than the drag of CORONAS-I at the same height due to a higher level of solar activity in 2000.

The latter is corroborated by the behavior of the upper atmosphere density in Cycle 23 at the heights of 400, 370, and 300 km (Fig. 7). The density at these heights was calculated using the relations

$$\rho(h) \approx \rho_0 \exp[G(L)]; \quad (4)$$
$$\rho_0 = n_0 M / N_a,$$

where $\rho_0$, g/cm$^3$ is the atmosphere density at 120 km (the height where the diffusion-driven separation of atmospheric components begins), $n_0$, cm$^{-3}$ is the particle number density at this height, $M$ is the mean molecular mass, $N_a = 6.0238 \cdot 10^{23}$ mole$^{-1}$ is the Avogadro constant, and $G(L)$ are the semi-empirical coefficients of the MSIS model [11, 12].

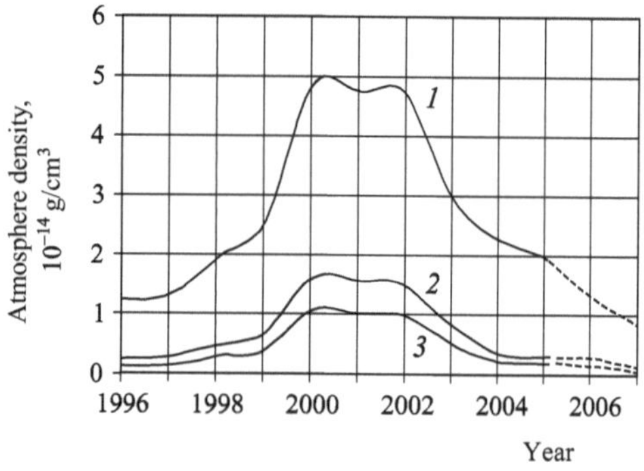


**Fig. 7** Density variations of the Earth atmosphere in Cycle 23 on April 1 in the period 1996–2007 at the heights of 300 km (1), 370 km (2), and 400 km (3). The *dashed* parts of the curves are the density variation forecast

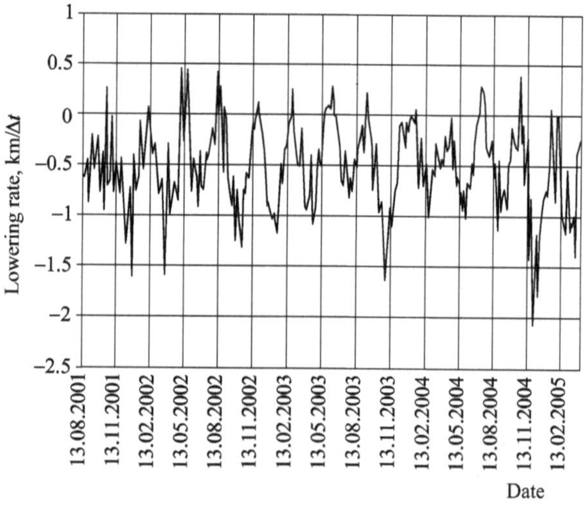


**Fig. 8** Variations of the perigee decay rate of the CORONAS-F satellite in five-day intervals ($\Delta t$) according to the orbital data of the NASA Goddard Space Flight Center for the period from November 2001 through March 2005

The particle number densities were calculated using the mathematical model of the upper atmosphere WMAO 01 [22].

## *Evolution of Orbital Parameters of the CORONAS-F Satellite*

Figure 8 illustrates the variations of the parameter $\Delta h_p/\Delta t$, km/day—the perigee lowering rate of the CORONAS-F orbit, which is equal to an average of 3 km/month.

In addition to the variations of the orbit caused by the nonsphericity of the Earth, Fig. 8 also shows smooth cyclic variations with a period of three months and nonperiodic peak-shaped variations associated with geoeffective events on the Sun that cause magnetic storms in the Earth magnetosphere. During these events, the satellite additionally loses 1–3 km in altitude; however, the pattern of decrease in the orbital height outside these events remains unchanged.

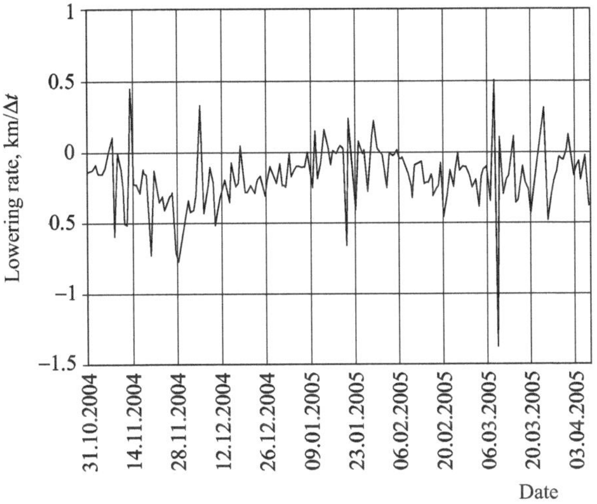


**Fig. 9** Variations of the perigee decay rate of the CORONAS-F satellite as inferred from one–day orbital data of the NASA Goddard Space Flight Center for the period from October 2004 through March 2005

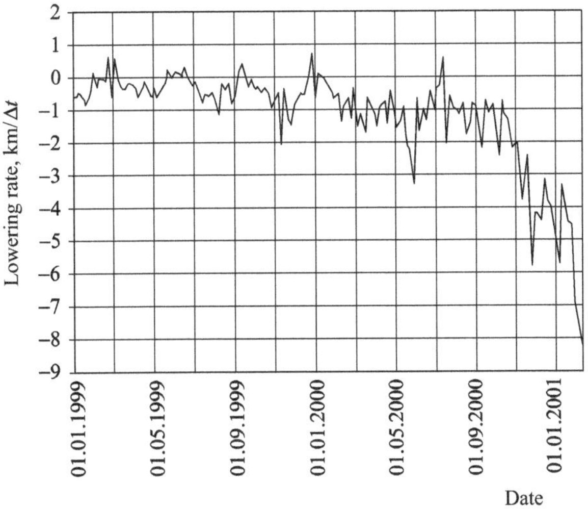


**Fig. 10** Variations of the perigee decay rate of the CORONAS-I satellite in five–day intervals ($\Delta t$) according to the orbital data of the NASA Goddard Space Flight Center for the period from January 1999 through January 2001

Figure 9 illustrates in more detail the variations in the perigee decay rate of the CORONAS-F satellite in the periods of high solar activity and strong geomagnetic disturbances in October–November 2004 and 2005 as inferred from one-day orbital data.

The lowering rate of the CORONAS-I satellite is shown in Fig. 10. The perigee decay rate was of the order of 3 km/month during a time interval in 1999. It was important to assess the conditions under which an appreciable loss of altitude of the satellite orbit occurs. One can readily see that the character of decrease of the perigee height of the CORONAS-I satellite changed in November 1999 against the background of increasing solar activity and the associated changes in the atmosphere. The lowering rate doubled, and after one year—in November

2000—the altitude of the satellite orbit began to decrease abruptly. In February 2001 (at an altitude of about 160 km), the CORONAS-I satellite was shut down and ceased to exist in early March 2001. For the time when the satellite orbit started to decay abruptly, the computed density of the atmosphere at the heights of the CORONAS-I orbit became critical reaching $\sim 10^{-14}$ g/cm$^3$ [5, 23].

## *Prediction of the CORONAS-F Lifetime*

Our prediction of the orbital lifetime of the CORONAS-F satellite in its orbit around the Earth is based on the following procedure. First, the orbital data of the CORONAS-I satellite are used to determine the critical density of the atmosphere at which important variations in the orbital height of the satellite begin. Then, we use this critical density to determine the orbital scenario of the CORONAS-F satellite [5]. We find the altitude of the atmospheric layer with the critical density known for satellites of the type considered, but for the conditions of minimum solar activity and, at the same time, we go on keeping track of the perigee height of the satellite. In such a way, we find the time when the naturally descending satellite reaches dense atmospheric layers.

Figure 7 shows the computed variations of the density of the Earth atmosphere at the perigee heights of the CORONAS-I and CORONAS-F satellites in the period 1996–2006. Table 5 gives the numerical values of variations of the atmospheric density ($\lg \rho$) at the CORONAS-I and CORONAS-F perigee heights in 1996–2006 for April 1 of the year under consideration (see Fig. 7).

One can see that, according to the forecast, CORONAS-F will reach the layer with a critical atmospheric density of $10^{-14}$ g/cm$^3$ at a height of $h_p \approx 300$ km. Note that the mathematical model of the atmosphere employed here uses as input parameters the seasonal and daily density values during the most unfavorable periods of the year and day (April 1 and daytime). At the height considered here, the density of the atmosphere varies by a factor of 1.5 depending on the season and by a factor of 2 from day to night.

It was interesting to compare the result obtained with the estimate of the orbital lifetime $\tau_L$ of CORONAS-F from the rate of change of the orbital period. For this purpose, we used the relations from [18]

$$\tau_L \approx -eT \cdot F(e)/\Delta T, \tag{5}$$

$$e \approx 3H/a.$$

For the chosen quiet period of solar and geomagnetic activity in early August 2004, the orbital eccentricity is $e \approx 0.026$; the orbital period is $T \approx 92.4$ min; the rate of change of the orbital period is $T/\Delta T \sim -0.003$ min/day; the function of the satellite lifetime is $F(e) \sim 0.85$ km; $H \sim 60$ km; $h_p \sim 400$ km; and the major semi-axis of the ellipse is $a = 6778$ km. The orbital lifetime of the CORONAS-F satellite since August 2004 as computed from Eqs. (2) and (3) is

**Table 5** Variations of the atmospheric density at the perigee heights

| CORONAS-I | | | CORONAS-F | | |
|---|---|---|---|---|---|
| Year | $h_p$, km | lg $\rho$ | Year | $h_p$, km | lg $\rho$ |
| 1996 | 481 | −15.55 | 2002 | 460 | −14.27 |
| 1997 | 481 | −15.49 | 2003 | 427 | −14.37 |
| 1998 | 475 | −15.11 | 2004 | 398 | −14.44 |
| 1999 | 463 | −14.91 | 2005 | 362 | −14.42 |
| 2000 | 425 | −14.07 | 2006 | 326 | −14.18 |

equal to about 720 days. For satellites of the same type as CORONAS-F, the errors in the lifetime estimates based on the change of the orbital period lie within 10–15 %. Thus, the lifetime of the CORONAS-F satellite is equal to almost two years beginning from August 2004. This result virtually coincides with the lifetime inferred above by analyzing the variations of the atmospheric density at the perigee heights.

The estimate of the CORONAS-F orbital lifetime obtained above is based on the assumption that, in the period close to the minimum of the solar cycle, the satellite flight conditions will be sufficiently favorable, i.e., a quiet geomagnetic field and low solar activity.

## *Low-orbit Flight of the CORONAS-F Satellite: The Dependence of Orbital Parameters of the Satellite on the Solar Activity Conditions*

Due to natural orbital evolution, any satellite lacking a propulsion system sooner or later undergoes the low-orbit flight phase, which determines the deorbit mode. The CORONAS-F satellite had the low-orbit phase of its flight at altitudes of ∼ 300–400 km [5, 6, 23].

The low-orbit flight of a spacecraft is determined not only by its design characteristics and mass but also by the state of the Earth upper atmosphere, which, in turn, depends on the level of solar activity.

The altitude of the critical density of the upper atmosphere for the active breaking of a satellite decreases by 100–150 km from the maximum to the minimum of solar activity. Active breaking of a satellite followed by its deorbit takes place at altitudes of ∼ 400–450 km and ∼ 300–350 km in the periods of maximum and minimum solar activity, respectively.

Above, we estimated the lifetime of the CORONAS-F satellite for the period from August 2004 onwards based on the forecast of atmospheric density variations at the heights of the perigee of the satellite orbit and on the rate of change of the orbital period. Under the conditions of low solar activity and quiescent geomagnetic field, both methods yield similar values. The CORONAS-F satellite had to operate in orbit nearly two years beginning from August 2004. Subsequent solar and

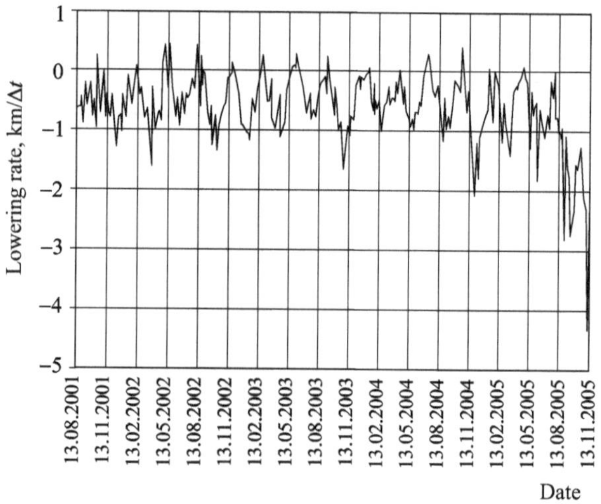


**Fig. 11** Variations of the perigee decay rate of the CORONAS-F orbit based on the NASA GSFC orbital data for five-day intervals ($\Delta t$) for the period from November 2001 through December 2005

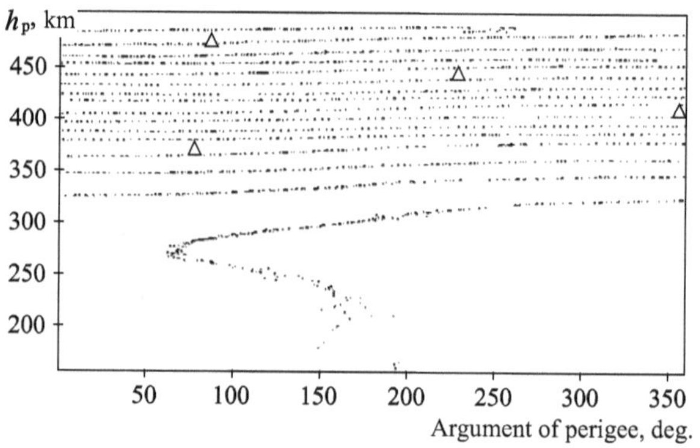


**Fig. 12** Variations in the height and perigee argument of the CORONAS-F orbit: the *triangles* mark successively the beginning of years 2002, 2003, 2004, and 2005

geomagnetic activity events showed that the heliogeophysical parameters used in our computations differed from the really observed ones. Below, we analyze the events on the Sun and in the geomagnetic field during the period from August 2004 through December 2005.

Figures 5 and 11 illustrate the orbit lowering and variation in the perigee decay rate of the CORONAS-F satellite. Note that significant breaking of the satellite began when it reached the upper atmospheric layer with a density of $\sim 10^{-14}$ g/cm$^3$. The plot of the perigee decay rate of CORONAS-F in Fig. 11 shows conspicuously (the same as Fig. 10) the time when the satellite reached the critical density layer (approximately on August 13, 2005).

Figures 12 and 13 show the variation in the height and perigee argument of the CORONAS-I and CORONAS-F orbits during the final phases of their flight. The satellites entered dense atmospheric layers and ceased to exist in February 2001 and in early December 2005, respectively. An additional inversion of the perigee argument of CORONAS-F is apparent beginning at an altitude of 250 km.

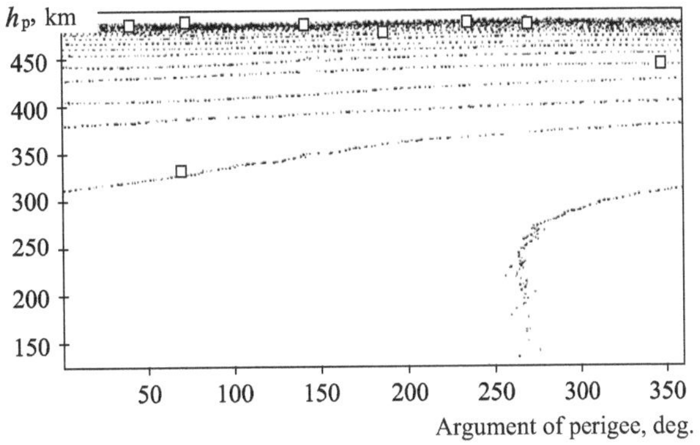


**Fig. 13** Variations in the height and perigee argument of the CORONAS-I orbit: the *squares* mark successively the beginning of years from 1994 through 2001

The problem of its origin requires further investigation, which lies beyond the scope of our paper

## Solar and Geomagnetic Activity in 2004–2005 and the Upper Atmosphere

When forecasting conditions in the Earth upper atmosphere during the "low-orbit" stage of flight of the CORONAS-F satellite, we used for comparison the parameters for the previous, odd (21st) solar activity cycle. Solar flares during the similar period (with respect to the epoch of minimum activity) were few in number and had low importance. A total of 18 large solar flares and one extreme flare occurred in 1984–1985. During the period of 2004–2005, there were a total of 62 large flares and two extreme events (an X7.5 flare in January 2005, which was characterized by the highest flux of protons with energies >100 MeV in that cycle, and an extreme flare of importance X17.1 in September 2005). For more details, see chapter "Experiment with the SPR-N Instrument on board the CORONAS-F Satellite: Polarization, Temporal and Spectral Characteristics of the Hard X-Ray of the Solar Flares".

Judging by its properties, the current 23rd cycle of solar activity is medium–high [24]. The most powerful flare events of Cycle 23 occurred in 2003 [25]. We, therefore, considered the time interval from August 2004 through August 2006 to be a period of rather quiescent Sun. At the beginning of this period, the layer of the upper atmosphere with a density critical to the drag of the CORONAS-F satellite was located at a height of $\sim$ 300 km [5]. However, there were a few intervals in 2004 and 2005, when the activity in the Sun was abnormally high, and an extremely powerful X-ray burst (X17.1/3B) was recorded in September 2005.

The period under consideration proved to be abnormally active and totally different from the corresponding period of the previous odd Cycle 21. More than 200 X-ray flares of high importance occurred. They were accompanied by ejections of coronal plasma, which triggered powerful (significant) geomagnetic disturbances

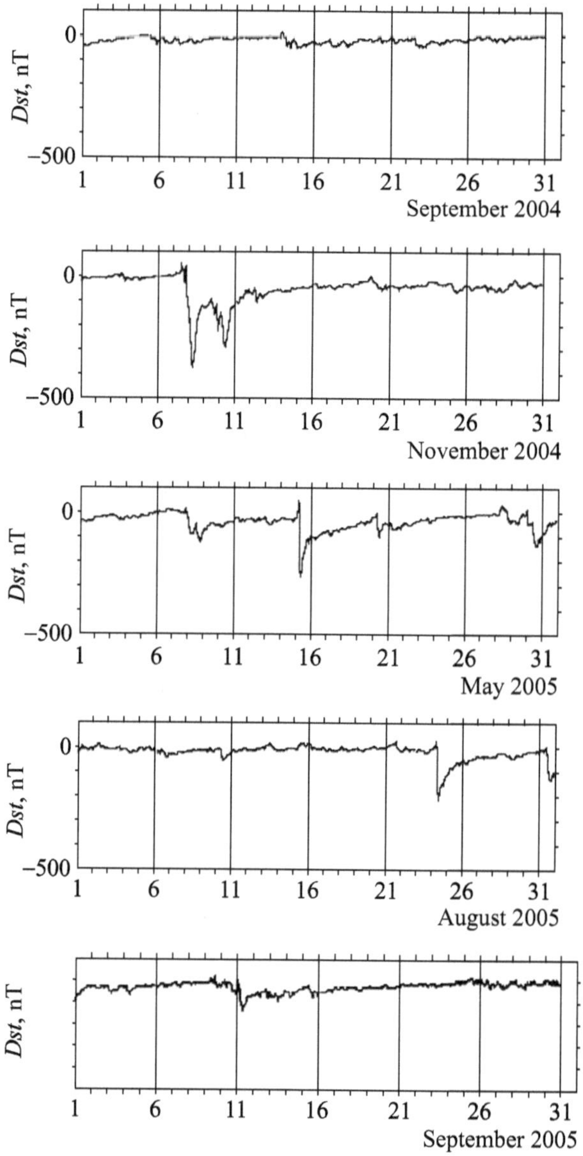


**Fig. 14** Characteristic *Dst*-variations during the period 2004–2005

in the near-Earth space, such as the events of August and November 2004 and those of January, May, August, and September 2005.

The geomagnetic field was also extremely disturbed during the period from August 2004 through December 2005. Figure 14 shows the characteristic *Dst*-variations. The variations of September 2004 are typical of quasi-quiescent

**Table 6** *Dst*-variations and the planetary geomagnetic activity index $A_p$

| Date | *Dst*, nT | $A_p$, nT |
|---|---|---|
| 24—28.08.04 | 126 | 195 |
| 10—11.11.04 | 370 | 206 |
| 16—20.01.05 | 121 | 91 |
| 20—22.01.05 | 105 | 94 |
| 08.05.05 | 127 | 64 |
| 14—17.05.05 | 263 | 87 |
| 29—31.05.05 | 138 | 93 |
| 12.06.05 | 106 | 35 |
| 23.06.05 | 97 | 48 |
| 12.07.05 | 90 | 48 |
| 23—25.08.05 | 216 | 105 |
| 10—14.09.05 | 147 | 101 |
| 11.11.05 | 147 | 105 |

conditions. The *Dst* deviations were equal in absolute value to $\leqslant$ 40 nT. In November 2004 and in May and August 2005, five sudden-commencement storms of different intensities ranging from −200 nT to −370 nT were observed. What these disturbances had in common was their duration. Thus, the November storm lasted for > 80 h, and the May storm (from the maximum deviation of ~ −260 nT to ~ −40 nT) lasted four days. It is important to emphasize here that ~ 48 h after the main phase of these storms, the perigee of the CORONAS-F orbit decreased substantially (by 600–1000 m), while during the entire active lifetime of the satellite under quiescent conditions, the perigee decay rate was $\Delta h_p \approx 3$ km/month.

Table 6 gives an idea of the state of the geomagnetic field in the period from 2004 through 2005 based on *Dst*-variations and the planetary geomagnetic activity index $A_p$.

It should be noted that, in spite of their outstanding intensity, the events of 07.09.2005 in the Sun did not cause any substantial *Dst*-variation. On the other hand, the index of planetary geomagnetic activity $A_p$ was at that time equal to ~ 100 nT.

The analysis of *Dst*-variations suggests that the plasmasphere was permanently "compressed" during the period under consideration, which might tell on the state of the Earth atmosphere on a planetary scale [26].

It is evident that such an active geomagnetic field had to tell on the atmospheric density. Figure 15 shows two curves describing the density variations of the Earth upper atmosphere in the period from August 2004 through 2005 calculated using the WMA01 mathematical model of the Earth upper atmosphere developed at IZMIRAN [9].

Curve *1* in the right-hand part of the figure represents the atmospheric density calculated from actually observed parameters of solar and geomagnetic activity and curve *2* shows the density in August 2004 based on the forecasted quiet geomagnetic conditions and low solar activity.

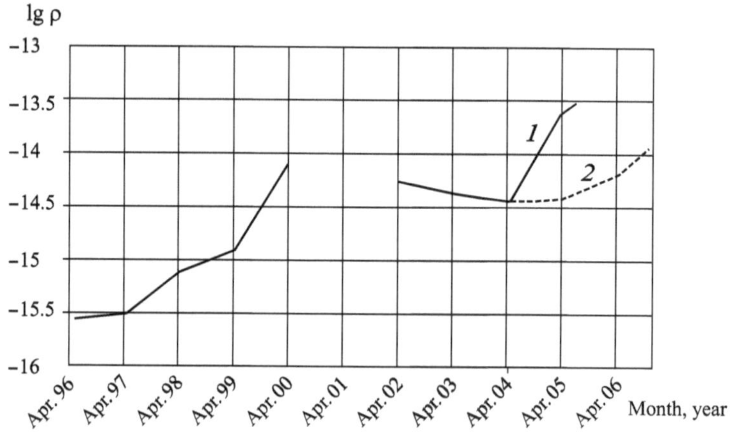


**Fig. 15** Variations in the Earth atmospheric density (lg $\rho$) at the perigee heights of the CORONAS-I (*left*) and CORONAS-F (*right*) satellites on the day of April 1 for the period from 1996 through 2006

Curve *1* is drawn taking into account the real solar and geophysical parameters and represents the actual density of the Earth upper atmosphere at the satellite perigee heights. Curve *2* represents the density forecast obtained in [6] under certain assumptions concerning the solar activity conditions.

The density of the upper atmosphere during the period under examination proved to be twice as high as it was during a quiescent period. This factor determined the orbital lifetime of the CORONAS-F satellite.

The results of our analysis of the CORONAS-F orbit show that very powerful solar events and the accompanying phenomena in near-Earth space and in the Earth upper atmosphere affected substantially the duration of the low-orbit flight phase of the satellite in the period 2004–2005. These solar events were impossible to predict in the late epoch of decline of the solar cycle close to the minimum.

On the other hand, the reliability of forecast of the Earth upper atmosphere depends directly on the forecast of solar activity.

For quite objective reasons (the lack of a fundamental physical theory of solar activity), a fully reliable forecast of solar activity and its effect on near-Earth environment is unavailable.

With the present-day state of the art in solar-terrestrial physics, the problem of medium-term forecast of solar activity and its geophysical effects can be solved by creating a space-borne monitoring system aimed mainly at providing comprehensive, real-time information on solar activity and the state of the inner heliosphere of the Solar System (down to the Earth orbit) [27].

# Mathematical Model of the Earth Upper Atmosphere WMA01 and Its Use for Updating the Available Models

The mathematical model of the upper atmosphere WMA01 was developed at IZMIRAN for processing and analyzing new experimental spacecraft data on the structure and state of the Earth upper atmosphere and ionosphere and their dependence on solar activity, as well as for the medium-term forecast of the Earth environment used in operational space flight control (change of trajectory, deceleration, and deorbit of spacecraft, or other preventive measures) [9].

This section contains a brief description of the algorithm and examples of refinement and use of the mathematical model of the Earth upper atmosphere WMA01, as well as its principal features, which are to be taken into account when updating the atmospheric models available.

The model makes it possible to evaluate the state of the Earth atmosphere in interactive mode with allowance for both past and future variations of solar and geophysical conditions.

## *WMA01 Atmospheric Model*

The control (input) parameters of WMA01 are the solar radio flux at a frequency of 2800 MHz $\{F_{10.7} = F \cdot 10^{-22}\ \mathrm{W \cdot m^{-2}(Hz)^{-1}}\}$ and the geomagnetic activity—the planetary activity index $A_p$ (the amplitude of the magnetic field disturbance in nT). The model output is the number density of the upper atmosphere (in $\mathrm{g \cdot cm^{-3}}$), the temperature of the neutral atmosphere $T(h)$, the concentrations (in $\mathrm{cm^{-3}}$) of the components—molecular nitrogen $N_2$ and oxygen $O_2$, atomic oxygen O, nitrogen N, helium He, hydrogen H, and argon Ar, and the mean molecular weight of the atmosphere in the specified region of the Earth or along the given flight path of the spacecraft.

The concentration of the neutral atmosphere components is calculated using the following relations:

$$n_i(h) = n_{120i} \cdot \exp[G_i(L)] \cdot D(h) \cdot [T_{120}/T(h)]^{1+\alpha}; \qquad (6)$$

$$D(h) = [T_{120}/T(h)]^{\gamma} \cdot \exp[-\alpha\gamma\zeta(h)]; \quad \gamma = m_i \cdot g_{120}/\chi R_g T_\infty;$$

$$g_{120} = g/(1 + 120/R)^2; \quad \chi = T_{120}/(T_\infty - T_{120});$$

$$\zeta(h) = (h - 120) \cdot (R + 120)/(R + h); \quad T_\infty = \overline{T}_\infty[1 + G(L)],$$

where $n_i$ and $m_i$ are the concentration and molecular weight of the $i$th component; $n_{120i}$ is the concentration of the $i$th component at the height of 120 km; $T_{120}$, $T$, $T_8$, $\overline{T}_\infty$, and $K$ are the atmospheric temperatures at 120 km, at an arbitrary height, at the height of the exosphere, and the mean exospheric temperature, respectively (the relations for $T_{120}$ and $T$ are the same as for $T_8$); $\alpha$ is the coefficient of thermal diffusion, $\alpha = -0,4$ for H and He and $\alpha = 0$ for O, $O_2$, N, $N_2$; and Ar; $g$ and $g_{120}$, $cm \cdot s^{-2}$ are the accelerations due to gravity at the sea level and at the height of 120 km; $R_g = 8.314 \cdot 10^{-3}$ $g \cdot km^2/(mole \cdot s^2 \cdot deg)$ is the gas constant; and $R$, km is the radius of the Earth.

The function $G_i(L)$ involves the same coefficients as the empirical atmospheric model MSIS90 [11,12] and expresses the dependence of the atmospheric parameters on local time $LT$, latitude $\varphi$ , longitude $\lambda$, colatitude $\theta$, universal time UT, 10-cm solar radio flux ($F_{10.7}$), geomagnetic activity $A_p$, and exospheric temperature $T_\infty$?. In other words, the character of the height profile—its inclination, gradient, etc.—of the $i$th atmospheric component is determined by the $i$th $G_i(L)$.

Diffusive equilibrium is assumed above 120 km. In model calculations of the neutral atmosphere parameters, the boundary condition is usually specified at a height of 120 km, where the concentration of atmospheric constituents $n_{120i}$ is assumed to be equal to the global mean concentration at 120 km [19,28].

The input model parameters are given in two program files: (1) the file of control parameters and the file of coordinates.

The file of control parameters includes the day, month, year, $F_{10,7}$ for the preceding day, $(F_{10,7})_{90}$ averaged over the three preceding months, geomagnetic activity $A_p$, starting time of computations (UT, h:min:s), minimum and maximum heights, and step in height (km). Such a record of the control parameters was designed to describe the atmospheric structure in the specified regions of the Earth.

The coordinate file involves the time, latitude, longitude, and height of every significant part of the path in order to describe the atmospheric conditions along the flight trajectory (orbit).

It is to be noted that the described model is sensitive to the parameters given in the control or coordinate file. A certain drawback of WMA01 (as well as of all present-day atmospheric models) is that the empirical coefficients used in calculations are normalized by the data on atmospheric composition obtained by mass-spectrometric measurements, as well as by the drag data for different satellites for several tens of years. The dependence of these coefficients on external factors is one of the reasons why the present-day models of the Earth upper atmosphere need correction.

## *The Algorithm and Software Package for Numerical Computation of the Atmospheric Parameters*

The WMA01 model is realized in the form of application modules written in the FORTRAN language. The particular version of the model under consideration works in the MS DOS operational system. The Grafor package is used for graphical representation of the results.

The model consists of the following modules:

- CMN—computation of the atmospheric parameters;
- oure—output preparation of the computation results;
- egrp—displaying or printing out the chosen parameter in a graphical form.

Each module receives and processes the initial data in the file form according to its designated purpose and issues the resulting files for subsequent computations. The data are transferred from module to module in unformatted files. The control is exercised by the batch file opk.bat, which takes account of the model structure and interaction of the particular modules.

The database of solar and geomagnetic activity indices is supplemented with data from IZMIRAN and from the web sites of the World Data Centers:
ftp://ftp.ngdc.noaa.gov/STP/GEOMAGNETIC_DATA/INDICES/Kp_Ap/;
ftp://ftp.ngdc.noaa.gov/STP/SOLAR_DATA/SOLAR_RADIO/FLUX/;
http://www.swpc.noaa.gov/today.html—Space Environment Center (SEC), Boulder, USA;
http://www.izmiran.rssi.ru—Solar and Geophysical Forecast Center, IZMIRAN.

The data on solar and geomagnetic activity entered in the corresponding files of the atmospheric model under consideration cover the period 1961–2001.

At the heights of 0–80 km, the activity indices are constant and are written as follows: $F_{10.7} = 150$, $(F_{10.7})_{90} = 150$, $A_p = 4$. Below 72.5 km, the atmospheric constituents O, H, and N are not calculated; i.e., they are equal to 0. These are the features of the MSISE–90 empirical model atmosphere coefficients used in the WMA01 model [11, 12].

## *Examples of Simulation of Space–Time Variations in the Atmosphere over Specified Regions of the Earth*

The choice of the model input parameters depends on the particular task. Two classical procedures can be applied in the course of modeling.

- A column of the Earth atmosphere in a specified region, e.g., in the range of latitudes $\varphi = 60°$ S–60° N and longitudes $\lambda = 0°$–360° at the heights of 70–300 km is simulated at a given time (UT, h:min) with a step of 5° in latitude, 15° in longitude, and 10 km in height. Such a model allows us to compute the atmospheric parameters at any minimal height down to the ground. The universal and local times are related by the expression $\mathrm{LT} = \mathrm{UT} + \lambda/15$, where $\lambda$ is in arc deg.
- The atmospheric parameters are determined along the spacecraft trajectory with the significant path sections specified.

Figure 16 shows an example of the calculated density cross-section of the neutral atmosphere along the equator (the indices of solar and geomagnetic activity are arbitrary and correspond to quiet conditions). Other calculation results are illustrated in Fig. 20 and are given in Table 8.

The results of test calculations of the density and temperature of the atmosphere show that the parameters determined using the WMA01 model are similar both in magnitude and in daily, altitude, and latitude variations to those obtained from other models.

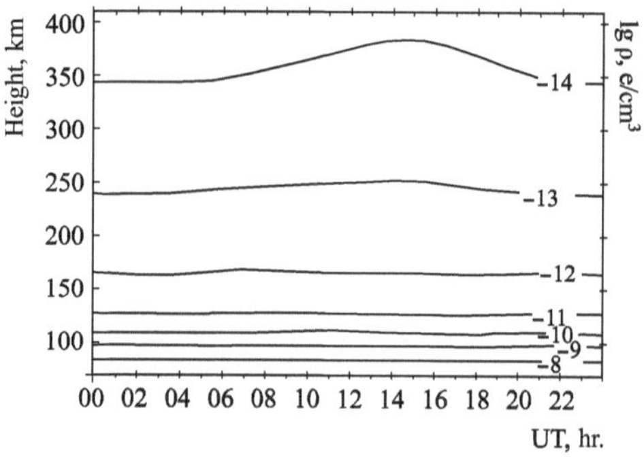


**Fig. 16** A computed density cross-section of the neutral atmosphere along the equator. The left scale is the atmosphere cross-section height; the right scale is the log density of the neutral atmosphere (08.01.2001, $F_{10.7} = 156$, $A_p = 5, 0$)

The comparison of the values obtained in terms of the model under consideration and the MSIS90 model reveals a discrepancy by a few per cent. For example, under moderate activity conditions ($F_{10.7} = 178$, $A_p = 56$) at 12.00 UT on 6 April 2000, the values of $\rho$ computed at the points with coordinates $\varphi = 0°$ and $\lambda = 0°$ at a height of 250 km were equal to $1.306 \cdot 10^{-13}$ (MSIS90) and $1.314 \cdot 10^{-13}$ (WMA01) $g \cdot cm^{-3}$; i.e., the difference made up $\sim 1\,\%$.

## *Deorbiting the Mir Orbital Complex. Solar and Geomagnetic Activity: The Response of the Earth Upper Atmosphere*

A good illustration of a successful use of the WMA01 mathematical model of the Earth upper atmosphere was deorbiting of the Mir space station.

The final stage of flight of the Mir orbital complex coincided with the epoch of maximum of the activity Cycle 23. The forecast of the upper atmosphere and the choice of the deorbit time were based on continuous analysis of solar activity for the period from December 2000 through March 2001 with the use of observation data from the world solar and geophysical observatories. Figure 17 illustrates the distribution of the solar and geophysical observatories over the world. Their coordinates are given in Table 7.

The basic parameters of solar and geomagnetic activity as observed and forecasted at these observatories are:

- Moscow—current solar radio emission at the frequency of 3000 MHz (10.0 cm), variations of geomagnetic field components, forecast of the solar radio flux $F_{10.0}$, and planetary geomagnetic activity index $A_p$;
- Ottawa—current solar radio emission at the frequency of 2800 MHz (10.7 cm);
- Boulder—forecast of the solar radio flux $F_{10.7}$ and $A_p$-index;
- San Vito, Holloman, Ramey, Learmonth—optical observations of the Sun, including $H_{\acute{\alpha}}$ (6563 Å) and white–light images (visible spectrum), and solar magnetograms.

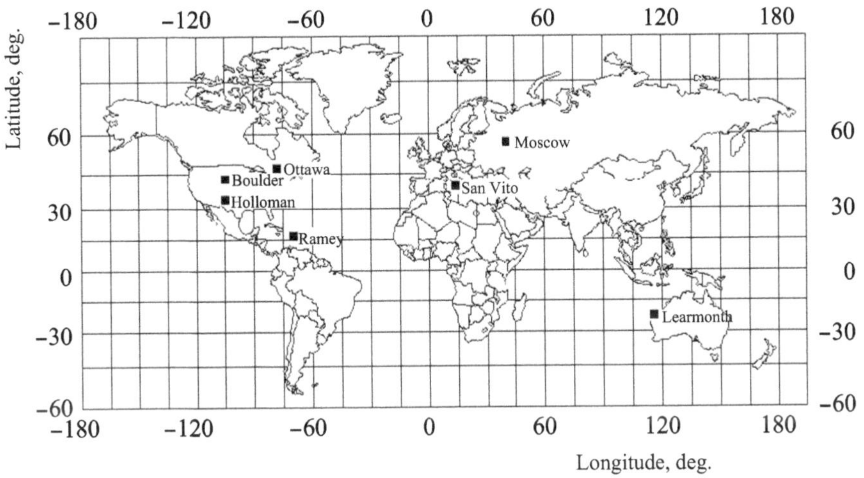


**Fig. 17** The distribution of solar and geophysical observatories over the world

**Table 7** Coordinates of the world solar and geophysical observatories

| Observatory | Country | Coordinates |
|---|---|---|
| Moscow | Russia | 55° N 37° E |
| Ottawa | Canada | 45° N 76° W |
| San Vito | Italy | 41° N 18° E |
| Boulder | USA | 40° N 105° W |
| Holloman | USA | 32° N 106° W |
| Ramey | Puerto Rico | 18° N 67°W |
| Learmonth | Australia | 21° S 115°E |

The variations of solar activity during part of Cycle 23 (from September 2000 through May 2001) are illustrated in Fig. 18.

From December 2000 to February 2001, we received the daily forecast of solar activity issued by the Space Environmental Center. The time interval under consideration involved the autumnal (September) and vernal (March) equinoxes. The solar active events observed in the equinox periods are known to produce strong geophysical effect [11, 12, 19].

In the first decade of March 2001, the Sun showed signs of the development of new active regions. This was to result in the increase of solar activity. Hence, we anticipated a change from the favorable, quiet period to the increase of $F_{10.7}$ and the onset of an unfavorable period characterized by strong disturbances of the atmospheric temperature and density. Under such conditions, the process of deorbiting the Mir station could face certain difficulties.

Figure 19 shows the variations of solar and geomagnetic activity during 11–31 March 2001. One can readily see two periods, call them favorable and unfavorable. In the favorable period (15.03.01–17.03.01), the activity indices were not high—$F_{10.7} = 136$ and $A_p = 3$ (quiet conditions); in the unfavorable period

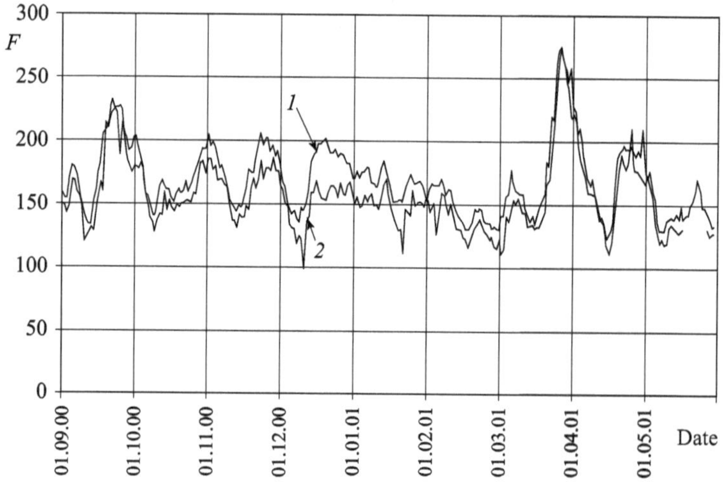


**Fig. 18** Variations of the solar radio flux $F$ at the frequencies of 2,800 MHz (10.7 cm) (1—Ottawa) and 3,000 MHz (10.0 cm) (2—Moscow) in the period 01.09.2000–31.05.2001

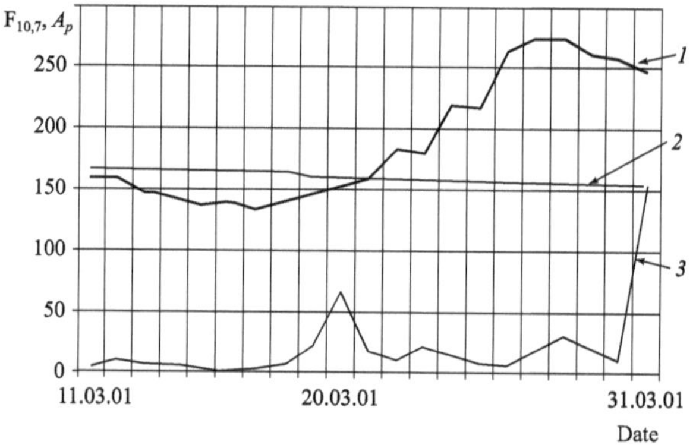


**Fig. 19** Variations of the solar radio flux $F_{10.7}$ for the current day (1) and for the preceding 90 days (2) and the planetary geomagnetic activity index $A_p$ (3) during the period from 11.03.2001 through 31.03.2001

(31.03.01), $F_{10.7} = 246$ and $A_p = 155$ (strongly disturbed conditions). On the day of deorbiting the Mir station (23.03.01), these parameters were $F_{10.7} = 180$ and $A_p = 21$ (moderate disturbance).

A distinctive feature of the unfavorable period under discussion was that the basic parameters of the solar and geomagnetic activity recorded at that time were the highest for the past two years. E.g., in addition to the aforesaid, the area of active regions on the solar disk $S_p$ was $\sim$ 4000 m.v.h. (millionths of the visible hemisphere) compared to $S_p \sim 390$ in the favorable period.

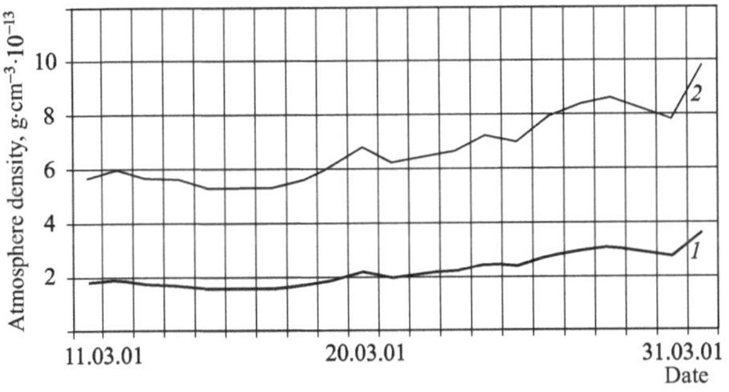


**Fig. 20** Variations of atmospheric density at the fixed heights of 220 km (1) and 180 km (2) during the period from 11.03.2001 to 31.03.2001

**Table 8** Predicted and observed values of the atmospheric density

| Height, km | Observed values $F_{10.7} = 180$; $AF_{10.7} = 158$; $A_p = 21$ | Predicted values $F_{10.7} = 145$; $AF_{10.7} = 158$; $A_p = 20$ | Deviation from the forecast |
|---|---|---|---|
| 220 | $2.189 \cdot 10^{-13}$ | $1.876 \cdot 10^{-13}$ | 15 % |
| 180 | $6.648 \cdot 10^{-13}$ | $6.019 \cdot 10^{-13}$ | 10 % |

The response of the atmosphere to variations of solar activity in the period from 11.03.01 to 31.03.01 is illustrated in Fig. 20. The figure shows the atmospheric density variations calculated using the WMA01 model with fixed parameters of solar and geophysical activity. Thus, the atmospheric density in the unfavorable period exceeds the quiet–time values almost by a factor of 1.5 at the heights of 220 km ((56 %)and 180 km (46 %) and almost by a factor of 2 at 300 km.

On the other hand, the increase of the calculated density corresponding to the predicted parameters of solar and geophysical activity on the day of deorbit of the Mir station (23 March 2001) at the propulsion switch–on point (35.4° N, 24.59° E) on the trajectory crossing the equator at 07:57 UT at 4.5° W was 15 % and 10 % at the heights of 220 km and 180 km, respectively (see Table 8).

The model under discussion was used to assess the conditions in the upper atmosphere at the final stage of flight of the Mir orbital station. This, along with other measures, allowed the experts of the Mission Control Center-M (Korolev, Moscow Region) to simulate eventual emergency situations on the orbit and along the deorbit path, including the effect of solar and geophysical activity on the Earth upper atmosphere. Thus, the risk factor was minimized, and the Mir station was submerged successfully in the calculated region of the Pacific [9].

The WMA01 model was also used to estimate the atmospheric density at the orbital heights of CORONAS-I and CORONAS-F satellites under different solar and geophysical conditions. A forecast of the orbital parameters and potential lifetime of the CORONAS-F mission was developed for the period 2004–2006 based on variations of the atmospheric density at the orbital heights depending on heliogeophysical conditions [5].

## Solar Active Events During the Flight Period of the CORONAS-F Mission

The CORONAS-F satellite was launched on 31 July 2001, i.e., in the mid-maximum of Cycle 23 (the second component of the physical 22-year solar cycle) when 5.3 years of the cycle had already passed. The evolution of Cycle 23 showed the scarcity of our knowledge of the regularities that govern the solar periodicity. Cycle 23 was a medium-intensity solar cycle. It fitted perfectly well in the scheme of evolution of the previous cycles similar in the pace of development and height, for which the Wolf number value was $80 < W^* \leqslant 130$.

This was the first solar cycle when the integral activity of solar processes began to decrease after nearly 60 years of high activity (Cycles 18–22). On the other hand, in spite of a dramatic general decrease of the number of solar flares, there were intervals of high and extremely high flare activity during the flight period of the CORONAS-F mission. The on board equipment recorded 143 major solar flares. Thirteen of these events were X-ray flares of importance $> 5$ and 4 flares had importance $\geqslant 10$, including three outstanding events of 28.10.2003 (X 17.4), 04.11.2003 ($X > 17.5$), and 07.09.2005 (X 17.1).

Table 9 provides the principal characteristics of the major flares of importance $\geqslant 3$ Cycle 23. Twenty of these events (marked bold) occurred during the flight of the CORONAS-F satellite. There were two clearly defined time intervals when the flare–generating active regions were mainly observed—before and after June 2003. The most important active regions of Cycle 23 that produced major flares during the flight of the CORONAS-F satellite (August 2001–May 2005) are listed in Table 10.

### *Brief Characteristics of the most Significant Flare Events*

2001 The first major event recorded after the launch of CORONAS-F was an X5.3 flare that occurred on August 25 in the sunspot group **AR9591** (S12L295; $X_1^{5.3} + M_9 + C_{49}$; XRI = 6.67) and was accompanied by $\gamma$-ray emission [29]. At the end of September, the visible solar disk was traversed by a large sunspot group **AR9632** (S18 L268, $X_1^{2.6} + M_1 + C_9$; XRI = 2.71), which produced two flares of importance X2.6 and M7.6. The former caused an intensive proton event in near-Earth space on September 24 (12,900 $cm^{-2} \cdot s^{-1}\ sr^{-1}$). In the next rotation (October 22–25), the same sunspot group **AR9672** (S18L268, $X_2^{1.3} + M_5 + C_{16}$; XRI = 4.01), again, produced a series of large solar flares (M6.7, X1.2, M6.9, and X1.3), which triggered a minor proton event on September 22–23 and a moderate magnetic storm on October 28.

One of the most intensive proton events (31,700 $cm^{-2} \cdot s^{-1}\ sr^{-1}$) and a severe magnetic storm of November 5–6 followed the major flare event of importance X1.0/3B, which occurred on November 4 in the sunspot group **AR9684** (N06L136; $X_1^{1.0} + M_1 + C_{11}$; XRI = 1.21).

**Table 9** The major (importance $\geqslant 3.0$) solar flares of Cycle 23 that produced the most intensive emission flux in the range of 1–12.5 keV

| Location | Year/month/day | Importance | $\tau$ | Coordinates | AR |
|---|---|---|---|---|---|
| **1** | **2003/11/04** | **X > 17.5 $\gamma$ n$^0$** | **11$^m$ (X28)** | **S19W83L286** | **10486** |
| 2 | 2001/04/02 | X > 17.5 | 5$^m$? (X22$^+$) | N19W90L152 | 9393 |
| **3** | **2003/10/28** | **X17.2 $\gamma$ n$^0$** | | **S16E08L286** | **10486** |
| **4** | **2005/09/07** | **X17.1 $\gamma$** | | **S12E89L229** | **10808** |
| 5 | 2001/04/15 | X14.4 | | S20W85L001 | 9415 |
| **6** | **2003/10/29** | **X10.0 $\gamma$** | | **S15W02L286** | **10486** |
| 7 | 1997/11/6 | X9.4 $\gamma$ GLE | | S18W63L352 | 8100 |
| 8 | 2006/12/05 | X9.0 $\gamma$ | | S07E791009 | 10930 |
| **9** | **2003/11/02** | **X8.3 $\gamma$ n$^0$** | | **S14W56L286** | **10486** |
| **10** | **2005/01/20** | **X7.1 $\gamma$ GLE** | | **N15W56L177** | **10720** |
| 11 | 2006/12/06 | 6.5 $\gamma$ | | S06E63L009 | 10930 |
| **12** | **2001/12/12** | **X6.3$\gamma$** | | **S09E16L020** | **9727** |
| **13** | **2005/09/09** | **X6.2$\gamma$** | | **S10E66L229** | **10808** |
| 14 | 2000/07/14 | X5.7 $\gamma$ | | N22W07L320 | 9077 |
| 15 | 2001/04/05 | X5.6 $\gamma$ | | S21E31L001 | 9415 |
| **16** | **2003/10/23** | **X5.4** | | **S21E88L286** | **10486** |
| **17** | **2005/09/09** | **X5.3 $\gamma$** | | **S11E74L229** | **10808** |
| **18** | **2001/08/25** | **X5.3 $\gamma$ n$^0$** | | **S17E34L296** | **9591** |
| 19 | 1998/08/18 | X4.9 $\gamma$ | | N33E87L034 | 8307 |
| **20** | **2002/07/23** | **X4.8 $\gamma$** | | **S13E72L204** | **10039** |
| 21 | 2000/11/26 | X4.0 | | N18W38L352 | 9236 |
| 22 | 1998/08/19 | X3.9 $\gamma$ | | N32E75L035 | 8307 |
| **23** | **2003/11/03** | **X3.9 $\gamma$** | | **N08W77L291** | **10488** |
| **24** | **2005/01/17** | **X3.8 $\gamma$** | | **N15W25L177** | **10720** |
| 25 | 1998/11/22 | X3.7 $\gamma$ | | S27W82L029 | 8384 |
| **26** | **2003/05/28** | **X3.6 $\gamma$ n$^0$** | | **S07W20L182** | **10365** |
| **27** | **2004/07/16** | **X3.6** | | **S11E35L044** | **10649** |
| **28** | **2001/12/28** | **X3.4 $\gamma$ GLE** | | **SEL018** | **9767?** |
| 29 | 2006/12/13 | X3.4 GLE | | S06W24L009 | 10930 |
| 30 | 1998/11/28 | X3.3 $\gamma$ n$^0$ | | N17E32L182 | 8395 |
| **31** | **2002/07/20** | **X3.3$\gamma$ n$^0$** | | **S13E90L204** | **10039** |
| **32** | **2002/08/24** | **X3.1$\gamma$** | | **S02W81L298** | **10069** |
| **33** | **2002/07/15** | **X3.0 $\gamma$** | | **N19W01L012** | **10030** |

Note: $\tau$ is the saturation time of the soft X-ray detectors on board the GOES-9 and GOES-10 satellites at a threshold of $X > 17.5$ during the maximum phase of extreme solar flares

A major flare event occurred on November 21 in the sunspot group **AR9704** (S17L270; $X_1^{1.1}$ | $M_3$ | $C_{44}$; XRI = 2.61). It comprised a low-importance flare C4.7 and a large eruptive filament accompanied by a bright coronal mass ejection (CME) and type II and IV dynamic radio bursts. This event caused one of the most severe magnetic storms in Cycle 23 (November 24–25) and a large proton event (18,900 $cm^{-2} \cdot s^{-1}\ sr^{-1}$) on November 22–24. The last intensive flare recorded in 2001

**Table 10** The most important flare-generating active regions observed during the CORONAS-F flight time (August 2001—December 2005)

| No. | AR | CMP | Coordinates | $S_p^{max}$; class | Flares by importance | FER date (hour) |
|---|---|---|---|---|---|---|
| 1 | 9561 | 28,5.08.01 | S19L295 | 740; FKI; $\delta$ | $X_1 + M_9 + C_{49}$ (1) | 24–25 (36) |
| 2 | 9608 | 11,5.09.01 | S27L109 | 1050; FKC; $\delta$ | $M_{13} + C_{27}$; (3) | 9 (19); 16–18 (69) |
| 3 | 9632 | 26,4.09.01 | S19L272 | 790; DKI; $\delta$ | $X_1 + M_2 + C_9$ (1) | 23–24.09 (22) |
| 4 | 9661 | 17,3.10.01 | N15L357 | 800; EKI; $\delta$ | $X_2 + M_2 + C_{16}$ | 19.10 (14) |
| 5 | 9672 | 24,0.10.01 | S18L268 | 590; DKI; $\delta$ | $X_2 + M_5 + C_{16}$ (4) | 22–25.10 (72) |
| 6 | 9690 | 11,3.11.01 | S18L027 | 1420; FKC; $\delta$ | $M_{18} + C_{49}$ (1) | 7–9.11 (80) |
| 7 | 9704 | 20,2.11.01 | S17L270 | 620; DKI; $\delta$ | $X_1 + M_3 + C_{44}$ (2) | 22.11 (7) |
| 8 | 9715 | 30,3.11.01 | N04L137 | 990; EKC; $\delta$ | $M_4 + C_{28}$ (2) | 27–29.11(39) |
| 9 | 9733 | 14,3.12.01 | N15L313 | 560; EKI; $\delta$ | $X_2 + M_4 + C_{20}$ (2) | 11–13.12 (54) |
| 10 | 9893 | 9,0.04.02 | N15L215 | 490; EKI; $\delta$ | $M_4 + C_{17}$ (1) | 14–16.12 (58) |
| 11 | 9961 | 25,7.05.02 | S22L331 | 480; EAI; $\delta$ | $X_1 + M_2 + C_{10}$ (2) | 20.05 (5) |
| 12 | 10017 | 29,1.06.02 | S18L235 | 610; EKI; $\delta$ | $X_1 + M_6 + C_{16}$ (2) | 2–5.07 (65) |
| 13 | 10030 | 15,9.07.02 | N18L012 | 1350; EKC; $\delta$ | $X_2 + M_6 + C_{40}$ (4) | 11–12(34); 15–18(59) |
| 14 | 10039 | 28,6.07.02 | S12L204 | 940; FKC; $\delta$ | $X_3 + M_4 + C_{28}$ (4) | 20–23(51); 2–4(46) |
| 15 | 10044 | 28,2.07.02 | S21L210 | 780; FKC; $\delta$ | $M_6 + C_{22}$ (2) | 26.07 (22) |
| 16 | 10069 | 17,8.08.02 | S08L298 | 1990; EKC; $\delta$ | $X_2 + M_{17} + C_{55}$ (5) | 16(12); 20–22(48); 23–24(17) |
| 17 | 10314 | 15,0.03.03 | S15L061 | 500; EKI; $\delta$ | $X_2 + M_6 + C_{23}$ (2) | 17.03 (39) |
| 18 | 10338 | 21,6.04.03 | N18L285 | 250; EAI; $\delta$ | $M_9 + C_{31}$ (2) | 23–24(36); 26–27(39) |
| 19 | 10365 | 26,6.05.03 | S07L182 | 880; DKC; $\delta$ | $X_2 + M_9 + C_{17}$ (5) | 27–29(25); 31–2(56) |
| 20 | 10375 | 7,7.06.03 | N13L022 | 1250; FKC; $\delta$ | $X_3 + M_{29} + C_{45}$ (6) | 9–12(62); 14–15(42) |
| 21 | 10386 | 22,3.06.03 | S07L188 | 310; DKC; $\delta$ | $X_1 + M_3 + C_9$ (2) | 15–17 (47) |
| 22 | 10484 | 24,0.10.04 | N03L354 | 1750; EKC; $\delta$ | $X_2 + M_{16} + C_{28}$ (3) | 19(13); 25–27(58) |
| 23 | 10486 | 29,3.10.03 | S17L283 | 2610; FKC; $\delta$ | $X_7 + M_{16} + C_{16}$ (9) | 22–24(59); 27–29(59); 1–5(85) |
| 24 | 10488 | 28,4.10.03 | N08L291 | 1750; FKC; $\delta$ | $X_3 + M_7 + C_{17}$ (2) | 2–4.11(41) |
| 25 | 10501 | 19,6.11.03 | N03L002 | 410; DKI; $\delta$ | $M_{11} + C_{20}$ (2) | 17–18(31); 20(22) |
| 26 | 10537 | 12,8.01.04 | N04L009 | 270; DKC; $\delta$ | $M_7 + C_{10}$ (3) | 5–7(55) |
| 27 | 10540 | 19,1.01.04 | S16L285 | 350; FKO; $\delta$ | $M_5 + C_{13}$ (2) | 17–20 (62) |
| 28 | 10564 | 25,0.02.04 | N14L160 | 870; EKC; $\delta$ | $X_1 + M_1 + C_{20}$ (2) | 26.02(20) |
| 29 | 10646 | *11.07.04 | N13L169 | 220; DAI; $\beta$ | $M_4 + C_{11}$ (3) | 12–13(36) |

(continued)

**Table 10** (continued)

| No. | AR | CMP | Coordinates | $S_p^{max}$; class | Flares by importance | FER date (hour) |
|---|---|---|---|---|---|---|
| 30 | 10649 | 19,0.07.04 | S10L044 | 510; FKC; $\delta$ | $X_6 + M_{10} + C_6$ (10) | 13–14(29); 15–17(55) |
| 31 | 10652 | 23,3.07.04 | N08L345 | 2010; FKC; $\delta$ | $M_{17} + C_{59}$ (3) | 20–22(36); 24–25(20) |
| 32 | 10656 | 12,2.08.04 | S13L083 | 1360; FKC; $\delta$ | $X_2 + M_{23} + C_{94}$ (5) | 13–14(48); 17–18(42) |
| 33 | 10691 | 29,2.10.04 | N14L133 | 270; DAO; $\delta$ | $X_1 + M_8 + C_{25}$ (2) | 30.10 (13) |
| 34 | 10696 | 06,4.11.04 | N08L026 | 910; EKC; $\delta$ | $X_2 + M_{13} + C_{37}$ (7) | 3–4(44); 6–7(39); 9–10(9) |
| 35 | 10715 | 03,6.01.05 | N05L339 | 350; DKC; $\delta$ | $X_1 + M_4 + C_{13}$ (1) | 30–1(26) |
| 36 | 10720 | 15,7.01.05 | N13L179 | 1630; EKI; $\delta$ | $X_5 + M_{19} + C_{65}$ (8) | 14–15(17); 16–17(9); 18–20(39) |
| 37 | 10786 | 7,6.07.05 | N13L056 | 420; DKC; $\delta$ | $X_1 + M_{14} + C_{41}$ (3) | 13–14.07(22) |
| 38 | 10808 | 14,3.09.05 | S10L229 | 1000; EKC; $\delta$ | $X_{10} + M_{20} + C_{47}$ (13) | 7–10.09(76); 12–15(72) |
| 39 | 10826 | 3,5.12.05 | S03L249 | 490; EKC; $\delta$ | $M_4 + C_{13}$ (2) | 2.12 (18) |

Note: CMP is the time of passage through the central meridian of the Sun; $S_p^{max}$ is the maximum sunspot area in millionths of the visible solar hemisphere; FER is the period of flare energy release; the column "Flares by importance" contains in brackets the number of major flares of X-ray importance $\geqslant$M5; * indicates the sunspot group that appeared at W33 on 11 July 2004

occurred in the sunspot group **AR9733** (N14 L313; $X_2^{6.2} + M_4 + C_{20}$; XRI = 8.99) on December 13 and had a significant effect on the daytime Earth ionosphere.

2002 After the dramatic events mentioned above, the activity in the Sun was not high until the second decade of July 2002, when two large flare-producing sunspot groups appeared on the visible disk: **AR10030** (N18 L012; $X_3^{3.0} + M_5 + C_{40}$; XRI = 7.33) and **AR10039** (S12 L204 $X_3^{4.8} + M_4 + C_{28}$; XRI = 10.33). The disturbances generated by the flares in these two active regions affected mainly the day-side ionosphere. In the next rotation, in August 2002, the flare activity was associated with the sunspot group **AR10069** (S10L298; $X_2^{3.1} + M_{17} + C_{55}$; XRI = 8.16). In the period from August 20 to August 24, this group produced four large flares, the most intensive of which (X3.1 on 24.08) was the source of a proton event (317 $cm^{-2} \cdot s^{-1}\ sr^{-1}$).

On September 5, a major flare event occurred in the decaying sunspot group **AR10102** (N08L019). It consisted of a flare of importance C5.2 and a large filament eruption, which resulted in a full-halo CME. The interplanetary shock wave generated in this event reached the Earth at 16:38 UT on September 7 and caused a strong magnetic storm in near-Earth space. The medium-importance flares and the filament eruption observed during September 27–29 were the most likely source of a severe magnetic storm on 1–2 October 2002.

**2003** The next period of enhanced flare activity started on 14 March 2003, as a rapidly developing sunspot group **AR10314** (S15L061; $X_2^{1.5}+M_6+C_{23}$; XRI = 4.23) appeared in the central zone of the southern hemisphere. All flares of high and medium importance recorded in the period under discussion occurred in that group during 42 h on March 17–19.

The end of May 2003 (27–29.05) was marked by a series of major solar flares in the sunspot group **AR10365** (S07 L182; $X_3^{3.6}+M_9+C_{33}$; XRI = 7.91), which caused magnetic storms in near-Earth space and a series of proton events of low and medium intensity. The period of 9–12 June 2003 was characterized by high flare activity in the sunspot group **AR10375** (N12 L022; $X_3^{1.7}+M_{29}+C_{45}$; XRI = 11.42).

The concentration of flare events in the cycle under consideration reached its maximum in the period from October 10 to November 5, 2003, when three large sunspot groups were passing simultaneously over the visible solar disk. One of the groups, having the cycle largest area, was observed in the southern hemisphere and the other two groups, in the northern hemisphere [30]. The period started with the appearance of the sunspot group **AR10484** (N03 L354; $X_2^{1.2}+M_6+C_{28}$; XRI = 5.73) from behind the eastern limb on October 17. Soon after its appearance (on October 19) it produced a flare of importance X1.1. The emergence of a new powerful magnetic flux on October 26 increased significantly the group area and resulted in two more major flares (X1.2/2N M7.6/2N). After that, the fully evolved sunspot group disappeared behind the western limb on October 29. A large sunspot group **AR10486** (S17 L283; $X_7^{>17.5}+M_{16}+C_{16}$; XRI = > 62.56) that evolved on the invisible side of the Sun appeared on the disk on October 21. On October 23, it produced flares of importance X5.4/1B and X1.1/1N and on October 24, a flare of importance M7.6/1N. A new powerful magnetic flux emerged in this group on October 24–25, 2003 increasing its area nearly twice and producing flares of importance X1.2/3B (26.10), M5.0/1F and M6.7/1F (27.10). Another magnetic flux that emerged on October 27–28 made the group area reach a cycle record value ($Sp$ = 2610 m.v.h.) and resulted in the third intensive flare (X17.3/4B) in the cycle on October 28.

The disturbances from the aforementioned flares caused a minor geomagnetic storm on October 28, which transformed to a severe storm (the strongest one in the cycle) on October 29–31 when the disturbance from the last flare reached the Earth. The auroras accompanying that storm were observed in the South California and South Greece. The same flare generated high-energy solar protons that arrived at the Earth and caused a very intensive proton event (29,500 $cm^{-2} \cdot s^{-1}\ sr^{-1}$).

The next period of energy release in the active region under discussion began on November 02, 2003 with a flare of importance X8.3/2B and continued on November 4 with a flare of importance $X > 17.5$/3B ($\tau = 11$ min), which produced the cycle most intensive flux of soft X-rays. This flare occurred near the western limb of the Sun and had no significant effect on geomagnetic conditions, but large solar proton events were recorded in near-Earth space with the maxima on November 2 and 4, 2003. The third sunspot group **AR10488** (N08 L291; $X_3^{3.9}+M_7+C_{17}$; XRI = 8.57) formed at the center of the northern hemisphere on October 27. In spite of a rapid evolution, it produced only moderate flares until November 3, when the flares of importance X2.7 and X3.9 took place. In the second rotation, significant geoeffective events occurred only in the first of the aforementioned sunspot groups, **AR10501** (N03 L002; $M_{11}^{9.6} + C_{20}$; XRI = 4.03). The disturbance from a complex flare event on November 18, when the optical flare of importance 2N was accompanied by two X-ray bursts (M3.2 and M3.9), caused a severe geomagnetic storm, which was but a bit weaker than that observed in October.

2004 The next significant burst of flare activity was associated with the sunspot group **AR10649**(S10L044; $X_6^{3.6} + M_{10} + C_6$; XRI = 14.57), which appeared on the visible disk on 11 July 2004. All major and moderate flares occurred in that group during the period from July 13 through July 18; however, they had little effect on near-Earth space and caused but sudden ionospheric disturbances on the day side and a minor magnetic storm on July 17. In the first decade of July 2004, it was established by the method of helioseismic tomography that a large sunspot group was evolving on the back side of the Sun. On July 16, it appeared on the visible disk as **AR10652** (N08L345; $M_{17}^{9.1} + C_{59}$; XRI = 4.75). While passing over the disk, this group of area equal to 2010 m.v.h. produced 17 flares of importance M, including three major flares. The geophysical effect from these flares was very strong on July 25 and 27 when the storm values approached the threshold for a severe magnetic storm. A new magnetic flux emerged rapidly on August 10–11, 2004 in the sunspot group **AR10656** (S13L083; $X_2^{1.8} + M_{23} + C_{94}$; XRI=8.67). It resulted in the occurrence of four major (one of importance X1.0) and twelve moderate flares during 54 h on August 13–15. The other significant flares (one of importance X1.8 and five of importance M) occurred during 56 h on August 17–19. None of these flares caused noticeable geomagnetic disturbances, their effect on the Earth environment being reduced to sudden ionospheric disturbances.

A seven-day-long period of significant flare activity started on November 3, 2004, in the sunspot group **AR10696** (N08L026; $X_2^{2.5} + M_{13} + C_{37}$, which had appeared in the vicinity of the eastern limb on November 1, 2004. During 96 h from November 3 to November 7, this group produced, among others, two significant geoeffective events: a 2N optical flare on November 6, which "combined" three X-ray bursts M9.3, M5.9, and M1.4, and an X-ray flare of importance X2.0 on November 7. The disturbances from these flares caused a severe magnetic storm in near-Earth space during November 9–11 and a medium-strong proton event. The second interval of enhanced activity was observed on November 9–11, when two major flares (M8.9 and X2.5) occurred in this AR during 9 h.

2005 r. The very first minute of 2005 was marked by a flare of importance X1.7/2B in the sunspot group **AR10715** (N05L339; $X_1^{1.7} + M_4 + C_{13}$; XRI = 2.69). Thc flarc was accompanied by a coronal mass ejection, which reached the Earth 35 h later and caused a minor magnetic storm. On January 10, a new group **AR10720** (N09L177; $X_5^{7.1} + M_{19} + C_{65}$; XRI = 21.5) appeared at the eastern limb in the northern hemisphere and began to evolve rapidly from January 12. For 17 h of January 14–15, 11 flares of importance ($X_2 + M_9$) were recorded in this sunspot group, including four major flares. The second enhancement of flare activity was recorded during January 16–17 and lasted 9 h ($X_1 + M_3$, 1 major flare), and the third one lasted 39 h from January 18 to January 20 ($X_2 + M_4$, 3 major flares) and ended with a flare of importance X7.1/3B, which was second in intensity to the event of November 4, 2003. The flux of charged particles with the energies more than 100 MeV generated in that flare was the largest one recorded since October 1989. The flux of highest-energy (>1 GeV) protons from that flare caused a large increase in the counting rate of ground-based neutron monitors ($\sim$ 105 % at the Moscow cosmic rays station). That was the largest flare in cosmic rays for the past two cycles. All major flares were geoeffective and resulted in one severe and two moderate magnetic storms in near-Earth space accompanied by large proton events.

On July 1, the sunspot group **AR10786** (N13L056; $X_1 + M_{14} + C_{41}$) appeared from behind the eastern limb in the north hemisphere. Geoeffective flares occurred in that active region when the sunspot group was at the center of the disk. Two X-ray flares of importance M4.5 (on July 7) and M2.8 (on July 9) were accompanied by the full-halo coronal mass ejections that caused moderate geomagnetic storms on July 10–11 and 12–13. As the group was approaching the western limb, it produced 11 flares of importance M and one flare of importance X during 54 h (July 12–14). Three of them were major flares (M5.0 on 13.07 and M9.1 and X1.2 on 14.07). The result of the major flares was a solar proton event, which lasted till the beginning of July 17. There is indirect evidence that significant flares in the active region under discussion continued to occur until the last low-intensity solar proton event was recorded three days after the sunspot group had disappeared behind the western limb.

In September 2005, the solar activity increased unexpectedly and reached an extreme level. The region that appeared on the visible disk was one of the most active ones not only in the cycle but also all over the history of solar observations. A large sunspot group of the previous rotation **AR10798**(S12L216) that had produced a flare of importance M5.6 on August 22 resulting in a strong geomagnetic storm on August 24–25 grew threefold in area while staying on the invisible side, and its activity increased dramatically. Powerful coronal mass ejections indicative of major flare events had been recorded in the yet invisible sunspot group **AR10808** (S10L229; $X_{10} + M_{20} + C_{47}$) a few days before September 7 when it appeared from behind the eastern limb. The appearance of this group on the visible disk was accompanied by the fourth intensive X-ray flare (B X17.1/3) in the cycle. Then, until September 10, the group produced seven other major flares, six of which had the X-ray importance X. Such a high flare activity caused a minor and a large magnetic storms in near-Earth space on September 9 and September 10–14, respectively, as well as an intensive solar proton event. The second period of high flare activity in the AR under discussion was observed during September 12–13. The observers recorded three major flares during 39 h. Two of them belonged to the X-ray class higher than X. The disturbances from these flares caused a moderate geomagnetic storm on September 14–15. The previously recorded proton event lasted at a moderate level till September 16. The last major flares observed in this AR occurred on September 15 and 17. Altogether, the group produced 11 flares of importance X and 22 flares of importance M and ranked fourth in its flare potential since the beginning of solar X-ray observations.

In December, the most significant flare events occurred at the very beginning. On December 2, two major flares of importance M7.8/1B and M6.5/1N and two moderate flares were recorded during 18 h in the medium-size group **AR10826** (S03L249; $M_4 + C_{13}$) in the south hemisphere. In spite of a favorable position of the flares, their effect on the Earth reduced to a mere electromagnetic impact, i.e., the impact of solar radiation (mainly UV and soft X-rays) on the Earth ionosphere. Such impacts usually result in sudden ionospheric disturbances, which cause radio blackouts at short waves.

## *Solar Activity in Cycle 23 and Near-Earth Space*

The evolution of Cycle 23 followed a scenario typical of the normal activity cycles; i.e., the most intensive flare events occurred in the rising and declining phases. Some particular evolution features and the pattern of active regions characteristic of Cycle 23 might be indicative of a change in the magnetic field generation mode in the solar convection zone. In this case, it is possible that the solar activity is entering a period of medium-high and low cycles that may last for about 50–70 years (5 or 6 cycles). If the number of active regions decreases, the number of coronal holes and, therefore, the number of recurrent high-speed solar wind streams will increase, resulting in a relative increase in the occurrence of recurrent small geomagnetic disturbances in near-Earth space. A significant decrease in the number of flare events will reduce the number of usually more intensive sporadic geomagnetic disturbances and increase the periods of quiet geomagnetic conditions. A reduced emergence of the magnetic flux will decrease the modulation zone of galactic cosmic rays in interplanetary space, increase their intensity in near-Earth space and, accordingly, increase the radiation background not only in the epoch of minimum but also throughout the cycle. The sporadic heating of the Earth atmosphere will decrease significantly resulting in further pollution of near-Earth space (unfavorable conditions for the removal of space debris from low orbits) and, probably, in other adverse consequences.

## Conclusion

The space weather forecasts become increasingly demanded in various spheres of the human activity. The first thing we need to produce a reliable forecast of geophysical conditions is the information on active processes in the Sun. It takes quite a long time (1–4 days) for a magnetic cloud ejected from the Sun to reach the Earth. Therefore, a short-term forecast of the space weather is sufficiently reliable provided that high-quality observations of the solar corona (preferably, at different angles) and cosmic ray variations are available.

In this chapter, we have described the method of quantitative estimation of the neutral density of the Earth upper atmosphere using the absorption measurements of solar short-wave radiation on board the CORONAS-F satellite.

The model calculations of the concentration of atomic oxygen [O] have been compared with the CORONAS-F absorption measurements in the range of 304 Å. It is shown that, under quiet conditions, the model values agree closely with experimental data. Under disturbed conditions, the experimental concentrations of [O] are by 50 % higher than the model ones. If we consider the total concentration of the atomic oxygen and molecular nitrogen, the experimental measurements and model estimates for the disturbed conditions are comparable.

The results obtained show that the method suggested above for comparing the simulated and experimental atmospheric parameters can be used successfully to correct the existing models of the Earth upper atmosphere.

A method is proposed for estimating the lifetime of low-orbit satellites and for predicting the atmospheric density variations at the perigee heights taking into account the heliogeomagnetic conditions. The method is based on the mathematical model of the Earth upper atmosphere, WMA01, developed at IZMIRAN.

The analysis of the orbit evolution of the CORONAS-I and CORONAS-F satellites has shown that the lifetime of CORONAS-F at the low-orbit flight stage was affected essentially by the outstanding solar events that occurred in the period immediately preceding the cycle minimum and by the accompanying processes in near-Earth space and in the Earth upper atmosphere (2004–2005).

## References


1. Oraevsky, V.N., Sobelman, I.I.: Comprehensive studies of solar activity on the CORONAS-F satellite. Astron. Lett. **28**(6), 401 (2002)
2. Kotov, Yu.D., Yurov, V.N., Boldyrev, S.I.: The scientific tasks and characteristics of the scientific equipment of the Russian CORONAS–FOTON space project. Kosmichna nauka i tekhnologiya (Space science and technology) **9**(5/6), 53–59 (2003)
3. Zhitnik, I.A., Boyarchuk, K.A., Bugaenko, O.I., et al.: Effects of the absorption of solar XUV radiation by the Earth's upper atmosphere at altitudes of 100–500 km in the X-ray solar images obtained on board the CORONAS-I (TEREK telescope) and CORONAS-F (SPIRIT X-ray complex) satellites. Solar Syst. Res. **37**(4), 296–301 (2003)
4. Zhitnik, I.A., Bougaenko, O.I., Delaboudiniere, J.-P., et al.: Spirit X-Ray Telescope/Spectrohelio-meter Results. "Solar Variability: from Core to Outer Frontiers". ESA SP-506. **2**, 915–918 (2003)
5. Boldyrev, S.I., Ivanov-Kholodny, G.S., Kolomiitsev, O.P., et al.: Evolution of the Orbit of the CORONAS-F satellite and a prediction of its lifetime. Solar Syst. Res. **39**(6), 501–507 (2005)
6. Boldyrev, S.I., Ivanov-Kholodny, G.S., Kolomiitsev, O.P., Osin, A.I.: Specifics of the orbital evolution of the CORONAS-F satellite at the final stage of its flight. Solar Syst. Res. **41**(5), 420–424 (2007)
7. Boldyrev, S.I., Ivanov-Kholodny, G.S., Kolomiitsev, O.P., Osin, A.I.: Influence of solar activity on variations in the density of the Earth's upper atmosphere. Geomagnetism Aeronomy **51**(4), 546–549 (2011)
8. Zhitnik, I.A., Kuzin, S.V., Sobelman, I.I., et al.: Main results of the SPIRIT experiment on board the CORONAS-F satellite. Solar Syst. Res. **39**(6), 442–452 (2005)

9. Oraevsky, V.N., Ivanov-Kholodny, G.S., Kanonidy, Kh.D., et al.: The upper atmosphere response to the solar geophysical variations on a final stage of flight MOF "Mir". Acta Astronautica. **53**(1), 75–84 (2003)
10. Kolomiitsev, O.P., Egorov, I.B., Surotkin, V.A.: Solar eclipse effects in the ionosphere F–region. Modelling research. Proc. Russian Acad. Sci. Phys. Ser. **64**(9), 1886–1891 (2000)
11. Hedin, A.E.: MSIS–86 thermospheric model. J. Geophys. Res. **92**, 4649–4662 (1987)
12. Hedin, A.E.: Extension of the MSIS thermosphere model into the middle and lower atmosphere (MSISE–90 atmospheric model). J. Geophys. Res. **96**, 1159–1172 (1991)
13. Ivanov-Kholodny, G.S., Nikol'sky, G.M.: The Sun and the ionosphere. In: M.: Nauka. p. 455 (in Russian) (1969)
14. Kolomiitsev, O.P., Zhitnik, I.A., Ivanov-Kholodny, G., et al.: Preliminary results of comparison of satellite measurements to numerical simulations of some parameters of the Earth's upper atmosphere. Solar Syst. Res. **40**(4), 326–330 (2006)
15. Akasofu, S.-I., Chapman, S.: Solar-Terrestrial Physics, p. 901. Clarendon, Oxford (1972)
16. Ibid.
17. Ratcliffe, J.A.: Physics of the Upper Atmosphere, p. 586. Academic, New York (1960)
18. King-Hele, K.: Theory of Satellite Orbits in an Atmosphere. Butterworths, London, p. 165 (1964)
19. Handbook of geophysics / Rev. Edition, p. 570. Macmillan, New York (1960)
20. Hargreaves, J.K.: The Upper Atmosphere and Solar–Terrestrial Relations: An Introduction to the Aerospace Environment, p. 298. Van Nostrand Reinhold Company, New York (1979)
21. Bruzek, A., Durrant, C.: Illustrated Glossary for Solar and Solar-Terrestrial Physics, p. 224. D. Reidel Publishing Co., Dordrecht (1977)
22. Boyarchuk, K.A., Ivanov-Kholodny, G.S., Kolomiitsev, O.P., et al.: Response of the midlatitude ionosphere to extreme solar events in October–November 2003. Geomagnetism Aeronomy **45**(1), 80–86 (2005)
23. Boldyrev, S.I., Ivanov-Kholodny, G.S., Kolomiitsev, O.P.: The problem of predicting the lifetime of low–orbital satellites. In: Procedings of the "Scientific session MEPhI–2006", M.: MEPhI, vol. 7. pp. 30–31 (in Russian) (2006)
24. Ishkov, V.N.: Properties of the current 23rd solar-activity cycle. Solar Syst. Res. **39**(6), 453–461 (2005)
25. Ishkov, V.N.: Evolution and flare productivity of active regions in October–November 2003. Solar Syst. Res. **40**(2), 117–124 (2006)
26. Afonin, V.V., Kolomiitsev, O.P., Mizun, Yu.G.: Satellite measurements of electron temperature and its behavioral characteristics in the region of the main ionospheric valley. Geomagnetism Aeronomy (in Russian) **18**(3), 290–292 (1978)
27. Kuznetsov, V.D., Boldyrev, S.I., Zaitsev, A.V.: The forecast of space weather in near-Earth space. In: Procedings of the "Scientific Session MEPhI–2007", M.: MEPhI, vol. 7, pp. 30–32 (in Russian) (2007)
28. Fatkullin, M.N., Zelenova, T.I., Kozlov, V.K., et al.: Empirical models of the midlatitude ionosphere. M.: Nauka, p. 256 (in Russian) (1981)
29. Kuznetsov, S.A., Kudela, K., Myagkova I.N., Yushkov B.Yu. X-ray and gamma-emission solar flare catalogue obtained by SONG on board CORONAS-F satellite. In: Proceedings of ISCS "Solar Variability as an input to the Earth's Environment" ESA SP-535, pp. 683–686 (2003)
30. Veselovsky, I.S., Panasyuk, M.I., Avdyushin, S.I., et al.: Solar and heliospheric phenomena in October–November 2003: causes and effects. Cosmic Res. **42**(5), 435–488 (2004)

# On-Board and Ground-Based Complexes for Operating the Science Payload of the CORONAS-F Space Mission

**A.I. Stepanov[1], D.V. Lisin, V.D. Kuznetsov, A.N. Afanas'ev, A.I. Osin, and J. Schwarz**

To ensure reliable operation of the science payload of the CORONAS-F satellite and to exercise its flexible control in the course of realization of the research program, an on-board and a specialized ground-based control complexes (GCCs) were designed and manufactured at the Pushkov Institute of Terrestrial Magnetism, Ionosphere, and Radio Wave Propagation (IZMIRAN). A demand for such systems arose because the service facilities of the satellite basic platform were unable to satisfy the requirements of the unique scientific experiments, i.e., an efficient on-line control of the variety of scientific instruments, managing large amounts of scientific information, etc.

The on-board control complex (BCC) was designed for acquisition, storage, and transmission of the telemetry information, the reception of commands from the mission control center, their storage on board and delivery to the scientific instruments at a fixed time.

The task of the GCC was to generate and transmit executive commands to the satellite, to receive the telemetry information, and to perform its first-level processing, archiving, and delivery to the user. Operated together, the GCC and BCC ensured precise timing of the telemetry received.

The development of GCC and BCC, as well as the solution of a wide scope of problems associated with the information support of the CORONAS-F project was realized at the IZMIRAN center for Space Information Technologies (Fig. 1). The

[1]Deceased

A.I. Stepanov · D.V. Lisin (✉) · V.D. Kuznetsov · A.N. Afanas'ev · A.I. Osin
Pushkov Institute of Terrestrial Magnetism, Ionosphere, and Radio Wave Propagation RAS (IZMIRAN), Troitsk, Moscow, Russia
e-mail: lisindv@izmiran.ru

J. Schwarz
German Remote Sensing Data Center, National Ground Segment (DLR), Neustrelitz, Germany

V. Kuznetsov (ed.), *The Coronas-F Space Mission*, Astrophysics and Space Science Library 400, DOI 10.1007/978-3-642-39268-9_18,

**Fig. 1** IZMIRAN Center for Space Information Technologies (*left*). Four-element antenna-feeder device (AFD) for the 137-MHz range (*right*)

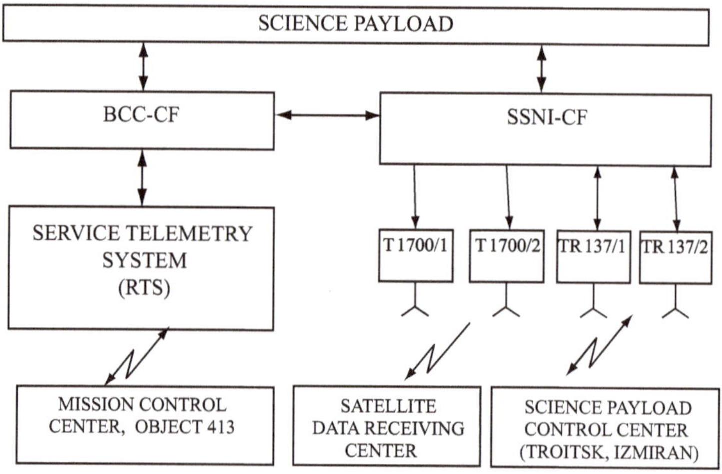


**Fig. 2** Block diagram of the on-board service systems of the CORONAS-F science payload

Center combined various elements of the information support of the project that did not only ensure proper operation of the CORONAS-F science payload but also saved expenses on manufacturing the instruments and testing all systems involved, i.e., the on-board scientific equipment, BCC, and GCC.

## On-Board Control Complex

The basic on-board control complex for the CORONAS-F mission was designed and manufactured at IZMIRAN. It comprised a block of commands and commutations BCC-CF, which operated the science payload via the spacecraft service systems; a data acquisition system SSNI-CF including a 32-bit processor, 128-MB memory, and an integrated command radio line (CRL); the TR137/1 and TR137/2 transceivers of the CRL, which received commands from the Earth at a rate of 4-kbit/s and transmitted the service information to the Earth at a rate of 64-kbit/s;

and the T1700/1 and T1700/2 telemetry transmitters with a transmission rate of 1-Mbit/s.

Figure 2 represents the block diagram of the on-board service systems of the CORONAS-F science payload.

## Block of Commands and Commutations

The block of commands and commutations (BCC-CF) was intended to switch on the scientific devices, control their operation modes via the spacecraft's service system, and establish commutation between the devices. The BCC-CF was based on diode-relay modular circuits. It received 130 commands of five types, including 19 clock pulses and 6 power busses and generated 276 output commands of three types (10 clock pulses, 42 power busses, and 11 general commands for 21 devices). The block was supplied with hot-spare power busses and power sources for the TR137 transceivers. The power consumption was 1.3-W (pulses up to 90-W in the command issue mode). The block had the mass of 9.8-kg and the dimensions of 430 × 434 × 100-mm.

### *Data Acquisition System*

The data acquisition system SSNI-CF was intended for acquisition, storage, and transmission of telemetry information from the science payload to the Earth and for reception of commands from the Control Center, their storage on board and delivery to the scientific devices at a fixed time. The main characteristics of the SSNI-CF are: 20-MHz 32-bit computer with an Intel-compatible processor, 128-MB mass memory, simultaneous servicing of 19 devices, interface performance rate for each device equal to 120-kbit/s, downlink transmission rate equal to 1-Mbit/s, uplink reception rate equal to 4-kbit/s, digital and analog inputs, digital interface with the transmission of time codes to the instruments; synchronization of the satellite time with UT accurate to 1-ms without GPS receiver; on-board processing and compression of data; firmware realization of functions; complete double cold sparing; possibility of in-flight reprogramming of the on-board computer and device controllers via CRL. The mass of the block including the radiation shield is 10-kg; the energy consumption is 8-W.

### *The Scanning Transceivers of the CRL and the Telemetry Transmitter*

The scanning transceivers of the CRL (TR137-1 and TR137-2) (Fig. 3), in conjunction with SSNI-CF, received commands during the satellite communication

**Fig. 3** Command radio line scanning transceivers (TR137-1 and TR137-2) (*left*). Telemetry transmitters (T1700-1 and T1700-2) (*right*)

sessions with GCC, issued the command approval and signal information; were recording the signal level in the ≪receiver≫ mode and the output power in the ≪transmitter≫ mode. The main parameters are: working frequency range—137–138 MHz; scanning pitch—25 kHz; mass—407 g; transmitter: output power—5 W; downlink transmission rate—64 kbit/s; FSK-modulation; receiver: sensitivity—1.2 μV; uplink reception rate—up to 19.6-kbit/s; FSK-modulation.

The telemetry transmitter (Fig. 3) ensured the transmission of scientific information to the ground. It had the working frequency of 1.701 GHz; the output power of 10 W; the downlink transmission rate of 1-Mbit/s; PSK-modulation; the mass of 500-g; and the energy consumption of 30-W.

## Check-and-Test Stand

Unique check-and-test equipment and a test bench (Fig. 4) were designed and manufactured for testing the CORONAS-F science payload. The tests were carried out at IZMIRAN and at the Plesetsk launch ground.

## Ground-Based Control Complex

The GCC comprised a four-element antenna-feeder device (AFD) for the range of 137 MHz with 12-dB amplification (see Fig. 1); a 137-MHz autotracking device with a remote loading block; TP137 transceiver with a 70-W signal amplifier; a bit synchronizer and a facility for loading information to the personal computer (PC) at a rate of 64-kbit/s; a PC-based CRL unit, a ground GPS receiver generating precise time signals; an operator's workstation on the basis of PC; local computer network and Internet facilities; and no-break power-supply sources of 5.4-kW total power.

The firmware facilities of the GCC were used to receive orbital data from the NORAD system; to compute the communication zones for 2 weeks onward; to schedule the control and data transmission sessions; to receive data on spatial orientation of the satellite from the SRT-K stellar detector; to compute the field of vision of the on-board aerials with allowance for their blockage and actual

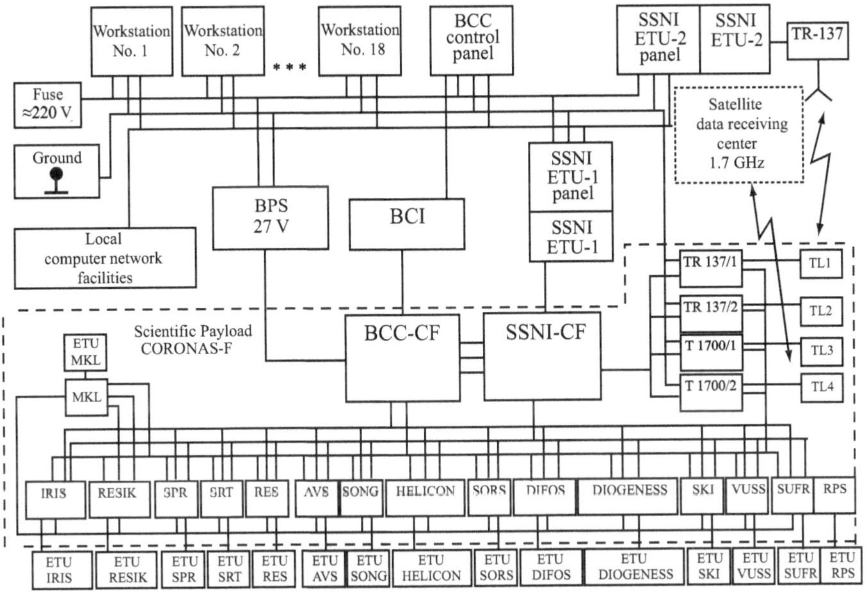


**Fig. 4** Check-and-test equipment and bench for testing the scientific equipment

SC orientation during the communication session; to receive command lists from experimenters, to control the switch on/off of the on-board transmitters taking into account the aerial fields of vision along a given orbit; to generate commands for calibration of the scientific instruments, etc.; to sort commands from the general list and to refer them to the on-board time; to convert the commands received to the binary code and to load them to CRL and farther, to the spacecraft; to keep GCC operation protocols; to receive telemetry data and realize their first-level processing and delivery to the users through the Internet; to keep full archives of telemetry information, and to execute other operations.

The scheme of the ground segment of the CORONAS-F information supply system is represented in Fig. 5.

The CORONAS-F service systems were controlled from the Mission Control Center (CF-MCC). An information exchange channel was established between the MCC SC-CF and the Information and Computing Center at IZMIRAN to ensure on-line operation. It was used to transfer the service formats and telemetry between the MCC SC-CF and the science payload control group.

Within the period of orbital operation of the CORONAS-F mission, the MCC IZMIRAN held 819 communication sessions, transmitted 6.908-MB of information to the on-board computer, and received and processed more than 112-GB of scientific data. All telemetry information received was processed promptly and delivered to the experimenters within an hour after each communication session. Owing to the internal radio line for control of the science payload, the traditional problem of the ≪shortage≫ of executive instructions was solved successfully. For

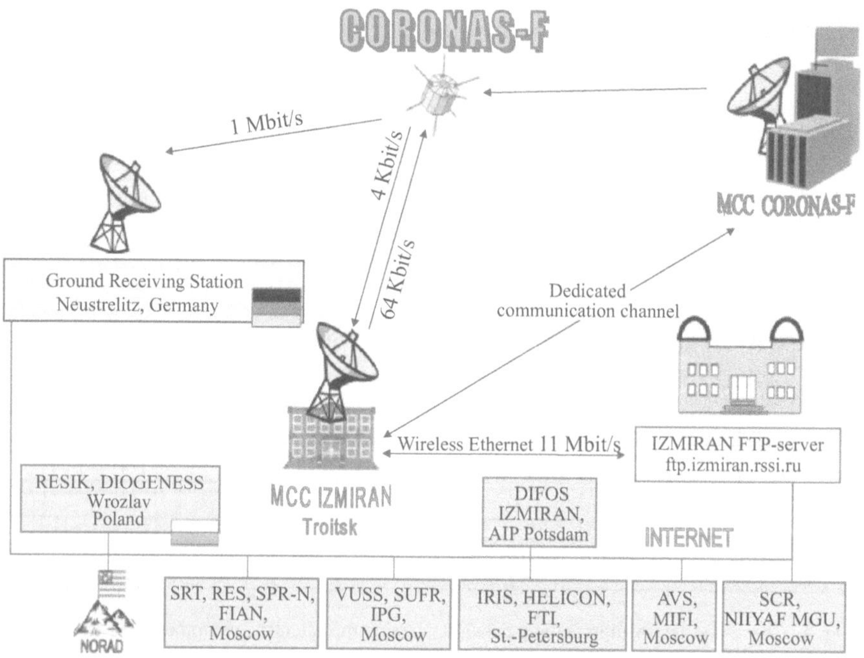


**Fig. 5** Scheme of the ground segment of the CORONAS-F information supply system

the first time, the investigators could issue on-line commands without limitation and could update and modify the experiment software in the course of observations.

At present, the MCC IZMIRAN is used as a command center for the space missions operating within the frames of the Federal Space Program.

## Software Facilities

The MCC IZMIRAN comprised five personal computers with special-purpose software facilities that executed real-time control of the antenna rotator for pointing at SC during the communication sessions; formed the transport layer of the communications protocol in the control radio line; received and analyzed the telemetry transmitted from SC as a feedback in the required format; maintained the communications protocol in the MCC closed local net and performed centralized loading of instructions to the antenna and CRL management PC from the operator's workstation; analyzed the operation of ground systems via the self-control PC; received time signals from GPS and performed time synchronization both in the closed and in the open local computing network (LCN) of the Mission Control Center; prepared executive orders in accordance with the claims received and other commands generated at MCC; computed the ballistic characteristics and synchronized the time on board with the universal time to an accuracy of 1 $\mu$s;

computed the SC orientation in space on the basis of data received from the science payload; planned the communication and telemetry transfer sessions for 2 weeks.

## Control of Orbital Characteristics

The orbital characteristics of the CORONAS-F satellite were controlled using the set of orbital parameters (SOP) available on the server of the NASA Goddard Space Flight Center, Orbital Information Group. In the process, the syntax and correctness of every new SOP were analyzed. The SOP obtained were used at MCC IZMIRAN to compute the SC trajectory; to operate automatic tracking of the object of the 137-MHz antenna CRL; to plan the SC control and 1.7-GHz telemetry transfer sessions to determine the zones where the on-board aerials are shadowed by the SC structure elements during the planned telemetry sessions and to generate switch-on commands for the T1700/1 and T1700/2 transmitters in the zones of consistent reception; to refill the database of SC orbital parameters during the flight period; to transmit these data to experimenters for referencing scientific information to the SC trajectory. As a result of this work, the experiment teams received pinpoint information on the orbital motion of the satellite in the standard format approximately once a day.

## Timing Control

Synchronization of the scientific data obtained and the SSNI operations with the universal time was performed using a specialized complex of soft and hardware facilities on board and on the ground. The on-board segment comprised an inner 32-bit SSNI timer with a counting rate of 1 μs, which operated from a quartz oscillator, and a timer generating the on-board time code; a specialized hardware circuit transmitting successive time codes from SSNI-CF to all instruments of the science payload every millisecond; and a specialized hard-software circuit to insert the blocks of current on-board time codes into the general telemetry flow in the 1,700 MHz radio channel at a rate of 1-Mbit/s. The ground segment comprised a high-precision timer with a 0.1 μs counting rate and synchronization with UT based on the data from SC navigation systems; a marker of precise ground time for telemetry blocks; a program for synchronizing the on-board and ground time scales taking into account the lag of radio propagation on the SC-Earth trajectory according to SC orbital data.

The system described above enabled the timing update in each telemetry session (i.e., 2–3 times a day) and allowed us to make a timing table that was used to convert the on-board time to UT for the control purposes and scheduling the experiments.

Thus, in the course of the flight tests, the SSNI on-board time was referred to the ground time scale with an accuracy of 1–5 μs and an average update frequency

1–2 times a day. This information was sufficient for mathematical processing and solution of the science payload managing problems.

## Conclusion

The on-board and ground-based control systems developed at IZMIRAN worked without failure during the entire operation of the satellite. They ensured an expeditious and flexible control of the CORONAS-F scientific equipment in the course of the experiments. The BCC spare set was not brought in operation, which proved a high reliability of the system. The scientific information obtained was processed promptly and delivered to the experiment teams.

The authors would like to express their sincere gratitude to the developers of the AUOS-SM science platform of the CORONAS-F satellite—the staff of the Yuzhnoe Design Bureau (Dnepropetrovsk, Ukraine)—and personally to V.I. Dranovsky, A.M. Popel', and A.A. Karapetyants, who ensured reliable functioning of the satellite's service systems. We also thank A.D. Ryabov (Space Research Institute), V.I. Kozlov, I.A. Egorova, and N.F. San'ko (Roskosmos), as well as the staffs of the Plesetsk launch site and Russian Space Forces for a fruitful cooperation in realizing the CORONAS-F Project.

# CORONAS-F: Infrastructure and Organization of the Information Exchange

A.I. Osin, E.P. Trushkina, and A.A. Freizon

The success of space missions depends to a large extent on their information support. The infrastructure and the routine work on receiving telemetry information, passing it to the computing centers, preprocessing, and delivery to the users (participants of scientific experiments) must be maintained around the clock all over the spacecraft flight period that usually lasts several years. All these functions must be performed promptly because the operating modes of the on board scientific equipment are scheduled on the basis of data received from the spacecraft. Nowadays, the information support of space missions is unthinkable without developing and managing a special information resource (WEB site) in the Internet. The WEB site of the mission is a means of coordination of the efforts of all participants of a scientific experiment and, simultaneously, a valuable tool for making the results obtained available to the scientific community.

The means for automatic transmission and preprocessing of telemetry data from the CORONAS-F science payload comprise the receiving centers equipped with antennas and transceivers, communication channels, access and preprocessing servers, and the appropriate software facilities. The skilled personnel must ensure regular maintenance and management of all elements of this infrastructure for a long time, sometimes, on a 24-h basis.

Besides the standard receiver of telemetry signals, RTS, the CORONAS-F satellite was furnished with a more efficient data acquisition system (SSNI) developed at IZMIRAN. The RTS telemetry was passed to IZMIRAN from the CORONAS-F Mission Control Center (Krasnoznamensk, Moscow Region) via a dedicated direct channel (64 kbit/s). The Information and Computing Center (ICC) at IZMIRAN insured the reception and preprocessing of data and maintained communication, including a wireless channel (11 MB/s), with operators at the IZMIRAN Center

A.I. Osin (✉) · E.P. Trushkina · A.A. Freizon
Pushkov Institute of Terrestrial Magnetism, Ionosphere, and Radio Wave Propagation, Russian Academy of Sciences (IZMIRAN), Troitsk, Moscow, Russia
e-mail: osin@izmiran.ru

V. Kuznetsov (ed.), *The Coronas-F Space Mission*, Astrophysics and Space Science Library 400, DOI 10.1007/978-3-642-39268-9_19,

for Space Information Technologies (CSIT) charged with the overall coordination of the project. The SSNI telemetry was received by the German Remote Sensing Data Center at Neustrelitz. Then, the data in HDF format were uploaded to the FTP server ftp.nz.dlr.de. The software package installed on the IZMIRAN server spotted new data files on the remote server and ensured their transmission via the FTP to IZMIRAN. To realize a preliminary processing of data, it was necessary to know the satellite's orbital parameters, which were automatically and regularly received from the OIG server (NASA/GSFC). The format conversion and preprocessing of data were done without human attendance by the programs run in the compatibility mode (Wine) on the server under OS Linux. The processed data were stored in the corresponding folders on the FTP server, from which they were regularly downloaded by the users. The process of delivery and preprocessing of telemetry was automatically displayed in real time on the project server (http://coronas.izmiran.ru). The same site opened access to the complete archive of the satellite's orbital parameters. The process of delivery and treatment of the telemetry information was fully automated. The operators had only to keep an eye on the process interfering occasionally to clear up the troubles. The processed data were available to the users on the IZMIRAN server about an hour after their appearance on the DLR server in Germany. After preliminary processing, the original telemetry files were archived on compact disks CD-R.

## Processing of SSNI Telemetry

The main operations on the delivery of the original SSNI data from the satellite and their preprocessing were performed by a set of the script programs in the Bash, Perl, and C languages:

- **autoftp**—the delivery of telemetry from remote receiving points every 30 min 24 h a day;
- **procssni**—preprocessing of SSNI telemetry every 10 min 24 h a day;
- **tle, tle–coronas**—the delivery of the satellite's orbital parameters (TLE—Two Line Elements) every 4 h 24 h a day;
- **purge–coronas**—automatic cleaning of the working directories (the processed files were deleted automatically every 3 weeks);
- **gees, plot–gees**—the reception of orbital parameters and graphical display of the dynamics of the satellite orbit on the mission WEB site;

as well as by two Wine-run programs (Windows)—**hdf2bin, proc_kf**, which converted HDF into BIN and performed preprocessing of the telemetry session.

The procedures of delivery and preliminary processing of the telemetry were registered in several logs:

**auto.log**

2005-01-15 11:10 CHECK ftp.nz.dlr.de/op/coronas
2005-01-15 11:40 CHECK ftp.nz.dlr.de/op/coronas

2005-01-15 11:40 BEGIN coronas_19480.hdf.gz 26924703
2005-01-15 11:43 DONE coronas_19480.hdf.gz 26924703
2005-01-15 12:10 CHECK ftp.nz.dlr.de/op/coronas
2005-01-15 12:40 CHECK ftp.nz.dlr.de/op/coronas

Data spotted and download to IZMIRAN started. The download of 27 MB (typical size of one session) took about 3 min.

**autoftp**

11:40:00 CHECK DATA at ftp.dlr.de
11:40:05 DOWNLOAD DATA STARTED
11:40:11 MAIL -> CORONAS OPERATOR: "ORBIT 19480 DLR DOWNLOAD STARTED"
11:43:19 DOWNLOAD FINISHED (27 MB, 137 KB/sec)
11:43:25 MAIL -> CORONAS OPERATOR: ORBIT 19480 DLR DOWNLOAD FINISHED (26924703 bytes)

The operator is notified by e-mail on the download start and finish, session number, data size, and transmission rate.

**procssni**

11:50:00 CHECK DISK SPACE
gzip -d $hdf.gz *(unpacking of hdf.gz -> hdf)*
HDF2BIN *(format conversion, wine)*
PROC_KF *(preprocessing, wine)*
zip *(archiving of instrument data)*
aaa_NNNNN.zip -> FTP:/CORONAS/F/data/SSNI/aaa *(uploading to FTP server)*
*.log,*.cm1,ut_kbv.dat -> FTP *(service files)*
zip hdf *(archiving of original data)*
11:52:00 MAIL -> CORONAS OPERATOR ("ORBIT $orbit PREPROCESSING FINISHED")

Listed below are the stages of preliminary processing of data of an ordinary session. The information on data download and processing was also displayed on the CORONAS-F WEB site (http://coronas.izmiran.ru/icc):

2005-01-15 08:40 ORBIT 19480 DLR DOWNLOAD STARTED
2005-01-15 08:43 ORBIT 19480 DLR DOWNLOAD FINISHED (26924703 bytes)
2005-01-15 08:52 ORBIT 19480 PREPROCESSING FINISHED

The process was fully automated. A full cycle of data processing from spotting the data on the remote server ftp.dlr.de (Germany) to the appearance of processed data on the FTP server at IZMIRAN and corresponding notifications by e-mail and on the Project site takes about 12 min (11:40–11:52). Taking into account the 30-min delay between the successive starts of autoftp, the time interval between the appearance of data on the DLR server and the end of processing at IZMIRAN might reach 45 min (30 min on the average). The maximum time of a cycle under most unfavorable conditions was about 55 min.

## Processing of Service Telemetry

The service telemetry (RTS) was transmitted to IZMIRAN via a dedicated channel from the Mission Control Center at Krasnoznamensk (Fig. 1). All operations on tracing, transmitting, integrity control, and notification of the sender and recipient of data about the availability of new data files were performed using the software facilities of the communication server. The same channel was used for automatic transmission (including the conversion and acceptance acknowledgement) of formalized orders for operation of scientific instruments from CSIT IZMIRAN to the CORONAS-F Mission Control Center (Fig. 2).

The software comprised the following modules:

- **mccget**—delivery of service telemetry from MCC to IZMIRAN;
- **mccput**—sending orders from IZMIRAN to MCC;
- **mccack**—processing the acceptance acknowledgements;
- **mccchk**—check of integrity;
- **mcclib**—library of functions;
- **crc**—computing checksum;
- **mcctmi, tnatmi**—reducing raw data to a common format

While the process of RTS transmission to IZMIRAN was unattended, the preliminary processing of the service telemetry could not be done automatically for various reasons and was performed later by operators. Preprocessing of the service telemetry was carried out with the aid of the RTSDATA package developed at IZMIRAN in 1993–1994 when the launch of the CORONAS-I satellite, the first one of the CORONAS series, was under preparation. After preliminary processing, the telemetry files were archived on CD-R disks. The RTSDATA package involves the following programs:

- RTS—the basic program for the data structure recovery, timing, and composition of output file (RTS);
- RTSERR, RTSERROR—detection and statistics of the processing errors;
- RTSTIME—timing checkup;
- RTSMKU—structure recovery checkup in MKU channel;
- RTS2TXT—conversion to the text format;
- RTSDEV—data selection and composition of output files of the particular instruments.

The data processed were stored on CD-R disks. They were also uploaded to the IZMIRAN server where they were available via FTP during 3 weeks.

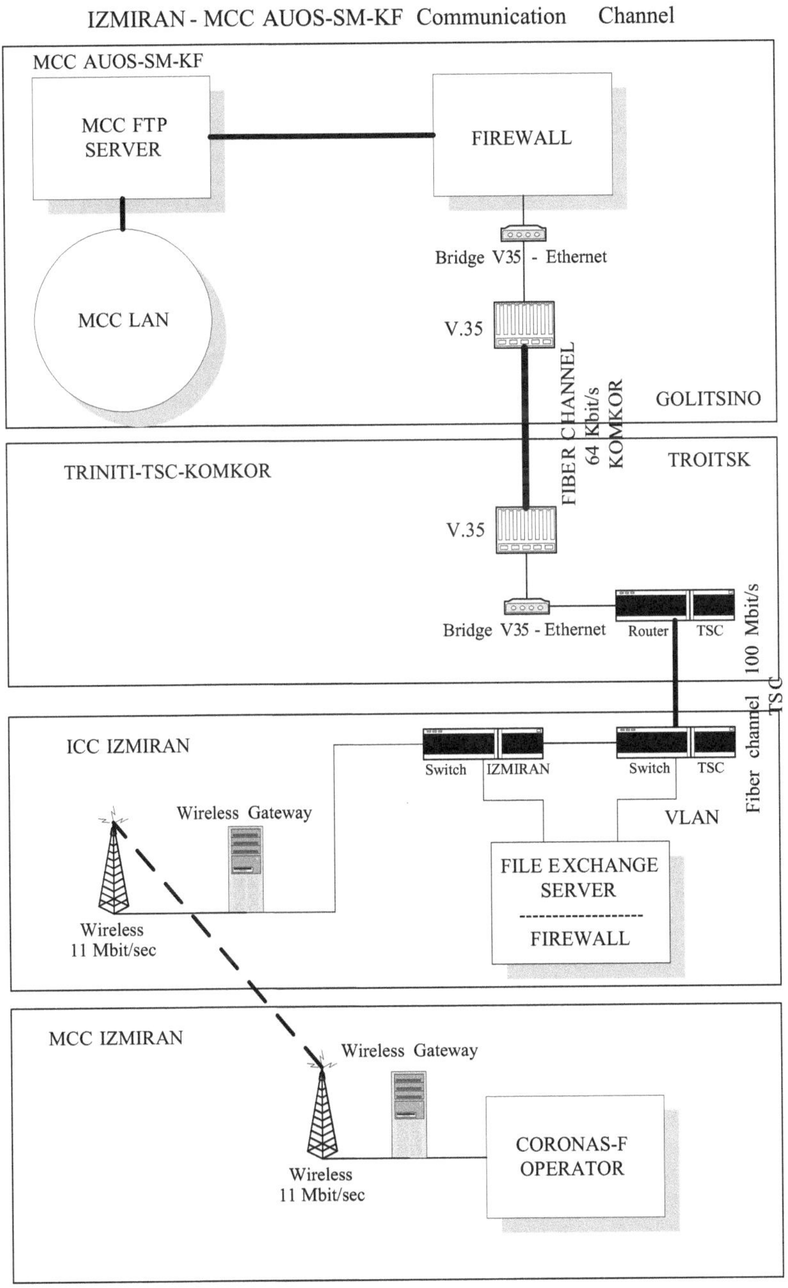


**Fig. 1** Structure of the information exchange channel between the AUOS-SM-CF Mission Control Center and IZMIRAN (ICC and CSIT)

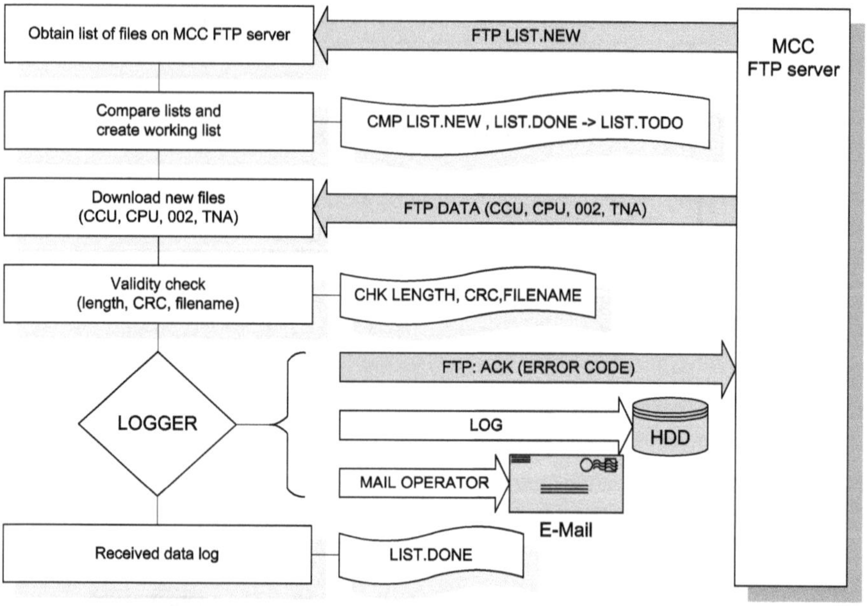


**Fig. 2** Block diagram of the delivery of telemetry data from the AUOS-SM-CF MCC to IZMIRAN (MCCGET). Files transmitted to IZMIRAN: CCU—management communication session, CPU—diurnal management program, 002—starting traffic conditions in the 002 format, TNA—scientific telemetry. The structure of transmitted files: LENGTH (file length, 4 bytes), DATA, CRC (checksum, 2 bytes). The file length and check sum are transmitted in the net format, i.e., the higher-order bytes are passed first

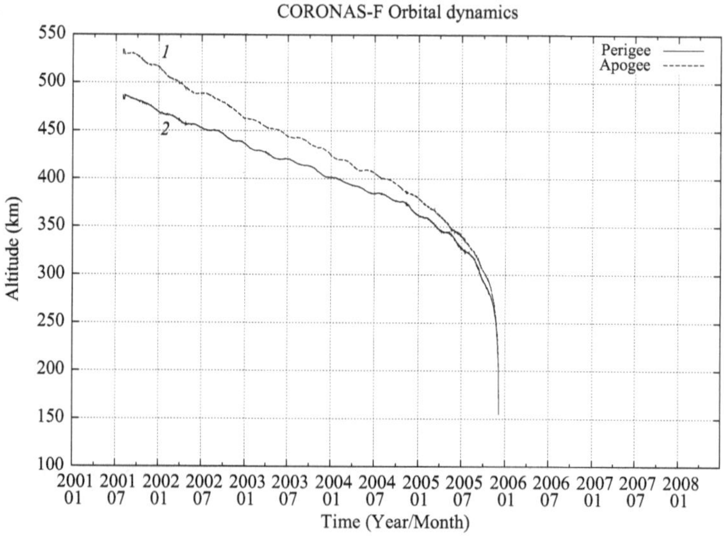


**Fig. 3** Monitoring of the CORONAS-F orbital parameters (apogee 1 and perigee 2)

## Internet Resource of the CORONAS-F Mission

The Internet resource http://coronas.izmiran.ru was established to keep the project participants and the broad scientific community informed of the details of realization of the mission. Most of the documents on the site are in two languages: Russian and English. They provide both a general description of the objectives and technical aspects of the project and detailed information on the particular instruments installed on board the satellite, their specifications, and authors and designers of the scientific experiments. A number of services are available in the ICC section, e.g., the mapping of the satellite current position over the Earth, graphical representation of the satellite orbit since the launch time (Fig. 3), access to the common database of orbital elements (TLE), as well as a dynamically updated log of acceptance and automatic preprocessing of telemetry information.

# Organization of a Unified Data Archive and Accessories for Processing Solar Images

**A.I. Osin and E.P. Trushkina**

The orbital solar observatory CORONAS-F was launched in late July 2001. A vast database from 12 onboard instrumental complexes was accumulated for 3 years of the satellite operation time. The data processing information was displayed on the project site http://coronas.izmiran.ru. The bulk of data received from CORONAS-F amounted to 100 MB/day. On-line processing of such amounts of data was quite a challenge. The preprocessed CORONAS-F data were, then, subjected to additional processing by the principal investigators at the corresponding institutions, including the calibration, space-time binding, clearing, etc. The resulting data are stored at the participant institutions in the form of profiles, spectrograms, and images. The stored "raw" data are the main source of reliable information for further investigations, but they usually need further processing to be visualized. Besides, a full cycle of processing of the scientific data takes much time and can even drag on for years under financial stringency and shortage of personnel. The final information is stored at different institutions that do not possess unified means of access, analysis, and visualization of data. This hampers significantly the use of the entire database by individual researchers.

## Organization of a Unified Data Archive

An attempt to create a unified data archive in the form of network resource supplied with all necessary means of data access and visualization was made at IZMIRAN—one of the head participants of the CORONAS-F project. The archive was set up within the program of creating a virtual laboratory on the basis of a powerful server

A.I. Osin (✉) · E.P. Trushkina
Pushkov Institute of Terrestrial Magnetism, Ionosphere, and Radio Wave Propagation RAS (IZMIRAN), Troitsk, Moscow, Russia
e-mail: osin@izmiran.ru

V. Kuznetsov (ed.), *The Coronas-F Space Mission*, Astrophysics and Space Science Library 400, DOI 10.1007/978-3-642-39268-9_20,

with additional software installed, such as IDL 6.0, IDL Data Miner, MatLab 7 R14, and MatLab Webserver. The idea was to organize a centralized data resource to ensure access to experimental databases. There is no need that the entire data archive of the project be gathered at one place. The data of the particular experiments carried out on board a scientific spacecraft may be stored on the servers of the participant institutions; however, a successful use of such a distributed archive requires a certain unification and coordination of the data structure. The proposed scheme of organization of a unified data archive involves the creation of a virtual laboratory, which would provide certain on-line facilities for the search and visualization of data. All operations can be executed in interactive mode via the Internet by invoking the IDL/MatLab procedures through the CGI-interface on the archive server. The work on the ≪Virtual Archive≫ was supported by the Russian Foundation for Basic Research, project no. 05-07-90092.

## Preimaging Facilities

The visualization tools were tested using the data of the SRT-K X-ray telescope (FIAN). The data processed were available on the web site of the Lebedev Physical Institute in FITS format; however, they were difficult to use, particularly, to the scientists working in allied fields. The final processing of solar images in FITS format comprises various procedures, including spatial orientation, removing artifacts, choosing appropriate colors, etc. This job is rather complicated, involves manual operations, and can only be done by a team of investigators.

In order to simplify and expedite the preparation of solar images obtained from the CORONAS-F satellite to the preliminary analysis and presentation, an experimental workstation (virtual laboratory) http://vlab.izmiran.ru/coronas-f/srt-c/ was established at IZMIRAN (see Fig. 1).

The workstation is based on the TCP/IP, HTTP, SSI, and CGI technologies and uses Apache Web Server (Linux), Perl, and IDL 6.0. The users of the virtual laboratory run the image regime by entering the corresponding parameters in HTML form and passing them to the invoked CGI scenario, which, in turn, transfers the set of the parameters to IDL procedures. The set of tools of the virtual laboratory makes it possible:

- to choose the data file by specified parameters, e.g., the wavelength, year, month, and day of observation;
- to display the original image in achromatic mode on a reduced or full scale;
- to remove automatically some typical defects, such as the horizontal lines that result from the information transfer losses;
- to remove the surplus background brightness and lateral light;
- to rotate the image in accordance with the FITS-file header information;
- to view the image in arbitrary colors with correction of the contrast;
- to view the image in the format determined by a specified set of parameters.

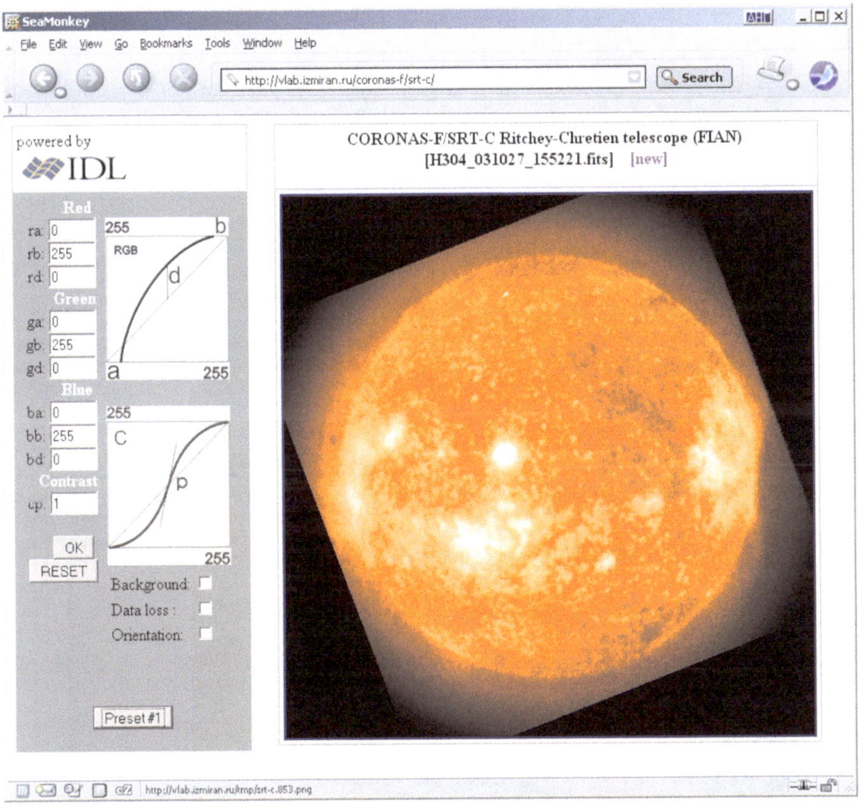


**Fig. 1** A screenshot of the CORONAS-F/SRT-C image processing toolkit

To automate these procedures, a number of programs for processing in FITS format and visualizing original solar images were developed using the software available on the server (IDL 6.0).

The preparation of images comprises operations based on the information contained in the FITS-file headers, e.g., the solar radius and image rotation angle. These parameters are necessary to fix the center of the Sun and realize the image inversion with respect to the vertical axis passing through the center. The original image often contains various defects. The program simulates the profile of these defects and subtracts it from the original data (Fig. 2). The next stage of processing is to fix the coordinates of the solar center. For this purpose, the position of the border of the Sun is estimated approximately by recognizing the local maximum of the emission intensity. The last operation in the preliminary processing of an image is its inversion relative to the vertical axis through the solar center and its rotation by the angle specified in the FITS-file header (see Fig. 3).

For better discrimination of the image details and increasing the contrast, the intensity values are additionally recalculated with empirical coefficients,

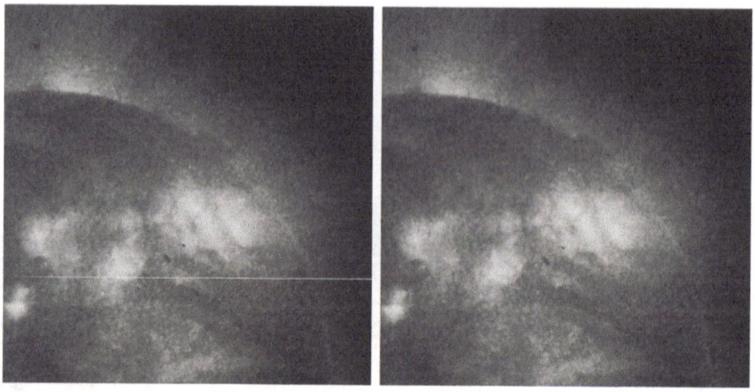

**Fig. 2** An example of removing image defects

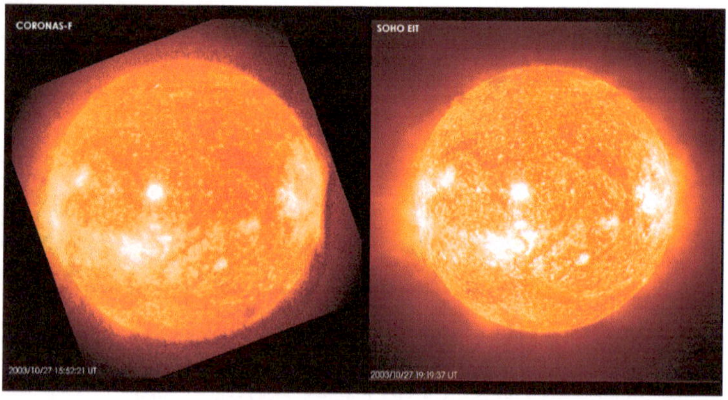


**Fig. 3** Comparison of the solar images from CORONAS-F and SOHO

and appropriate color palette is chosen for each of the four wavelengths. The optimal palette is obtained by fitting the intensity of the basic colors. A thorough selection of all parameters made it possible to obtain an image suitable for preliminary analysis and presentation of the results.

# Conclusion

Effective operation of the CORONAS-F solar observatory was the result of joint effort of a large team of scientists and engineers and a broad cooperation between Russian, Ukrainian, and West-European institutions under the leadership of the Pushkov Institute of Terrestrial Magnetism, Ionosphere, and Radio Wave Propagation of the Russian Academy of Sciences. The work was coordinated by the Council on Space of the Russian Academy of Sciences and the Russian Federal Space Agency.

The participants of the CORONAS-F Project are grateful to the management of the Council on Space of the Russian Academy of Sciences, the Russian Federal Space Agency, the National Academy of Sciences and National Space Agency of Ukraine, as well as to all organizations responsible for the launch and control of the satellite, for successful realization of the CORONAS-F mission.

Russian and Ukrainian technicians at the Plesetsk launch ground before the launch of the CORONAS-F satellite

V. Kuznetsov (ed.), *The Coronas-F Space Mission*, Astrophysics and Space Science Library 400, DOI 10.1007/978-3-642-39268-9,

The third satellite of the CORONAS Project prepared for launch—CORONAS-PHOTON—will continue the study of solar activity and its terrestrial impacts in the activity cycle 24 (2007–2018).

Troitsk, Moscow, Russia V.D. Kuznetsov
30 July 2007

# Index



V. Kuznetsov (ed.), *The Coronas-F Space Mission*, Astrophysics and Space Science Library 400, DOI 10.1007/978-3-642-39268-9,